DISCOVERING THE UNIVERSE

CHARLES E. LONG

North Hennepin Community College

HARPER & ROW, PUBLISHERS

SAN FRANCISCO

CAMBRIDGE
HAGERSTOWN
NEW YORK
PHILADELPHIA

LONDON
MEXICO CITY
SÃO PAULO
SYDNEY

1817

Sponsoring Editor: Malvina Wasserman
Project Editor: Molly Scully
Production Coordinator: Marian Hartsough
Text Design: Nancy Benedict
Illustrator: Dick Cole
Color Insert Art: Catherine Brandel
Copy Editor: Carol Dean
Cover Design: Nancy Benedict
Color Separations: Focus 4
Compositor: Bi-Comp, Incorporated
Printer and Binder: R. R. Donnelley & Sons

DISCOVERING THE UNIVERSE

Library of Congress Cataloging in Publication Data

Long, Charles E 1943-
Discovering the universe.

Includes bibliographies and index.
1. Astronomy. I. Title.
QB43.2.L66 520 79-25200
ISBN 0-06-044034-1

The Cover In the foreground is Kitt Peak National Observatory, photo by Gary Ladd. The middle ground is the Pleiades and Nebulosity in Taurus, courtesy of California Institute of Technology and Carnegie Institution of Washington. The background is Lagoon nebula in Sagittarius, courtesy of Kitt Peak National Observatory and Cerro Tololo Inter-American Observatory.

This book is dedicated to my parents,
Margaret and Bernard Long,
who first showed me the stars,
and to my wife, Phyllis,
who helped me write about them.

CONTENTS

THE STUDY OF DYNAMICS AND THE INVENTION OF THE TELESCOPE 90

LIGHT AND THE SPECTROSCOPE: ACCOMPLISHING THE IMPOSSIBLE 124

THE SUN, THE DAZZLING MYSTERY: A PRELIMINARY LOOK 138

INVESTIGATING THE SUN IN THE TWENTIETH CENTURY 162

THE INNER SOLAR SYSTEM: A VIEW OF OUR NEIGHBORHOOD 196

THE OUTER SOLAR SYSTEM: JUPITER AND BEYOND 250

THE SUPERNOVA PUZZLE 340

THE MILKY WAY 368

THE GALAXIES: ISLANDS OF STARS 394

COSMOLOGY, THE EDGE OF THE UNIVERSE, AND LIFE 444

PREFACE

This book can be used as a text in a one- or two-quarter or one-semester astronomy course at community colleges or four-year institutions, but it can also be interesting and useful to the general reader. The reader I have in mind needs no science or mathematics background; I tried to anticipate and clear away any roadblocks that could hinder your understanding and enjoyment of the universe by giving explanations that are as clear and complete as possible.

THE GOALS OF THIS BOOK

I hope this book will have these effects:

1. You will acquire the habit of looking regularly at the sky, and will be able to appreciate the objects in the universe with deeper understanding.
2. You will understand better what physical theory is. You will study scientific observations and the resulting theories, and will learn about their rise and fall as accepted theories.
3. You will acquire the concepts and habits of thought so that you can read other works and follow new astronomical discoveries.

STYLE AND STRUCTURE

Both style and structure are used to seek these goals. I tried to write prose that is easy and pleasant to read. While this meant sacrificing elegance and conciseness in favor of a relaxed manner and fuller explanations, I tried not to sacrifice precision or real scientific reasoning.

I agree with Arthur Koestler when, in *The Act of Creation,* he wrote, "To derive pleasure from the art of discovery, as from the other arts, the student must be made to re-live, to some extent, the creative process." Further on, he wrote, "The traditional method of confronting the student not with the problem but with the finished solution means depriving him of all excitement," to which I would add, "and deprives him or her of valuable insight."

To this end I structured the text as a series of problems—puzzling observations with which astronomers have struggled. You learn how astronomers grappled with each problem in various ways, and how the tension was released when a satisfactory theory was obtained. Furthermore, where appropriate, you see how and why certain theories were eventually discarded.

Your appreciation of a science is that much richer if you can comprehend the twists and turns which were required to produce our present body of theory. In addition, by knowing how theories have lived and died and how our current ideas have come about, you can acquire an informed view of current knowledge and theory. And, overall, you gain an appreciation of science, what originality and creativity have gone to advance it, and what an achievement the scientific manner of looking at the physical universe is.

ASTRONOMERS ARE HUMAN, TOO

As an important factor in the attempt to capture and maintain the reader's interest, I tried to emphasize that astronomical research is performed by a wide range of fascinating human beings. The universe is only a part of the story of astronomy; another part consists of the struggles, successes, and failures of people to come to terms with the universe.

GETTING DOWN TO BASICS

Because you may have no previous knowledge of the sky, I begin, in Chapter 2, with the basics—the apparent motions of the stars, sun, moon, and planets. My students enjoy this material, perhaps because it helps them see the sky with new understanding, and also it makes them eager to learn the reasons for these motions.

Thus, Chapter 3 begins a study of the procession of great ideas that have been proposed to explain the heavens. This takes us from the monumental geocentric theory eventually to the mind-bending ideas of general relativity and modern cosmology.

The topics I selected to describe our evolving conceptions of the universe are those that promote as continuous a flow of ideas as possible. I have integrated such topics as instrumentation and physics into the narrative when possible so that their connection to astronomy will be apparent.

SPECIAL FEATURES

I would like to point out a number of features that should make your study of the universe easier and more enjoyable. You can easily see and enjoy quite a few of the heavenly objects and events we discuss. I have suggested some of these in the text; see also the Readings section at the end of Chapter 2.

The text is liberally cross-referenced so that you can turn back and review previous concepts. The index is complete and may be used to locate particular topics of interest.

An important task involved in learning a science is to master the vocabulary. All words defined in the text are printed in **boldface** type. The Glossary near the end of the book should also prove useful in your study of these definitions.

The illustrations were done with particular skill and care. Photographs worthy of study were chosen. Notice that there is a collection of color plates in the middle of this book. These show objects that are especially beautiful and instructive when seen in color.

There are occasional puzzles in the text. They are intended to give you something to think about at some of the critical points in the discussion. Try to solve them as best you can before reading the solutions, which are found in Appendix 2. In addition, exercises are offered at the end of each chapter to help you test your understanding of the most important topics.

Each chapter (except Chapter 1) ends with a section called "Summing Up," which highlights the main points of the chapter.

The text incorporates numerous boxes and marginal notes. These are used to present material you might not choose to read the second time through the chapter. I hope you will find them interesting and informative.

Each chapter has an annotated list of readings, provided in case you want to know more about at least some of the fascinating objects, people, and ideas we discuss. I have chosen the readings to be at the right level for the user of this book.

The appendixes contain helpful information, such as Appendix 1, which will give you a start toward an enjoyment of the constellations, and other appendixes which give summaries of data.

I hope you enjoy discovering the universe.

Charles E. Long

ACKNOWLEDGMENTS

I would like to give special thanks to the following people to whom I am indebted for their support and invaluable criticisms: Emilia Belserene, Herbert Lehman College; Wayne Smiley and Robert Warasila, both of Suffolk County Community College; Gladwin Comes, Broward Community College; Leo Standeford, Mankato State University; David Linton, Parkland College; Anthony Lazzaro, California State College; Francis Lestingi, SUNY-Buffalo; William Penhallow, University of Rhode Island; Michael Chriss, College of San Mateo;

Peter Brancazio, Brooklyn College; Charles Whitney, Harvard University; Charles Hathaway, Kansas State University; and Thomas Harrison, North Texas State University. I would also like to thank Robert Morris, Robert Schadewald, and Patricia Pennington for their help. Dr. Michael Shurman, University of Wisconsin-Milwaukee, was an early inspiration to me. My wife, Phyllis, was my best reader for clarity of description, and Mrs. Pat Probst did a magnificent job typing the manuscript. And finally, I would like to thank Carol Dean for her outstanding copyediting of the manuscript and Molly Scully, Nancy Benedict, and the staff of Harper & Row for their excellent work producing this book.

PHOTO CREDITS

Page 3 (upper) NASA, (lower) Anglo-Australian Observatory; p. 4, Lick Observatory; pp. 55, 73, 79, 91, 94, 101, 104, Yerkes Observatory; p. 110 (upper) Anglo-Australian Observatory, (lower) Mount Wilson and Palomar Observatories; pp. 116, 119, NASA; p. 133, Hale Observatories; p. 140, Mount Wilson and Palomar Observatories; p. 142, Hale Observatories; p. 145, Yerkes Observatory; p. 150, NASA; p. 151, Yerkes Observatory; p. 152, Lick Observatory; *Color Plates:* Plates 3—12, NASA; Plate 13, Darryl Roberson; Plates 14, 15, NASA; Plate 16, California Institute of Technology and Carnegie Institution of Washington; Plates 17, 18, Kitt Peak National Observatory and Cerro Tololo Inter-American Observatory; Plates 19–22, California Institute of Technology and Carnegie Institution of Washington; p. 163, Kitt Peak National Observatory; p. 171, Hale Observatories; p. 176, Mount Wilson and Palomar Observatories; p. 177, NASA; pp. 179, 184, Yerkes Observatory; pp. 190, 192, 193, 197, 198, NASA; pp. 202, 204, 206, Lick Observatory; p. 207, Mount Wilson and Palomar Observatories; pp. 207, 208, Lick Observatory; pp. 208–218, 223, 224, 226, NASA; p. 228, Hale Observatories; p. 230, NASA; pp. 233, 234, Lick Observatories; p. 235, Yerkes Observatory; pp. 236–240, NASA; p. 241, Lowell Observatory; pp. 242–247, NASA; p. 252, Yerkes Observatory; p. 253, Hale Observatories; pp. 255, 256, 257, 259, NASA; pp. 262, 265, 268, Lick Observatory; p. 271, Hale Observatories; p. 272, Lick Observatory; p. 273, Yerkes Observatory; p. 274, Lick Observatory; pp. 276, 277, Yerkes Observatory; p. 291, Hale Observatories; pp. 293, 301, Yerkes Observatory; p. 306, Lick Observatory; p. 316, Yerkes Observatory; pp. 319, 320, 321, 329, 330, 334, 335, Lick Observatory; p. 336, Hale Observatories; p. 341, courtesy Charles A. Whitney; pp. 342, 344, Lick Observatory; p. 346, Hale Observatories; p. 347, Lick Observatory; p. 348, Cerro Tololo; p. 365, Lick Observatory; pp. 370, 372–374, Yerkes Observatory; p. 375, courtesy Charles A. Whitney; pp. 376, 377, Anglo-Australian Observatory; pp. 379–381, Lick Observatory; p. 382, Hale Observatories; p. 389, Bell Laboratories; p. 390, National Astronomy and Ionosphere Center, Cornell University; pp. 395, 398, Yerkes Observatory; p. 398, courtesy Charles A. Whitney; p. 399, Lick Observatory; p. 400, Kitt Peak National Observatory—Cerro Tololo Inter-American Observatory, p. 413, Hale Observatories; pp. 414, 415, Lick Observatory; p. 416, Kitt Peak National Observatory—Cerro Tololo Inter-American Observatory; pp. 417, 418, Lick Observatory; p. 419 (upper), Mount Wilson and Palomar Observatories; p. 419 (lower), Kitt Peak National Observatory—Cerro Tololo Inter-American Observatory; p. 420, Lick Observatory; p. 421, Kitt Peak National Observatory—Cerro Tololo Inter-American Observatory; p. 422, Hale Observatories; p. 427, Mount Wilson and Palomar Observatories; p. 427, Kitt Peak National Observatory—Cerro Tololo Inter-American Observatory; p. 428, Lick Observatory; p. 431, Hale Observatories; p. 433, National Radio Astronomy Observatory; p. 434, California Institute of Technology; p. 436, Kitt Peak National Observatory—Cerro Tololo Inter-American Observatory; p. 437, Lick Observatory; p. 440, Kitt Peak National Observatory—Cerro Tololo Inter-American Observatory; pp. 441, 442, Hale Observatories; p. 459, courtesy Joseph Weber.

DISCOVERING THE UNIVERSE

ASTRONOMY AND ASTRONOMERS

What is the most spectacular sight nature offers? Which view is the most awesome, the most breathtaking? The thundering waters of Niagara Falls? The Grand Canyon's depths? The Alps? If you ask, you will receive these and many other answers. And yet I find that very few people name what may be the most awesome view of all—the sky viewed on a very clear, moonless night from a very dark location. No words can express the mingled terror and awe that can grip us when we truly experience the infinite depths of space, all hung with blazing stars. For a moment, we seem taken out of ourselves, suspended over the brink of eternity. Just for an instant we are stripped of the centuries of civilization that cover and insulate us from nature and we stand primitive before the universe. One senses something of the impact such a view had upon the ancient philosopher Chalcidius, who wrote: "No man would seek God nor aspire to piety unless he had first seen the sky and the stars." Afterward, we may contemplate the experience. Our reaction may or may not be the same as Chalcidius', but it can lead us to ask, "What is the meaning of all this? What is my place in the scheme of things? Why am I here?"

In times past, most people could not afford to travel to the Grand Canyon or the Alps, but anyone could step outside and see the true splendor of the night sky. Today that backyard view is denied most of us, particularly city dwellers. Light pollution in and near cities, where most of us live, blots out the splendor. We must travel far from a city to witness it. And the problem grows worse, decade by decade.

This book is intended to be an aid to those who wish to know more about the universe. It is concerned with astronomers, those who have observed and thought about the objects in the sky, and also is concerned with astronomy, the results of their labors. Ours is an age largely deprived of the truly dark night sky but enriched in other ways. We can purchase binoculars or a telescope and investigate the sky on our own. Large, expensive telescopes have been built; we shall discuss what they reveal and see some of the pictures they have taken. And we live in the space age; technology and we taxpayers have made it possible to send space probes to other planets, probes that send back wonderful pictures, closeups of other worlds. We have even sent representatives out into that awesome sky, and they have returned and told us what it is like.

1.1 WHY DO ASTRONOMERS TAKE UP ASTRONOMY?

Observatories cannot be heated, because the heat escaping through the slit in the dome would ruin the view. Why do so many observers endure the cold and the occasional boredom of long nights of painstaking observation? Some astronomers never look through a telescope or at the sky professionally. Rather, they study observers' data and try to derive an explanation of what the observers saw. Why do so many theorists spend long, anxious, frustrating hours juggling numbers, performing endless computations, risking failure (for success is given to very few), and agonizing over some of the most complex, difficult problems known?

The reasons are many, and each person has several, no doubt. Perhaps a few astronomers merely seek financial reward; to them it's just another job. A few others, perhaps on an ego trip, seek a discovery that will bring them fame. But most astronomers are seeking something else. They are genuinely dedicated to their task, and they care about heavenly objects. They have observed the night sky and want to understand that overwhelming experience.

The rewards dedicated astronomers seek are many. They seek a kind of escape, a kind of freedom. Looking down upon your hometown from an airliner at 30,000 feet (ft), you can feel a sense of release from the trivial cares in which you can become enmeshed when walking those tiny streets. How much greater is your freedom when, at least mentally, you wander among the stars.

Figure 1.1 Saturn and its rings.

Astronomers seek themselves, too. As all of us do from time to time, they wonder what their place is in the scheme of things and about their very existence. They seek clues among the stars.

They seek the great and the small. They enjoy pondering huge galaxies and tiny atoms. They desire to comprehend the vast distances in our universe and to come to know the tiny details on the surfaces of the planets. They want to stretch their sense of time to encompass the long lives of the stars and the even longer past and future life of the universe, as well as to probe the split-second events within atoms, events which, in the last analysis, cause so many astronomical phenomena.

They seek beauty—the beauty of Saturn's rings (Figure 1.1), the beauty of hundreds of thousands of stars in a globular cluster (Figure 1.2), and the beauty of an emission nebula, a gigantic glowing cloud of gas (Figure 1.3). And, too, they seek the beauty of a really satisfying theory. (See the color plates in this book.)

And finally, among other motivations, they seek to satisfy their curiosity. They wonder. "Men were first led to the study of science, as indeed they are today, by wonder," wrote Einstein. And they seek that moment, that supreme moment, when they at last can say, "Aha! I understand!"

Figure 1.2 Omega Centauri, a globular cluster.

Figure 1.3 The Great nebula in Orion, an emission nebula.

1.2 THE AHA

The first requirement of a scientist is that he be curious. He should be capable of being astonished and eager to find out.

Erwin Schrödinger

There is a legend that the great ancient Greek scientist Archimedes was in the bath and thinking about a problem that had puzzled him for some time. Suddenly, out of the blue, the solution came to him. He became quite excited, to say the least, for he ran naked through the streets shouting, "Eureka" (I have found it).

Some people call it the "eureka phenomenon." I call it "the aha," but whatever you call it, it is one of those moments that all creative people, not just astronomers, seek. We shall meet some cases of the aha in this book. The aha is one of the five links that often lead to a theory, the goal of science. The links are curiosity, observation, puzzlement, the aha, and a theory.

Curiosity is the first step. It draws us to look at something and wonder about it. Most of us share this trait.

Observation is the next step. We must watch the phenomenon carefully, exactly, patiently, and repeatedly. Most people will see something in the sky and say, "I wonder why that happens?" and then go inside and watch television. The professional observer, if he or she is curious enough, will begin a program of study which may require years of painstaking, tedious instrument building, observation, and record keeping. We shall meet a number of such observers in chapters to come. They are a breed apart.

Next, if a theory is to result, must come puzzlement. The data are assembled, but they don't make sense. Someone, perhaps the observer or some other person, becomes upset, annoyed, frustrated, but also intrigued by the interesting and odd results of the observations. This person may pace, wring his or her hands, and lose sleep over the problem. It sounds, and it can be, painful, but it is often a necessary condition if a theory is to be created.

This anguish starts the researcher's subconscious mind working on the problem. It is in the subconscious mind that all great ideas seem to originate. As William James put it, the idea is to "get your mind whirling and see what happens." Most people have had the experience of sensing the subconscious at work, perhaps just before sleep. Poincaré said that, at such times, "Ideas rose in clouds. I felt them collide." A poet describes ". . . a subconscious linking of ideas, of emotional tensions, of imagery, of memories."

And then, if a person is very fortunate, the subconscious ideas fall together and, as if by a miracle, the solution pops into the conscious mind. The more labor spent in stirring up the unconscious, the more powerful this experience, the aha, will be. Those who have known it most exquisitely report that it can be a peak experience as wondrous as any religious, artistic, or mystical experience, even as wondrous as falling in love. Physicist Fred Hoyle says: "You are high about three days." Richard Feynman, another physicist, reports: "You go wild. It's ecstacy!" Max Born, also a physicist, recalls: "I shall never forget the thrill I experienced when I succeeded." And Albert Einstein reports: "The years of anxious searching in the dark with their intense longing, their alternations of confidence and exhaustion and the final emergence into light—only those who have experienced it can understand that."

Finally, after the emotions have calmed, the observer uses the new insight to work out a theory, which shows how the observations can be explained in a simple, beautiful way. This leads the observer to say, "Yes, now I see why. Now it makes sense."

1.3 SCIENTIFIC THEORY

We all seek a buildup of tension at times in our lives. We seek it in the hope that it will be released again. Music builds to a dissonance to be resolved in a consonance. A football game builds tension and release. Riddles and puzzles have the same function.

In science, the tension comes from the problem, and the release from the theory. A **theory** is an explanation of a problem, an explanation that allows us to see connections between seemingly unconnected facts and to see them in a way that makes sense and gives pleasure.

It is important to know that in science the word "theory" has a meaning somewhat different from its meaning in everyday conversation. You hear people say, "Oh, your idea is just a theory." They mean that your idea is a guess, a hypothesis with little to back it up. In science, an initial guess is called a **hypothesis** or **speculation.** Only after the idea has been fully worked out, only after it is seen to explain many of the observations satisfactorily, and only after it has stood up to the critical analysis of other workers in the field is the word "theory" conferred upon it.

In this book we study some of the theories that have evolved throughout the centuries to explain astronomical phenomena. We shall observe many things about theories. One of the most important is that there has been a progression of theories. This progression is one of the most interesting features of science. As theory follows theory, each succeeding generation finds two things: not only deepening satisfaction as increasingly satisfactory answers are found, but also a kind of dissatisfaction as each new theory brings forward new questions. It seems that the need for new theories never ends. Like a child asking, "Why?" after each increasingly profound answer is obtained, science always delves more deeply into the mysteries of our puzzling world.

As we study the theories of astronomy, we shall also observe theories being discarded. A theory can be discarded for either of two reasons. The first is that an observation has been made that contradicts the theory. This can kill a theory, especially one that is rather new and not well integrated into the minds of scientists. However, even a lack of agreement with observations may not dislodge a long-standing, highly successful theory. In the absence of something better, scientists avoid using the theory where it is seen not to apply.

The most interesting way a theory can be displaced is when a new, better theory takes its place. We shall see a number of examples of such revolutions in thought as we study the chapters that follow.

One could ask here, What makes one theory better than another, competing one? The new theory should explain the observations in such a way that we can rationally understand them more clearly than before. Perhaps the new theory can even explain certain observations the older one could not or which even contradicted the older theory. But even more importantly, the new theory should be more beautiful than the older one.

To those who believe that science deals only with cold, hard facts and reasoning, this last statement may seem astonishing. Yet, as we shall see exemplified in Chapter 3 and elsewhere, aesthetic appeal is almost as important in science as it is in art. One should respond to a great theory by saying not only, "Oh, I see it clearly now," but also, "What a wonderful explanation." "Search for simplicity and beauty," this is the creed of a scientist. According to Newton, "truth is ever to be found in simplicity and not in the multiplicity and confusion of things." Paul Dirac (one of the most eminent of twentieth-century physicists) stated: "It is more important to have beauty in one's equations than to have them fit the experiment." However, a theory that contradicts experiment or observations for too long a time will eventually have to be discarded, no matter how beautiful.

1.4 WHY WOULD A NONSCIENTIST STUDY ASTRONOMY?

Be intent on things above rather than on things of Earth.

Colossians, 3

The physicist Paul Dirac had just been awarded the Nobel Prize. He was asked by a reporter whether any of his work was explainable to the nonscientist. He thought for awhile and then answered, "No." Now, the writer of a science book for nonscientists has to feel otherwise. He has to feel that it is important that as many people as possible share his love for his field and understand as much of it as they can.

This book has a number of goals. Here are some of the effects a book like this one might have.

The reader may look at the sky more often and enjoy the sights there. Even in a large city, many celestial objects are easily visible and becoming familiar with them can be immensely rewarding. When one is outdoors, the sky is half of the view. Why not look at it at least once each day?

The reader may come to know the universe better. Most other subjects except astronomy are concerned with things on our earth. In astronomy, one studies the earth and everything else.

An acquaintance with the universe can help one, in some small way, come to know oneself better.

The reader may obtain a deepended insight into the process of science—how it works and what it means.

Last, at least some readers may be moved to use the tools (concepts, theories, ways of thinking) learned in this book to read other books, pursue a closer study of the sky using the naked eye, binoculars, or a telescope, and become better able to follow and enjoy future progress in astronomy.

But enough of this, let us turn to Chapter 2 and observe the wonders of the sky.

EXERCISES

1. What, at least in the author's opinion, is the most spectacular sight nature has to offer?

2. What is a theory, in the sense in which it is used by scientists? What is the difference between a scientific hypothesis and a scientific theory?

3. Describe the five links that can lead to a theory.

4. Which two criteria are often used in deciding between competing theories?

WHAT'S GOING ON UP THERE? THE APPARENT MOTIONS OF THE STARS, SUN, MOON, AND PLANETS

In 134 B.C., Hipparchus, now recognized as the greatest astronomer of ancient Greece, noticed a star in the sky, a star he believed had not been visible before. This inspired him to draw a map of the stars. The result was the first good star chart, one that showed the positions of 850 stars. From then on, any suspected new stars could be checked against Hipparchus' chart. Hipparchus spent much of his life observing and recording the positions and motions of the stars, sun, moon, and planets.

Hipparchus lived from about 190 to 120 B.C.

Today we call a star that suddenly flares up a *nova*. We will discuss this phenomenon in Section 12.8.

The purpose of this chapter is to gain an acquaintance with some of the basic findings made by astronomers like Hipparchus and his predecessors. We discuss the motions in the sky that can be observed by any careful, patient person using the naked eye over a period of several years.

One point concerning the organization of this and the next two chapters should be made before we begin our study of the sky. In this chapter we deal only with observations, that is, we describe *appearances,* the apparent motions of the stars, sun, moon, and planets. For example, we make statements such as "The sun rises in the east," a description of the apparent motion of the sun.

Such observations have, throughout time, led to the question, Why does this occur? Having seen some phenomenon, one begins to search for an explanation. This might be called the "scientific impulse," for it is one of the starting points for all science. Throughout this chapter, as we describe certain observations, perhaps you will find yourself wondering what the explanations for them might be. We do not answer such questions in *this* chapter, but instead gather the explanations into the chapters that follow. Thus the apparent motions are described in this chapter and explained in the following chapters. This order should be helpful, because the explanations of the observations will not be clear until the observations themselves are clearly understood.

The reader should look at the sky as often as is convenient. Try to observe and enjoy as many of the phenomena we discuss as time allows.

With these preliminaries out of the way, let us turn our attention to the splendors of the sky.

2.1 THE APPARENT MOTIONS OF THE STARS

As discussed in Chapter 1, the two types of explanations given in science are called hypothesis and theory. See Section 1.3 for the distinction.

Polaris, the North Star One's first reaction to the staggering beauty of a truly dark, unpolluted night sky is one of awe. Then bewilderment

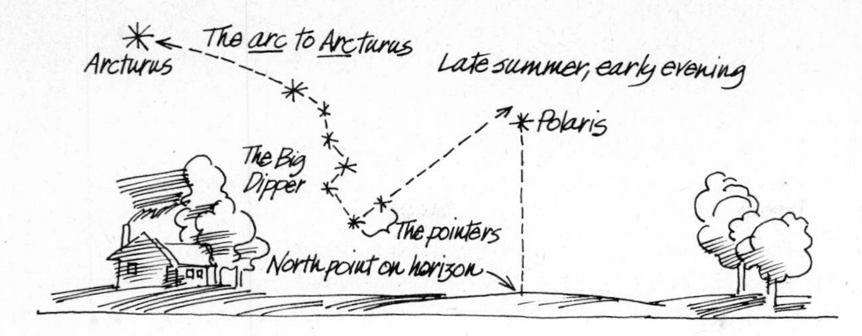

Figure 2.1 The Big Dipper as a guide to Polaris and Arcturus. From most locations in the continental United States, the Big Dipper is a circumpolar group; also it is always found somewhere in the northern sky. Once the Big Dipper has been found, one can locate Polaris by extending a line from "the pointers," as shown. One may also extend the curve (or, arc) of the Big Dipper's handle to locate the orangish star Arcturus. (Arcturus is not always above the horizon.)

sets in. So many stars! Early astronomers imposed order on the scattered stars by grouping them, in their imagination, into what we call **constellations.** Gradually, these groupings and the brighter stars acquired the names we use today. Many find it a pleasure to become familiar with at least some of these constellations and brighter stars. (See Appendix 1.)

Many of the star names we use derive from names coined by Arab astronomers. Many of these names sound enchanting. Consider, for example, "Mizar" or "Aldebaran."

Some people are surprised the first time they notice that the stars move. Slowly but surely they march across the sky. Loosely speaking, they rise in the eastern sky and set in the western sky. You can easily perceive this motion by selecting an observation point and watching a star as compared to some stationary object. Return to the same place every quarter hour or so and observe the star's change in position.

This chapter describes observations. Thus, here, as elsewhere in this chapter, a statement such as, "The stars move," should be understood to be short for, "The stars *appear to* move." We do not yet wish to concern ourselves with the question of whether or not they really do move.

If you were to watch enough stars in this way, you would eventually notice a single star which, of all the thousands visible, is not moving. (We assume that the reader lives in the northern hemisphere.) This star's name is Polaris. It is also called the Pole Star or the North Star. Figure 2.1 shows how to locate Polaris. Polaris is always found in the same place in the sky, right over the north point on the horizon. Since it does not move, it is always at the same angle above the north point on the horizon when viewed from the same location. Learning to find Polaris is the first step in learning to navigate by the stars. The point on the horizon directly below Polaris marks north. When you are facing Polaris, west is on your left, east on your right, and south behind you. The stars provide a compass in the sky.

Actually, Polaris moves. It traces out a small circle in the sky. However, it is not easy for a casual observer to notice this motion, which was discovered in the third century B.C. For simplicity, we ignore the motion of Polaris and consider it stationary.

Angles in the Sky Three commonly used terms will be defined next: angular separation, angular diameter, and altitude. Determining the actual (or linear) distance between two stars is very difficult. (See Section 12.1.) On the other hand, the angular separation of two stars (or of any two points in the sky) is easy to measure. (Box 2.1 and Figure 2.2 provide a review of angles.) Choose two stars. Sight along a long, straight stick pointed at one star. Sight along a second stick at the second star. Now have an assistant measure the angle formed by the sticks. The resulting angle is known as the **angular separation** of the two stars. See Figure 2.3. The angular separation of any point on the **horizon** (the circle at which the ground seems to meet the sky) and the **zenith** (defined as the point directly overhead) is 90°.

One's actual horizon may be distorted from the ideal by hills or other obstructions. A true horizon may be seen from a ship at sea or on very level land.

BOX 2.1

A REVIEW OF ANGLES

An *angle* is formed whenever two straight lines meet. The size of an angle, often expressed in degrees, indicates how widely spread the two lines are. If the two lines meet at a right angle, they are said to be perpendicular. Two such lines form a 90° angle, as shown in Figure 2.2a. If a 90° angle is placed in a circle with its corner at the circle's center, it marks off a quarter circle, as in Figure 2.2b. A half circle is marked off by a 180° angle. One says that a full circle represents 360°.

Now, if one were to divide up the quarter circle in Figure 2.2b into 90 parts, each part would represent 1°. Often angles even smaller than 1° are measured. One degree is further divided into 60 parts. Each part is called a *minute of arc*, or $\frac{1}{60}$ of a degree. (One should be careful to include the words "of arc" to avoid confusion with minutes of time.) See Figure 2.2c.

In making very precise measurements, it may become necessary to divide the minute of arc. A *second of arc* is $\frac{1}{60}$ of a minute of arc, or $\frac{1}{3600}$ of a degree.

This summary is worth remembering: 60 seconds of arc make 1 minute of arc; 60 minutes of arc make 1°; 360° make one full circle.

A round object's actual diameter, measured in, say, meters or feet, is known as its **linear diameter.** A related concept is **angular size** or, in the case of a round object, **angular diameter.** For example, if you point two sticks at the opposite edges of the full moon, the angle you measure is the moon's angular diameter. The result is about $\frac{1}{2}$° (or 30 minutes of arc).

The word "altitude" as used by mountain climbers represents a different concept having the same name.

The **altitude** (also known as the **elevation**) of an object in the sky is its angular separation from the horizon. Point one stick at the object and the other stick at the point on the horizon directly below the object. The angle made by the sticks is the object's altitude. An object's altitude cannot exceed 90°.

Travelers and the Altitude of Polaris As viewed from any one location on earth, Polaris appears to be stationary. In other words, the altitude of Polaris is constant as viewed from any given location. However, even by Hipparchus' time it had been known for centuries that as one travels north the altitude of Polaris increases. Not only is the position of Polaris affected; the entire sky seems to slide up in the north and down in the south as one travels north. Stars formerly just barely visible above the southern horizon slide down and disappear. It was apparent that, if one could go far enough north, Polaris would be at the zenith, that is, directly overhead.

Similarly, as one travels south, the altitude of Polaris decreases and new southern stars are seen as the sky slides upward in the south. Travelers who continue far enough south eventually see Polaris drop down to the horizon and then disappear.

This observation, that the altitude of Polaris increases as one travels north, is important for determining the shape of the earth and for celestial navigation. (This topic is discussed further in Section 3.1.)

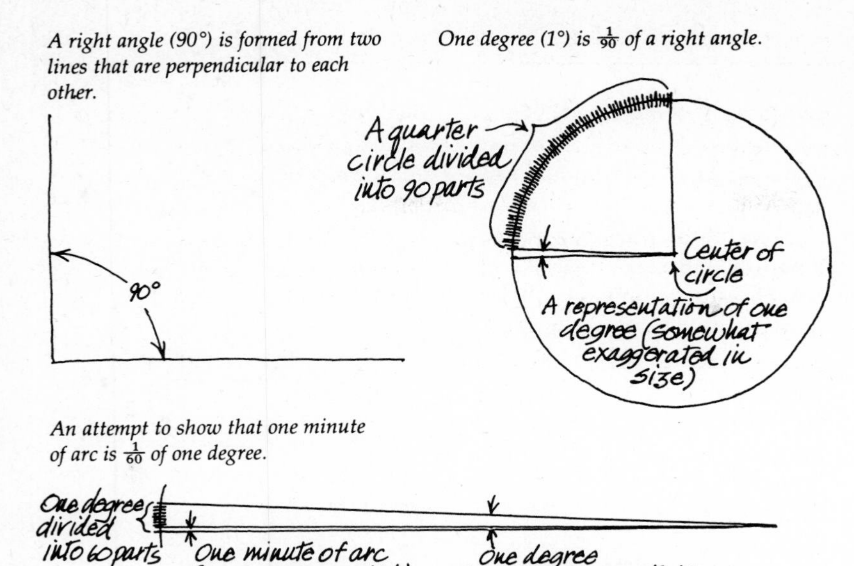

Figure 2.2 Various units used to describe the size of angles. Both 1° and (especially) 1 minute of arc are too small to be shown accurately in such a small drawing. In order to represent 1 minute of arc by two lines that meet at one end and are 1 centimeter (cm) (about the length of a fingernail) apart at the other end, the lines would have to be about 34 meters [37 yards (yd)] long.

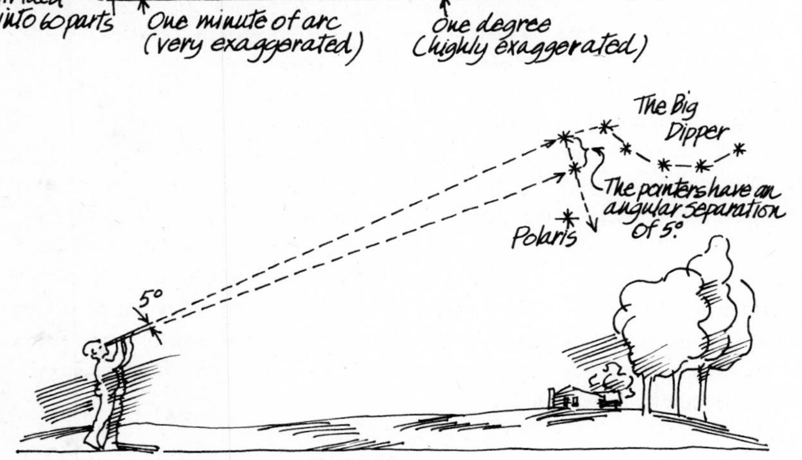

Figure 2.3 How to measure angular separation. This is an instructive, if impractical, method. One points sticks at two objects and measures the angle between the sticks. Several practical methods have been developed, but we need not know them for this book. This figure illustrates angular separation, *not* linear separation.

The Apparent Motions of the Other Fixed Stars The grouping of stars into the major constellations we use today is among the oldest of all traditions. It is evidence that the stars do not appear to swarm rapidly around in the sky like a cloud of gnats. Observation shows that the stars do move and that they march across the sky in an orderly parade, always staying in formation. The Big Dipper, for example, has not changed appreciably in shape since the earliest known drawing.

Consider any two stars in the sky. Since all stars stay in formation, these two stars will not drift closer together or farther apart as they move across the sky. More concisely, one says that *the angular separation of any two stars is constant.* This is why the stars are often called **fixed stars.** Very careful measurements made over long intervals of time finally have revealed that the stars do drift relative to each other, but only very, very slowly. This **proper motion** of the stars is discussed in Chapter 12. Now solve Puzzle 2.1 on the next page.

PUZZLE 2.1

WHAT PATH DOES A STAR FOLLOW?

Think about a star not very far from Polaris in the sky. What is the shape of its path? Also, which direction, clockwise or counterclockwise, does it travel on that path? [Remember, you now know that (1) all stars except Polaris move, (2) the angular separation of any two stars is constant, and (3) stars move generally upward in the eastern sky and generally downward in the western sky.] (Think through each puzzle. Perhaps a sketch will help. When you are finished, study the solution given in Appendix 2 even if you are confident of your own solution. My solution may give you another point of view.)

PUZZLE 2.2

A RACE BETWEEN TWO STARS

Imagine a line drawn horizontally to the left from Polaris. Suppose two stars are on this line, one being further from Polaris. In a race, which will return to the line first? Will it be a tie? The solution is found in Appendix 2.

In this chapter, the phrase "a star close to Polaris" merely implies that the angular separation of Polaris and the star is small. The star may actually be much farther from us or nearer to us than Polaris. One cannot perceive depth when viewing the sky.

We use the concept of the celestial sphere in this chapter to help describe appearances. We will face the question of the reality of the celestial sphere in subsequent chapters.

Having studied Puzzle 2.1 and its solution, we now know that the stars travel in circles centered on Polaris. Now solve Puzzle 2.2.

Circumpolar Stars The two stars we followed in Puzzle 2.2 are examples of circumpolar stars. A **circumpolar star** is one situated so close to Polaris that, as viewed from your location, it never reaches the horizon. It never sets because its path is such a small circle.

Stars That Are Not Circumpolar Stars that are not circumpolar rise somewhere in the eastern half of the sky and set somewhere in the western half. Visualize the paths they follow. Many people find this somewhat difficult to do (but see below). It is true, but hard to visualize, that the further a star is from Polaris the less time it spends above the horizon and the more time below. Also, some stars are so far south that observers in the United States never see them. These stars never rise above the horizon.

The Celestial Sphere Ancient astronomers devised a convenient method for visualizing the motions of the stars. They invented the celestial sphere. When you stand out in a flat field, the earth looks like a flat disk. It always appears that you are at the center of the disk. Further, it appears that the sky is a dome overhead and that the dome curves down to touch the ground at the edge of the disk. The edge of the disk is called the **horizon.** After locating Polaris, you can locate the north point on the horizon directly below it. Then, as you face north, you can locate the east point on your right, the west point on your left, and the south point directly behind you. See Figure 2.4.

(A word of clarification: Phrases such as "in the east" and "in the eastern sky" usually mean "somewhere in the eastern half of the sky, not too far above the horizon." On the other hand, phrases such as "at the east point on the horizon" and "due east" refer to *one point on* the horizon, the point 90° east of the north point.)

The **celestial sphere** is conceived to be an extension of the dome described earlier. Imagine that the disk of the apparently flat earth is surrounded by a snugly fitting sphere. The stars move as if they are attached to this sphere.

Various parts of the celestial sphere are named much like the parts on a globe of the earth. It has a north pole, called the **north celestial pole.** (Refer to Figure 2.5.) As luck would have it, there is a conveniently placed star marking this location, Polaris. (Actually, Polaris is not located exactly at the north celestial pole. However, it is close enough for us to ignore this approximately 1° discrepancy.)

Now imagine a line drawn from the north celestial pole through the center of the disk and extended until it strikes the celestial sphere again. This locates the **south celestial pole.** (The region of the sky near the south celestial pole is visible only to those who live in the southern hemisphere. It happens that there is no star to mark the south celestial pole.) The line connecting the two poles is called the **axis.**

Now we can begin to visualize the motions of the stars more easily. The celestial sphere appears to rotate about its axis from east to west as indicated by the arrow in Figure 2.5. The celestial sphere carries the apparently attached stars with it, causing them to rise in the east and set in the west. The motion of a representative star is shown in Figure 2.5. Notice that, unless you live at the earth's equator, the stars rise on a slant rather than straight up from the horizon.

The **meridian** is an imaginary line in the sky that starts at the north celestial pole, runs through the zenith down to the south point on the horizon, and then on to the south celestial pole. The meridian is stationary relative to the observer; it does not rotate with the celestial sphere. One significance of the meridian is that, when a star or other celestial object crosses the meridian, it has reached its greatest angle above the horizon; that is, it has reached its maximum altitude.

The **celestial equator,** which is analogous to the earth's equator, is an imaginary circle dividing the celestial sphere into two halves, a northern half and a southern half. If you start at either celestial pole and mark off a quarter of a circle (90°) in any direction, you will come to the celestial equator. As viewed from your location on earth, the celestial equator (supposing it were somehow made visible) intersects the horizon at both the east point and the west point. The point where it cuts the meridian is of interest. If the altitude of Polaris is high at your location, this intersection point will be low in the south. If Polaris is low, this point will be high.

At this point the reader should consider the celestial sphere and how the stars move. How do circumpolar stars move? Why is it that the further a star is from Polaris the less time it spends above the horizon? Why are some stars never above the horizon? Figure 2.6 has been drawn as an aid. If a three-dimensional model of the celestial sphere is available, it is well worth consulting.

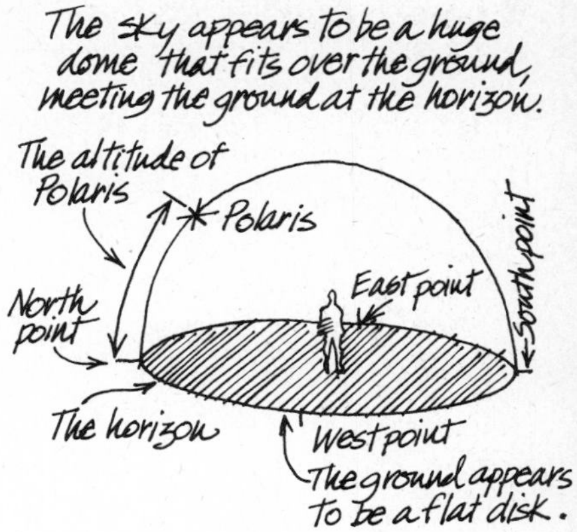

Figure 2.4 The world's largest domed stadium. This concept, that the sky appears to be a dome over what appears to be a flat earth, is the basic idea behind the celestial sphere, shown in Figure 2.5. One can imagine that Polaris and all other stars appear to be attached to the dome. The observer always seems to be located at the center of the apparently flat disk.

The Rotational Period of the Celestial Sphere How long does it take the celestial sphere to rotate once? You can measure this yourself by observing how long it takes a given star to travel its circular path

Figure 2.5 **The concept of the celestial sphere.** The reader should imagine the view from inside the sphere where the observer is shown standing. How will the stars appear to move if each is attached to the celestial sphere rotating about its axis as shown?

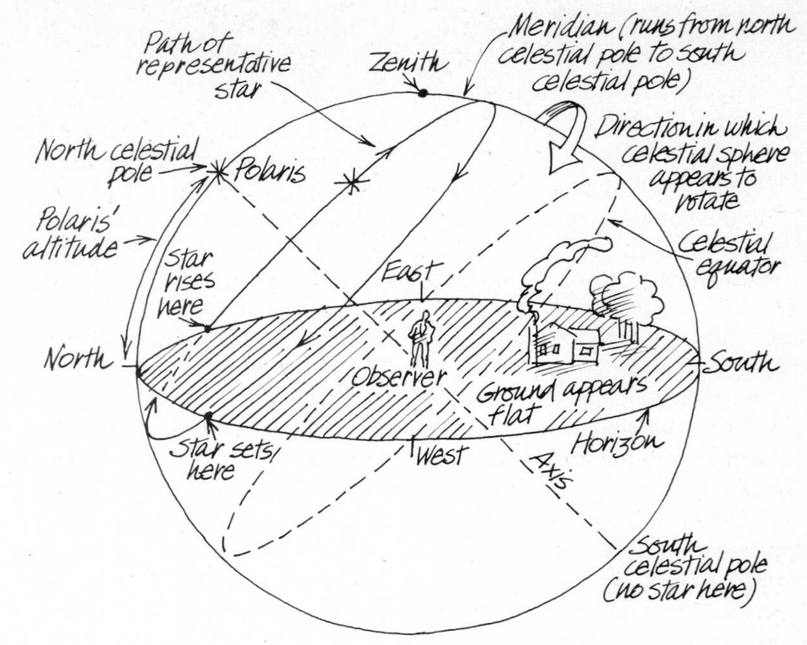

Figure 2.6 The motions of stars. The celestial sphere is an aid to visualizing the motions of stars. Stars move as if attached to the celestial sphere. The figure illustrates the statement, The further a star is located from the north celestial pole (i.e., from Polaris), the less time it spends above the horizon.

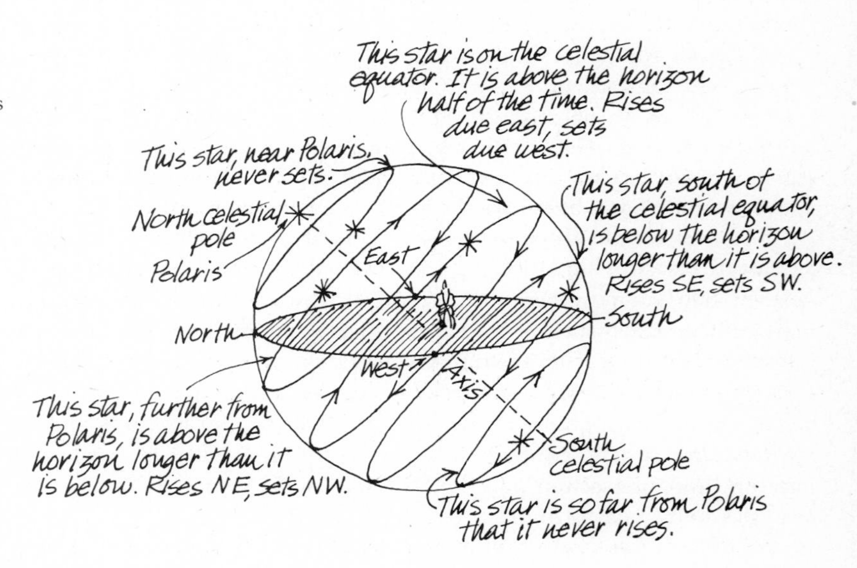

once. A casual observer would say it takes 24 hours (hr), but a careful observer would find that the celestial sphere rotates once every *23 hr, 56 minutes* (min), a number the reader may wish to remember. An observer equipped with the appropriate instruments would obtain a result closer to 23 hr, 56 min, 3.6 seconds (sec). This length of time is called a **sidereal day.**

The stars do not quite keep step with our clocks which run on the 24-hour solar day, discussed in Section 2.2. The fact that the stars take 4

PUZZLE 2.3

STAR WATCHING AND THE SIDEREAL DAY

Suppose each evening you look out an east window when your clock reads exactly 11:00 P.M. (There is nothing significant about 11:00 P.M. Choose any convenient time.) Suppose that tonight you see a bright star just above the eastern horizon at that time. Then you close the curtains until tomorrow at 11:00 P.M., when you look again. Will the star be higher above the horizon? Will it be at the same place? Will it be invisible because it has not yet risen? (Check your solution with the one in Appendix 2.)

---

min less than 24 hr to complete one circuit makes star watching more interesting. To understand some of the consequences of the "missing four minutes," as some call it, solve Puzzle 2.3.

The same kind of reasoning that answered Puzzle 2.3 also leads one to conclude that the stars in the western sky are a little nearer the horizon each day at the same time. The overall effect is that, each succeeding evening at the same time, new constellations gradually appear above the eastern horizon while old ones gradually sink below the western horizon. The result is an interesting, slow parade of stars.

Consider once more the star just above the eastern horizon tonight at 11:00 P.M. Tomorrow it will be higher at 11:00 P.M., and the next night higher yet. Night by night it will creep across the sky until, at 11:00 P.M., it will be in the western sky. How long will you have to wait until this star starts repeating? In other words, how long from tonight must you wait until you again see the star just above the eastern horizon at 11:00 P.M.? Careful observation will show that, night by night, each star actually advances slightly less than 1° per day if always observed at the same time. In one year, it is found, the star advances 360° or one full circle. Thus, the parade of stars lasts one year before it starts repeating. If the constellation Orion is rising tonight at midnight, it will again be rising at midnight one year later.

2.2 THE APPARENT MOTION OF THE SUN

Warning: This book will encourage you to view many celestial objects, but you must *never* look at the sun directly with your naked eye or through a telescope, binoculars, smoked glass, photographic negatives or anything else. Your eye is very sensitive and easily damaged. A safe way to observe the sun is described in Box 8.1. Ophthamologists frequently see patients who have suffered permanent damage to their retinas from glancing at the sun.

Since this book is intended to be an introduction to astronomy, we do not give every interesting detail on every subject. In discussing the sun, for example, we describe its motion without delving into some of the finer details. Those who desire more information should consult a more advanced text.

One learns in childhood that the sun rises in the east and sets in the west. In fact, unless it is circumpolar, every heavenly object rises in the eastern sky and sets in the western sky. There are no exceptions. If you see something rising in the west, it might possibly be an airplane or an artificial earth satellite, but it isn't the sun, moon, a planet, or a star.

Perhaps it cannot be overemphasized that in this chapter we are describing apparent motions. We do not mean to assert or deny that the observed motions are real motions.

The Daily Motion of the Sun The motion of the sun through the sky may seem complex unless it is broken down into two parts. These parts may be called the daily motion and the motion in declination.

The term **daily motion** refers to the part of the sun's motion that it repeats each day. Each day it rises in the eastern sky—higher and higher above the horizon until it reaches the meridian. We base our system of time on the sun, because it has such a powerful influence on our activities. Clocks are set to 12 o'clock noon when the sun reaches the meridian. Times before noon are labeled A.M. for *ante meridiem* (Latin for "before the meridian"). After noon, the sun drops toward the western horizon. Times occurring after noon are labeled P.M. or *post meridiem* ("after the meridian"). Because our clocks run on a 24-hr system based on the sun's daily motion, the sun returns to the meridian once every 24 hr. (Note, not every 23 hr 56 min, but every 24 hr.)

This is one of several places in which we reluctantly omit fine details. Although, on the average, the sun returns to the meridian once every 24 hr, there are daily discrepancies. These tend to cancel out in the long run.

The Sun's Motion in Declination At the same time it is performing its daily motion, the sun is also moving in another way. This motion, which may be called the sun's **motion in declination,** takes a year to complete. We can cancel out the daily motion by agreeing to observe the sun's position each day at noon. That way, the sun will always be on the meridian when we observe it. We will then find that it is at a slightly different place on the meridian each day.

Suppose we begin observing the sun in late March. On March 20 we find the noon sun right on the celestial equator. (The date may vary slightly, but we will not concern ourselves with such details.) Were we to watch the sun in its daily motion on March 20, we would find that it followed the celestial equator that day, rising at the east point on the horizon and setting at the west point. Because of this, the sun spends half of its time above the horizon and half below. We have 12 hr of daylight and 12 hr of night. Since it is spring and day and night are equal in length, this day is called the **spring** or **vernal equinox.** (*Equinox* means "equal night" in Latin.) (See Figure 2.7 and Table 2.1.)

A careful observation at noon the next day would reveal that the sun has moved a bit north of the celestial equator. This trend continues for three months until, on June 21, the sun reaches its maximum distance north of the celestial equator. The day on which the sun stops moving north is called the **summer solstice** (from the Latin for "sun stationary"). The sun is $23\frac{1}{2}°$ north of the celestial equator that day. It rises in the northeast and sets in the northwest. It spends the maximum amount of time above the horizon. We have our longest day and shortest night. We have warm weather at this time for several reasons. In the northern hemisphere the days are long, so the sun has more time to heat the land and oceans. Also, the land and oceans have less time to cool off during the short nights. Moreover, the sun is high in the sky during much of the day, so that its rays come down at the least possible

The seasons are reversed in the southern hemisphere. In late June the southern hemisphere has short days, long nights, and cold weather.

View from outside of celestial sphere

View seen by the observer in figure on far left, i.e., one who is on earth, facing south

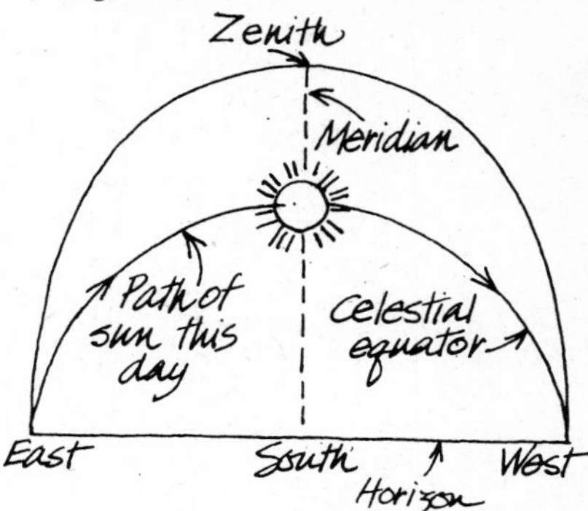

March 20 Spring equinox **or** *September 22 Autumnal equinox*

On both days the sun is on the celestial equator.

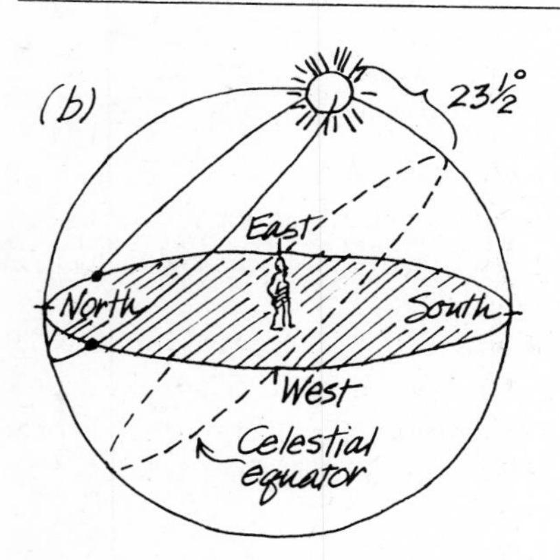

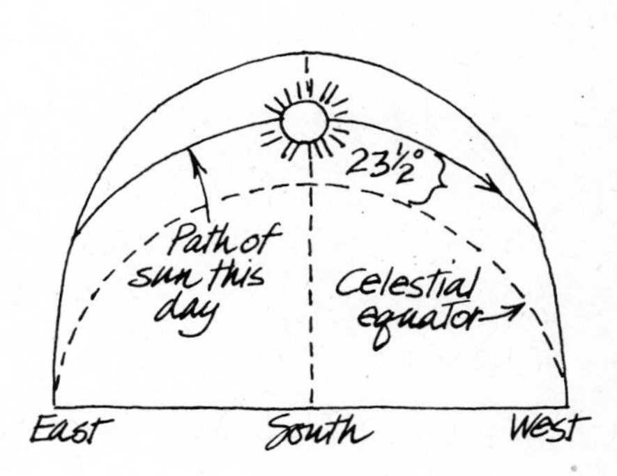

June 21 Summer solstice

The sun is $23\frac{1}{2}°$ north of the celestial equator.

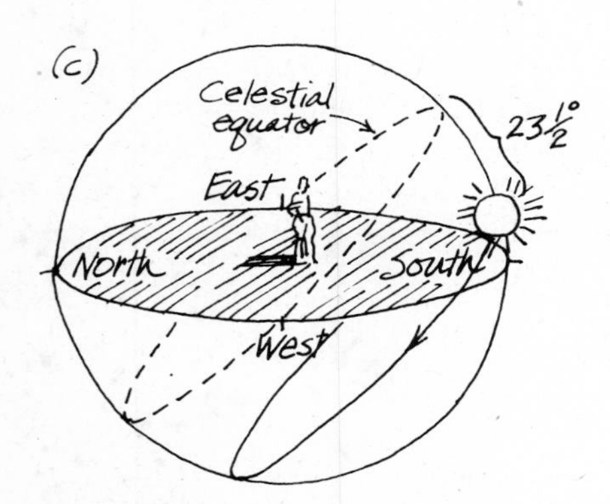

December 22 Winter solstice

The sun is $23\frac{1}{2}°$ south of the celestial equator.

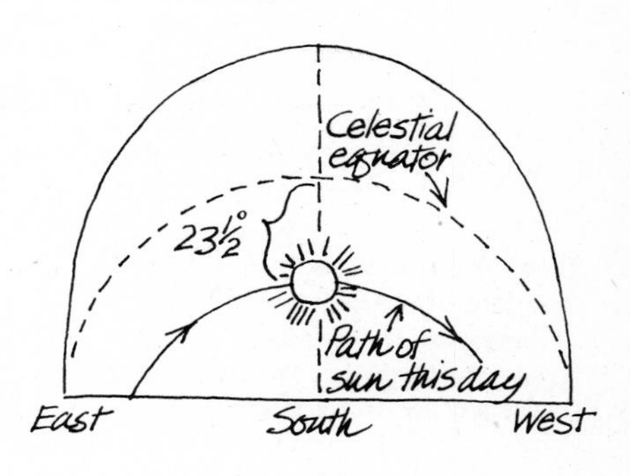

Figure 2.7 The sun's motion in declination. Four important positions of the sun are shown as viewed from outside the celestial sphere (left) and as viewed from the earth (right). The sun's path for each day is shown. The sun's position at noon (on the meridian) is indicated. Starting with the spring equinox, the sequence of views is (a), (b), (a), (c). Study this figure along with Table 2.1.

Table 2.1 The Sun's Motion in Declination

Date	View in Figure 2.7	Position of sun	Name of event	In the northern hemisphere: Length of daylight	Length of night
March 20	(a)	On celestial equator; moving farther north each day	Spring (or vernal) equinox	12 hr	12 hr
June 21	(b)	$23\frac{1}{2}°$ north of celestial equator; not moving north or south today	Summer solstice	Longest daylight	Shortest night
September 22	(a)	On celestial equator; moving farther south each day	Fall (or autumnal) equinox	12 hr	12 hr
December 22	(c)	$23\frac{1}{2}°$ south of celestial equator; not moving north or south today	Winter solstice	Shortest daylight	Longest night

PUZZLE 2.4

A RACE BETWEEN THE SUN AND A STAR

Suppose that stars are visible during the day. What if a star and the sun are on the meridian one day at noon. Which will return first to the meridian the next day? (The answer is given in Appendix 2.)

Each observer on earth can compare the motions of the heavenly bodies to a framework that is stationary relative to the observer. This framework consists of the horizon and meridian. When we use the phrase "motion with respect to an observer on earth," we mean that the observer watches the motion of the object as it rises, crosses the sky, and sets. The observer relates this motion to the fixed framework.

slant. This makes the sun's rays most effective in heating the earth's surface. (Usually the hottest weather follows the summer solstice by about six weeks. Any object requires time to heat up when heat is applied.)

After June 21 the sun begins moving south, back toward the celestial equator, day by day. In three months it reaches the celestial equator again. September 22 is called the **fall** or **autumnal equinox.** It is a repeat of the spring equinox, one half year before, except that now, day by day, the sun is moving south.

The sun continues moving south for three months until, on December 22, things look rather bleak. On this day the sun is $23\frac{1}{2}°$ south of the celestial equator. Fortunately, it goes no further but stops its southerly motion. This day is known as the **winter solstice.** If we were to follow the sun all day, we would find that it rose in the southeast and set in the southwest. We have our shortest daylight hours and longest night. Short days and the slanting rays of the sun combine to produce our coldest weather. (The earth continues to cool for about six weeks after the solstice.)

The Apparent Motion of the Sun with Respect to the Fixed Stars The combined daily motion and motion in declination of the sun describe the apparent motion of the sun as viewed by an observer on the earth. It is convenient to have yet one more way of describing the movement of the sun. Up to now we have discussed the motion of the sun with respect to an observer on earth. Another convenient way of describing the motion of the sun is to compare its apparent motion with respect to the fixed stars. There is nothing new here; it is just another way of describing the sun's motion. As an introduction to this topic, solve Puzzle 2.4.

We use the phrase "to the right" as a rough equivalent for "to the west," and "to the left" for "to the east." These phrases apply to an observer in the northern hemisphere looking toward the south at objects in the vicinity of the meridian. The reader should get used to the correct terminology (east and west) because the rough equivalents will be dropped later on.

In Puzzle 2.4 we assume the stars are visible by day. Actually, the sun causes so much glare in the atmosphere that it drowns out the weak starlight. Yet we know that the stars are really there during the day. The brighter stars appear when the moon completely covers the sun, that is, during a total solar eclipse. Astronauts can see the stars by day once they get above our atmosphere.

Now for the point at which we have been aiming: We can express the result of the race in the previous puzzle in two *equivalent ways.*

1. With respect to an observer on earth, the star takes 4 min less than the sun to return to the meridian. As a result, the next day the star is about 1° further west (to the right) of the sun.

2. Statement 1 is equivalent to saying: As compared to the star, the sun has fallen behind. Each day the sun will be about 1° farther to the east (to the left) of the star, as shown in Figure 2.8. One can say that *the sun moves from west to east* (from right to left) *with respect to the fixed stars.*

This last sentence would be nonsense if it had not included the phrase "with respect to the fixed stars." Ordinarily one assumes that motions are referred to a framework fixed in the earth. Speed limit signs really should say, "90 kilometers per hour (km/hr) *with respect to the ground,*" but everyone knows what is meant. If a motion is referred to some other framework, a phrase such as "with respect to" or "as compared to" should appear. You will also see it put this way: "The sun moves through the fixed stars from west to east."

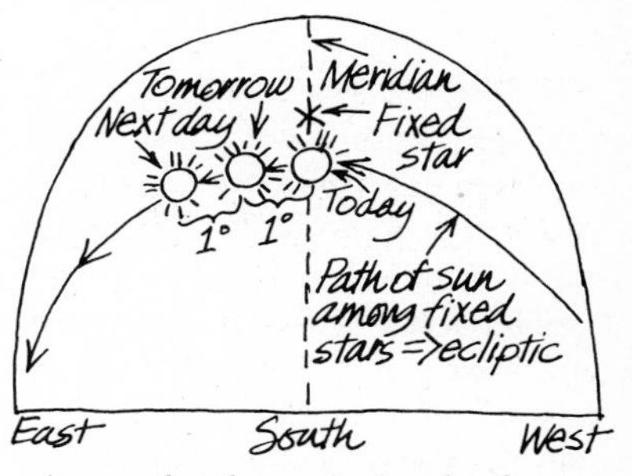

Assume that the sun is viewed each day only when a certain star is on the meridian.

Day	*Time*
Today	*12:00 Noon*
Tomorrow	*11:56 A.M.*
Next day	*11:52 A.M.*

Each day the sun seems to have shifted its position as compared to the fixed star. It is progressively farther to the east (left) of the star each day.

Figure 2.8 The motion of the sun among the fixed stars. As explained in the solution to Puzzle 2.4 (see Appendix 2) the sun falls behind a fixed star by slightly less than 1° each day. One says, "The sun travels from west (right) to east (left) among the fixed stars." At the the same time, both the sun and the star continue to rise in the east and set in the west each day. The sun merely falls behind the swifter star.

If the sun is lined up with a certain fixed star today, it will be slightly less than 1° to the east of it the next day. It will move about 1° further each day. How long will it take before the sun once again lines up with the same star? Since the sun covers slightly less than 1° per day, it takes slightly more than 360 days to cover 360° or a full circle. The exact result is that the sun covers a full circle with respect to the fixed stars in 365.26 days or one year. A **year** is defined as the time required for the sun to return to a particular place among the stars in the constellation Pisces. This motion of the sun among the fixed stars is called its **annual motion.**

You may have noticed the connection between the missing 4 min described in Section 2.1 and the 1° per day motion of the sun among the fixed stars. The reason a star advances by about 1° or 4 min each day at the same time of day is that our clocks run on 24-hr sun time.

As the sun moves through the fixed stars each year, it follows the same track. For example, it always passes close to Regulus, the brightest star in the constellation Leo (the Lion), and the star Spica, the brightest star in Virgo (the Virgin). The path the sun traces out each year among the fixed stars is called the **ecliptic.** On the celestial sphere, the ecliptic is a large circle intersecting the celestial equator at two points. The ecliptic is tilted at an angle of $23\frac{1}{2}°$ with respect to the celestial equator.

BOX 2.2

ASTROLOGY

In the beginning, astrology and astronomy were largely practiced by the same people. (Hipparchus, the astronomer, was a well-known exception.) For thousands of years before the Christian Era, many people studied the stars and planets (astronomy) and tried to figure out how the stars affected the lives of humans (astrology). The two professions were not seen as distinct for thousands of years. Tycho Brahe, the great Danish observer of the sixteenth century (see Chapter 5), was in large part motivated to measure celestial motions because he wanted to further refine his astrological predictions.

The break between astrology and astronomy began not long after Tycho's day. Gradually the gulf between the two fields widened until today the separation is complete.

The break between the two subjects began in the sixteenth century at about the same time as the scientific revolution. Since the time of ancient Greece, no really major advances had occurred in science, although some interesting work had been done during the Middle Ages. As science was reborn, scientists slowly began to recognize that a good theory must meet certain criteria that astrology did not meet. Among these criteria was predictive ability. A good theory concerning the motion of the sun, for example, should be able to predict accurately where the sun will be in the sky at some future specified date. Although astrologers had made a few successful predictions, astronomers began to realize that all too many astrological predictions were either wrong or were so vague that one could never be certain whether they had come to pass or not. Every so often scientists have performed careful scientific experiments to test astrological predictions. If astrology were ever to pass such a test, it would be dramatic news indeed. However, astrology has failed all such tests. Astronomers today classify astrology as a superstition.

Many of the rules of astrology were laid down thousands of years before the Christian Era. Since that time, slow but steady changes have occurred in the sky. The ecliptic has undergone a very gradual shift in position with respect to the fixed stars. When astrology was formulated, the sun passed through only the first 12 constellations listed in Table 2.2. A change in the sun's path now causes it to pass through Ophiuchus in addition to the other 12 constellations. In addition, this shift in the sun's path causes it to pass through the various constellations at times markedly different from those observed in the early days of astrology. Some astrologers have attempted to improve their predictions by taking this knowledge into account, but no improvement has been found in careful scientific tests.

See Figure 2.9. It may be thought of as attached to and moving with the celestial sphere.

Table 2.2 lists the 13 major constellations the sun passes through each year. The band in the sky centered on the ecliptic in which these constellations lie is called the **zodiac.** See also Box 2.2.

2.3 THE APPARENT MOTION OF THE MOON

We describe the apparent motion of the moon from three different points of view: motion with respect to an observer on earth, motion with respect to the fixed stars, and motion with respect to the sun.

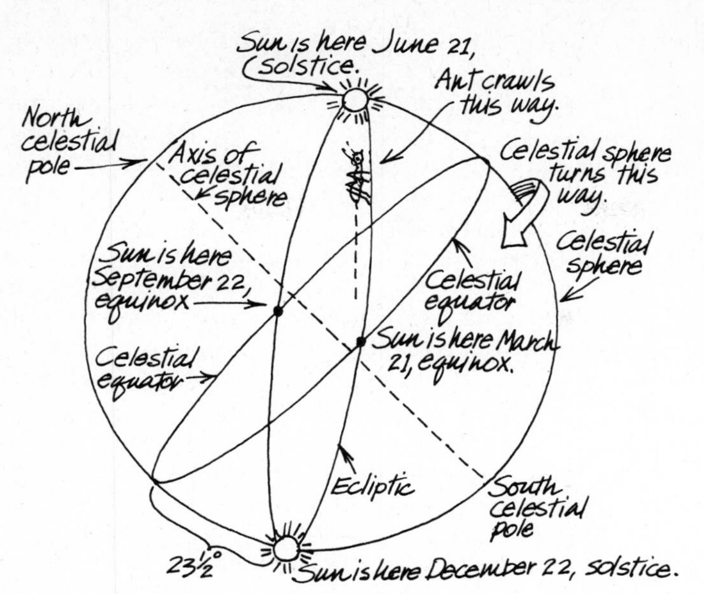

Figure 2.9 The ecliptic on the celestial sphere. As far as appearances are concerned, one might imagine that the sun behaves like an ant crawling on the celestial sphere. It takes one year to complete the trip. This motion carries it north and south of the celestial equator at different times of the year. At the same time, the celestial sphere spins around once each 23 hr, 56 min carrying the sun with it. This spinning causes the sun's daily motion.

Table 2.2 The Constellations of the Zodiac[a]

Constellation name	*Approximate date when constellation is on the meridian at midnight*[b]	*Approximate date when sun is in the midst of the constellation*[c]
Sagittarius (archer)	July 7	January 7
Capricornus (goat)	August 8	February 8
Aquarius (water carrier)	August 25	February 25
Pisces (fishes)	September 27	March 27
Aries (ram)	October 30	April 30
Taurus (bull)	November 30	May 30
Gemini (twins)	January 5	July 5
Cancer (crab)	January 30	July 30
Leo (lion)	March 1	September 1
Virgo (virgin)	April 11	October 11
Libra (scales for weighing)	May 9	November 9
Scorpius (scorpion)	June 3	December 3
Ophiuchus (serpent holder)	June 11	December 11

[a] These are the constellations the sun passes through once each year. The sun spends varying amounts of time in each one mainly because of the varying widths of the constellations on the ecliptic. One may locate the approximate position of the sun on one's birthday by referring to the closest date in the last column. Do not be surprised if the result differs from tables in astrological sources. See Box 2.2.
[b] The constellation is best viewed for several months on either side of this date.
[c] What is the relationship between these dates and the ones in the previous column? Why is this so?

PUZZLE 2.5

A RACE BETWEEN THE SUN, THE MOON, AND A FIXED STAR

Suppose that the sun, the moon, and a fixed star are on the meridian at one time. In what order will they cross the finish line tomorrow? (See Appendix 2.)

The Apparent Motion of the Moon with Respect to an Observer on Earth If the moon is on the meridian right now, later it will set in the west and still later it will rise in the east. On the average, it returns to the meridian about once every 24 hr, 50 min. As a result, if the moon is on the meridian (i.e., directly above the south point on the horizon) at 10:00 P.M. tonight, it will return to the meridian at about 10:50 P.M. tomorrow and at about 11:40 P.M. the night after that. One can say that the moon crosses the meridian an average of about 50 min later each night.

Much as the times of rising and setting of the sun vary widely, so do those of the moon. It is correct to say that, *on the average*, the moon rises 50 min later each day and sets 50 min later each day. But, at times, the lag is much more or much less than 50 min.

The Apparent Motion of the Moon with Respect to the Fixed Stars Now we are in a position to predict the outcome of a three-way race. Work out Puzzle 2.5.

One conclusion you can draw from this puzzle is that the moon travels among the fixed stars from west to east (right to left). It travels among the fixed stars in the same direction the sun does. When discussing motion with respect to an observer on earth, we say that the moon moves more slowly than the sun. (It lost the race.) But, when motions are compared to the fixed stars, the moon is swifter than the sun. Measurement shows that, on the average, the moon travels 13° on the celestial sphere per day, as compared to the sun's 1° per day. A careful observer can notice this rapid motion of the moon with respect to the fixed stars after only an hour or two. Thirteen degrees per day works out to roughly $\frac{1}{2}°$ per hour. This means that the moon seems to move through the stars by an amount equal to its own apparent diameter each hour.

Since the moon moves 13° per day on the celestial sphere, it travels completely around the celestial sphere in a relatively short time. If it appears to be near a certain star now, it will travel around the celestial sphere and "visit" that star again in about 27 days, 8 hr. This is known as the **lunar sidereal period.** (*Lunar* means "of the moon." *Sidereal* comes from the Latin word *sidereus,* or "starry." *Period* in science often means "interval of time until something starts to repeat." Thus we have the "starry period of the moon.")

The sun travels along the ecliptic among the fixed stars year after year. The motion of the moon among the fixed stars is not quite as simple. We will be content to know that the moon, in its travels, stays close to the ecliptic. It is always found no more than 5° north of or south of the ecliptic. This means that it also passes through the same constellations (the zodiac) that the sun does. The moon does not pass through exactly the same *parts* of these constellations from one pass to the next.

The fact that the moon stays near the ecliptic on the celestial sphere has several implications. For example, like the sun, the moon is never found near the northern stars, such as the Big Dipper or Polaris. As viewed from the United States, the moon always crosses the meridian in the southern half of the sky.

The Apparent Motion of the Moon with Respect to the Sun A third way to think about the motion of the moon is to consider how it catches, or appears near, the sun in the sky, leaves the sun, and later returns. As we saw in the solution to Puzzle 2.5 (See Appendix 2), the sun falls behind a star by 1° per day, while the moon falls behind a star by 13° per day. Thus the moon travels about $13° - 1° = 12°$ farther to the east of the sun each day. If the moon appears to be near the sun today, the length of time required for it to return to the sun again is called one **lunar synodic period.** How long is that? The moon gains 12° per day on the sun and so takes about $360° \div 12 = 30$ days to catch up to it once more. More precisely, one lunar synodic period equals about 29 days, 13 hr.

We may say that the moon returns to the sun about once each month. This is no coincidence. On many calendars, the month is, at least roughly, based on the moon's motion. The words "month" and "moon" come from the same root. Many nations base their calendar on the moon. On our calendar, the date of Easter is reckoned by using the first full moon in spring.

The Phases of the Moon During one synodic period, the moon undergoes a gradual change in appearance from a crescent to a disk and back to a crescent. One says that the moon passes through various **phases.** A related fact, which the casual observer might not notice, is that the moon always has the same phase when it is in the same position relative to the sun.

We next describe the moon's phases during one synodic period. (The reader may easily observe the results we are about to describe.) Since both the sun and the moon appear to move, we find it convenient to observe the moon each day just after the sun has set. This will allow us to concentrate on the relationship between the moon's phase and the position of the moon relative to the sun.

Let's begin our observations on a day when the newspapers and calendars say it is the day of the **new moon.** We find that the moon is not visible at sunset. On this day it is nearest the sun in the sky. Before sunset, the moon is lost in the intense glare caused by the sun, and at sunset it is also setting. We will be unable to see the moon for about two days until is has moved far enough from the sun. By watching each evening just after sunset we will at last glimpse the moon in the west. It appears to be a slim crescent, and sets not long after the sun does. As shown in Figure 2.10, the horns of the moon always point away from the sun, and the bright side of the moon is always toward the sun.

As we continue to view the moon at sunset each evening, we note two things. Each day it moves further toward the east. It also appears to increase in width. One says the moon is *waxing* (growing). This process

Figure 2.10 The phases of the moon.

New moon: *The moon is so near the sun in the sky that it is not visible because of the sun's glare. The new moon sets as the sun does.*

Several days after new moon: *The moon appears to have moved away from the sun toward the east. The horns (points) of the crescent moon always point away from the sun. By the same token, the illuminated side of the moon is always toward the sun.*

First quarter moon: *The moon is about 90° east of the sun. The moon looks like a half circle, bright side toward the sun. Some calendars and almanacs mistakenly draw the first quarter moon as a crescent. The first quarter moon crosses the meridian at sunset.*

Full moon: *The moon is about 180° from the sun. It looks like a full circle and rises at about sunset.*

Last (or third) quarter moon: *The moon is approaching the sun day by day. It is now about 90° west of the sun and crosses the meridian at about sunrise. Its shape has become a half circle.*

Several days before the next new moon: *The moon is a slim crescent as it approaches the sun. Soon it will be a new moon and the cycle will begin anew.*

continues until, about one week after the new moon, the **first quarter moon** occurs. At sunset we in the northern hemisphere now face south to view the moon. It is near the meridian at sunset, and so about 90° east of the sun. At first quarter, the moon is shaped like a capital "D". There is a temptation to call this the "first half moon," but this is not the traditional name. During this time it is easy to verify that the moon is often visible in broad daylight by looking for it before sunset.

The daily progression of the moon eastward away from the sun continues until, about two weeks after the new moon, the moon is in the region of the sky opposite the sun. This is the time of the **full moon.** The moon now looks like a complete circle and rises in the east at about sunset.

After the full moon, we can no longer look for the moon at sunset because it now rises after sunset. We can change our strategy by observing the moon at *sunrise.* As the sun is about to rise in the east, we see the full moon setting in the west. Day by day the moon continues to travel toward the east relative to the sun, but now it is approaching the sun. We also notice that the moon is gradually *waning* (thinning).

About three weeks after the new moon, or about one week after the full moon, the moon is near the meridian at sunrise. It looks like a half circle again, but now like a *backward* letter "D." This phase is called the **third quarter,** or **last quarter.** As we continue to observe at sunrise, we see the moon getting closer and closer to the sun each morning as it thins down to a slim crescent. About five days after the third quarter, it is so close to the sun that the narrow crescent is lost in the twilight. Finally, about one month (actually 29 days, 13 hr) after the previous new moon, the moon is new once more, that is, it appears to be close to the sun in the sky. We have observed a complete cycle of the phases of the moon.

Ordinarily, one loses track of the moon during the new moon. It is somewhere within 5° of the sun in the sky but not visible. Once in a rare while, however, observers get a rare treat. The moon passes directly in front of the sun at the new moon. This splendid event is called a solar eclipse and is discussed further in Section 8.2.

2.4 THE APPARENT MOTIONS OF THE PLANETS

The sun and the moon do not look like fixed stars, and they do not *behave* like fixed stars; they *travel among* the fixed stars. Careful observation with the naked eye also reveals five objects that *do* look like stars but which move with respect to the fixed stars. These are the five planets, Mercury, Venus, Mars, Jupiter, and Saturn. (The Greek word *planetes* means "wanderer.") The planets discovered after the invention of the telescope will be discussed in Chapter 11.

How can one tell if an object that looks like a star is really a planet? The surest method is to consult the monthly sky chart in the latest issue

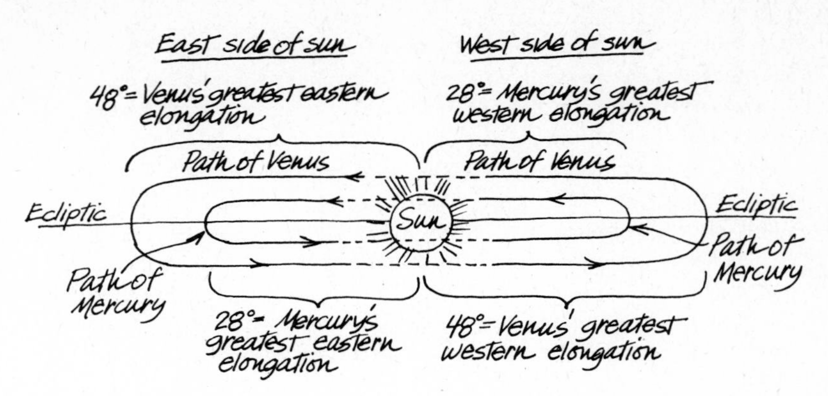

Figure 2.11 The apparent motions of the inferior planets. Each inferior planet travels slowly to the east side of the sun, turns around, and then travels quickly to the west side. The paths are not drawn solid near the sun to indicate that the naked eye cannot see the planets there. The eye cannot perceive depth in the sky, and so one cannot determine whether an inferior planet passes behind or in front of the sun.

of a magazine such as *Astronomy* or *Sky and Telescope.* (See also this chapter's "Readings" section.)

The Inferior Planets: Mercury and Venus The five planets easily visible to the naked eye can be conveniently divided into two groups according to their apparent motions. The **inferior planets** are Mercury and Venus, while the **superior planets** are Mars, Jupiter, and Saturn.

The words "inferior" and "superior" are not used to indicate the value of each planet. They are used more nearly in the sense of "nearer" and "farther." This usage will become clearer in subsequent chapters.

The inferior planets are distinguished by the fact that they are the sun's constant companions. Mercury and Venus are always found in the sky not very far from the sun. They move from one side of the sun to the other but never leave its vicinity completely. Venus, for example, moves to the east (left) of the sun until it reaches an angular separation of about 48° from the sun. This position is called its **greatest eastern elongation.** Then it rapidly moves back and reappears on the western (right) side of the sun. It eventually reaches its **greatest western elongation** of about 48°. Next, Venus returns to the sun, and the cycle begins again. Mercury performs a similar oscillation from one side of the sun to the other. However, Mercury moves faster than Venus and has maximum elongations never exceeding about 28°. Mercury appears to stay even closer to the sun than Venus does. See Figure 2.11. Among other things, this figure shows that the inferior planets remain quite close to the ecliptic as they perform their motions.

Because of their behavior, the best time to look for Mercury or Venus is just before sunrise or just after sunset. (See Figure 2.12.) When either inferior planet is to the east of the sun, it sets after the sun does. If it is far enough from the sun, one can view it in the west in the evening at that time. It is an "evening star." Similarly, when it is west of the sun,

An inferior planet as evening star. *When the planet is on the east side of the sun, it sets after the sun; it is seen in the west after sunset. The planet's path relative to the sun is shown as a dashed line. Each night the planet is further advanced along the path. The planet travels away from the sun slowly and returns quickly.*

An inferior planet as morning star. *When the planet is on the west side of the sun, it rises ahead of the sun; it is seen in the east before sunrise. The planet advances daily along the dashed line. It travels quickly away from the sun and returns slowly. (Keep in mind that a phrase such as "travels away from the sun" means "the angle between the planet and the sun as observed from earth increases."*

Figure 2.12 An inferior planet as evening star and morning star.

it rises ahead of the sun in the eastern morning sky as a "morning star." Venus is very easy to locate when farthest from the sun. It is very bright, third after the sun and the moon in brightness. Its angular separation from the sun periodically becomes large enough so that it may be viewed in a completely dark sky. Mercury, on the other hand, is always seen during twilight, even at maximum elongation. A beginner will need some patience and guidance during a first attempt to locate Mercury. (See also Section 10.3.)

A planet is said to be in **conjunction** with the sun when its angular separation from the sun reaches a minimum. The planet is then nearly lined up with the sun. The planet is not usually visible to the naked eye at conjunction, because it is lost in the sun's glare. Two types of conjunction have been defined for inferior planets. **Inferior conjunction** occurs while a planet is crossing from the east side of the sun to the west side. **Superior conjunction** occurs for an inferior planet when it is crossing from the west side of the sun to the east side.

Each inferior planet goes through a cycle as it moves with respect to the sun. If, for example, we begin watching the planet when it is at its greatest eastern elongation, the sequence is: greatest eastern elongation, inferior conjunction, greatest western elongation, superior conjunction, and back to the greatest eastern elongation. Then the cycle begins to repeat once more. The length of time required for an inferior planet to execute one such cycle is known as its **synodic period.** Mercury's synodic period is much shorter than Venus' (see Table 2.3). Perhaps that is why Mercury was named for the fleet messenger of the gods. Table 2.3 also shows that each inferior planet takes less time to travel from greatest eastern to greatest western elongation than vice versa.

Table 2.3 Some Data Concerning the Motions of the Planets As Viewed from Earth

Planet	Synodic period		Inferior planets		Approximate time planet spends executing retrograde motion during each synodic period (months)[a]	Approximate time required for planet to traverse the celestial sphere once (years)[a]
	Expressed in days	Expressed in years	Approximate time from greatest eastern to greatest western elongation[a]	Approximate time from greatest western to greatest eastern elongation[a]		
Mercury	116	0.32	45 days	71 days	1	1
Venus	584	1.60	5 months	15 months	1	1
Mars	780	2.14			2	2
Jupiter	399	1.09			4	12
Saturn	378	1.03			4	30

[a] These numbers can only be considered rough averages. The actual numerical value in any individual instance may deviate from the number given.

The Superior Planets: Mars, Jupiter, and Saturn The superior planets periodically go into conjunction with the sun, but they also appear to leave its vicinity and go to the region of the celestial sphere farthest from the sun. A planet in the region of the sky farthest from the sun is said to be in **opposition.** When a planet is in oppostion, it rises as the sun sets because it is very nearly in the direction opposite the sun as viewed from earth. While in opposition, a planet crosses the meridian at midnight. It is important to remember that superior planets go into opposition with the sun while inferior planets do not. (One never observes Mercury or Venus on one's meridian at midnight.) An additional interesting fact is that each superior planet looks brighter during opposition than at other times.

If we observe the motion of a superior planet with respect to the sun starting at the time of conjunction, we will find that it first moves to the west side of the sun and may be seen rising ahead of the sun in the morning. Its angular separation from the sun continues to increase until opposition occurs. At such a time the planet is visible all night, rising at sunset and setting at sunrise. Next the planet begins to return to the sun, its angular separation decreasing. If watched every evening just after sunset, the superior planet will appear farther west, closer to the sun, each night until the next conjunction occurs. The length of time required for this entire cycle is known as the superior planet's synodic period. See Table 2.3.

The Apparent Motions of the Planets with Respect to the Fixed Stars All planets are characterized by the fact that they move with respect to the fixed stars. Each planet stays near the ecliptic as it performs this motion. If one watches a planet from one week to the next, one will *usually* observe that it travels from west to east among the fixed stars during that time. (The need for the word "usually" is explained below. Recall that the sun and the moon *always* travel from west to east with respect to the fixed stars.) This overall eastward drift of the planets eventually carries them completely around the celestial sphere. Table 2.3 gives the *average* time each planet takes for one trip around the celestial sphere.

Each inferior planet accompanies the sun as it travels with respect to the fixed stars, sometimes leading it and sometimes lagging behind. As a result, after the sun has completed its yearly trip around the celestial sphere, the inferior planet also has returned to the same general region of the celestial sphere. One can say that an inferior planet takes *roughly* one year to travel around the celestial sphere as it accompanies the sun.

The average motion of each superior planet also carries it in an eastward direction on the celestial sphere. Mars takes an average of two years to complete the circuit, Jupiter about 12 years, and Saturn about 30 years. With respect to the fixed stars, the average motion of Mars is more rapid than Jupiter's, and Jupiter in turn moves more rapidly than Saturn. The sun takes only one year to make the circuit and, as a result, appears to catch up to the superior planets. During one year, Saturn makes little progress eastward among the fixed stars, so that the sun catches up to it in a little more than one year. This helps us to understand why Saturn's synodic period is only a little longer than one year. (Why is Mars' synodic period so long?)

The Direct and Retrograde Motions of the Planets with Respect to the Fixed Stars We now consider the most puzzling feature of planetary motions. Imagine a very ancient astronomer observing a bright starlike object in the sky for several months. He notices at first that it is changing its position with respect to the fixed stars, moving farther east, each night. He calls this wandering star a "planet" and its motion **direct motion.** He assumes that it will continue this eastward motion forever. But during the last few nights the planet stops its eastward drift among the fixed stars. Has it become a fixed star? No, for now the astronomer observes that it has turned around in its tracks and begun to back up. It is now traveling from east to west among the fixed stars. The planet is performing **retrograde motion.** Each night the astronomer marks the position of the planet on a star chart. The path the planet follows among the fixed stars can then be traced out, as in Figure 2.13. See also Figure 2.14.

The retrograde motion continues for some time. Then the planet stops and resumes its normal direct motion, west to east among the fixed stars.

The above description of direct and forward motion applies equally well to inferior and superior planets. However, the inferior planets differ from the superior planets in this respect: The inferior planets perform retrograde motion while near the sun in the sky, but the superior planets perform retrograde motion when in opposition to the sun. An inferior planet performs retrograde motion during an interval of time beginning before and ending after its inferior conjunction. During the rest of its synodic period, it is performing direct motion,

Whether a planet is performing direct or retrograde motion with respect to the fixed stars, an observer watching the planet in the sky *always* sees it travel across the sky from east to west. Each planet *always* rises in the east, crosses the meridian, and sets in the west. Direct and retrograde motion refer to the motion of the planet *with respect to the fixed stars.*

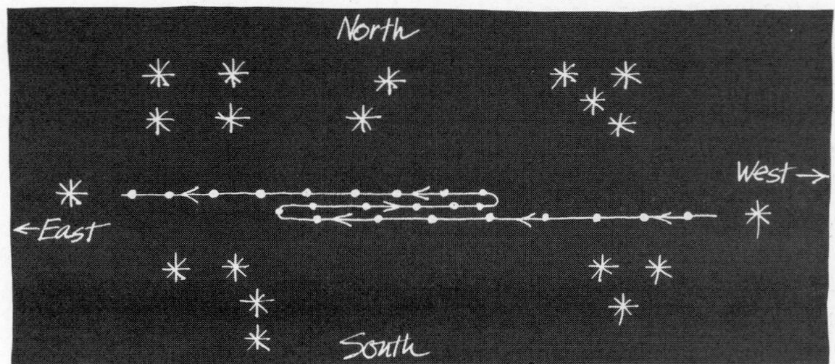

Figure 2.13 A planet performs retrograde motion. The changing position of a planet is noted on a map of the sky. (Observe that here as on most star maps, east is on the left and west on the right, the opposite of a land map. A star map is made to be held up for comparison with the sky.)

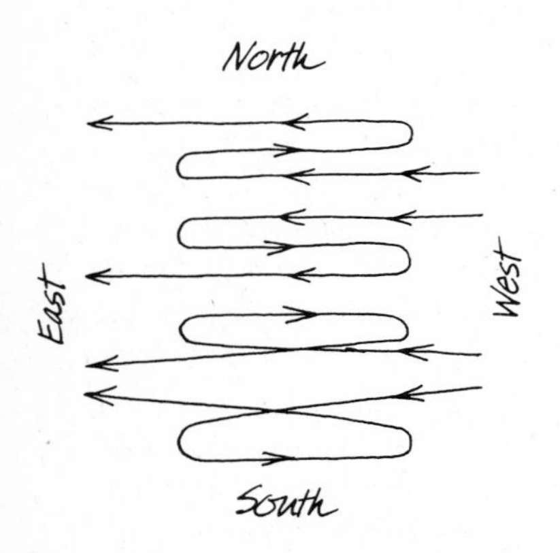

Figure 2.14 The variety of shapes of retrograde loops. The shape of a retrograde loop executed by a planet depends on whether the planet is drifting north or south (or first one and then the other) during its retrograde motion. We have not emphasized the north-south motions of the planets in this text.

that is, it is traveling from west to east with respect to the fixed stars.

It is not surprising that each inferior planet performs retrograde motion while near the sun in the sky. An inferior planet is *always* found relatively near the sun. However, it is an unexpected fact that each superior planet performs retrograde motion only during the time it is in opposition. The superior planets perform direct motion as their angular separation from the sun increases. The retrograde motion of a superior planet never occurs until its angular separation has reached a maximum. Then and only then does it turn around and back up relative to the fixed stars.

SUMMING UP

The objects in the sky appear to move in a number of interesting ways. The stars move as if affixed to a rigid celestial sphere. The sun exhibits a motion composed of its daily motion and its motion in declination. The moon's motion is somewhat similar, except that it performs one cycle in roughly a month rather than a year as in the case of the sun. The inferior planets remain relatively close to the sun in the sky, while the superior planets execute both conjunctions and oppositions with the sun. All planets usually move from west to east with respect to the fixed stars, but each planet periodically performs retrograde motion. A superior planet executes retrograde motion only when in opposition to the sun and always looks brightest when doing so.

In the next chapter we begin to investigate the theories developed to explain the observations we have studied in this chapter.

EXERCISES

1. Define the following terms carefully and exactly: horizon, degree, minute of arc, second of arc, angular separation, angular diameter, altitude, fixed star, circumpolar star, celestial sphere, celestial pole, meridian, zenith, celestial equator, equinox, solstice, ecliptic, lunar sidereal period, lunar synodic period, conjunction, opposition, inferior conjunction, superior conjunction, direct motion, retrograde motion.

2. What is unusual about the apparent motion of the star Polaris as viewed by the casual observer?

3. How do the stars near Polaris in the sky appear to move?

4. As one watches any two stars in the sky move, what is significant about their angular separation?

5. Describe the connection between the location of an observer and the altitude of Polaris.

6. Look up the latitude of your location on a map. As explained in Section 3.1, the altitude of Polaris as viewed from your location is equal to your latitude. What is the maximum angular separation a star can have from Polaris and still be circumpolar as viewed from your location? (This question is a little difficult. Draw a good diagram. Answer: An angular separation equal to your latitude.)

7. Draw a sketch of the celestial sphere and show a flat earth inside it. Label the parts of the celestial sphere (meridian, celestial equator, etc.). Show how the rotation of the celestial sphere explains the motions of selected stars.

8. What feature of a star's location on the celestial sphere determines the length of time the star remains above the horizon?

9. Suppose a star sets at 10:00 P.M. tonight. At what time will it set tomorrow night? Will it be visible at 10:00 P.M. tomorrow night?

10. Describe the daily motion of the sun.

11. Describe the sun's motion in declination and its relation to the length of our daylight and to the seasons.

12. Describe the apparent motion of the sun with respect to the fixed stars. How is this motion related to the year?

13. If the moon is on the meridian tonight at 11:00 P.M., what time will it be on the meridian tomorrow night?

14. Describe the motion of the moon with respect to the fixed stars.

15. Describe the location of the moon in the sky at sunset when it is new moon, first quarter moon, and full moon. Where is the moon in the sky when it is third quarter moon and the sun is rising?

16. What is the difference between the apparent motions of a fixed star and a planet?

17. What is the major difference between the apparent motion of an inferior planet and that of a superior planet?

18. Which of the following planets can be seen rising at sunset: Mercury, Venus, Mars, Jupiter, Saturn?

19. Is Mars ever visible all night? Is Venus?

20. What is the approximate angular separation of a superior planet from the sun when the planet is performing retrograde motion? How does its brightness then compare with its brightness at other times?

21. Does Venus ever rise in the west? Does Saturn? (If you answered "yes" to either of these questions, you need to restudy the concept of motion with respect to the fixed stars. No celestial objects ever rise in the west.)

22. Suppose someone tells you, "Retrograde motion means backing up, so when a planet performs retrograde motion it goes from west to east." Help this confused person with a clear discussion emphasizing the concept of motion with respect to the fixed stars.

23. Suppose that the full moon, Jupiter, and a star are on the meridian tonight at midnight. In what order will they finish the race tomorrow night? (Notice that Jupiter is in opposition to the sun.) Answer: Jupiter, star, moon.

READINGS

Those wishing to learn the constellations should look at Appendix 1 for a quick start. Much better, and highly recommended, is

The Stars, by H. A. Rey, Houghton Mifflin, Boston, 1970.

For more information on watching the sky without a telescope, read

Whitney's Star Finder, by Charles A. Whitney, Knopf, New York, 1974 (revised 1977), and *Naked Eye Astronomy*, by Patrick Moore, Norton, New York, 1965.

Information on the positions of the planets and other interesting occurrences in the sky may be obtained in the following magazines which also contain timely articles on current astronomical research and developments.

Sky and Telescope, published by Sky Publishing Corporation, 4a Bay State Road, Cambridge, Mass.

Astronomy, published by Astromedia Corporation, 411 E. Mason Street, Milwaukee, Wisc.

A monthly guide to the sky for naked-eye observers, one that features drawings of what to look for each day and which is very useful, reasonably priced, and highly recommended is

Abrams Planetarium Sky Calendar, Abrams Planetarium, Michigan State University, East Lansing, Mich.

A detailed guide to events during the year as well as a wealth of information may be had in

Astronomical Calendar, published annually by Guy Otewell, Department of Physics, Furman University, Greenville, S.C. Write for details.

For more on astrology by a noted astronomer who is interested in the subject, see

"Interview with George Abell on Astrology and Astronomy," by R. Reis, *Mercury*, March/April 1976.

PTOLEMY'S GEOCENTRIC THEORY: BRILLIANCE OR NONSENSE?

Most early theories were in fact geocentric, that is, earth-centered. A geocentric theory sometimes known as the Ptolemaic theory is the principal subject of this chapter.

Aristarchus of Samos (about 320 to 250 B.C.), as he is often known, was once accused of being irreligious because of his views.

Ptolemy does acknowledge that Hipparchus not only contributed his excellent observations but also was an important contributor to the theory. He describes Hipparchus as a "diligent and truth-loving man."

I find the observations described in the last chapter interesting in themselves. It is enjoyable merely to watch the varied motions in the sky from year to year. But, almost inevitably, one is not satisfied to stop there. One seeks understanding. This can be provided by a theory. From one point of view, a theory answers the question, What would these motions look like if I viewed them from a point in outer space?

The first theory we investigate is often called the geocentric theory. Other theories preceded it. For example, Aristarchus proposed that the sun, not the earth, was at the center of the universe. The earth, he said, revolved around the sun. The idea was intriguing but not worked out in great mathematical detail. As the geocentric theory became more and more successful, Aristarchus' idea was pushed into the background. (But see Chapters 4 and 5.) The geocentric theory became so sophisticated and mathematically precise that it superseded all competition.

Introducing the Geocentric Theory The geocentric theory was worked out by a number of astronomers, among whom Hipparchus was prominent. These scientists lived during a period of about 400 years that was one of the high points of civilization. Most of them lived in the flourishing Mediterranean empire originally founded by Alexander the Great (who died in 323 B.C.)

Ptolemy (usually pronounced tahl'-a-mee) lived in Alexandria, Egypt, from about 100 to 170 A.D. He wrote a book that preserved and improved upon the ideas of Hipparchus and the other geocentric theorists, bringing the geocentric theory to a highly refined state. His book is usually known as *The Almagest.* It was preserved by Arab scholars who studied and copied it during Europe's Middle Ages. Ptolemy did not always specify the originator of each individual idea. But for convenience, the geocentric theory is often referred to as "Ptolemy's theory," although this is an oversimplification.

As we describe the principal features of the geocentric theory as found in *The Almagest,* the reader is asked to keep an open mind. Digest the theory first and then evaluate it. Some authors see the theory as a historic obstacle to astronomical progress (one author even calls it "nonsense"), while others see it as a brilliant advance in understanding. It is hoped that the reader will take a fresh look at the subject. See Box 3.1.

BOX 3.1

SCIENTIFIC CHAUVINISM

In the past, there have been those who believed that their own age or their own culture was far superior to any other. Some are surprised to find ancient Egyptian drawings that are "amazingly lifelike." Others believe that certain ancient cultures were too primitive and childlike to build such monuments as the Egyptian pyramids. Some feel justified in asserting that ancient astronauts must have built the monuments. It seems they cannot admit that geniuses have lived in every age and in every culture.

A related type of chauvinism can crop up in some scientific circles. At times, there has been a tendency to dismiss the ideas and accomplishments of earlier peoples as naiveté or even nonsense. Implied in this attitude is that one's own age has found the theories that are *the* truth and which will stand up for all time henceforth. Many an age has had a tendency to see itself as the final high point of scientific history.

There have been reactions to this tendency from time to time. In our age, many find it possible to appreciate the brilliance and important contributions of former workers. Furthermore, we should recognize the strong probability that in 100 years many of our own ideas and theories will be seen as interesting but obsolete stepping-stones in scientific history. People then may read the later chapters of this book and think, "How quaint."

Before considering some of the details of the geocentric theory, we list three of the main characteristics underlying all that follows in this chapter.

1. The theory is geocentric (from *geo,* "earth," and *centric,* "centered"). The earth is assumed to be at the center of the universe with all heavenly objects revolving around it.

2. It is assumed that the apparent motions of the heavenly objects are real motions. For example, not only does the sun *appear* to move across the sky from east to west each day, it really does so.

3. All the observed motions are explained using circles. It is helpful to keep in mind that the ancient scientists thought highly of the circle. We have lost that bias today. We no longer feel a strong aesthetic attraction to that "perfect shape," the circle. But listen to Ptolemy: "Whosoever thinks my theory is too artificial should remember that motion in a circle corresponds to the divine nature. To reduce all apparent irregular motions to motion in a circle may well be called a feat. Thereby, the ultimate goal of science is attained."

3.1 THE EARTH IN THE GEOCENTRIC THEORY

Most of the assertions in this chapter should contain the phrase "according to the geocentric theory," but that would be wearying, so we often omit the phrase.

According to the geocentric theory, the earth is at the center of the universe, and all objects revolve around it. The earth is at rest, not rotating: it is stationary. Ptolemy pointed out that, if the earth were to move any significant distance from the center of the celestial sphere, the

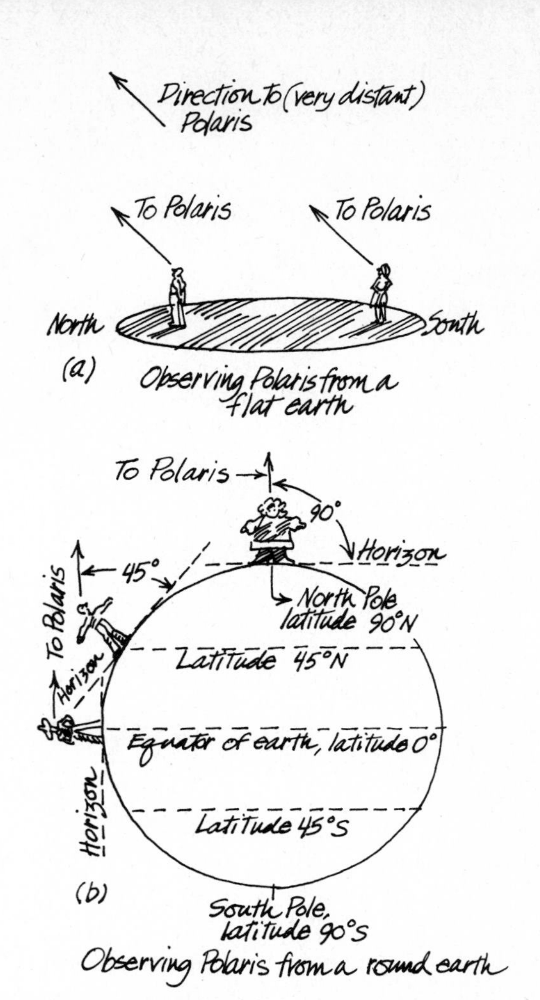

Figure 3.1 The altitude of Polaris and the shape of the earth. As shown in (a), all observers on a flat earth would see the distant Polaris at roughly the same angle above the horizon (i.e., at roughly the same altitude). The situation on a round earth is shown in (b). Each of the three observers finds that the angle from Polaris to his horizon equals his latitude.

constellations would seem to change shape as we get nearer to or farther from them. No such effect had ever been observed.

The Round Earth, Polaris, and Latitude The geocentric theory also states that the earth is round. I was told in grade school that Columbus was the first to decide that the earth is round. I was misled. Actually, most educated people throughout history have thought that the earth is round. We are interested in the reasoning Ptolemy used to decide that the earth is round. The earth looks flat. Ptolemy gathered evidence and from that *deduced* that the Earth is round. We will encounter many other instances in which people have found indirect, ingenious methods to reach a conclusion.

The ancients found a number of indications that the earth is round. One of the clues is astronomical and was mentioned in Section 2.1: The altitude of Polaris increases as one moves north. Scientists well before Ptolemy's time had pointed out that this is evidence that the earth is round. On a flat earth, the altitude of Polaris would change little as one traveled north. If the earth is round, the observation is nicely explained. See Figure 3.1.

In this analysis, we assume that Polaris is not close to the earth but rather beyond the sun, moon, and planets.

This fact has led to a system for describing how far north or south one is on earth. Each location in the northern hemisphere is assigned a **latitude** equal to the altitude of the north celestial pole as viewed from that location. At the earth's equator, the north celestial pole appears to be on the horizon; thus the equator is assigned a latitude of 0°. At the earth's north pole, the north celestial pole is at the zenith; the latitude of the north celestial pole is assigned a latitude of 90° north. Minneapolis, Minnesota, lies halfway between the equator and the North Pole; its latitude is 45° north. Locations in the southern hemisphere are similarly assigned south latitudes equal to the altitude of the south celestial pole.

Navigators in the northern hemisphere who wish to know their location north or south of the earth's equator measure the altitude of Polaris and apply a small correction to allow for the fact that Polaris is not quite at the north celestial pole; the result is the navigator's latitude. See Figure 3.2.

The reader might find it interesting to try this. Estimate the altitude of Polaris from your location by any method, no matter how rough. Then look up the latitude of your location on a map. Is there rough agreement?

Because 1° of latitude on the earth corresponds to about 111 km of distance, you can estimate your distance from the equator by multiplying your latitude by 111 km [or 69.1 miles (mi)]. For example, the latitude of Minneapolis, Minnesota, is 45°, so it is 111 × 45 or about 5000 km (3110 mi) from the equator.

A resident of Minneapolis uses a sextant to measure the altitude of Polaris. She finds that it is 45° above the horizon and concludes that she lives at a latitude of 45°N in the northern United States. She then returns to her snow shoveling.

A resident of the Florida Keys (southern tip of Florida) measures the altitude of Polaris and finds that it is only 25° above the horizon. She concludes that she lives at a latitude of 25°N in the far southern United States. She then returns to her winter lawn mowing.

Figure 3.2 Using Polaris to determine latitude.

3.2 THE CELESTIAL SPHERE IN THE GEOCENTRIC THEORY

The stars appear to move as if they are attached to a huge, rigid celestial sphere. (See Section 2.1.) The geocentric theory simply asserts that the celestial sphere is real. All the motions of the stars discussed in Section 2.1 are elevated to the status of reality. The stars are fixed in relation to each other because they are attached to the celestial sphere. The varied motions of the stars described earlier are simply explained in this way. Polaris does not appear to move much because it lies so close to the north celestial pole. The other stars appear to move as the earth-centered celestial sphere rotates once every 23 hr, 56 min.

PUZZLE 3.1

THE HORIZON AND THE DISTANCE OF THE STARS

Study Figure 3.3. Following Ptolemy's theory, the earth is shown round and placed inside the celestial sphere. The man on the earth represents an observer and the dashed line his horizon. The drawing implies that an observer can see *less* than half of the celestial sphere, which, as we know, is incorrect. Modify the figure so that any observer on the earth will more nearly be able to see half of the celestial sphere. (Compare your answer with the solution in Appendix 2.)

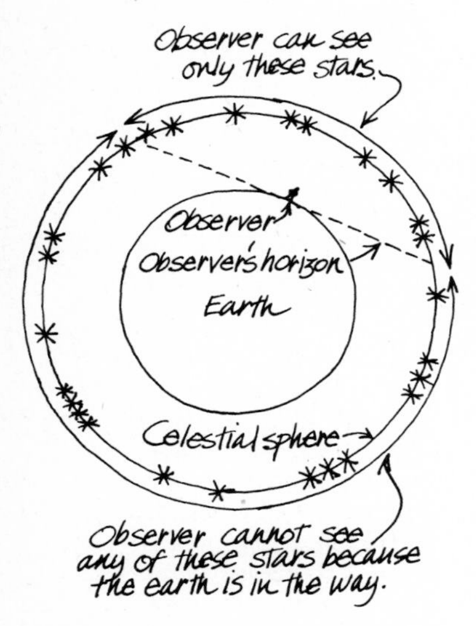

Figure 3.3 Puzzle 3.1 asks, "How can this figure be improved?"

In Ptolemy's theory, all stars are attached to the celestial sphere, and so all stars are the same distance from the earth. Ptolemy added that the distance to the stars must be very large compared to the diameter of the earth. The key to this conclusion is the horizon. An observer standing in a large, level field notices that *the horizon always cuts the celestial sphere in half.* No matter when the observation is made, the observer will find that just half of the celestial sphere is above the horizon. If one draws a circle on a model of the celestial sphere connecting the stars on the horizon at any one time, it will be found that the circle divides the celestial sphere into two equal parts. Now anticipate Ptolemy's reasoning by solving Puzzle 3.1. (See also Figure 3.3.)

3.3 THE SUN IN THE GEOCENTRIC THEORY

As in the case of the stars, the theory really contributes nothing very novel to our understanding of the motion of the sun, except to say that the motion is real. The reader may wish to review Section 2.2 at this point.

How might it affect our seasons if the ecliptic were identical with the celestial equator?

According to the geocentric theory, the sun revolves around the stationary earth. One can think of the sun as being "carried along" with the celestial sphere. This causes the sun's daily motion, its rising and setting. At the same time, it "crawls along" the celestial sphere in the direction opposite that in which the celestial sphere turns. This motion with respect to the fixed stars carries the sun 1° per day along the celestial sphere from west to east and is the cause of its annual motion.

The yearly north-to-south motion, which we call the sun's motion in declination, is due to the $23\frac{1}{2}°$ tilt of the ecliptic with respect to the celestial equator. The annual motion of the sun along the ecliptic carries it north of the celestial equator in summer and south of the celestial equator in winter. Study Figures 2.9, 3.4, and 3.5 carefully.

Figures 2.9 and 3.5 describe the same motions in the same way but from slightly different points of view. The main difference is that Figure 2.9 shows the sun actually on the celestial sphere, whereas the theory (Figure 3.5) assumes that the sun is closer to us than the stars are. Some writings on the geocentric theory imply that the motion of the sun is somehow affected by the motion of the celestial sphere—that it is carried along with the celestial sphere. That is why it moves around the earth in almost the same amount of time that the stars do.

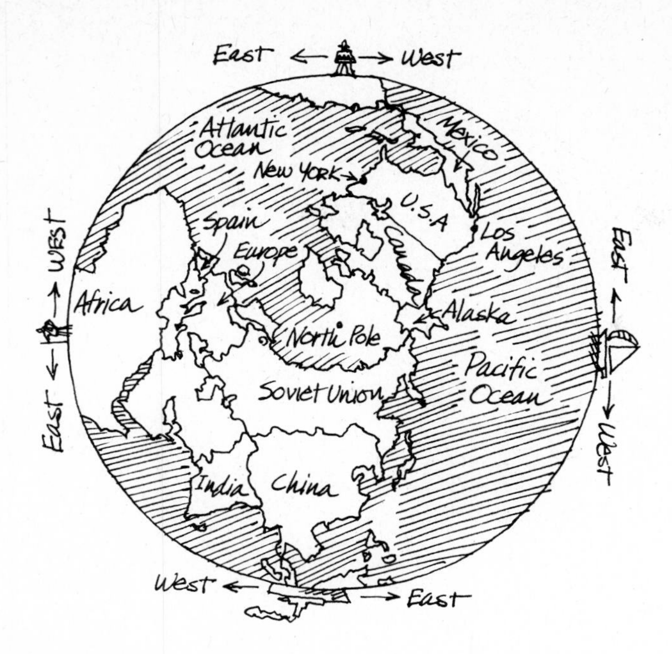

Figure 3.4 The earth as viewed from above its north pole: an aid in interpreting east and west in later illustrations. Unless otherwise stated, all figures that show the earth as viewed from outer space give the view from a rocket ship in space as it flies over the North Pole. We are not accustomed to looking at maps from this vantage point, so the reader should become familiar with this illustration. Note in particular the directions east and west as given for each of the four persons shown.

This last aspect of the theory may seem a bit mysterious. How can the sun be affected by the celestial sphere if the two objects do not touch one another? It may be helpful to think of the celestial sphere and its contents as "the universe." The whole universe, except the earth, rotates once in 23 hr, 56 min. The sun, moon, and planets participate in this rotation; when they execute their own additional motions with respect to the fixed stars, this should be thought of as a departure from the common spinning of the universe.

3.4 THE MOON IN THE GEOCENTRIC THEORY

You may wish to review Section 2.3.

According to the geocentric theory, the moon revolves around the stationary earth. Ptolemy assumed that the moon is closer to the earth than the sun is. It was known that during a solar eclipse the moon passes between the sun and the earth. Again, the theory explains the observations of the moon's motion by asserting that what is seen is real. The moon is carried along with the celestial sphere, which causes it to rise, move across the sky from east to west, and set. At the same time, the moon crawls along the celestial sphere from west to east, in a

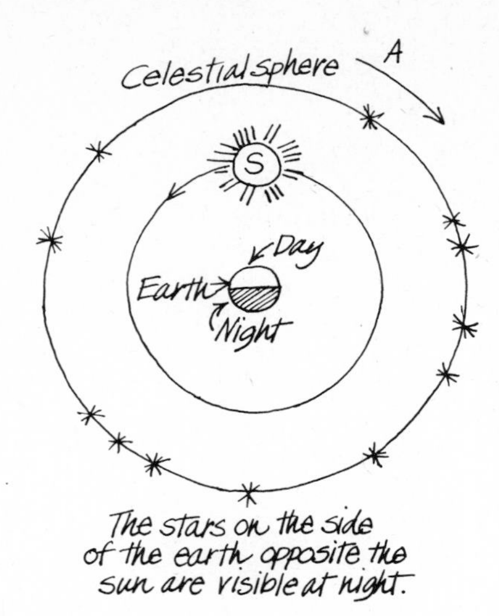

Figure 3.5 The motion of the sun in the geocentric theory. According to the geocentric theory, this is the view of the earth, sun, and stars that would be obtained by a viewer in outer space, above the earth's north pole. (We simplify this and similar illustrations by ignoring the fact that the ecliptic is inclined with respect to the earth's equator.) The reader should practice the following two ways of thinking about such figures. Both are instructive.

Method 1: Such illustrations are primarily intended to show the motion of an object (here, the sun) with respect to the fixed stars. With this method, the stars' motion is temporarily forgotten. Mentally keep the stars at rest. Then, mentally make the sun go around the earth as shown by the arrow on the sun's path. The sun should go around the earth in one year. This makes it revisit each constellation once a year. The sun should move 1° per day along its path. This is what is meant in the text by, "The sun crawls along the celestial sphere." Check that the sun's arrow shows a motion among the fixed stars from west to east, as it should. See Figure 3.4.

Method 2: A further stretch of the imagination applied to this figure can demonstrate the motion of the sun *and* the stars as viewed from the earth. Mentally keep the *earth* at rest. Then imagine the rest of the figure, sun *and* stars, turning around the earth in the direction shown at A. This explains the rising in the east and setting in the west of the sun and fixed stars. (Verify these directions by referring to Figure 3.4.) Figure 3.5 should turn around the stationary earth once every 23 hr, 56 min. This will cause a star to return to an observer's meridian in that time. To complete the mental animation of the figure, imagine the sun's symbol cut out and placed on the back of a trained ant. The ant has been instructed to crawl along the sun's path at a rate of slightly less than 1° per day so that it returns to its original location *on the paper* after one year. Imagine, then, the whole page turning around the stationary earth clockwise (east to west) once every 23 hr, 56 min as the ant-sun crawls slowly along the page counterclockwise. Notice how the ant-sun is "carried along with the celestial sphere."

PUZZLE 3.2

THE MOTION OF THE MOON WITH RESPECT TO THE FIXED STARS

Draw a figure like Figure 3.5, only draw it for the moon: First draw a large circle and label it "celestial sphere." Draw a small circle at the center labeled "earth." Draw an earth-centered circle between the earth and the celestial sphere, to represent the moon's path, and place the moon on it. Draw an arrow on the moon's path indicating a counterclockwise motion. For convenience, we let the stars remain at rest temporarily. Now imagine the moon traveling around its path once. How long should this trip take? Be sure to read the solution in Appendix 2 when you are done.

direction opposite the motion of the celestial sphere. The speed is such that it travels 13° per day from west to east with respect to the fixed stars. Now solve Puzzle 3.2.

A good theory about the phases of the moon had already been given by Aristotle roughly 500 yr before Ptolemy's time. We postpone a detailed discussion of this topic until the next chapter, Section 4.7. (See also Figure 3.6.)

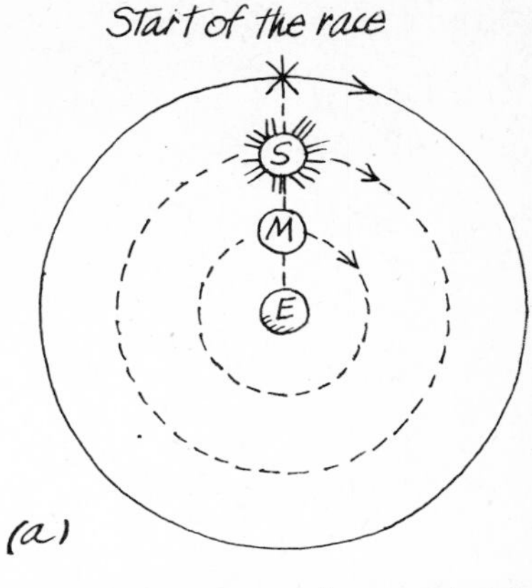

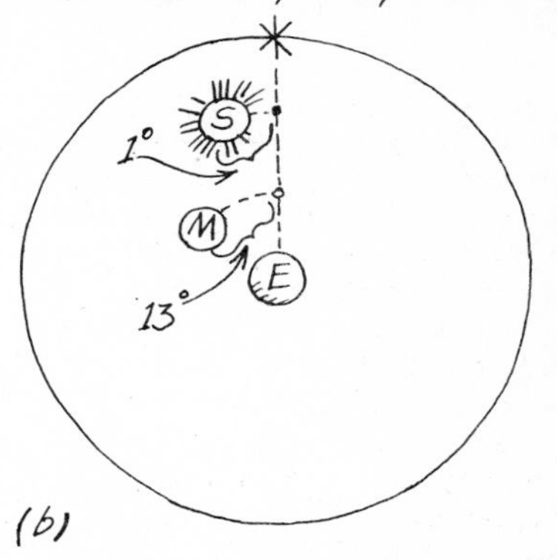

Figure 3.6 A race between the sun, the moon, and a star. In Chapter 2 (see Puzzle 2.4) we analyzed a race. Here is Ptolemy's view of what happens as viewed from outer space. In (a) we note that the observer (dot on the earth) sees the racers lined up on his meridian. All three objects move toward the west, set, and rise again in the observer's east. The race ends when the star wins by crossing the meridian first (b). Another way to explain the result is that the sun and the moon have traveled from west to east *with respect to the fixed star.*

3.5 THE CONCEPT OF THE EPICYCLE

To this point the geocentric system has not proposed any striking novelties. It has merely taken appearances as reality. We see that the theory explains the motions of the stars, sun, and moon using nothing but circles, a persuasive point to the ancients.

But now the theory reaches an interesting stage. The motions of the planets do not appear to be straightforward circular motions around the earth. Can the theory use circles to explain the motions of the inferior planets from side to side of the sun and the equally peculiar motions of the superior planets? The way in which the theory solves this problem is fascinating.

The beauty of the solution Ptolemy employs is that it uses only one concept, the epicycle, to explain the motions of all five planets. A theory using a different concept for each planet would be less attractive. The various motions of the inferior and superior planets are thereby seen to have a similar basis.

The basis of the concept is to combine rotary and forward motion. Figure 3.7 shows a woman holding a ball. In (a) she swings the ball in a circle at a speed of 2 meters per second. Near the top of the path, the ball is moving forward at 2 meters per second. Near the bottom of the path, the ball is moving backward at a speed of 2 meters per second. In (b) the woman holds the ball steadily while walking forward at a speed of 1 meter per second. Finally, in (c) she swings the ball as in (a) but does so while walking as in (b). The resulting motion of the ball is the point of interest. Near the top of its path the ball travels ahead of the woman, having a combined speed of $2 + 1 = 3$ meters per second with respect to the ground. Near the bottom of its path, however, it is traveling backward with respect to the ground because it has a forward motion of 1 meter per second due to the walker and a backward speed of 2 meters per second due to the rotary motion. The result is that, near the bottom of the path, the ball moves *backward* at 1 meter per second with respect to the ground. A combination of rotary and forward motion can result in an object moving alternately forward and backward. Note that, overall, the ball makes forward progress—it stays near the woman.

Now then, in the geocentric theory the concept of the epicycle is used to combine rotary and forward motion. We next define three terms. A **deferent** is a circle centered on the earth. An **imaginary point**

Ball goes forward at
2 meters per second
when here.

(a)

Ball goes back-
ward at
2 meters per
second when
here.

A woman stands still and moves a ball in a circle at a speed of 2 meters per second. This is a rotary motion.

The woman walks forward at a speed of 1 meter per second. She holds the ball still with respect to herself.

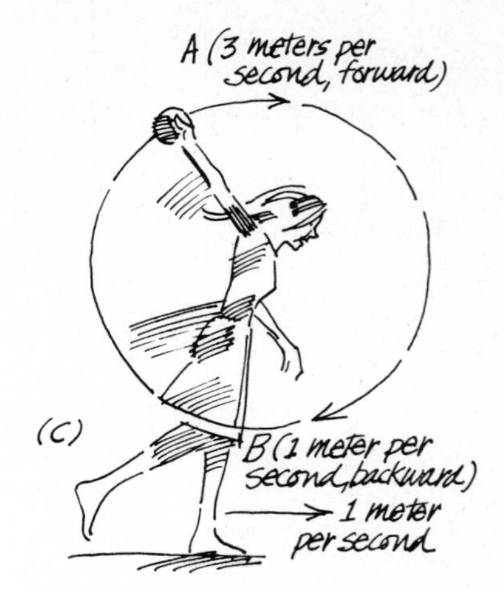

The woman walks at a speed of 1 meter per second while moving the ball in a circle at a speed of 2 meters per second. This is forward plus rotary motion.

Figure 3.7 A combination of forward and rotary motion. This illustration shows how a combination of forward and rotary motion can cause an object to go forward and backward alternately. It is the idea behind the epicycle in Ptolemy's theory.

is a point on the deferent that travels at a steady speed around the earth, staying on the deferent. An **epicycle** is a circle drawn with the imaginary point as its center. As the imaginary point moves along the deferent, the epicycle moves with it in such a manner that the imaginary point remains at the epicycle's center. (See Figure 3.8.)

The final step is to imagine a planet on the epicycle. The planet travels along the epicycle at a steady speed. The rotary speed of the planet on the epicycle is greater than the forward speed of the imaginary point on the deferent. Now compare the motion of the ball in Figure 3.7c with the motion of the planet in Figure 3.8. The forward-walking person is analogous to the forward-moving imaginary point. Imagine the view obtained by an observer on the earth. When the planet is far from the earth, the observer sees it going forward (performing direct motion). When the planet is inside the deferent, nearest the earth, it is observed to be backing up (performing retrograde motion).

A nit-picking point. The analogy between Figures 3.7 and 3.8 is not exact. The motion of the imaginary point, which we are calling "forward motion," is along the deferent and therefore also "rotary." Because the deferent is so large, the analogy is not misleading.

The resulting motion of a planet is shown in Figure 3.9. The planet alternately travels forward for a while and then backs up briefly. Overall, the planet travels around the earth in the forward direction with brief episodes of retrograde motion. Once again, the geocentric theorist has explained motion entirely in terms of circles. It was felt that this enhanced the beauty of the theory.

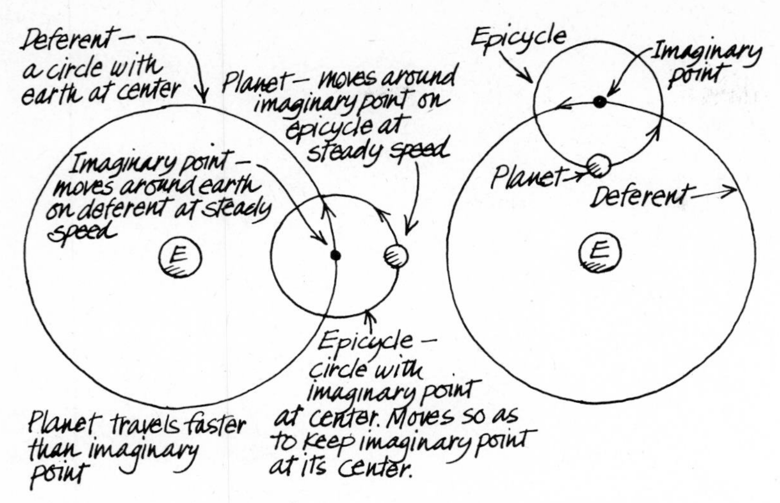

When the planet is in the position shown, outside its deferent, it will appear to be executing direct motion (west to east among the fixed stars) as viewed from earth.

The same planetary system as viewed at a later time. The planet has traveled to the inside of its epicycle. Observers on earth see the planet performing retrograde motion (east to west among the fixed stars) in this situation.

Figure 3.8 The epicycle used to explain planet motions. The forward motion of the imaginary point is combined with the rotary motion of the planet on the epicycle. The result is that the planet appears to move forward most of the time but to retrograde at regular intervals. See also Figure 3.9.

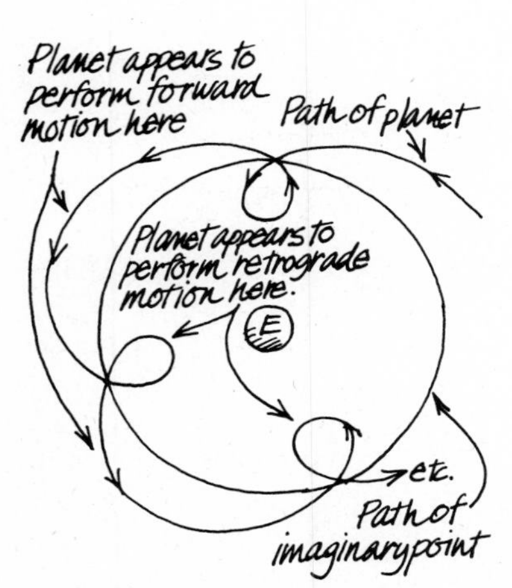

The motion of a planet according to Ptolemy. The planet in Figure 3.8 follows this sort of path. The view is from a spaceship hovering over the earth's north pole.

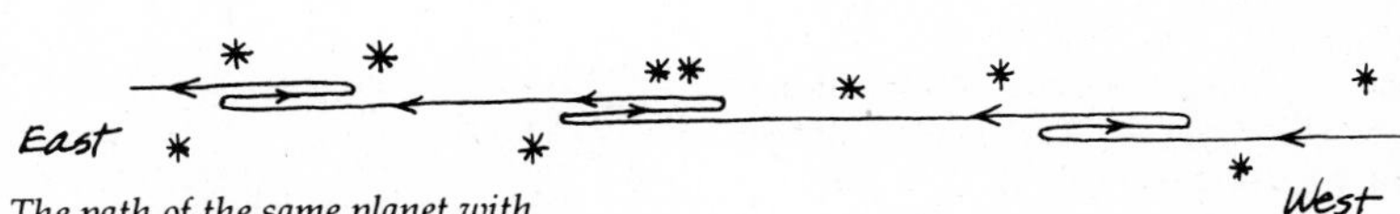

The path of the same planet with respect to the fixed stars as viewed by an observer on earth.

Figure 3.9 The motion of a planet as explained by Ptolemy. If a pencil were placed on the planet in Figure 3.8 and used to trace out the subsequent path of the planet, the result would be the looping curve shown at the left. This illustrates the motion of a planet as viewed from outer space according to the geocentric theory.

Not allowed! The observer on earth sees the inferior planet at a large angle from the sun. This never happens. Venus can be, at most, 48° from the sun as viewed from earth.

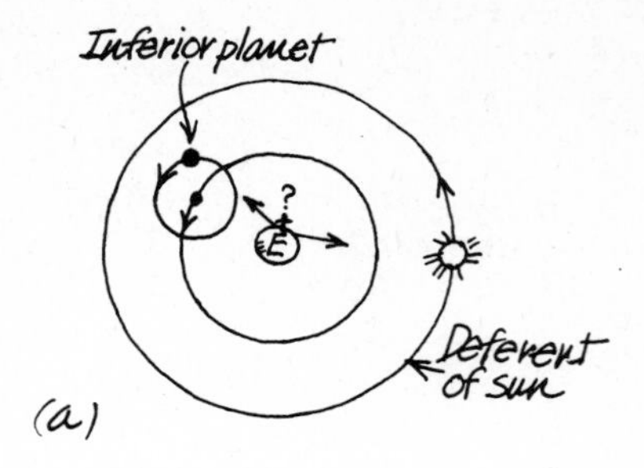

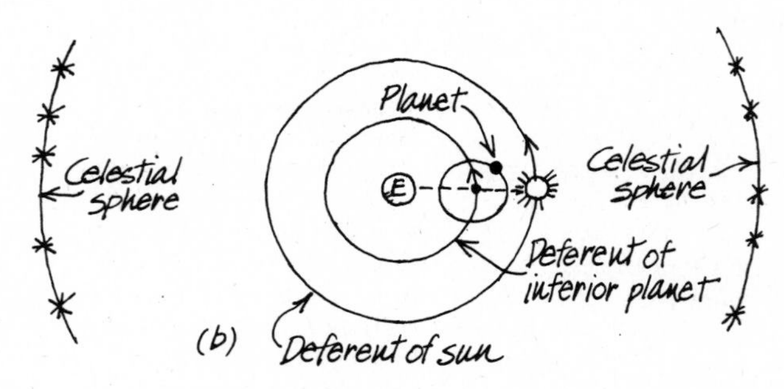

Figure 3.10 Inferior planets and the sun. As viewed from the earth, the inferior planets seem to stay near the sun in the sky. In Ptolemy's theory, this is assured by the assumption that for each inferior planet the imaginary point of the planet stays on a line from the earth to the sun as shown in (b).

3.6 THE INFERIOR PLANETS IN THE GEOCENTRIC THEORY

You may wish to review Section 2.4.

When applying the concept of the epicycle to the motion of the inferior planets, two additional assumptions were made. First, it was assumed that the inferior planets stay in the region between the earth and the deferent of the sun. This was simply accomplished by making the deferents of Mercury and Venus smaller than the deferent of the sun.

Second, Ptolemy assumed that the imaginary points of each inferior planet always remain on a line from the earth to the sun. As illustrated by Figure 3.10, this assumption was made necessary by the observation that, as seen from the earth, the inferior planets never stray far from the position of the sun in the sky. This means that, as the sun travels with respect to the fixed stars on its yearly path around the earth (as in Figure 3.10b), the imaginary point of the inferior planet also travels around the earth in one year.

It is the motion of the inferior planet on its epicycle that causes the inferior planet to appear to move from one side of the sun to the other.

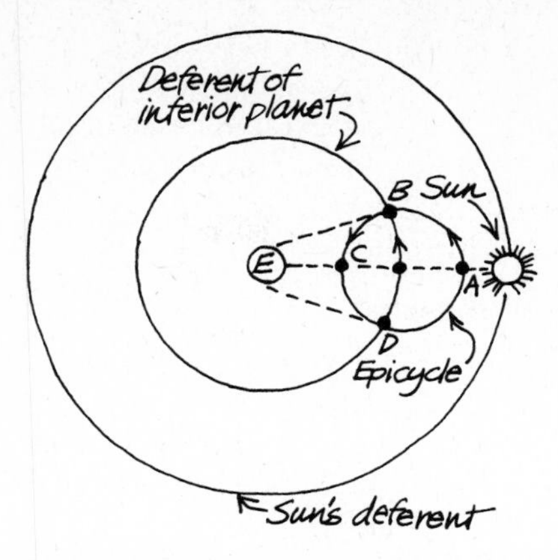

Figure 3.11 The geocentric theory of the motion of an inferior planet. At point A on the epicycle, the planet is said to be in superior conjunction. At B, the planet is at maximum eastern elongation. Next the planet seems to get brighter as it approaches the earth. At point C the planet is at inferior conjunction and is executing retrograde motion. At D it has reached maximum western elongation and thereafter appears to head back toward the Sun. Last, the fact that an inferior planet seems to travel rapidly from B to D and slower from D to B can be explained by the theory. From B to D the planet is closer to the earth, and so only *seems* to be going faster.

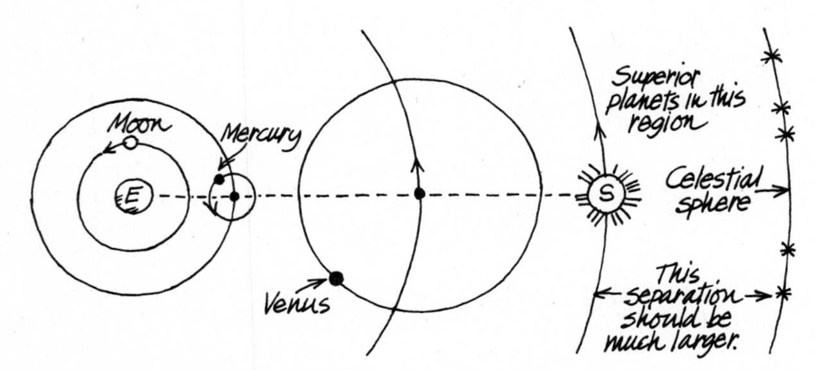

Figure 3.12 The inferior planets in the geocentric theory. Venus' epicycle must be drawn much larger than Mercury's epicycle. This will cause Venus' greatest elongation to be larger than Mercury's, as it should be. As shown later in Figure 3.16, the superior planets were placed in the region between the sun and the celestial sphere. This region is much larger than could be conveniently indicated in the illustration.

The inferior planet's synodic period, according to the theory, is the length of time required for it to travel its epicycle once. These details are explained further in the caption for Figure 3.11.

Finally, Figure 3.12 shows both of the inferior planets as envisioned by Ptolemy. Notice that Ptolemy places Mercury closest to the Earth and Venus farther away. He does not say why he chose to arrange these two planets this way. Ptolemy was aware that he had no real knowledge of the distances of the planets, and his choice of order was really a matter of preference. Some Greek astronomers used other orders.

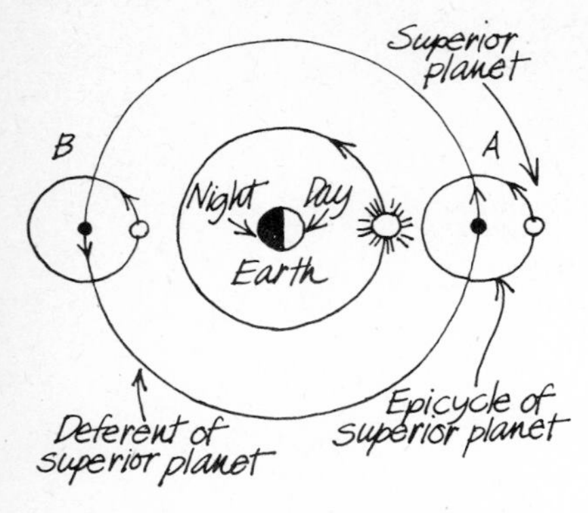

Figure 3.13 Two possible positions of a superior planet according to the geocentric theory. When the superior planet is in a position such as A, people on the earth observe that it is lined up with the sun in the sky, in conjunction. A planet in position B is said to be in opposition, because it seems to be in the part of the sky opposite the sun as viewed from earth. The theory maintains that here the planet is performing retrograde motion as compared to the fixed stars. It further explains that the superior planet appears brightest at this time because it is nearest the earth.

3.7 THE SUPERIOR PLANETS IN THE GEOCENTRIC THEORY

The motions of the superior planets are explained by the same device that explained the motions of the inferior planets—the epicycle. Again, two assumptions are needed to characterize the superior planets' motions. First of all, Ptolemy decided to place the superior planets beyond the Sun. He stated that the sun seemed to be a natural divider between the inferior and superior planets. This assumption implies that the deferents of the superior planets are larger than the deferent of the sun. The order chosen for the superior planets was Mars, next in distance from Earth after the sun, then Jupiter, and last, Saturn, the furthest known planet from the earth. Ptolemy probably chose this order by observing that, with respect to the fixed stars, Mars is the most rapidly moving planet of the three, with Jupiter next fastest and Saturn the slowest.

The imaginary point of each planet travels once around its deferent in the average time required for that planet to revisit (as it appears from earth) any given constellation. These approximate times are Mars 2 years, Jupiter 12 years, and Saturn 30 years. See Table 2.3. These times, then, are the average length of time required for each planet to revolve around the earth once with respect to the fixed stars. The sun revolves around the earth once a year with respect to the fixed stars, according to Ptolemy. Since this is a more rapid motion than the apparent motions of the superior planets, the sun appears gradually to catch up to each superior planet and then leave it behind. The result is that at times a superior planet appears to be near the sun but at other times appears to be far from the sun in the sky. In other words, the superior planets not only move into conjunction with the sun but also into opposition. For the superior planets, the speeds of the imaginary points are not tied to the motion of the sun, as is assumed for the inferior planets. (See Figures 3.13 and 3.14.)

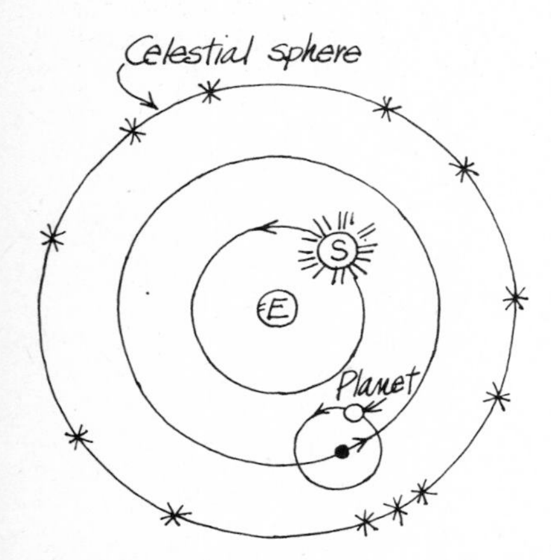

Figure 3.14 The motion of a superior planet according to the geocentric theory. The speeds of the imaginary point on its deferent and of the superior planet on its epicycle are dictated by the observed motions of the superior planets as given in Table 2.3. The resulting motion is shown in Figure 3.15.

On the other hand, the theory requires that the speed of each superior planet on its epicycle be strongly related to the sun's motion. The second main assumption needed to explain the superior planets' motions is this. A line drawn from a superior planet's imaginary point to the planet must always point in the same direction as a line drawn from the earth to the sun. Figure 3.15 illustrates this idea. This second assumption causes the superior planet to move at the proper speed so that it will perform retrograde motion at the proper time, that is, when it is in opposition to the sun as viewed from the earth. The result is that each superior planet moves in lockstep with the sun.

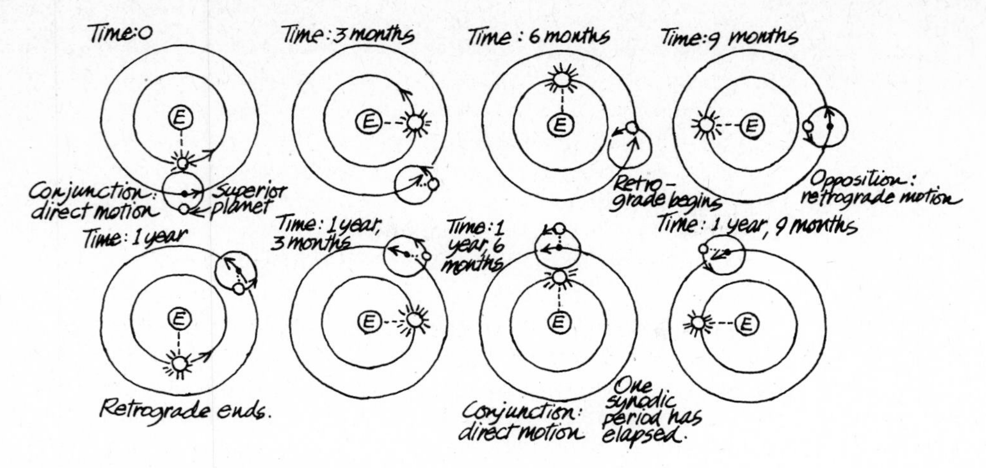

Figure 3.15 The motion of a typical superior planet according to the geocentric theory. Although the planet in this example does not move at exactly the same speed as any actual superior planet, its behavior is illustrative of Ptolemy's concept.

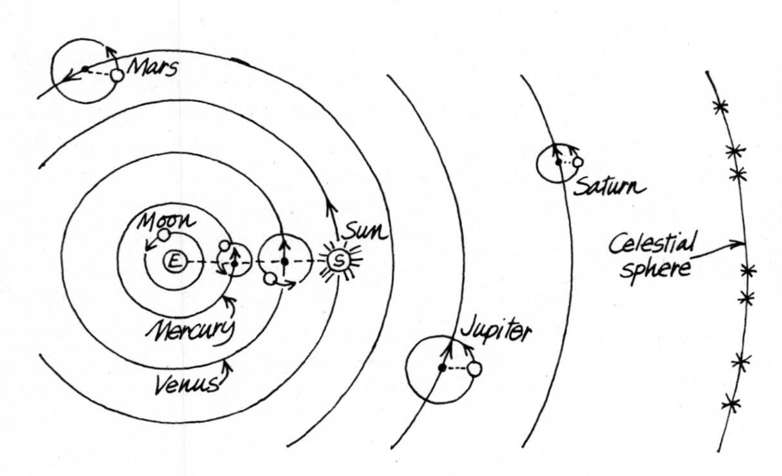

Figure 3.16 The geocentric universe: geocentric theory. This illustration shows the importance of the line from the earth to the sun in the geocentric theory. The imaginary point of each inferior planet must remain on this line. For each superior planet, the line from the planet's imaginary point to the planet must point in the same direction as this line.

Figure 3.16 displays the entire geocentric system according to Ptolemy. Ptolemy was quite aware that the order and placement of the planets in his scheme was a matter of taste. If he had wanted to, he could just as well have placed Venus beyond the sun. This could have been done by using an even larger epicycle for Venus so that its greatest elongations would still be of the correct size.

3.8 THE QUALITATIVE AND QUANTITATIVE ASPECTS OF A THEORY

Scientists use the words "qualitative" and "quantitative" in a certain sense. We pause here to illustrate the two ideas.

Loosely speaking, the **qualitative** aspects of a theory are those that can be expressed in words, and the **quantitative** aspects are those that require numbers and mathematics for a clear description. When describing a theory qualitatively, we may use thousands of words, analogies, sketches, and any other devices that give the reader an idea of the concepts of the theory and how they are related. This book is almost wholly devoted to qualitative explanations.

Many theories begin in an author's mind as qualitative concepts. The author may then strive to fill out the theory quantitatively by introducing numerical measurements and relating these by means of mathematical equations. When these equations lead to a quantitative prediction that can be tested experimentally, a crucial stage has been reached. If, when the test is conducted, it agrees with the prediction, scientists are compelled to give the theory serious consideration. If the test fails, something in the theory is wrong, and it is back to the drawing board for the author of the theory. Perhaps, by changing some part of the theory, it can be made to agree with the experimentation. Perhaps, on the other hand, the theory must be abandoned.

When quantitative aspects are introduced into a theory, the theory usually becomes much more definite. For example, "Mars exhibits retrograde motion because it travels on an epicycle," is a qualitative statement. A scientist encountering this statement would not necessarily be impressed. However, when the epicycle theory is used to make the prediction, "Mars will be 3.7° away from the star Regulus in exactly three months," we have a quantitative statement. If the prediction comes true, a scientist will be impressed by the theory. But we notice that, by producing a quantitative prediction, the author of the theory has taken a risk. Now the theory may easily be shown to be in error.

A student astronomer spends the bulk of his or her time learning the quantitative aspects of the field. The student may begin by reading a qualitative book like this one but then spends years intensively studying mathematics and physics in preparation for a thorough understanding of the quantitative aspects of astronomical theory. The readers of this book, most of whom will not pursue science professionally, should understand that our almost exclusively qualitative study of astronomy glosses over the very important quantitative aspects of the subject. Fortunately, we can nevertheless derive a great deal of understanding and pleasure from looking at and thinking about the universe in a qualitative way.

3.9 THE QUANTITATIVE ASPECTS OF THE GEOCENTRIC THEORY

We have described some of the major qualitative features of the geocentric theory. In the view of those who authored the theory, it was satisfactory on a qualitative level. To give only one example, the superior planets look brightest during retrograde motion because at such a time the planet must be traveling on that part of its epicycle nearest to the earth.

As Ptolemy worked out the qualitative aspects of the theory, he also contemplated the important quantitative aspects. His task was to use the ideas we have described and to refine this vision of the universe in such a way that the theory would correctly explain the measured positions of each heavenly body. Ptolemy had to adjust the size of each circle in the theory and also the speeds of each of the imaginary points and the speeds of the planets on their epicycles. His goal was to do this in such a way that the theoretical positions of the heavenly bodies on the celestial sphere agreed as well as possible with previously observed positions of these bodies. A further test of the success of his quantitative work was to see how well the theory could predict the future positions of heavenly bodies.

To make the theory a quantitative success required a mountain of labor and a high degree of mathematical skill. Ptolemy reported that he could rely in large part on the quantitative theory of Hipparchus when working with the sun and the moon, but that even Hipparchus had been unable to make the theory successful quantitatively for the planets. Ptolemy found that the qualitative theory for the planets as we have described it was not fully adequate to produce quantitative success. He had to introduce complications. For instance, it was necessary to shift the deferent of each planet from its previously earth-centered location; the center of each planet's deferent had to be moved to a different empty point in space. After this and other modifications, Ptolemy at last found success. The geocentric theory, as presented in *The Almagest* and further refined by later workers, was used by astronomers both to explain the universe qualitatively and to calculate the positions of the planets quantitatively. For centuries thereafter, the geocentric theory was taught in schools and used by professionals.

3.10 THINKING IT OVER

In Chapter 2 we learned about the observed motions in the sky. This led us to wonder what sort of theory could explain these observations.

In this chapter we studied such a theory. Having expended the effort necessary to understand the theory, the reader is now in a pleasant situation: He or she may now act as a critic. After reading a book or hearing a record, we all enjoy analyzing our reactions. We should do the same with the geocentric theory. This chapter concludes with a series of questions designed to aid in such a critique. These are the kinds of questions a scientist asks himself or herself after studying a theory.

I have tried to refrain from submitting my own reactions to the theory. I hope the reader cannot tell what they are. The questions are not meant to lead the reader to any particular conclusion. We are seeking an independent judgment, unprejudiced by any ideas held before reading this book.

Did you find that the geocentric theory helped you understand the observations described in Chapter 2? Did you say, "Oh, now I see why that happens," while studying the theory? Which parts of the theory

Although we shall not mention it again, this process of judging a theory should be applied to each of the theories we meet in this text.

did you find most satisfying? Did the theory challenge any ideas you had held before studying this chapter? If so, has Ptolemy's theory changed your mind? Why or why not? Which features of the theory, if any, did you find less than satisfying? State clearly why you found these features unsatisfying. (Do not merely say that "everybody knows" that such and such an idea is wrong. Use reasoning to back up your claim.) Finally, if you are able to find someone who has reached a conclusion opposite yours, try to convert that person and have that person try to convert you.

SUMMING UP

According to the geocentric theory, the earth is at rest and is spherical. The stars are attached to a real celestial sphere and rotate with it. Ptolemy deduced that the distance to the stars must be enormous compared to the diameter of the earth. The sun is carried along with the rotation of the celestial sphere but, so to speak, crawls backward (from west to east) along the ecliptic 1° per day. The apparent motion of the moon is explained in a similar fashion.

The motion of each planet is explained by means of an epicycle and a deferent. Each inferior planet's imaginary point remains in line with the sun while a line from the center of a superior planet's epicycle to the planet is oriented in the same direction as a line from the earth to the sun.

While developing the quantitative features of his theory, Ptolemy found it necessary to make a number of modifications of his qualitative scheme. After the adjustments, the theory was used for centuries by astronomers.

In the next chapter, a rival to the geocentric theory is described, a competitor that appeared some 1400 years after Ptolemy's time.

EXERCISES

1. Define these terms: geocentric, deferent, epicycle, qualitative, quantitative.

2. What are the three main characteristics of the geocentric theory?

3. How does the earth move in the geocentric theory? (Trick question!)

4. How did Ptolemy deduce that the earth is round? How did he deduce that the stars are very distant compared to the size of the earth?

5. How do the axis of the earth and the axis of the celestial sphere lie in space in relation to one another?

6. How does the geocentric theory account for the observed motions of the sun?

7. Sketch the earth, a planet, and its deferent and epicycle.

8. How does the geocentric theory account for the retrograde motion of planets?

9. What two assumptions concerning the motions of the inferior planets are required in the geocentric theory? Why is the assumption concerning the line from the earth to the sun necessary?

10. What two assumptions concerning the motions of the

superior planets are required in the geocentric theory? Why is the assumption concerning the line from the earth to the sun necessary?

11. How does the geocentric theory explain why the superior planets look brightest during opposition?

12. Draw a sketch of the universe according to Ptolemy.

READINGS

Somewhat scholarly but very interesting information about almost all of the people discussed in this book may be found in the monumental

Dictionary of Scientific Biography, edited by C. C. Gillispie, New York, 1970, 14 volumes.

More on the geocentric theory as well as the matters discussed in Chapters 4 through 6 of this book is discussed in

This Wild Abyss, by Gale E. Christanson, Free Press, New York, 1978. The early Greek theories of astronomy are also discussed in

The Copernican Revolution, by Thomas S. Kuhn, Harvard University Press, Cambridge, Mass., 1957, chaps. 1–4.

ABOUT REVOLUTIONS: THE HELIOCENTRIC SYSTEM

Copernicus' book is usually referred to as *De Revolutionibus.* The full title is *De Revolutionibus Orbium Coelestium,* or roughly, *Concerning the Revolutions of the Celestial Bodies.*

Blinded by illness and age, 70-year-old Nicolas Copernicus lay on his deathbed. The year was 1543, relatively early in the Renaissance. Copernicus had spent his life as befits our idea of a Renaissance man—artist, linguist, doctor, statesman, church official, economist, author, and astronomer. It is said that, as he lay dying, two books were brought for him to hold. One resulted from the rediscovery of Greek learning and the ever-increasing number of printed books; it was Ptolemy's *The Almagest,* printed in the original Greek. During his life Copernicus had studied the great text, which he now held, only in translation. The other book Copernicus held as he lay dying was his own work, *De Revolutionibus,* which had just been published. It was the greatest astronomical text since Ptolemy's. In it, Copernicus explained his own theory of the universe. He could not have guessed that the scientific revolution his book initiated would be honored 500 years after his birth (1973 was known as "The Year of Copernicus"). Our lives have been transformed by the scientific age his book helped, in large measure, to create.

4.1 THE COPERNICAN ERA

Copernicus lived from 1473 to 1543, an age of discovery, innovation, and revolution. A great deal happened during his lifetime. Columbus first sailed in 1492, and by 1519 one ship of Magellan's fleet had sailed around the world. Mercator invented the Mercator projection, a map-drawing technique that greatly aided navigation. Artists and musicians discovered new, exciting means of expression. It was, for example, the age of Michelangelo, of the artist-mathematician, Dürer, and of the ultimate "Renaissance man," Leonardo da Vinci. It was an age of reform and revolution in religion, witnessing, for example, Luther's 95 theses in 1517, King Henry VIII of England's break with Rome, and the founding of the Catholic order of priests known as the Jesuits. The revolutionary studies of the human anatomy by Andreas Vesalius were published in 1543, the same year Copernicus' book was published. Mathematics developed as important advances were made in algebra and trigonometry. We next study one of the most important ideas to come out of the Renaissance, Copernicus' heliocentric theory.

4.2 DOUBTS CONCERNING PTOLEMY'S GEOCENTRIC THEORY

The geocentric theory must be counted among one of the most successful theories in history. It had no important competition from the time of Ptolemy (about 150 A.D.) until Copernicus published his theory in 1543. Yet, before 1543, others had expressed doubts concerning the theory. Some of these doubts derived from observations made after the time of Ptolemy by Arab astronomers. These observations showed that the geocentric theory was not quite as accurate as it first had seemed. Attempts were made to improve it but were largely unsuccessful.

Certain qualitative aspects of the geocentric theory might seem dissatisfying, and there are a number of questions that the theory cannot answer. Perhaps, while reading the previous chapter, the reader would have liked to ask Ptolemy some of these questions:

1. Why do the planets revolve around imaginary points? What is so significant about an empty point in space that a planet would circle around it?

2. Why do the imaginary points of Mercury and Venus always line up with the sun? What causes this curious phenomenon?

3. Why is it that the superior planets go into retrograde motion only when in opposition to the sun? In other words, why should the line from the sun to the earth have any influence on the position of a superior planet on its epicycle?

Ptolemy's followers could have given no satisfying answer to such questions. They would have had to reply, "That is the way nature is constructed." One should hasten to add that, to this day, there has never been a theory in any field that was able to answer *all* questions. A theory can be satisfying on one level even if it is incomplete on a deeper level. It is such unanswered questions that spur scientists to devise ever more comprehensive, more satisfying theories. There are many phenomena to which today's scientists must respond, "That is the way nature is constructed."

Copernicus felt there was a need for a new theory of the heavens. His main concern was that Ptolemy's explanation was so complex as to be unbelievable. Copernicus was dissatisfied with the large number of epicycles in Ptolemy's theory, and Ptolemy's device of moving the deferents off-center struck him as artificial and unconvincing. He deeply believed that a better, simpler, and more beautiful theory must be possible. It is said that, when a king of Spain learned the geocentric theory, he exclaimed, "I wish the Creator had asked for my opinion; I would have recommended something simpler!" A similar emotion (see Box 4.1) drove Copernicus to spend much of his life searching for a better theory.

BOX 4.1

AESTHETICS: SCIENTIFIC FAITH

Science is sometimes thought of as a field dealing only with facts. Scientists are supposed to use only cold, hard logic. Their conclusions must not be based on personal beliefs that cannot be substantiated. Their work is often contrasted with an artist's pursuit of such a nebulous, subjective thing as beauty.

A close look at the work of the major scientists shows this view to be only partially correct. Copernicus was led to doubt the geocentric theory because he found it insufficiently simple and beautiful. He said of Ptolemy's theory, "It would have been unworthy of the Creator to use so many circles in order to make the Sun, Moon, and planets move." He shared two profound, unprovable articles of faith with the other major scientists of every age. The first is that the universe is understandable to the human mind. Any collection of puzzling data must have an underlying basis of simple, understandable laws. Such scientists believe that the universe is neither capricious nor chaotic. As Einstein put it, "The eternal mystery of the universe is its comprehensibility."

A second article of a scientist's creed is this: The best explanation of observed phenomena is also the most beautiful. Every great theory provides more than a mere understanding of the phenomena. It should also have the same properties that enhance a work of art: unexpected elegance, simplicity, and symmetry. A great theory should make us exclaim both, "Yes, I now understand," and, "How wonderful!"

4.3 THE CHARACTERISTICS OF THE HELIOCENTRIC THEORY

Copernicus spent much of his leisure time speculating about a new theory. He searched for ideas among the ancient Greek authors whose books had been rediscovered by European intellectuals. He found that several Greeks had imagined the sun to be in the center of the universe as, for example, Aristarchus had suggested. Copernicus did not record any such event, but we might speculate that one day the clue he was searching for hit him as a sudden inspiration. "Aha," he may have shouted, "Exchanging the positions of the earth and the sun is indeed the answer. Perhaps I can eliminate epicycles. Beauty and simplicity will surely be the result."

Before describing Copernicus' heliocentric theory (Figure 4.1) in detail, we list some of its most important characteristics:

One should understand that the phrase "according to the heliocentric theory" is unstated but implicit in many of the sentences of this chapter.

1. The theory is heliocentric (sun-centered). The sun is at rest in the center of the universe. The earth revolves around the sun, as do the other planets.

2. Some of the observed motions in the sky are not real motions. For example, the sun does not really rise, rather it is at rest. The observed motion is caused by the rotation of the earth.

3. Like all preceding astronomers, Copernicus used circles to represent all motions.

Next we turn our attention to each of the heavenly bodies and see how Copernicus explained the observed motions that were described in Chapter 2.

NICOLAI COPERNICI

net, in quo terram cum orbe lunari tanquam epicyclo contineri diximus. Quinto loco Venus nono mense reducitur. Sextum denique locum Mercurius tenet, octuaginta dierum spacio circũ currens. In medio uero omnium residet Sol. Quis enim in hoc

pulcherrimo templo lampadem hanc in alio uel meliori loco poneret, quàm unde totum simul possit illuminare? Siquidem non inepte quidam lucernam mundi, alij mentem, alij rectorem uocant. Trimegistus uisibilem Deum, Sophoclis Electra intuentẽ omnia. Ita profecto tanquam in solio regali Sol residens circum agentem gubernat Astrorum familiam. Tellus quoque minime fraudatur lunari ministerio, sed ut Aristoteles de animalibus ait, maximam Luna cum terra cognationẽ habet. Cõcipit interea à Sole terra, & impregnatur annno partu. Inuenimus igitur sub hac

Figure 4.1 A page from *De Revolutionibus* showing the heliocentric system.

4.4 THE STARS IN THE HELIOCENTRIC THEORY

Copernicus saw no reason to disagree with Ptolemy concerning two ideas about the stars. He agreed that the celestial sphere is real. Copernicus was also convinced by Ptolemy's proof that the distance to the stars is very large compared to the diameter of the earth. In fact Copernicus went one step further. By assuming that the earth orbits (i.e., revolves around) the sun, he concluded that the stars must be extremely distant even as compared to the size of the earth's *orbit*. Otherwise, as Ptolemy had declared (Section 3.1), the constellations would seem to change shape noticeably as the earth approached or receded from them.

But here agreement ends. Copernicus assumed that the celestial sphere and all the stars attached to it are at rest. The observed motion of the stars—rising, moving across the sky, and setting—is not real motion. Actually, Copernicus claimed, the stars appear to be moving because the earth is rotating.

4.5 THE DEBATE OVER THE ROTATING EARTH

Ptolemy argued that the earth is round because, for one thing, the altitude of Polaris increases as one travels north. Copernicus agreed. But the many readers of *De Revolutionibus* were shocked by Copernicus' next claim about the earth—that it rotates (spins) on its axis once each 23 hr, 56 min. This, Copernicus felt, is a much more satisfactory hypothesis than the hypothesis of a stationary earth.

Imagine an observer in the United States facing eastward. He sees the stars rise in front of him, pass high above him to the south, and set behind him. If the stars in the east are not actually moving westward, toward him, then the observer must be actually moving eastward, toward the stars. Copernicus assumed that the earth rotates from the west toward the east. (See Figure 4.2.) Similarly, it is the eastward rotation of the earth that makes the sun, moon, and planets seem to rise, move to the west, and set.

The assertion that the earth is in motion caused a gradually widening wave of protest, which lasted over 100 years. Debates took place both in public and in print. Histories of astronomy describe these conflicts in detail.

For brevity, and for a change of pace, let us imagine a fictional debate between George, who defends the geocentric theory, and Helen, a follower of Copernicus. This debate is a composite of some of the actual (if slightly modernized) arguments used during the century

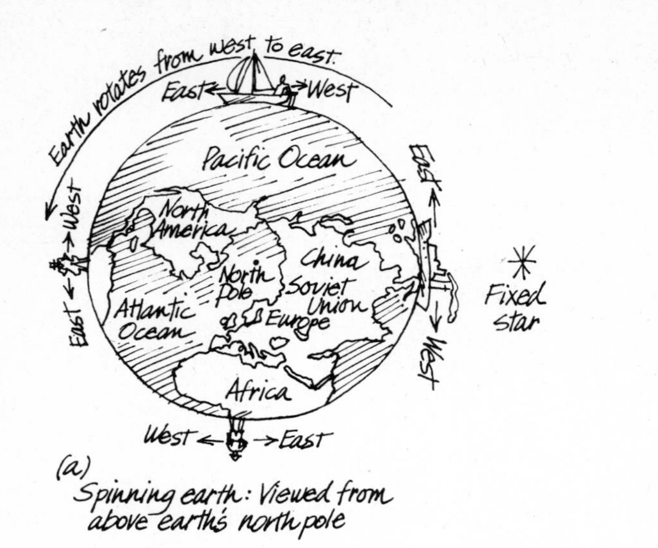

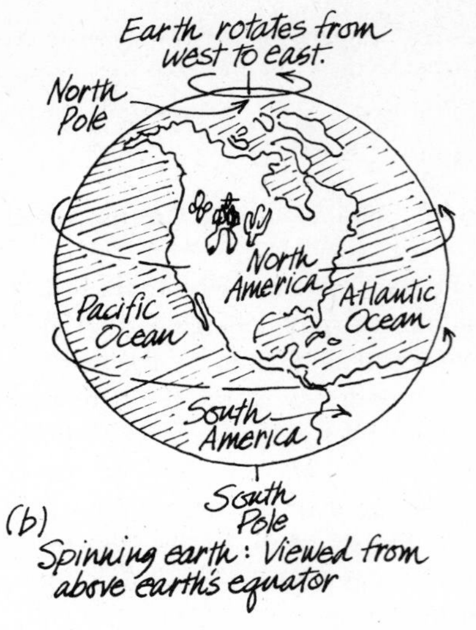

Figure 4.2 Rotation of the earth: heliocentric theory. The view from above the earth's north pole is shown in (a). For future reference, the reader should study the directions east and west for each observer shown. The rotation of the earth carries each observer toward his or her east. The rotation of the earth as viewed from above the earth's equator is shown in (b).

following Copernicus' death. (It is more rational and polite than some of the real debates.) The reader is asked to clear his or her mind of all prejudgments and imagine that he is scoring the debate as a fair-minded person living in the sixteenth century. Who has the best argument in what follows?

Resolved: That the earth rotates on its axis.

George: Ladies and gentlemen, I intend to argue against the resolution. Allow me to point out, initially, that the idea of a moving earth contradicts the opinions of the greatest minds of the past, those of Aristotle, Ptolemy, and the best thinkers of the present day. Furthermore it contradicts the opinion of the common people of every age. Simply put, a moving earth goes against common sense.

Helen: I welcome the opportunity to bring you the ideas of Copernicus, one of the greatest minds of our age. Yes, Copernicus contradicts the opinions of earlier authorities. But could it not be that, in our own age, Copernicus is an equal of these great men? Is it heresy to assert that he may have seen just a bit more clearly than these great sages? One must look at the *evidence* to decide which philosopher has the best argument. Common sense is a good guide in many cases but, in this case, common sense needs to be revised. In future ages, I predict, it will *become* common sense to say that the earth spins.

George: Besides violating common sense, the idea of a moving earth is a violation of religious belief. Copernicus' idea will undermine our religious life. The Bible clearly states that the *sun* moves. Witness: Ecclesiastes 1:5, "The Sun rises and the Sun goes down," and Psalms 19:6–7, ". . . the Sun . . . like a strong man, joyfully runs its course."

Helen: The charge that Copernicus was an enemy to religion is most false. Copernicus was an official of his church. He dedicated his book to the pope, and his work was supported by many famous church officials of his time. Copernicus read the Bible and tried to live a devout life by following its teachings. What shall we say then of such quotations as those given by my opponent? Surely the author of Ecclesiastes was not concerned with astronomy in his writing. Read the passage in context. He was giving an example of the sorts of events that repeat ceaselessly. It would have distracted from his point had he said, "The Earth rotates so as to cause the Sun to appear to rise and set." Furthermore, his readers would not have understood such wording. And is the quotation from the beautiful psalm to be taken as a scientific statement? It seems clear that the psalmist is using a figure of speech. When you say, "My uncle eats like a bird," you do not mean that your uncle *is* a bird. The author of the psalm is using poetic devices. Read the psalmist again. Does the author mean that the sun is literally a joyful, strong man? People like my opponent who take the Bible to be a mere science book do it a great disservice. The Bible is concerned with the meaning of life, the relationship between God and humanity, and the relationships between people. Finally, would it really affect your belief in God if the earth moved rather than the sun? Surely not. Look at Job 9:6, "He shakes the Earth out of its place." Do not place limits on what God can do.

George: Let's look at this question from a scientific point of view. Because the distance around the earth's equator is roughly 24,000 mi, and because the earth supposedly turns once in 24 hr, a person at the equator must be traveling at 1000 mi/hr! Yet we are asked to believe that he or she cannot feel such motion. My friends, I can close my eyes in a wagon and detect motion even if I am moving even at 5 mi/hr. Servant, beware you do not spill the milk—we are on a 1000 mi/hr wagon!

A debater in the sixteenth century might not have used this precise number for the circumference of the earth. Whatever number he used, it would not have changed the point of his argument materially. The diameter of the earth is about 13,000 km (7900 mi), and its circumference is about 40,000 km (25,000 mi).

Helen: Have you ever ridden on a sled over a very smooth, frozen lake? The motion is so even that you can barely detect it. Smooth motion is difficult to detect, and the earth turns *very* smoothly. Let me point out that Copernicus did *not* say that the Earth is rolling along a bumpy road. It is spinning in empty space, much like a spinning ball tossed in the air. It turns absolutely smoothly and evenly so that . . .

George: Pardon my impatience, but I wish to perform an experiment that will convince even the most committed Copernican. [He jumps into the air and comes down in the same place.] Do you see the

Figure 4.3 Tossing a ball: at rest and while moving. During the debate in the text, two experiments are conducted. They are illustrated here. In (a), Helen stands still and tosses a ball straight up into the air. In (b), Helen walks forward at a steady speed in a straight line and, while doing so, again tosses a ball straight up. The ball does not return to the floor *behind* her, rather, it continues the forward motion it has when it was in her hand. This forward motion causes it to remain directly above her hand throughout the flight.

meaning of what I have just done? I jumped off of the earth, giving it every chance to move at 1000 mi/hr under my feet. Of course it did not; the earth is at rest. If Copernicus were right, one could travel to the west by merely hopping a few times! If he were right a mother bird that left her nest could never return, because the tree would rush away at 1000 mi/hr! Furthermore, why don't we experience a destructive 1000 mi/hr wind?

Helen: The experiment is interesting but incomplete. Let me perform a complete experiment. Standing still, I toss a ball straight up. It comes down in my hand as expected. But now I again toss the ball straight up as I walk at a steady speed. Behold! It again returns to my hand rather than falling behind me. (See Figure 4.3.) It is the same way on earth. When one jumps off the earth, one continues his or her forward motion without stopping. One stays directly above the spot on earth that one left. When a bird leaves a tree, the bird does not suddenly stop. What would cause it to stop? The bird continues to move at the same eastward speed when on or off the tree. The air is part of the earth and travels with the earth and travels with the earth as it spins. Thus there is not a 1000 mi/hr wind. The jumping experiment has no value. The experiment *looks* the same whether the earth is moving or not.

George: Still no evidence *for* a turning earth, dear opponent? Of course not—because the earth is at rest. I leave you now with my last, and perhaps most convincing, proof. Imagine taking a ball of mud and spinning it until it goes 1000 mi/hr at its equator. What would happen? It would fly to pieces most spectacularly. I leave you with the question, Why doesn't the earth fly apart?

Helen: Dear opponent, you are trying to throw mud in our jury's eyes. If the earth does not spin, then, as you insist, the huge celestial sphere of stars must. But then imagine a star on the celestial equator. In only one day it must travel an enormous distance. It would have to travel at many millions of miles per hour. I put it to our jury: Which is more likely? A tiny earth turning once a day or a monstrous celestial sphere of stars whirling at incredible speeds? Which would be *more likely* to fly apart?

Had you witnessed this debate in the late sixteenth century, how would you have voted?

4.6 THE SUN IN THE HELIOCENTRIC THEORY

The sun appears to revolve around the earth once in 24 hr and to wander north and then south of the celestial equator once each year. Also, it seems to travel among the fixed stars along the ecliptic from west to east, taking one year for one complete circuit. There was much for Copernicus to explain. If the sun is really at rest, why does it appear to perform such motions? Copernicus' answer was that these observed motions were actually caused by the earth's motions.

Daily Motion The sun appears to return to the observer's meridian once every 24 hr. We have called this the sun's daily motion. According to Copernicus, it is actually caused by the rotation of the earth. The rotation (spinning) of the earth carries the observer and his or her meridian around and back to the stationary sun once every 24 hr. (See Figure 4.4.) Much of our life is regulated by this spinning of the earth.

We may observe that Copernicus gives the time for one rotation of the earth as 23 hr, 56 min with respect to the stars and as 24 hr with respect to the sun. This will be explored later in Puzzle 4.1.

Motion in Declination According to Copernicus, the earth revolves around the sun once each year. The path the earth travels on its journey around the sun is called its **orbit.** (We use the word "orbit" in the context of the heliocentric theory in preference to "deferent.") It is this revolution that causes the apparent annual motion of the sun which in turn causes our seasons.

Copernicus states that the axis of rotation of the earth does not lie at right angles to the plane of the earth's orbit. Rather, the axis is tilted at an angle of $23\frac{1}{2}°$. This tilt must be constant throughout the yearly journey; that is, the axis must always point in the same direction, toward the north celestial pole. This is why Polaris, which lies near the north celestial pole, appears to be nearly stationary. Figure 4.5 and Table 4.1 give the details of the sun's motion in declination according to the heliocentric theory.

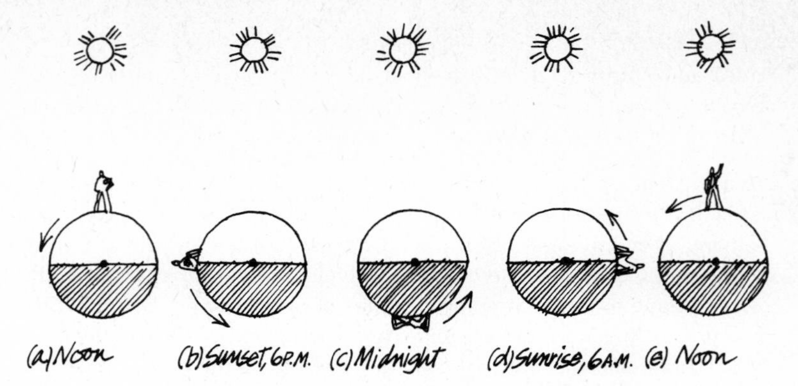

Figure 4.4 Viewing the daily motion of the sun from a rotating earth. In the heliocentric theory, the sun stands still. Its daily rising and setting are only an apparent motion caused by the rotation of the earth. In (a), the observer is in the middle of the day side of the earth. He sees the sun high in his sky, that is, on his meridian. The time is noon. In (b) it is 6 hr later. The earth has rotated one-quarter of a turn. The observer is about to enter the night side of the earth. He is getting his last look at the sun for the day; it is sunset. He sees the sun low in the sky, setting in his west. In (c) it is 6 hr later than in (b). The observer is in the middle of the night side of the earth; it is midnight. After 6 hr have elapsed, the earth has carried the observer to the position shown in (d). The observer sees the sun again as a new day begins; it is sunrise. He sees the sun low in the sky, rising in his east. In (e), the earth has rotated once with respect to the sun. It is again noon, 24 hr later than in (a). (For simplicity, we have suppressed the relatively slow motion of the earth around the sun. See Figure 4.6. In addition, we have described the situation for an observer on the equator at the time of equinox. You may wish to study Figures 2.7 and 4.5 and then think about sunrise and sunset times in your own location at differing times of the year.)

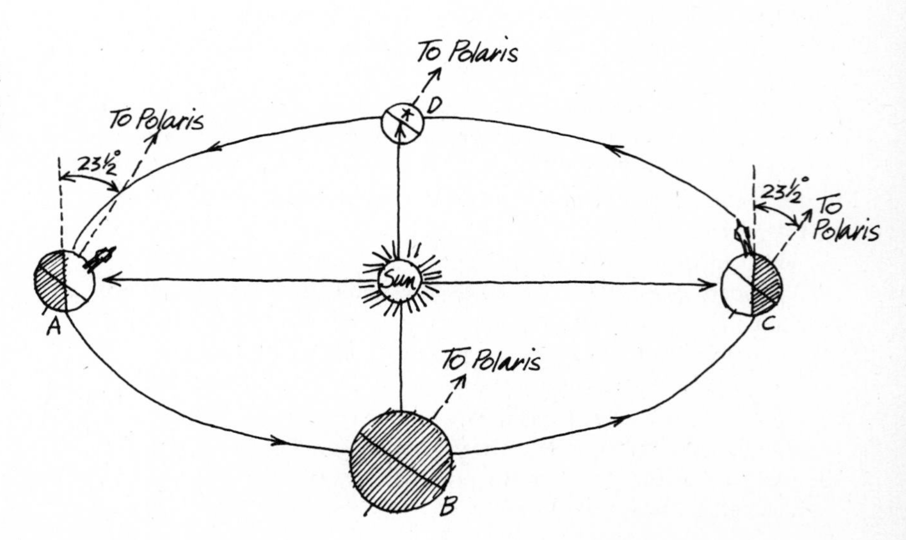

Figure 4.5 The heliocentric theory explains the sun's annual motion and the seasons. The view here is from a point in space just above the plane of the earth's orbit. See also Table 4.1.

Table 4.1 The Sun's Motion in Declination: The Heliocentric Theory

Position of the earth in Figure 4.5	*Date*	*Earth's axis points at*	*Earth's north pole is*	*Sun's rays at noon perpendicular to the ground at latitude*	*Every observer in North America*			*Name of this event*
					Sees the noon sun	*Experiences*	*Experiences the beginning of*	
A	June 21	Polaris	Always on day side	23½° north	23½° north of celestial equator	Long daylight	Summer	Summer solstice
B	September 22	Polaris	Entering night side	0°, equator	On celestial equator	Equal days and nights	Autumn	Fall or autumnal equinox
C	December 22	Polaris	Always on night side	23½° south	23½° south of celestial equator	Short daylight	Winter	Winter solstice
D	March 20	Polaris	Entering day side	0°, equator	On celestial equator	Equal days and nights	Spring	Spring or vernal equinox

Annual Motion Looked at from another point of view, the yearly trip of the earth around the sun can also explain the apparent motion of the sun among the fixed stars, the sun's annual motion. As the earth carries us to different points in its orbit, we see the sun from differing points of view, against differing starry backgrounds. (See Figure 4.6.)

You can check your mastery of Copernicus' theory as it applies to solar motion by attempting Puzzle 4.1. In it you are asked to explain the outcome of a race between the sun and a star using the heliocentric theory. Study the solution to the puzzle carefully. Do you find it more or less satisfying than the geocentric analysis given in Figure 3.6?

The followers of Copernicus felt that the heliocentric theory was more satisfying than the geocentric theory when applied to the sun's apparent motion. One no longer had to think of the sun as being "carried along" somehow by the whirling stars. Further, no mention needed to be made of the sun's "crawling along" the ecliptic in the reverse direction. The whole universe does not whirl around the earth. Rather, the earth rotates and revolves. The sun and the stars simply stand still. Others felt that the concept of a moving earth was too unbelievable to be used to explain anything else.

4.7 THE MOON IN THE HELIOCENTRIC THEORY

Ptolemy assumed that *everything* revolves around the earth. In Copernicus' view, the stars and the sun are at rest and the planets (the earth is considered a planet) revolve around the sun. Only the moon remains orbiting the earth in Copernicus' theory.

Copernicus sought a way to explain the motion of the moon in his theory. He had to explain (1) why the moon is observed to return to an

PUZZLE 4.1

THE RACE BETWEEN THE SUN AND A STAR: THE HELIOCENTRIC VIEW

In Puzzle 2.4, the reader was asked to predict the winner of a race between a star and the sun, starting and ending at the meridian. Then, in Figure 3.6, the outcome of the race was explained by using the geocentric theory. The reader should now draw a diagram explaining why the star wins the race, this time using the heliocentric theory. (Then see the solution in Appendix 2.)

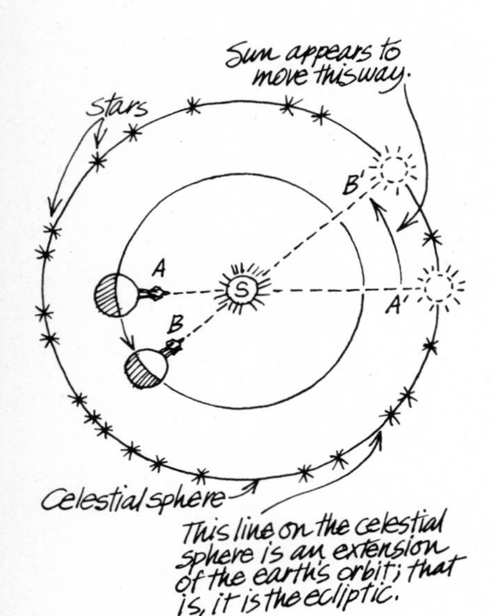

Figure 4.6 Why the sun seems to move from west to east among the fixed stars according to the heliocentric theory. As the earth revolves from A to B, the stationary sun appears to change its position with respect to the fixed stars from A′ to B′. The reader should check the directions east and west as shown in Figure 4.2a. Would the observer indeed describe the apparent motion of the sun as motion from west to east among the fixed stars? (For simplicity, the 23½° tilt of the earth's axis is suppressed here and in similar diagrams.)

observer's meridian once every 24 hr, 50 min, (2) why the moon is observed to travel west to east among the fixed stars, and (3) why the moon is observed to return to the same star once every 27 days, 8 hr (one lunar sidereal period). In one sense, Copernicus' explanation requires a bit more concentration on the reader's part than Ptolemy's did. The reason is that, in this theory, both the moon *and* the earth move.

Figure 4.7 shows how Copernicus explained the moon's motions. The earth's rotation carries the observer relative to the fixed stars in a circle once nearly every 24 hr. By then the moon has moved far enough in its orbit so that about another 50 min is required for the observer to, as it were, catch up to the moon. The moon now appears to be to the east of the star. That is, the moon appears to have traveled from west to east among the fixed stars. The moon continues on its orbit around the earth until, one lunar sidereal period later, it is once more lined up with the stationary star.

In Figure 4.7, we viewed the moon's motion from a position above the earth's north pole. Figure 4.8 shows how the moon moves as viewed by an observer positioned far above the stationary sun. Here the moon is shown moving through one complete lunar synodic period. The figure shows why the lunar synodic period is longer than the lunar sidereal period. Note the analogy to the idea behind the figure in the solution to Puzzle 4.1, found in Appendix 2.

In Chapter 2, the following observations about the moon's phases were recorded. The new moon is lined up with the sun. The first quarter moon is on one's meridian at sunset. The full moon rises at sunset. Finally, the last quarter moon is on the meridian at sunrise. It is not easy to see a pattern in these facts. But study Figure 4.9. Once you understand Copernicus' theory of the phases of the moon, you will perceive two of the advantages of having a good theory. First, the previous data are explained; they now make sense. Second, one need no longer memorize the various positions of the moon at each phase. One need only draw the appropriate sketch and figure out when to look for the moon at each phase. This is one of the marks of any good theory: It makes sense out of a set of observations. One idea may replace a burdensome set of seemingly disconnected data.

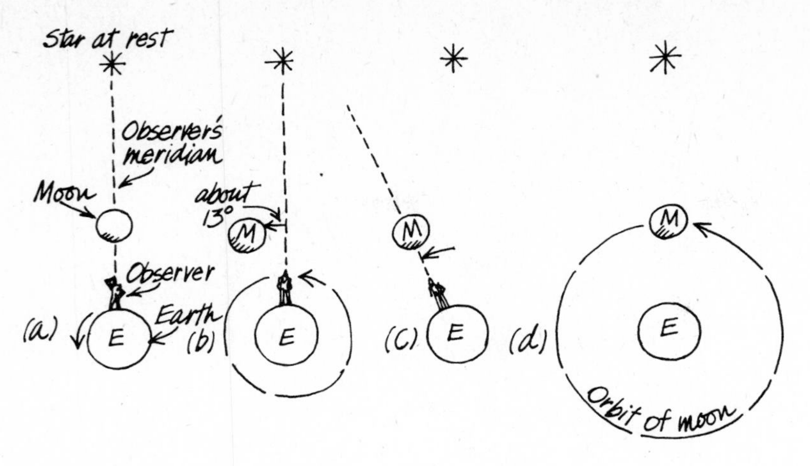

Start: Moon and stars are on observer's meridian.	*23 hr, 56 min after (a): Earth has rotated once. Star is again on meridian. Moon is now to the east of the star.*	*24 hr, 50 min after (a): The moon is again on the observer's meridian. The star is to the west of the meridian.*	*27 days, 8 hr after (a): The moon has gone completely around the earth and is lined up with the same star again. One lunar sidereal period has elapsed.*

Figure 4.7 The motion of the moon according to Copernicus. The motion of the moon around the earth plus the rotation of the earth on its axis can explain the apparent motion of the moon. Observe that the moon orbits the earth in the same direction (or sense) that the earth spins on its axis. The figure does not show that the moon's orbit lies in nearly the same plane as that of the earth's orbit.

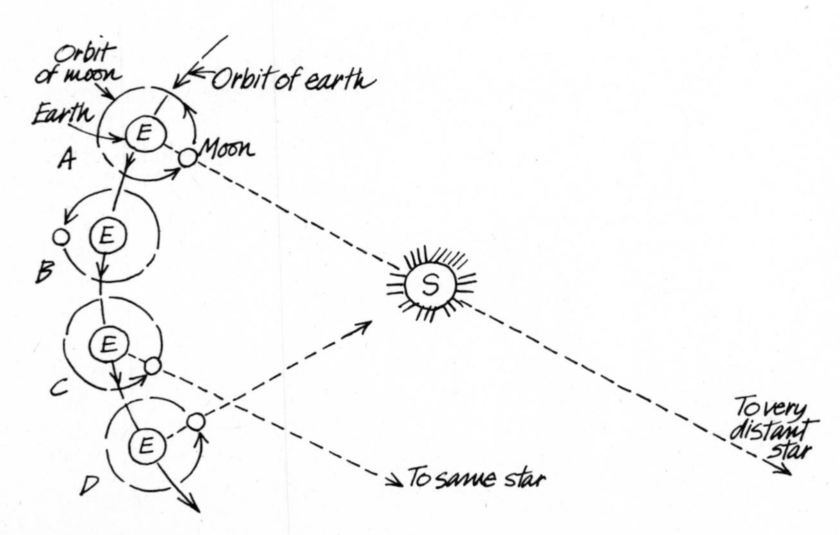

Figure 4.8 The motion of the moon as viewed from above the stationary sun. Position A shows a new moon. Position B, about 2 weeks later than A, illustrates the following full moon. At position C, the moon has orbited the earth once (with respect to the fixed stars). Thus, one lunar sidereal period has elapsed since position A. The next new moon does not occur until position D. There, the moon has once again lined up with the sun, and 29 days, 13 hr have elapsed since position A. That is, we have followed the moon's motion for one lunar synodic period.

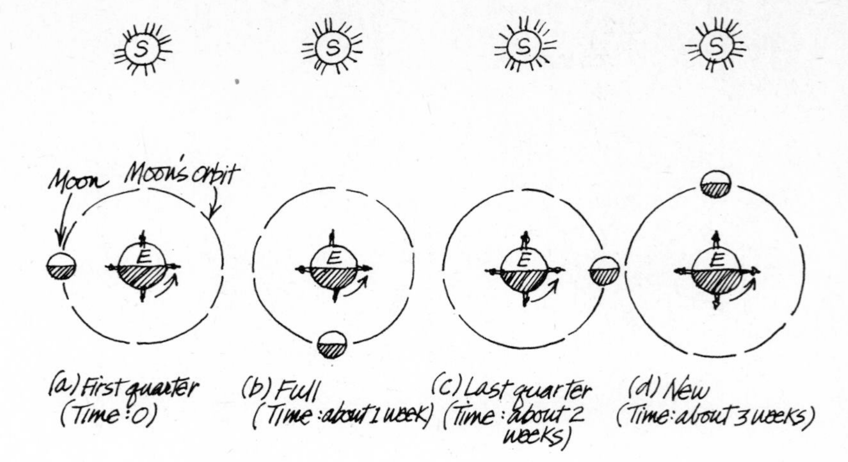

Figure 4.9 The phases of the moon. The four small dots on the earth represent observers. In (a), the observer on "top" of the earth (noon position) sees the first quarter moon for the first time that day. The rotation of the earth is turning her so that she sees the first quarter moon rising. (Keep in mind that the earth turns more quickly than the moon moves. The moon appears to be rising because of the earth's rotation. We have simplified the analysis by temporarily suppressing the orbital motion of the earth around the sun.) The evening observer sees the first quarter moon on her meridian, high in the sky. The midnight observer is getting her last look at the moon—the first quarter moon sets at midnight. Now study (b), (c), and (d) and verify these statements. The full moon rises at sunset, is on the meridian at midnight, and sets at sunrise. The last quarter moon rises at midnight, is on the meridian at sunrise, and sets at noon. The new moon, were it visible, would rise at sunrise, be on the meridian at noon, and set at sunset. By the way, the moon's orbit is tilted by 5° with respect to the plane of the ecliptic. This is why there usually is no eclipse at full or new moon. The earth, moon, sun lineup is ordinarily not exact.

4.8 THE INFERIOR PLANETS IN THE HELIOCENTRIC THEORY

We next discuss perhaps the most intriguing contrast between the geocentric and heliocentric theories: the explanation of the motions of the planets. Ptolemy used epicycles as a unifying assumption to explain the motions of all five planets. What did Copernicus have to offer?

Copernicus felt he had a much improved theory to offer. He found that, qualitatively, by imagining the sun at the center of the universe, he could eliminate epicycles from his theory. (But see Section 4.10.) His assumption was simplicity itself: Each planet revolves around the stationary sun in a circular orbit. We see that the heliocentric theory recognizes six planets. Earth has been added to the previous list. (Uranus, Neptune, and Pluto were yet to be discovered.)

When considering the inferior planets, Copernicus added one fur-

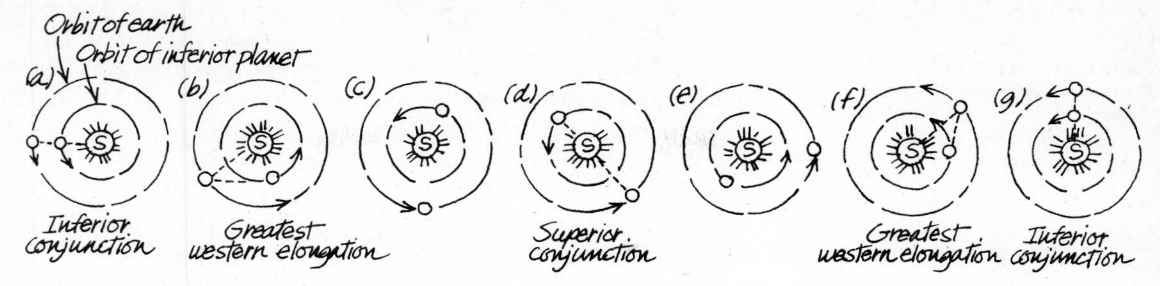

Figure 4.10 The motion of a typical inferior planet: heliocentric theory. At (a), the planet is in inferior conjunction with the sun as viewed from earth. As it passes the earth, it appears to be moving from east to west with respect to the fixed stars. This is retrograde motion. At (b) the planet has moved to the west of the sun and so is a morning star. At (d), the planet is said to be in superior conjunction. It next moves to the east of the sun and becomes an evening star.

ther simple assumption. The inferior planets have orbits smaller than the orbit of the earth. The two assumptions about an inferior planet (circular motion centered on the sun and an orbit smaller than earth's orbit) make it possible to explain the apparent motions of the inferior planets in a neat, qualitative (nonnumerical) way. In Figure 4.10 the motion of an inferior planet is shown as it travels through one synodic period. Looked at in one way, Copernicus adopted Ptolemy's idea about the motion of an inferior planet, but he shifted the center of the planet's epicycle to the sun. The looping motion of Venus as viewed from a position at rest in outer space has disappeared, and the planet now travels in a simple circle around a stationary center. Furthermore, the center of the circle is no longer an empty point in space. The planet circles a real object, the sun.

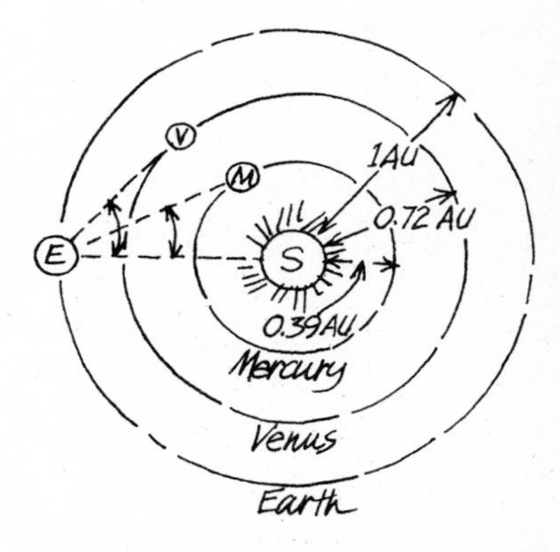

Figure 4.11 The orbits of the inferior planets according to the heliocentric theory. By assuming that the inferior planets travel in circular orbits around the sun, Copernicus found that he could determine the order of the inferior planets and even the relative size of each orbit. All he had to do was observe the angular separation of each inferior planet from the sun at maximum elongation (shown here). Then a simple calculation using trigonometry gives the radius of each orbit in terms of the radius of the earth's orbit.

Recall that Ptolemy had to assume that the imaginary points of Mercury and Venus stay lined up with the sun. This was viewed by some to be mysterious. In the heliocentric theory, as one author phrased it, "The mystery is no mystery." As viewed from earth, the inferior planets *appear* to move from side to side of the sun, never leaving the region of the sky near the sun because they have orbits smaller than the earth's. Figure 4.10 shows why the inferior planets always *seem* to stay near the sun in the sky. They simply cannot go to the side of the earth opposite the sun because they are confined to the region inside the earth's orbit.

Ptolemy preferred to assume that the order of the inferior planets, working outward from Earth was Mercury, then Venus, followed by the sun. Ptolemy would have granted that this was a somewhat arbitrary choice. On the other hand, if one is willing to accept Copernicus' assumptions mentioned above, all such arbitrariness automatically disappears. Mercury *must* be closest to the Sun, in the smallest orbit of all the planets. Venus *must* have a larger orbit. According to Copernicus, Venus is the inferior planet nearest Earth, contrary to Ptolemy's opinion. To see why the order of the inferior planets is not arbitrary in the heliocentric theory, refer to Figure 4.11.

Table 4.2 Data Concerning the Planets: Heliocentric Theory

Planet	*Radius of orbit, modern value (AU)*	*Sidereal period, modern value (earth years)*	
Mercury	0.39	0.24	(88 days)
Venus	0.72	0.62	(225 days)
Earth	1.00	1	(365.26 days)
Mars	1.5	1.9	
Jupiter	5.2	11.9	
Saturn	9.5	29.5	

The assumption that the planets orbit the sun fixes not only the order of the inferior planets but also the relative size of each orbit. One observes each inferior planet's greatest elongation and calculates the size of the respective orbit relative to the size of the earth's orbit. The larger the greatest elongation, the larger the orbit. Modern astronomers have given the average distance from the earth to the sun the name **astronomical unit** (AU). Thus one says that the radius of the earth's orbit is 1 AU. (Copernicus did not use this terminology.) Copernicus then was able to calculate that the radius of Mercury's orbit is about 0.36 AU and that the radius of Venus' orbit is about 0.71 AU. (The modern values are 0.39 and 0.72.) Copernicus did not have a reliable value for the actual distance of the sun from the earth in, say, kilometers. Therefore he did not know the *actual* distance of each planet from the sun. Yet he did have a way of calculating the *relative* size of each planet's orbit.

Besides determining the relative sizes of the orbits of the planets, Copernicus could also compute the relative speeds of the planets. He found that, when any two planets were compared, the planet with the smaller orbit moved faster in its orbit. Thus Mercury moves faster in its orbit, reckoned in kilometers per hour (or miles per hour), for example, than Venus. We can see why Mercury completes one orbit in a much shorter time than Venus; not only is Mercury the faster runner, it has the inside track.

We close this section by defining a new term used in the heliocentric theory. The **sidereal period of a planet** is the length of time required for a planet to revolve around the sun once, as *viewed from a fixed star*. Recall that the sun and the stars are at rest in this theory. Thus an imaginary line from the sun to a fixed star is a good reference line. The sidereal period of a planet may be thought of as the length of time required for the planet to return to such a line. One may also say that a planet's sidereal period is "the planet's year." Copernicus calculated the planets' sidereal periods (see Table 4.2) from the observations of earlier astronomers.

Copernicus calculated the sidereal periods from the measured synodic periods using a formula derived from his theory.

4.9 THE SUPERIOR PLANETS IN THE HELIOCENTRIC THEORY

Copernicus found that his theory explains the apparent motions of the superior planets just as neatly as in the case of the inferior planets. He merely had to assume that the superior planets travel in circular orbits larger than the earth's orbit. Figure 4.12 shows the motion of a typical superior planet during one synodic period. The theory easily explains why superior planets can go into opposition (while inferior planets cannot).

Based on the observations made and on his theory, Copernicus calculated the size of each planet's orbit and its sidereal period. Here, too, the order of the planets and the sizes of their orbits are not arbitrary but are fixed by observation and theory. (See Table 4.2.)

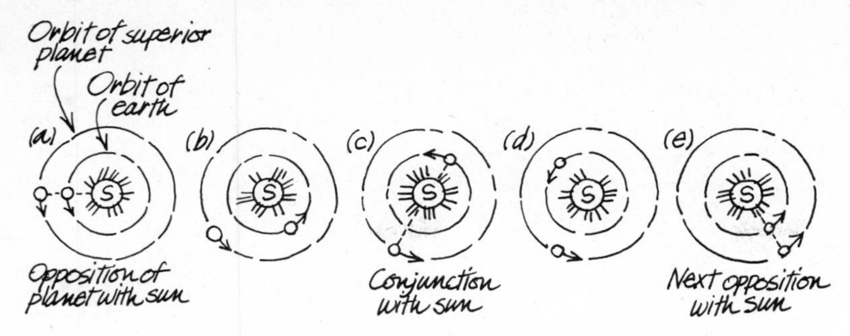

Figure 4.12 The motion of a typical superior planet: heliocentric theory. The swifter earth is shown here passing a typical superior planet and eventually overtaking it again. While studying this figure, imagine the view obtained by an observer on earth. In (a) the planet lies on the side of the earth opposite the sun and so is said to be in opposition. In (c) the planet is said to be in conjunction. In (e) the earth has again caught up with the superior planet. One synodic period has elapsed since (a) for this superior planet.

Perhaps the greatest challenge to the heliocentric theory was the seemingly erratic motion of the superior planets. In his qualitative considerations Copernicus met this challenge triumphantly. Recall the facts about the retrograde motion of the superior planets: (1) A superior planet reverses its normal progression periodically by traveling from east to west among the fixed stars. (Ptolemy explained this using epicycles. He said that the planet *really does* reverse its motion.) (2) The superior planet appears brightest during retrograde motion. (Ptolemy said that this is because the planet is nearest the earth at this time.) (3) Retrograde motion occurs for a superior planet only at opposition to the sun. (Ptolemy said that the planets' motions are geared to the sun's motion.)

The heliocentric theory's qualitative explanation of retrograde motion could not involve any actual reversal of the planet's motion. Copernicus had already asserted that all planets travel forward at all times. Copernicus said that the planets only *appear* to back up as the more rapidly moving earth passes them. Figure 4.13 describes an analogy involving a rapid car passing a slower one. A sophisticated observer would say that the faster car puts the slower car behind her as she overtakes it. A naive observer, riding in the faster car, might assume that her car is at rest. She therefore says that the other car has gone behind her because it has a reverse motion. Copernicus' contention was that believers in the geocentric system had taken a naive view when thinking about retrograde motion. The earth is not at rest, he claimed, but actually moving faster than the superior planet. As the earth overtakes the superior planet, the superior planet falls behind. Thus it merely *looks* as though it is backing up. Figure 4.14 is a sketch of the heliocentric point of view.

The words "naive" and "sophisticated" are used here as a proponent of the heliocentric theory might use them.

The more one thinks about this analysis, the more wonderful it becomes. For one thing, one can see that a superior planet must be

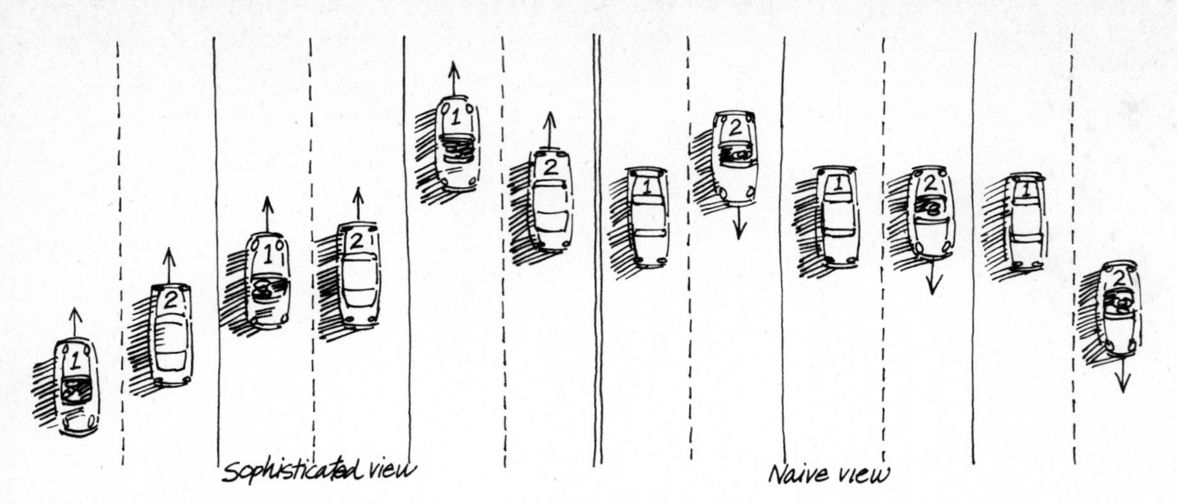

Figure 4.13 Two ways of looking at the relative motion of two cars. Imagine two observers in car 1. They both observe the behavior of car 2 and compare observations later. The sophisticated observer assumed that her car, car 1, was moving faster than car 2, so that the swifter car overtook the slower one. The naive observer in car 1 assumed that car 1 was at rest. To her, then, car 2 seemed to be going backward.

brightest during retrograde motion because it is nearest the earth at the time the earth passes it. Even more remarkable, the mystery of the simultaneous occurrence of the retrograde motion of a superior planet and its opposition disappears. A glance at Figure 4.12a clears away the mists. Aha! If retrograde motion is observed when the earth passes the superior planet, *of course* the superior planet must be on the side of the earth opposite the sun. Here, the heliocentric theory explains yet another fact that the geocentric theory handled by introducing a further assumption. The third unanswered question of Section 4.2 is not even relevant in the heliocentric theory.

4.10 THE ECSTACY AND THE AGONY: QUALITATIVE SUCCESS, QUANTITATIVE TROUBLE

The heliocentric universe, sketched in Figure 4.15, was an imposing conception. A proponent of the theory can list many qualitative reasons why this theory was an improvement over the geocentric theory. The heliocentric theory explains qualitatively all observed motions, and it does so simply and elegantly. There is no need for epicycles in the qualitative explanations. Not only is the order of the planets outward from the sun definite, but also the relative sizes of the orbits are fixed. Once one gets used to the idea of a rotating and revolving earth, simplifications follow. The heavens no longer whirl around the earth. The superior planets no longer circle on their epicycles in time with the

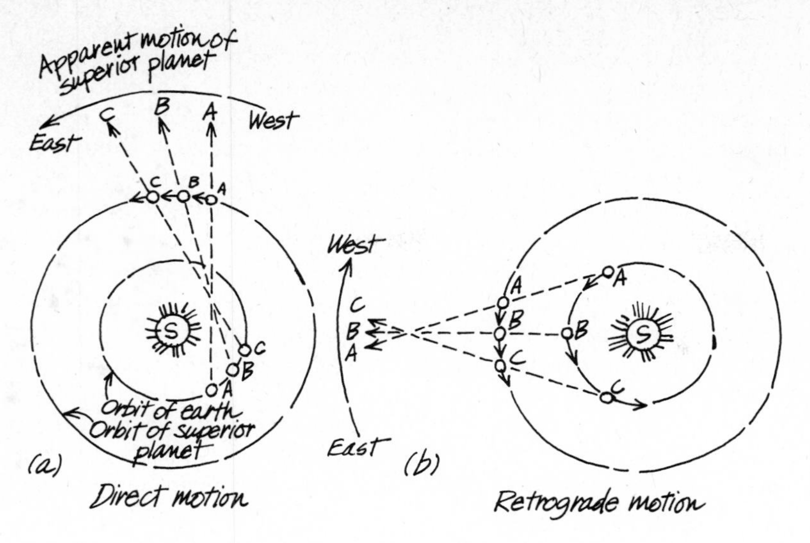

Figure 4.14 The explanation of direct and retrograde motion of a superior planet according to the heliocentric theory. In (a) and (b) the earth and the superior planet are shown in three successive positions. The situation when the superior planet is not near opposition is illustrated in (a). The planet appears to be gradually shifting its position from *west to east* with respect to the distant fixed stars. In (b), near opposition, when the earth is passing the slower superior planet, the superior planet appears to be gradually shifting its position from *east to west* with respect to the distant fixed stars, which is retrograde motion.

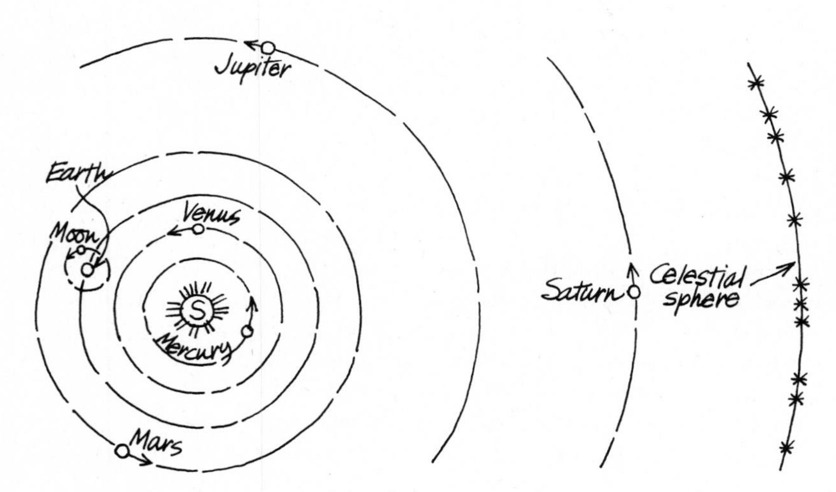

Figure 4.15 Copernicus' heliocentric universe. The further the planet is from the sun, the slower it moves. Such a drawing cannot show that each planet's orbit is slightly inclined (tilted) with respect to the earth's orbit. It should be emphasized that none of the inclinations is large. The solar system is flat, so to speak. The largest tilt of any orbit shown here is the 7° inclination of Mercury's orbit.

sun like mysterious clockworks. The unanswered questions in Section 4.2 have all been answered or—better—they have been defined out of existence. Copernicus could justly be proud of the qualitative features of his theory.

But when Copernicus turned his attention to the quantitative, that is, mathematical aspect of his theory, he ran into grave difficulties. His theory, as described above, did not explain the observed motions of the moon and the planets numerically. The simple, elegant theory failed as a practical theory. It could not accurately predict the future positions of the moon and the planets. Each planet was found to run ahead of or, at times, lag behind its predicted position.

In this chapter we have separated the qualitative aspects of the heliocentric theory from the quantitative ones. Most likely, Copernicus worked on both simultaneously. Most of Copernicus' book concerns his complex efforts to bring his theory into quantitative agreement with the observations.

Copernicus found, as had Ptolemy earlier, that he needed to modify his elegant, simple qualitative theory in order to bring it into line with quantitative observations. After much thought, he decided that he *had* to use epicycles in his theory. It was the only way he could think of to make the planets behave as they should. He conceived of an imaginary point traveling on each planet's orbit and the planet itself traveling on an epicycle around the point. The planets traveled slowly on their epicycles, so that they did not actually reverse their motion but did run ahead of or lag behind their imaginary points. When all was said and done, Copernicus had to console himself with this: His theory was superior because it required fewer epicycles than the current versions of the geocentric theory.

A major problem that retarded acceptance of the heliocentric theory was that, even when Copernicus used epicycles in his calculations, his theory gave predictions of planetary positions found to be no better than (and, in some cases, *worse* than) the predictions of the geocentric theory.

We now know that Copernicus' simple qualitative theory needed modification but that he had not found the *best* modification. The use of epicycles was, we now know, a mistake. Fortunately, history provided a pair of outstanding astronomers who collaborated to find the proper solution to Copernicus' quantitative troubles.

SUMMING UP

In the course of seeking a theory more beautiful than Ptolemy's, Copernicus constructed the heliocentric theory. In it, all objects except the moon orbit the sun. The celestial sphere is at rest and only appears to rotate because of the rotation of the earth. The apparent motion of the stationary sun is produced by the simultaneous rotation and revolution of the earth. The moon revolves around the earth relative to the fixed stars in 27 days, 8 hr. The inferior planets travel in orbits smaller than the earth's orbit, while the superior planets have larger orbits. The theory is particularly impressive in the manner in which it makes definite the relative sizes of planetary orbits and in its explanation of why the retrograde motion of superior planets can occur only at opposition.

Copernicus died leaving his theory in an uncertain state. It was, perhaps, a more beautiful conception than the geocentric theory but quantitatively it was not more successful. Looking back from our five hundred-year vantage point, we can see the importance of Copernicus' achievement. Once it became a quantitative success (a process described in the next chapter), the theory opened scientists' eyes. Ancient wisdom can be successfully challenged. Innovation can work. The technological and industrial revolutions that have so changed our societies have roots that can be traced back to the day Copernicus imagined the earth and the sun changing places.

EXERCISES

1. Define the following terms: heliocentric, orbit, rotation, revolution (the last two are defined only in the glossary), astronomical unit.

2. List the three unanswerable questions about the geocentric theory that are given in Section 4.2. Can you add any of your own? Does the fact that a theory cannot answer certain questions mean that it is a bad theory? Does it mean that a better theory might be possible? Why was Copernicus' dissatisfied with the geocentric theory?

3. What are the three main characteristics of the heliocentric theory?

4. How do the stars and the sun actually move according to the heliocentric theory? (Trick question!)

5. In what direction does the earth rotate and how long is one rotation relative to the fixed stars according to the heliocentric theory?

6. Suppose you were alive in the year Copernicus died and wanted to persuade someone to accept the heliocentric theory. What points would you raise? What points would you raise in order to argue against it?

7. If one jumps into the air and comes down in the same place, it neither proves nor disproves that the earth rotates. Explain why.

8. How does the revolution and rotation (on a tilted axis) of the earth explain (a) the sun's 1° per day apparent motion from west to east among the fixed stars, and (b) the sun's apparent motion in declination (thus the seasons).

9. Explain by means of a diagram why, in a race between the sun and a star, the sun loses by 4 min. Use the heliocentric theory.

10. Explain by means of a diagram why the lunar synodic period is longer than the lunar sidereal period. Use the heliocentric theory.

11. What time does the first quarter moon rise? What phase is it when the moon sets at sunrise? In what direction must one look to see the moon at noon if the moon is in its third quarter? (Answers: about noon; full; west.)

12. Which two assumptions did Copernicus make to explain the apparent motions of the inferior planets?

13. Why must Venus have a larger orbit than Mercury in the heliocentric theory?

14. What is the meaning of "sidereal period of a planet" in the heliocentric theory?

15. What additional assumption beyond circular orbits centered on the sun did Copernicus make in order to explain the apparent motions of the superior planets?

16. Use a diagram to give Copernicus' explanation of the retrograde motion of a superior planet. Why, in this theory, must such a planet exhibit retrograde motion only at opposition?

17. Draw two diagrams, one for the geocentric system and one for the heliocentric system, showing a superior planet performing retrograde motion. Explain why in each case the planet looks brightest during this time.

18. Use a diagram and the heliocentric theory to explain why Saturn (or, indeed, any superior planet) cannot go into opposition more than once each earth year.

19. Draw a sketch of the universe according to Copernicus.

READINGS

More on Copernicus and his work may be found in the works by Christanson and Kuhn cited in the "Readings" section at the end of Chapter 3.

RESCUING THE HELIOCENTRIC THEORY

Tycho Brahe had been raving in a fever all day. For the previous 20 years this Danish lord had been observing the motions of the heavenly bodies. He had far surpassed all previous astronomers in accuracy and completeness. Yet, as he lay dying, he mumbled over and over, "Don't let it seem that I have lived in vain." Why wasn't he satisfied that his observations would secure his fame? The reason was this. Led by his intuition and observations, Tycho had come to doubt the theories of both Ptolemy *and* Copernicus. This led him to invent a third theory, the Tychonic system of the world. He had pinned his hopes for fame on this theory and on his brilliant young German assistant, Johannes Kepler. Several days before Tycho died, he called Kepler to his bedside and told him of his last wish. Kepler was asked to promise that he would use Tycho's observations to show that Tycho's theory was the one, true theory. Under such circumstances, how could Kepler refuse? Nevertheless, as we shall see in this chapter, the outcome of Kepler's labor was by no means the one Tycho so fervently desired. Tycho's name lives on, but not precisely as he had hoped.

5.1 THE TROUBLE WITH THE HELIOCENTRIC THEORY

After Copernicus died in 1543, his ideas did not immediately attract many important followers. It is easy to see why. Although he had left a theory of great qualitative beauty, it had had no outstanding quantitative success. Also, in order to compete with the predictive ability of the geocentric theory, Copernicus had used epicycles. On quantitative grounds, few saw any compelling reason to reject the long-standing geocentric theory.

Looking back, we can understand why Copernicus had such quantitative problems. He was using very old data. The best observations available to him were those in Ptolemy's *The Almagest*, which had been written about 14 centuries earlier. Also he took the wrong path when he used epicycles.

Recall that the bulk of the observations in *The Almagest* were performed by Hipparchus who lived about 300 years before Ptolemy.

In spite of these problems, Copernicus laid the foundation of a successful theory. He could not know that it would require two additional lifetimes to complete his work. The story of this completion, which we

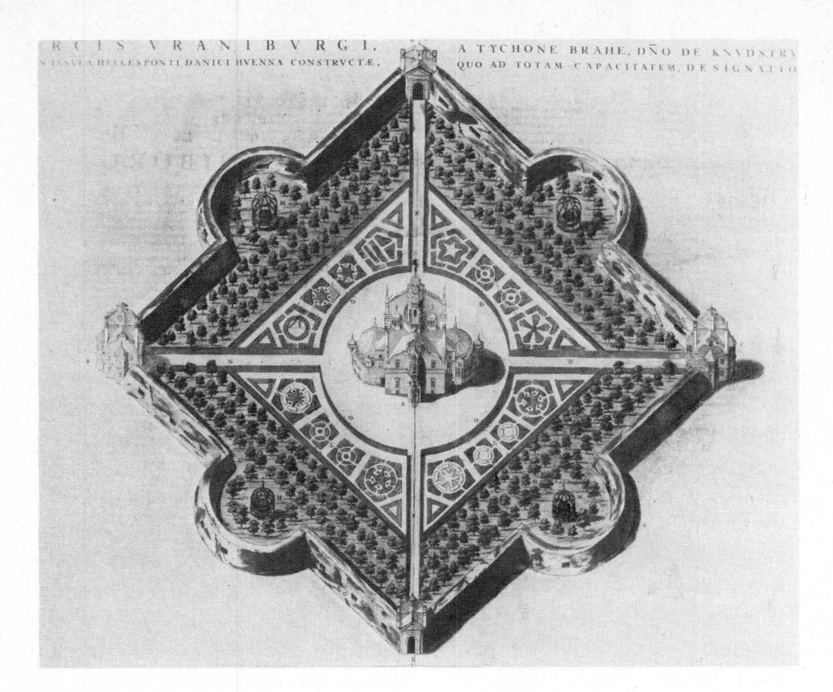

Figure 5.1 Uraniburg—castle of the heavens.

study in this chapter, gives us a glimpse of the inner workings of science. Science, as we have come to know it, was about to be born.

5.2 TYCHO BRAHE: THE ART OF THE OBSERVER

If the ills of the heliocentric theory were to be corrected, someone had to make a set of fresh, accurate, comprehensive observations of the motions in the heavens. Fate, or Providence, sent a man with the necessary qualifications and the passion to do the job: Tycho Brahe, (See Box 5.1 and Figure 5.1.) Tycho revolutionized the branch of science known as experimental or observational science, the art of the observer. A study of Tycho's methods shows them to be strikingly contemporary in many ways. His innovations helped create modern exact science.

Tycho's compelling need for accurate celestial measurements was complemented by the skills he brought to the task. He had to have funds. The king of Denmark gave him an island and generous funds to build his observatory and instruments, and hire assistants. Tycho designed instruments that were a vast improvement over previous ones. They were larger, for one thing, so that the degree marks on his scales were further apart. Also, the large size allowed the use of long, accurate pointers. Thus Tycho could read small subdivisions of an angle and point his instruments accurately at stars. His mechanical skills made it possible for him to build instruments that were large and yet rigid enough to be accurate. They were far more accurate than any astronomical instruments constructed previously.

BOX 5.1

TYCHO BRAHE: THE LIFE OF AN OBSERVER

Twins were born to the Brahe family on December 14, 1546. One of them died at birth. The other, Tycho, was destined to revolutionize observational astronomy. Tycho's life was eventful from the beginning. He was kidnapped by a rich uncle who raised Tycho as his own. Even threats of murder by his parents failed to get the child back.

Tycho conceived an early interest in astronomy. When his uncle detected it, he discouraged such an interest as not befitting a noble. When Tycho was sent to study law, a tutor was sent along to watch over him. Tycho studied the skies at night after the guardian had fallen asleep.

As if to assure Tycho's becoming an astronomer, three astronomical events occurred at various times in his life. Each of them renewed his interest in the work of the observer. The first occurred when he was 16. A partial eclipse of the sun had been predicted. Tycho was impressed when it actually came to pass. He later wrote, "[It] seemed godlike that men could measure the heavenly motions so accurately that they could predict their positions so far in advance." The theme "measure so accurately" was already occurring.

Then, in 1563, an almanac foretold the conjunction of Saturn and Jupiter. It bothered Tycho to observe that the predicted time of the conjunction turned out to be in error by a month. He saw that astronomy needed new, more accurate observations.

Then, from 1571 to 1572, Tycho became fascinated with chemistry and concentrated on chemical experiments. But, on November 11, 1572, as he was leaving his laboratory in the evening, he noticed a brilliant new star in the constellation Cassiopeia. Scarcely able to believe his eyes, he asked some nearby peasants if they saw it too. (They did.) This event sealed Tycho's devotion to astronomy. He observed the nova steadily until, in 1574, it became too dim to see.

Tycho decided to devote his life to observation. He was irked by the haphazard methods of the past. Furthermore, he wanted to help decide whose theory was correct and also hoped to improve the art of astrology. He wrote that, "If warned by Astrology, men could conquer the influence of the stars on themselves."

A piece of good luck came his way in 1576. In recognition of Tycho's growing fame, King Frederick II of Denmark granted him funds for the most luxurious observatory ever known. Tycho was given a whole island. He was to receive the rent from the tenants on the island. He was given funds to build instruments and an observatory, workshops, a windmill, a paper mill, 60 fish ponds, flower gardens, and a fountain. He was given servants. The king paid for Tycho's assistants and the artisans to build and maintain his castle-observatory with its lavish decorations and appointments. Tycho named it *Uraniburg,* castle of the heavens. See Figure 5.1. In return for this generosity, Tycho was commanded to observe the skies and govern the island. From 1576 to 1597 he observed the skies with skill and accuracy which far surpassed the work of any previous observer.

Another group of skills is required when it comes to actual observation. An accurate instrument is useless unless great care is exercised when using it. Tycho took enormous pains to eliminate sloppy or hasty observing. He knew the necessity of repetition. All star positions were measured over and over. He did not hesitate to repeat a measurement 20 times. He also was severely honest and, if any observations were at all doubtful, they were discarded and redone.

PUZZLE 5.1

ATMOSPHERIC REFRACTION

It is an interesting phenomenon that, when the sun appears to be on the horizon, its light is actually coming from below the horizon. The sun has already set but is still visible, so to speak. The reason for this is a phenomenon called **atmospheric refraction,** the bending of light by the earth's atmosphere. As illustrated in Figure 5.2, the atmosphere causes the light to bend downward, causing the sun to appear higher in the sky.

Light coming from the zenith is not refracted (i.e., bent) at all. Stars high in the sky appear at their correct places. The smaller an object's altitude (angle from the horizon), the higher above its correct location it appears. Objects nearest the horizon have their location raised the most. This is why the sun's disk appears flattened when it is near the horizon. Both the bottom and top edges of the sun are raised by atmospheric refraction, but the bottom is raised *more* than the top. The resulting flattened appearance of the sun is quite noticeable at sunset. (The sun is flattened by about 6 minutes of arc.)

Tycho was the first to allow systematically for atmospheric refraction in taking measurements of star positions. Now here is the puzzle. Suppose Tycho measures the altitude of two stars appearing one above the other. Suppose he finds that one star is 10° above the horizon and the other is 5° above the horizon. Should he conclude that the angular separation of the two stars is actually 5° or more or less than 5°?

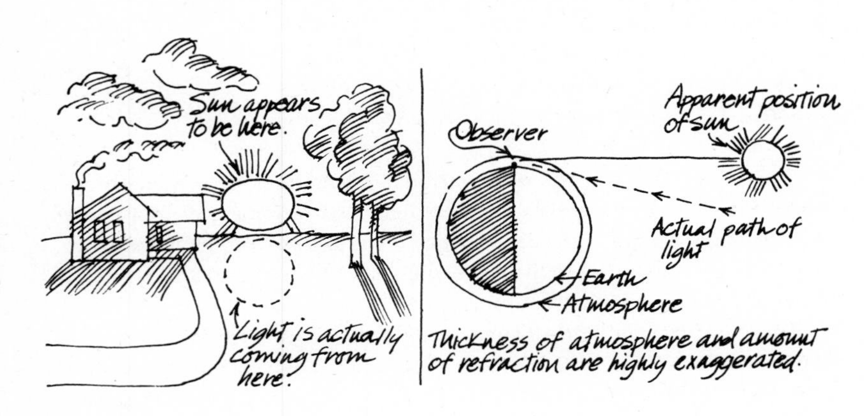

Figure 5.2 Atmospheric refraction.

Once careful measurements had been taken, a great deal of work still remained: data reduction. Many things had to be done to his data before they could be published. He had to eliminate random error, as much as possible. Any reported measurement is inevitably a bit larger or smaller than the actual value, and Tycho developed ways of averaging his data to reduce random error. He also attacked what is known as systematic error. For example, if the long pointer used for sighting a star sagged under its own weight, and sagged by varying amounts at various angles, Tycho took this into account and adjusted his readings accordingly. He was also the first to allow for the effects of atmospheric refraction in his measurements. (See Puzzle 5.1 and Figure 5.2.)

One of Tycho's greatest innovations was that he observed the motions of the sun, moon, and planets *continuously* for 20 years. Observers before his time had worked haphazardly, taking measurements only now and then. Tycho was the first to publish data on the night-by-night motions of the heavenly bodies.

Tycho revolutionized the standards of accuracy and completeness of astronomical observation. He worked in an era before the invention of the telescope. More recently, the telescope has been used to check his data on star positions. Tycho's best measurements of star positions are accurate to within 1 minute of arc. This is equal to the angular width of a pencil viewed from a distance of about 25 meters [about 27 yards (yd)]. Without the later invention of the telescope, Tycho's accuracy could not have been exceeded to any great extent.

Tycho's measurements of planet positions are accurate to about 4 minutes of arc. Because the planets move with respect to the fixed stars, one cannot repeat identical measurements of a planet over a series of years as Tycho could for stars.

Tycho left the world a rich heritage. His work was published after his death and contained observations of the sun, moon, and planets extending over 20 years. It also reported the positions of 777 stars. Each position was the average of his best 15 observations of that star. Tycho also had remeasured many astronomical data (such as the length of a year) with unprecedented accuracy. He had discovered several new irregularities in the moon's motion and showed that lunar trepidation, which Copernicus had tried to explain, did not exist. Perhaps these accomplishments can best be appreciated by noting that Hipparchus and Ptolemy's measurements had an accuracy of only 10 minutes of arc and were *much* less comprehensive.

5.3 AN ATTEMPT TO GRASP THE STARS: THE METHOD OF PARALLAX

One of the tasks of the observational astronomer is to make measurements. Unlike scientists in other fields, astronomers cannot come in contact with most of the objects they wish to measure. How far away are the heavenly bodies? This basic question has been and still is one of the most challenging astronomical puzzles.

The principal method of measuring the distance to an object one cannot reach is called the method of parallax. The **concept of parallax** may be illustrated this way. View a relatively nearby object against a more distant background. Now move sideways to a second observation point. Notice how the nearby object *appears* to have moved so that it is now lined up with a different place in the distant background. This apparent motion is called the **phenomenon of parallax** or merely **parallax.** An easy way to observe parallax is to hold up a finger against a distant background and wink first one eye and then the other. The finger then appears to move sideways. One should also do this experiment with the finger at various distances from the eyes. Observe that, the more distant the finger, the smaller its apparent motion.

This concept is used in determining the distance to a relatively nearby object. The technique is called the **method of parallax.** One measures the distance one has moved from one observation point to the other.

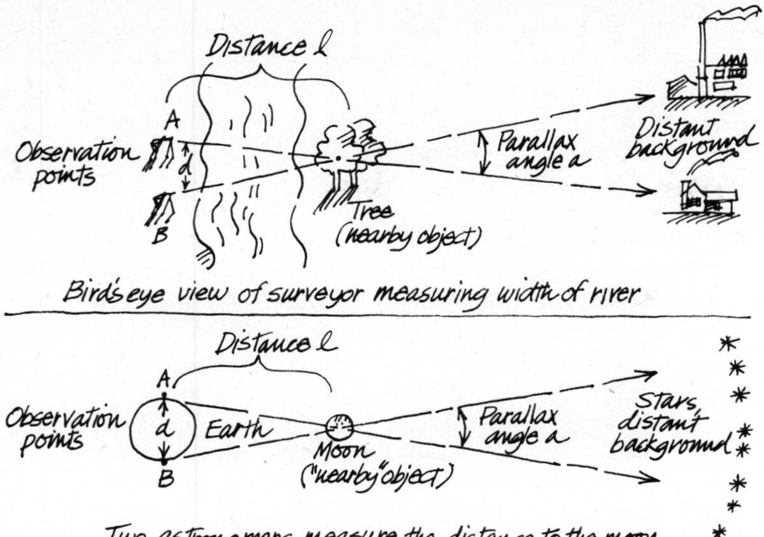

Figure 5.3 Using the method of parallax to measure distance. A surveyor measures the width of a river by measuring the distance to a tree on the opposite bank. The astronomers (one at A and one at B) measure the distance to the moon. The distant backgrounds above could not be drawn properly. They should be much farther away to the right. The parallax angle is measured by measuring the angular separation of the two positions, A and B, on the distant background.

One also measures the angular distance the object *appears* to have moved from one place on the background to the other. This angle is called the **parallax angle.** From these two measurements one can compute the distance to the relatively nearby object. For a given separation between observation points, the larger the parallax angle, the nearer the object. There is a formula for computing the distance, known to all surveyors and astronomers. Figure 5.3 shows how a surveyor and an astronomer use the method of parallax.

For those who are interested, here is the parallax formula. The distance R in meters to the relatively nearby object may be computed from

$$R = 57.3 \frac{d}{a}$$

where d is the distance in meters between the observation points and a is the parallax angle in degrees as shown in Figure 5.4. This form of the formula applies only if d is small compared to R.

Hipparchus had used the method of parallax to measure the distance to the moon. Ptolemy had given reasons for believing that the stars are very far away compared to the size of the earth. Copernicus had shown that, if the earth moves, the stars have to be very far away compared to the size of the earth's *orbit*. Tycho now wanted to go one step farther and measure *quantitatively* the distance to the stars using the method of parallax.

Tycho felt he should be able to measure the parallax angle of the nearest stars. By an ingenious method, he convinced himself that the stars appeared to have an angular diameter of about 1 minute of arc. (We now know that the actual angular sizes of stars are *much* smaller.) Based on this measurement, he estimated that, if the stars were as big as the sun, they should be no more than 30 AU away, three times

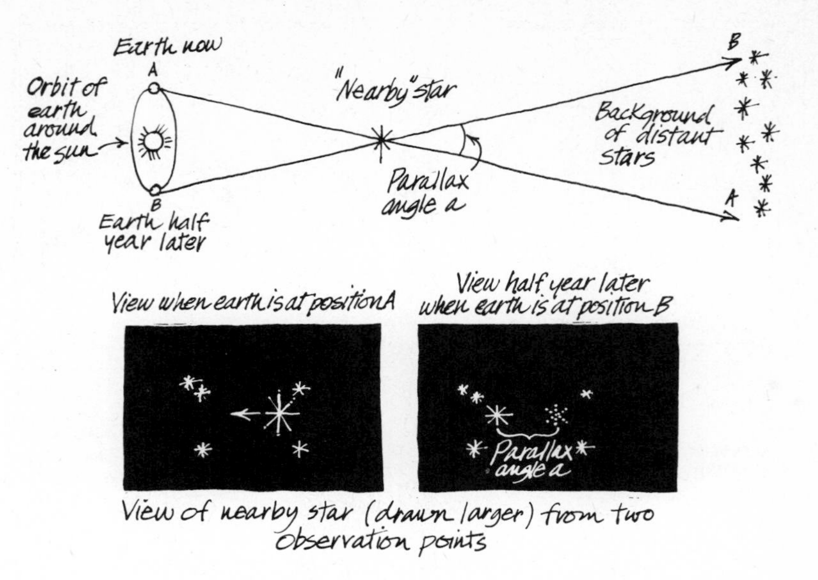

Figure 5.4 Measuring the distance to a nearby star using stellar parallax. So that all parallax data can be easily compared, astronomers have agreed to the following. The parallax angle reported for a star in a star catalog is the angle that would have been measured from two observation points 1 AU apart. This convention is followed no matter what the actual distance between the observation points was in the actual measurement.

Saturn's distance. This would be near enough to show a small but measureable parallax angle.

For purposes of clarity, we describe Tycho's search for stellar parallax in a way that makes it appear somewhat more systematic than it probably was.

He knew that the parallax angle would be small, so he needed a large distance between observation points. He decided to view the target star at intervals of one-half year, when the earth would be at opposite ends of its orbit. (See Figure 5.4.) He chose a number of bright stars for observation, assuming they would be among the nearest stars. Then, as time went by, he repeatedly compared the selected stars' positions with those of other stars in the same region of the sky.

Tycho found that no star ever exhibited noticeable parallax. There were at least three possible reasons:

1. Perhaps all stars are the same distance away. In other words, the celestial sphere is real.
2. Perhaps even the nearest stars are very distant. This would cause the parallax angle to be too small to measure with the naked eye.
3. Perhaps the earth does not revolve around the sun.

Tycho could not accept the first possibility, although he could not prove it wrong. If the second were correct the stars would have to be at least several thousand astronomical units away. Such distances seemed unimaginable to Tycho. Thus he chose the third possibility: Copernicus was wrong, the earth is at rest.

This intrigued Tycho. He liked the simplicity of Copernicus' qualitative ideas but, on the other hand, he had always found it difficult to believe that the heavy earth could move. Then, one day, he had one of those flashes of inspiration we have been calling aha's. He conceived the Tychonic system of the world. This theory steers a middle course between the views of Ptolemy and Copernicus. In it, the earth is at rest,

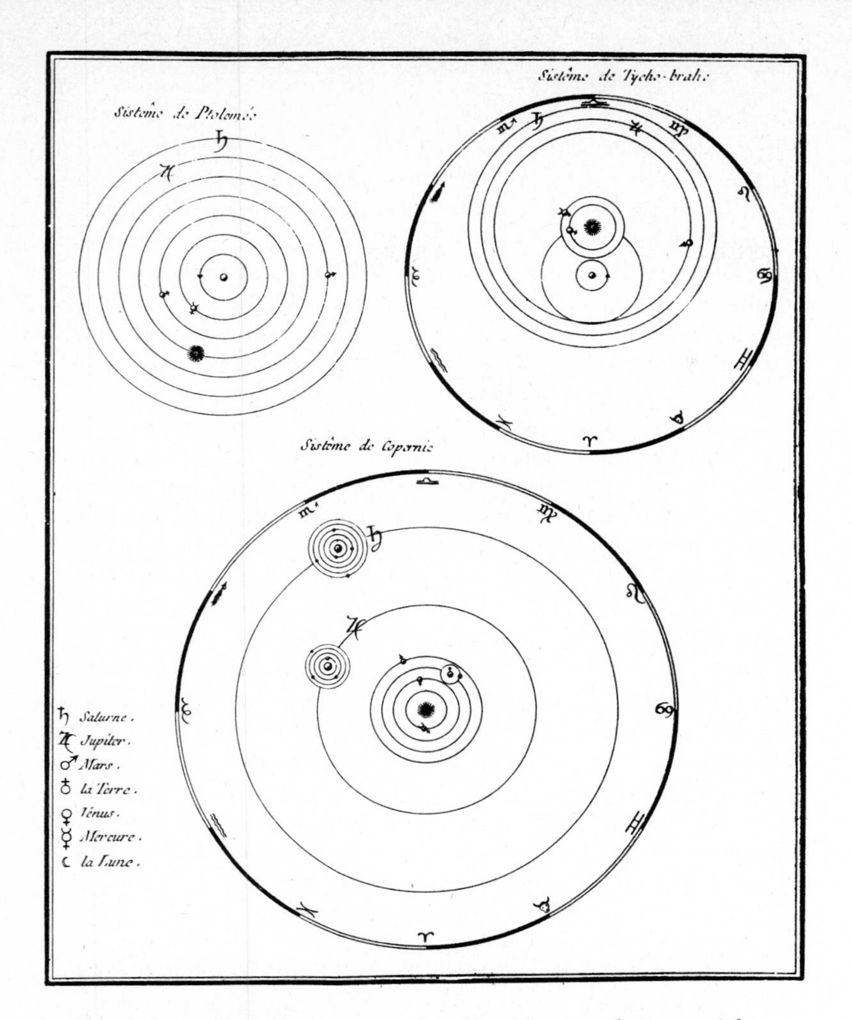

5.5 The three major theories of the solar system. Taken from an old French work, this drawing shows Tycho's theory at the upper right. Ptolemy's theory is shown without the epicycles.

the sun travels around the earth, but the planets travel around the sun. See Figure 5.5. Tycho's new theory could explain all the qualitative motions of the heavenly bodies. It also explained why stars showed no parallax: the earth is at rest, Tycho asserted.

Tycho attempted the necessary calculations that would make his theory a quantitative success. He soon met with the same types of problems that had plagued Ptolemy and Copernicus and realized that he needed help. He needed someone with sufficient mathematical skill who might be able to complete the theory by using Tycho's storehouse of observations. After reading a book titled *The Cosmic Mystery,* by Johannes Kepler, Tycho suspected that he had found the mathematician.

5.4 THE OBSERVER FINDS HIS THEORIST

Johannes Kepler, a German astronomer, was born in 1571, 28 years after the death of Copernicus, and died in 1630 at the age of 59.

Johannes Kepler believed in the heliocentric theory from the time it was shown to him, in private, by his science teacher. This belief put him in the minority among his fellow scholars. He debated the theory in public, giving these reasons for the sun being at the center of the solar system: (1) The sun seems to have a strong effect on the motions of the planets; it seems reasonable that the controlling body should be located at a central position. (2) The sun is the furnace that heats and lights the planets, so it should be centrally located. (3) The sun is a religious symbol of God, so it should have the place of honor. (4) Having the sun in the center leads to a more beautiful theory with fewer unanswerable questions.

Kepler was gifted in more than one specialty. During his university years, he had decided to become a Lutheran minister. The first job he obtained, however, was teaching mathematics. All this time, he had prayed that God would reveal to him the pattern underlying the spacing of the planets' orbits in the heliocentric theory. One day, during a mathematics lecture, he drew a diagram on the blackboard. Staring at it, he had a flash of insight. Aha, he suddenly saw the required pattern. The resulting theory was ingenious, interesting, and completely false, as it turned out. Yet that moment of false inspiration was one of the turning points in astronomical history for the following three reasons.

First, his discovery, published as *The Cosmic Mystery,* so thrilled Kepler, that he decided to devote himself to astronomy. Second, the discovery indicated to him that he needed more precise observations, both to confirm his cosmic mystery and to make successful the entire heliocentric theory. Last, the publication of this first book brought him to Tycho's attention. When religious persecution forced Kepler to leave the town where he was teaching, Tycho took him and his family in. Tycho became so convinced that the younger man possessed the necessary mathematical genius that Tycho, on his deathbed, made Kepler promise to prove the Tychonic system of the world. In this way, the precious observations came into perhaps the only hands that could prove Copernicus' theory a success. Thus it was that, at about the same time, an observer and a mathematician each experienced faulty aha's which brought them together. What followed was an enormous labor of love.

5.5 THE LABORS OF KEPLER: THE ART OF THE EMPIRICAL THEORIST

Briefly, when one searches through data looking for patterns in the numbers, one is said to be using the *empirical method.* This is in contrast to obtaining results by means of a theory that already exists, which is the *deductive method.*

One can imagine how eagerly Kepler set to work. He at last had possession of numerical data. Kepler began a long, patient search for the patterns he felt the data must contain. When he joined Tycho, Tycho was concentrating on Mars, the superior planet with the most irregular apparent motion. As a result, Kepler also became interested in Mars. Kepler concentrated his analysis on Mars for the next six years. Later he ascribed the choice of Mars to "an act of Divine guidance" because, as it turned out, of all the planets, Mars could best reveal the laws Kepler was to discover.

First Kepler reworked Ptolemy's theory of Mars based on the new data. Next, true to his word, he then worked on Tycho's theory. This difficult work finally convinced him that neither Ptolemy's nor Tycho's theories were satisfactory. With great relish, then, he turned to his pet project: finding the flaw in Copernicus' ideas and correcting it.

Correcting Copernicus' Theory This was no simple task. Kepler proceeded by trial and error. He would set up a trial orbit for Mars; this alone might take several weeks. Then he would find the appropriate data in the mass of Tycho's observations for comparison of this trial with the actual measurements. This might take another several days. When that comparison failed, he would change his assumptions and try again. It took six years before he finally found what he was looking for. The average person would surely have given up long before this or perhaps, once the difficulty was glimpsed, would never have even started. But Kepler was driven by a desire to uncover God's patterns and also to see Copernicus triumphant.

Kepler's account of his labors from 1600 to 1606 is quite complete, describing in dumbfounding detail the mistakes and wrong turns he took, as well as his successes. We can only briefly outline his procedure here.

First he cleaned house, so to speak, throwing away assumptions that did not appeal to him. He tossed out an assumption Tycho had made that orbits expand and contract. He also threw away the time-honored assumption that planets travel at a constant speed in their orbits. *His* planets would speed up and slow down. Finally, he renounced epicycles forever. We now know that each of these earlier ideas had to be discarded if Kepler was to be successful. To say the least, Kepler had excellent intuition. Then he set out into the unknown.

Trying Circles For the next five years Kepler tried circle after circle for Mars' orbit. He tried bigger circles, smaller circles, and circles more or less off-center, until he had made 70 trials and filled 900 large pages with his minute handwriting. To preserve accuracy, he usually performed arithmetic with six-digit numbers. All his computations were done by hand. Computers, electronic calculators, slide rules, and even logarithms lay in the future. A modern analysis has shown that Kepler copied three numbers from Tycho's data incorrectly near the start of the five years. But, near the end, he made some arithmetic errors which, miraculously, nearly canceled out the effects of the original mistakes.

After years of labor, Kepler found a circular orbit for Mars that agreed quite well with 10 of Tycho's measured positions for Mars. For these 10, theory agreed with observation with an error of no more than

2 minutes of arc. But when Kepler checked his results by comparing two additional observations of Mars with his theoretical positions, he found that these were both in error by about 8 minutes of arc. The discrepancy was not so large, only 8 minutes of arc, which is less than one-fourth of the angular diameter of the full moon. Many scientists would have been satisfied and published the theory. But to Kepler, the error might as well have been *8*°. He knew how careful Tycho had been and realized that the fault must lie in the theory and not in Tycho's observations. Yet, it must have required great integrity for Kepler to admit to himself that these two rather small discrepancies invalidated years of intense labor.

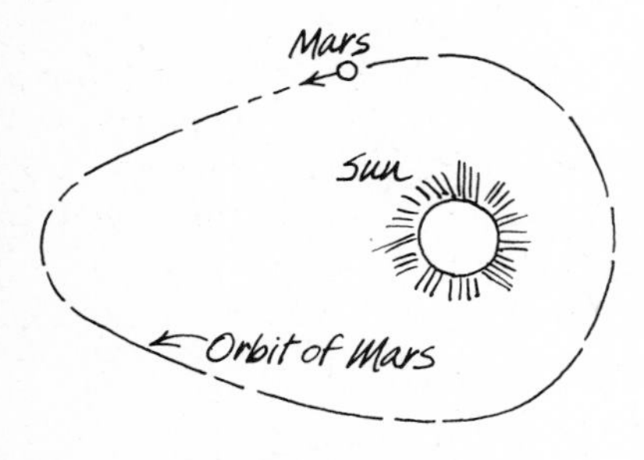

Figure 5.6 Kepler's egg-shaped oval. Kepler spent a whole year trying to prove that Mars had an orbit shaped something like the one shown here. The egg shape is eggs-aggerated here for emphasis. No such orbit was successful. (A serious article on Kepler's work in the *Journal for the History of Astronomy* (1974, 5:1–21) is entitled, "Keplerian Planetary Eggs—Laid and Unlaid, 1600–1605.")

Trying Eggs At this point, Kepler made a momentous decision. Ignoring the weight of all previous opinion, he wondered whether the problem might not lie in the time-honord assumption that the planets travel in circles. Only a few people in any age are able to rise successfully above the ingrained assumptions of past ages. Kepler, like Copernicus before him, was one of these. We can only wonder at the reserves of courage he drew upon to launch into the unknown. The question facing Kepler was, If not circles, what then?

Kepler spent the next year trying to find a successful egg-shaped orbit for Mars. The curve is like a deformed circle, a figure that is more pointed on one side than the other. (See Figure 5.6.) One egg shape after another failed to fit the observations. He did not like the egg, describing it as a "cart full of garbage" compared to a circle, but could think of nothing else to try. At one point during the year he wrote a friend that it was "as if Mars' orbit were a perfect ellipse. But regarding that I have tried nothing." At long last he came to the conclusion that the egg worked no better than the circle.

Trying Ellipses In a moment of desperation, he toyed with the idea of putting an epicycle onto an egg-shaped deferent but then dropped *that* madness. Then he noticed a mathematical formula for Mars' orbit. He tried, but could not understand what the formula implied. We now know that it implied that Mars' orbit is an ellipse. Finally Kepler stopped thinking about the formula and tried an ellipse as a last resort to see how well it would fit the data for Mars. Aha! It fit very well indeed. At long last he had achieved success. When he realized that his formula had implied an ellipse all along, he wrote, "Oh what a silly bird I have been!" and "It was as if I awoke from sleep and saw a new light."

The rest was easy (for a Kepler). He not only showed how splendidly the ellipse satisfied the data but also laboriously convinced himself that every other hypothesis failed. All that remained was to write his long book containing an account of his labors. He called it *Astronomia Nova*, or, *New Astronomy*. It is a book as significant as Ptolemy's *The Almagest* and Copernicus' *De Revolutionibus*.

Let us now look at this and other discoveries of Kepler.

BOX 5.2

THE ELLIPSE: COUSIN OF THE CIRCLE

After a long search, Kepler discovered the shape of each planet's orbit. Each is an ellipse. The ellipse is so closely related to the circle that, looking back, one wonders why Kepler did not turn to ellipses immediately after abandoning circles. This was partly due to bad luck and partly due to the fact that he had been attracted to egg shapes, which require only one point for their definition. An ellipse requires two points.

Figure 5.7 shows how to draw a circle. Choose one point and stick a thumbtack into the point. Loop a piece of string over the tack. Holding the string tight with a pencil point, trace out a circle.

The same figure illustrates a method for drawing an ellipse. Add another thumbtack and loop the string over both tacks. The pencil will now trace out an ellipse. Loosely speaking, an ellipse is a flattened circle.

Each of the two points of the ellipse marked by a tack is called a *focus* (plural *foci,* pronounced foe'-sī).

The amount of flattening of an ellipse is, in part, determined by the separation of the foci. As shown in Figure 5.8, if the foci are close together, the ellipse is nearly a circle. A circle may be thought of as an ellipse with both foci in the same place. The further apart the foci, the more flattened the ellipse appears. One says that, the further apart the foci, the greater the *eccentricity* of the ellipse. (This assumes that one uses the same loop of string each time.) An ellipse having an eccentricity of zero is a circle. The eccentricity of an ellipse defines its shape but not its size.

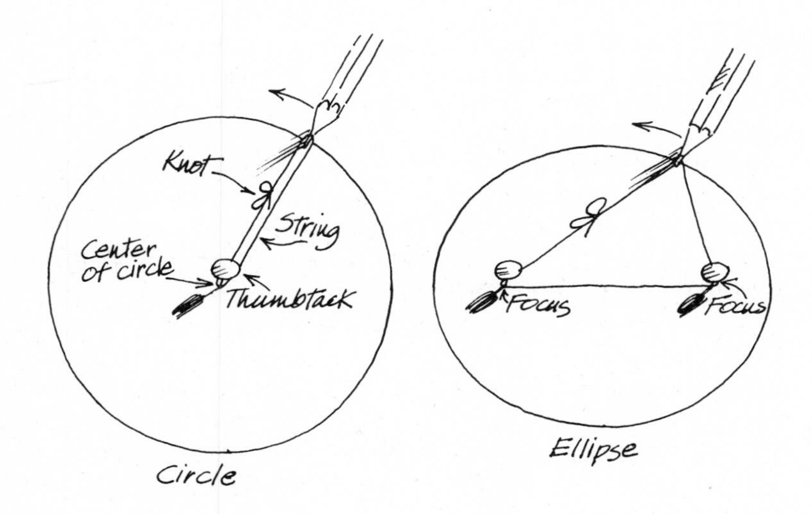

Figure 5.7 How to draw a circle and an ellipse.

5.6 KEPLER'S FIRST LAW

The two main discoveries found in *New Astronomy* have traditionally been called Kepler's first and second laws. The first law is the discovery that saved the heliocentric theory—the planets do not travel in circles. (Study Box 5.2 and Figures 5.7 and 5.8.)

Strictly speaking, Kepler's first law as stated here applies only to an object in orbit around a much more massive object in the absence of any other objects. We now recognize that the presence of other planets, for example, disturbs the orbit of Mars so that it is not exactly an ellipse.

Kepler's First Law

The path of each planet around the sun is an ellipse. The sun occupies one focus. (See Figure 5.9.)

Since Kepler's time it has been found that all astronomical objects executing closed orbits follow ellipses. This applies to comets (Section 11.8) orbiting the sun as well as to artificial satellites orbiting the earth. Armed with his first and subsequent laws, Kepler published tables, based on Tycho's data, that allowed astronomers to predict planetary positions with far greater accuracy than ever before.

As each planet orbits the sun, the point in its orbit at which it is nearest the sun is called the **perihelion** (per-e-hee'-lyun). The point in its orbit at which the planet is furthest from the sun is called the **aphelion** (e-fee'-lyun). See Figure 5.9. Hipparchus, in about 150 B.C., had already observed that the apparent size of the sun changes slightly during each half year. This was now explained by Kepler in terms of the earth's elliptical orbit.

The earth reaches perihelion each year on or about January 4, and aphelion on or about July 4. Residents in the northern hemisphere are often surprised to learn that the earth is nearest the sun during the cold month of January. The reason that the nearness of the sun makes little difference to the weather is that the earth's orbit is only slightly eccentric. It is nearly a circle having the sun almost at the center. Each year, the earth's distance from the sun varies by only about 3 percent of an astronomical unit. See Figure 5.9. Recall the discussion of the seasons in Sections 2.2 and 4.6.

Some planets have orbits somewhat more eccentric than most. (See Appendix 3.) Mercury, the innermost planet, and Pluto, the outermost planet, have the most eccentric orbits. Mars' orbit is also fairly eccentric, and that is why Kepler was lucky to choose it for study; it deviates markedly from circular motion. Venus' orbit is very nearly a perfect circle. Had Kepler concentrated on the motion of Venus, he might never have noticed that its orbit is really an ellipse.

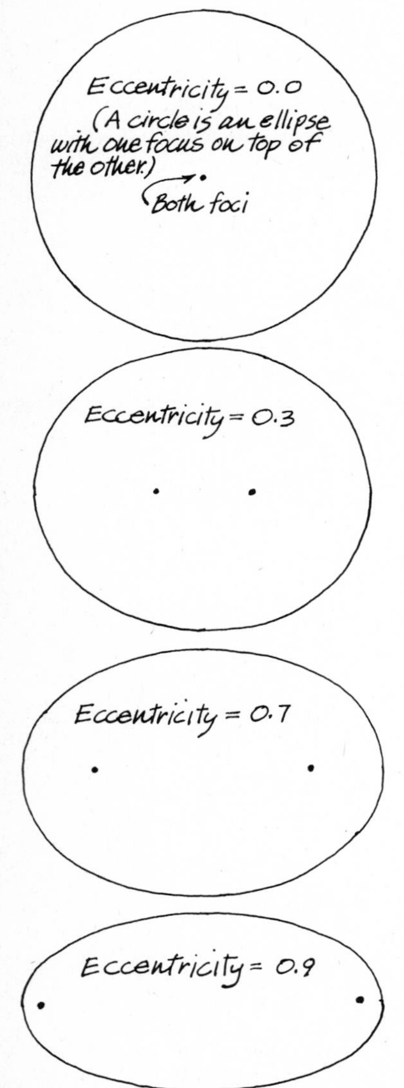

Figure 5.8 Foci and the eccentricity of ellipses.

5.7 KEPLER'S SECOND LAW

Of the many patterns Kepler found (or, in some cases only *thought* he found) in Tycho's data, only a few were of lasting importance. The second of these has to do with the speed of a planet in its orbit. Kepler had assumed that planets speed up and slow down in their orbits. By referring to Tycho's data, he could see that each planet moves along its orbit fastest when at perihelion and slowest when at aphelion. At other times, it is either speeding up (on the way to perihelion) or slowing down (on the way to aphelion). He felt that there must be a pattern or law governing the speed of a planet at each place in its orbit. The pattern he found is one of the most peculiarly formulated laws in science.

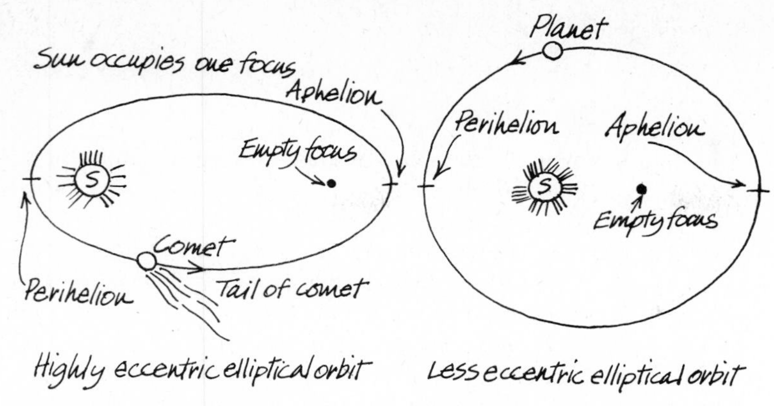

Figure 5.9 Kepler's first law. Two illustrations of elliptical orbits are shown. Many comets (Section 11.8) follow orbits that are highly eccentric (flattened). On the right a planet's orbit is illustrated. However, the eccentricity shown here is still exaggerated. A scale drawing of the earth's orbit would appear to the casual observer to be a nearly perfect circle with the sun very near the center.

Let us fantasize that Kepler found his second law in the following way. Suppose he drew a figure like Figure 5.10, showing a planet in four positions in its orbit. From A to B the planet is near perihelion and thus traveling rapidly. In one month, say, the planet travels a large distance along its orbit. From C to D it is near aphelion. There the planet travels more slowly and so, in an equal time interval (one month in this example), covers a smaller distance along its orbit. Did Kepler's inventive subconscious lead him to draw lines from the sun to each of the four positions of the planet? The result would be that two pieces of the ellipse have been marked off. They are shaded in the diagram.

Continuing our scenario, we imagine that Kepler, intrigued by this diagram, decided to investigate which of these regions has the larger area. It is as if the diagram showed the top of an elliptical cake. The question is, Which is the larger slice? or Which slice has the larger area? This was not an easy question to answer. It is easy to find the area of a rectangle (multiply length times width), but how does one find the area of an elliptical slice? Newton discovered a method for solving such problems, a method called integral calculus, but Kepler died 12 years before Newton was born. Kepler worked out his own method for finding such an area. Even though the method was based upon a fallacious idea, perhaps we are not surprised by now to learn that Kepler's method gave the correct result: The two shaded areas in Figure 5.10 are equal.

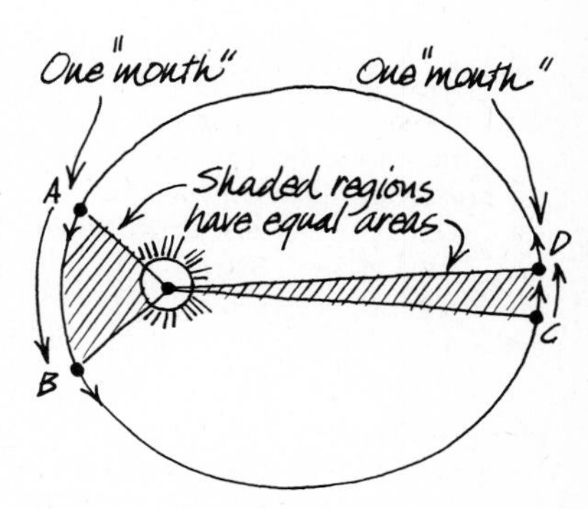

Figure 5.10 Kepler's second law. This diagram is used in the text discussion of Kepler's second law which states: A line drawn from the sun to a planet sweeps out equal areas in equal time intervals.

Kepler checked the result for other positions of the planet in its orbit and found that, if the planet traveled for, say, one month, it always defined a region equal in area to either of the original two areas. This is known as

Kepler's Second Law

A line drawn from the sun to a planet sweeps out equal areas in equal time intervals.

The above description of Kepler's thinking is highly simplified and idealized. As Kepler described his actual method of discovery, it was much more amazing and indirect. Once again, he made several mistakes which, as luck would have it, canceled out. It is even more bewildering to learn that Kepler actually discovered his second law *before* discovering his first law. This means he found the equal areas law before he knew the correct shape of the orbit!

No matter how he derived his second law, it agrees well with observations. It has been found to apply to any orbital motion. For example, a line drawn from the earth to an earth satellite also sweeps out equal areas in equal time intervals. Astronomers still use the second law to predict orbital speeds.

5.8 THE HARMONY OF THE WORLD AND KEPLER'S THIRD LAW

Early in life Kepler had conceived his pet project. It was to be called the *Harmony of the World* (*Harmonice Mundi*). In it he wanted to describe the ideas God had used in designing the universe. Assuming that God's interests were the same as his own, Kepler planned to base this all-embracing theory on mysticism, geometry, astronomy, and music.

The book was not completed until 1618. Kepler often worked on it during times of personal crisis. He found it soothing, he reports, to contemplate the glorious harmony he perceived in the cosmos. The resulting book is a staggering performance. Alas, it is not sufficiently persuasive. All of his grand theory has fallen by the wayside except for one additional pattern which is still considered important, Kepler's third law.

Here are the results of some modern determinations of the average orbital speed of certain planets: Mercury, 47.9 km/sec (29.8 mi/sec); Earth, 29.8 km/sec (18.5 mi/sec); Saturn, 9.7 km/sec (6.0 mi/sec). (See also Appendix 3.)

A few excerpts from Appendix 3: For Mercury, P = 0.241 earth years and R = 0.387 AU. For Earth, P = 1 earth year and R = 1 AU. For Saturn, P = 29.48 earth years and R = 9.54 AU.

Kepler's second law is concerned with the varying speed of a planet in its orbit. The third law, on the other hand, compares the speeds of the different planets with each other. Kepler knew that the larger a planet's orbit, the slower its average speed in its orbit.

He focused on the sidereal period (symbol, P) of each planet (the length of time required to complete one orbit as viewed from the sun; see Section 4.8) and the average radius (symbol, R) of each planet's orbit. (R is the average distance of the planet from the sun). The observations indicated that, qualitatively, the greater the average distance of a planet from the sun, the longer it takes for the planet to execute one complete orbit. But Kepler was not satisfied until he had found the exact quantitative law.

Kepler's third law can be stated in a compact formula. The formulation requires a familiarity with the ideas of squaring and cubing a number. Perhaps some readers would appreciate a review. When any

PUZZLE 5.2

USING KEPLER'S THIRD LAW

Here is an example of the kind of practical use that may be made of Kepler's third law. Suppose a new planet is found. After sufficient observation, astronomers determine that it orbits the sun once in, say, 8 years. Compute the average distance in astronomical units of the planet from the sun using Kepler's third law.

---

number, say P, is to be squared (symbol P^2, read "P squared"), one simply writes down the number twice and multiplies. Thus $P^2 = P \times P$. (The exponent 2 means "write down the number P *twice*.") For example, $4^2 = 4 \times 4 = 16$, and $10^2 = 10 \times 10 = 100$. Similarly, R^3 (read "R cubed") is a compact way of writing $R \times R \times R$; one writes down the number three times and multiplies. For example, $2^3 = 2 \times 2 \times 2 = 8$, and $10^3 = 10 \times 10 \times 10 = 1000$.

Now we can express

Kepler's Third Law

$$P^2 = R^3$$

(In order for the law to have this particularly simple form, P must be the sidereal period of the planet expressed *in earth years* and R must be the average distance of the planet from the sun in *astronomical units*.) The pattern Kepler found in the data for each planet, then, was that the sidereal period multiplied by itself results in the same answer obtained when the radius of the orbit is multiplied by itself twice. That is, for each planet $P \times P = R \times R \times R$. To a scientist, this number pattern has interest in and of itself, as does any pattern in nature. But the law also has practical use in astronomy. (See Puzzle 5.2.)

The reader may care to verify the third law for a few planets. For example, for Saturn, $P^2 = 29.46^2 = 868$ (rounded to three places), while $R^3 = 9.54^3 = 868$. Check.

5.9 A NEW AGE FOR ASTRONOMY

Kepler left the face of astronomy permanently changed. His powerful influence has been felt ever since in two major ways.

First, Kepler found the modifications of the heliocentric theory that had eluded Copernicus. Tycho had made observations so accurate that only a refined theory could explain them. Kepler performed the enormously difficult computations necessary to show that the heliocentric theory did fit the data. Gradually, all major astronomers were won over to the heliocentric theory because it provided by far the best method for producing accurate predictions of planetary positions. After Kepler, the heliocentric theory could no longer be ignored. At last, Copernicus was triumphant. Kepler composed a prayer in thanksgiving which ended, "Behold, I have completed the work to which Thou has called me. And

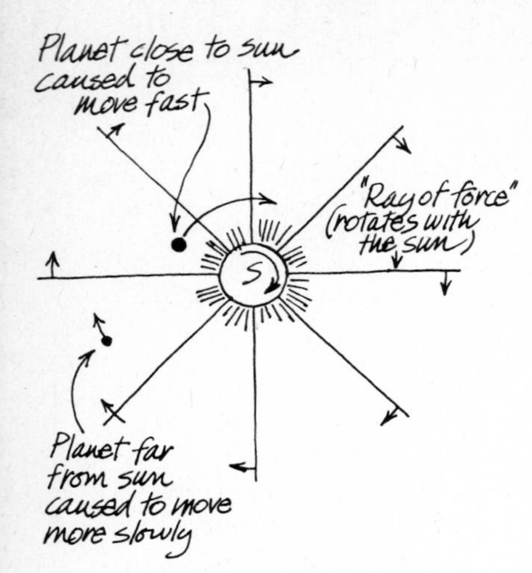

Figure 5.11 Kepler's paddlewheel model.

I rejoice in Thy creation whose wonders Thou has given me to reveal unto men."

Kepler's second innovation was only a tentative start toward a new physics, but it was an important start. All his life Kepler had asked, "How do the planets move?" His three laws were the foundation of an answer. But Kepler was a pioneer in a new field when he asked, *"Why* do the planets move that way?" As so often happens, a great discovery in a science led to new and even more profound questions. These questionings of Kepler's were the birth of a new era in the study of **dynamics,** the branch of science that seeks reasons for motion. Ptolemy and Copernicus had given one kind of explanation for observed motions, an explanation in terms of epicycles or simple circles. Kepler's contribution was not only to discover, for example, that orbits are ellipses, but to ask the deeper question, Why are they ellipses?

Kepler never did find satisfying answers to such questions, but he did have several interesting ideas. He noticed that each of his three laws gave a prominent place to the sun. This confirmed his opinion that the sun must have a great deal to do with the planetary motion.

He asked himself, "What keeps the planets moving?" His hypothesis was that the sun must send out rays of force emanating in straight lines. Assuming that the sun rotates (this guess was later discovered to be correct), he imagined that these rays of force somehow whip the planets along, keeping them in motion. (See Figure 5.11.) The more distant planets move more slowly because, at a greater distance from the sun, the "lines" are further apart and so somehow less effective. Kepler was not able to prove that $P^2 = R^3$ followed logically from this idea, however, and the hypothesis has never been revived.

Such ideas of Kepler are not important in themselves. However, it is important that he was the first to think about dynamics from a fresh viewpoint. The next chapter describes how the field of dynamics, guided in large part by Kepler's three laws, blossomed into a further major scientific advance.

SUMMING UP

Tycho left a legacy of accurate naked-eye observations of the locations and motions of the stars, sun, moon, and planets. His work was innovative in that he observed and recorded the motions of heavenly objects *continuously* for a lengthy period.

Having obtained Tycho's data, Kepler set to work to use it with the goal of perfecting the heliocentric theory. After many years and much travail, he was successful; the heliocentric theory became quantitatively valid. Kepler's most important results are embodied in what are known today as his three laws of planetary motion.

Kepler also began to investigate the reasons why the planets follow his laws. Although he made no real headway, we shall see in the next chapter how two men brought dynamics to a high level of success.

EXERCISES

1. Define the following terms: atmospheric refraction, the method of parallax, parallax angle, ellipse, focus of an ellipse, eccentricity of an ellipse, perihelion, aphelion.

2. List the skills essential to a scientific observer as exemplified by Tycho. Why was Tycho so much more accurate than earlier observers?

3. Explain, by use of a diagram, how the method of parallax can be used to measure the distance to an object one cannot reach. Now suppose that one observes two objects from the same pair of observation points. If object A exhibits a smaller parallax angle than object B, and if the two objects are in the same general direction, which object is more distant? Explain your answer by means of a diagram.

4. In Section 5.3, we listed three possible reasons to explain why Tycho was unable to observe any stellar parallax. What are they? Which did Tycho accept? Which would Copernicus have insisted on? Describe the theory derived from Tycho's choice.

5. What motivated Tycho to hire Kepler? What secret scientific motivation made Kepler accept?

6. What breakthrough enabled Kepler to at last rescue the heliocentric theory? State Kepler's first law.

7. State Kepler's second law. Draw a sketch to accompany the statement. Draw another sketch of a planet's elliptical orbit and place the sun at one focus. Label the orbit with the following words: aphelion, perihelion, moving slowest, speeding up, moving fastest, slowing down. Next, assuming the orbit to be the earth's, label the approximate position of the earth in the middle of each (northern) season of the year.

8. State Kepler's third law. What do P and R stand for? What units must they be expressed in? Choose the correct answers: The more distant a planet is from the sun on the average, the (faster? slower?) it travels in its orbit on the average and the (longer? shorter?) the time required to complete one orbit.

9. (a) Suppose a planet orbits the sun once each 27 years. What would be its average distance from the sun in astronomical units? Choose the answer from among these possibilities: 3; 9; 27; 81; 729; 19,683. (b) Suppose a planet were observed to be 16 AU from the sun on the average. How many years would it take for one orbit (4; 8; 16; 64; 128; 4096)? [Answers: (a) 9 AU, (b) 64 years.]

10. Why was the heliocentric theory more attractive for everyday use in practical astronomy than the geocentric theory after the work of Tycho and Kepler was completed?

READINGS

A splendid article on Tycho's life and work is

"The Celestial Palace of Tycho Brahe," by J. Christianson, *Scientific American*, February 1961, p. 118.

An enjoyable biography of Kepler containing much on Tycho (and Galileo) as well as details of Kepler's awesome labors is

The Watershed, by Arthur Koestler, Doubleday (Anchor Books), New York 1960.

An amusing article on the ellipse is found in the same issue as the above Christianson article:

"Diversions That Involve One of the Classic Conic Sections: The Ellipse," by Martin Gardner, *Scientific American*, February 1961, p. 146.

THE STUDY OF DYNAMICS AND THE INVENTION OF THE TELESCOPE

Galileo Galilei stood out under the stars one clear, dark night in 1610 and raised to his eye a hollow tube of lead, a tube fitted with a piece of glass at each end. See Figure 6.1. That night he became, in effect, one of the first space travelers. Up until then, one could only view the moon from a distance of roughly 400,000 km (roughly 250,000 mi). Galileo was one of the first to view the moon as if he were *30 times* closer, only about 13,000 km (8300 mi) away. The view was overwhelming in its beauty. From his new vantage point, Galileo could see valleys and mountains on the moon in splendid detail. The moon's apparent angular diameter was no longer $\frac{1}{2}°$ but 15°. Galileo accomplished this without actually leaving the earth. His tube of lead and glass had given him the power to view heavenly objects up close, as if he had traveled most of the way to their vicinity. This instrument, the telescope, opened up a frontier of space with an impact that has been duplicated since then only by actual space flight.

6.1 THE TWO ASTRONOMICAL FRONTIERS OF THE SEVENTEENTH CENTURY

By the early 1600s, the work of Tycho and Kepler had produced great success. Yet this success made it clear that two new astronomical frontiers were waiting to be explored. On the one hand, Tycho and Kepler had perfected the heliocentric theory, a satisfying theory concerning the paths that the planets follow. But this, as always, led to deeper questioning: *Why* do the planets move on these paths? Kepler wrestled with this question, but made little headway. (See Section 5.9.) The science of **dynamics** (the branch of physics that seeks reasons for motion) had made slow progress since the time of Aristotle (about 340 B.C.), but it was about to be reborn.

Tycho himself had arrived at the second frontier. Future naked-eye observers might repeat what he had done, but never significantly improve upon the accuracy or detail of his work. It was a frustrating limitation. Except for the sun and the moon, the celestial objects all look like points of light to the naked eye. Were humans meant to know no more of the heavenly bodies than their motions? Both Galileo Galilei and Isaac Newton explored these two frontiers with spectacular suc-

BOX 6.1

GALILEO GALILEI (1564–1642): FRESH THOUGHT

Galileo's father wrote, in a book on music, "Those who attempt to prove an assertion by simply relying on the weight of authority act very stupidly." Apparently, Galileo learned much from his father and, like his father, he spoke and wrote bluntly and plainly. Most importantly, he refused to accept something merely because it was written; he was attracted to fresh thought, "fresh" in the sense of being both new and disrespectful of authority.

Galileo's love of original thought led him to many important advances in science. Early in life he proposed the pendulum as a time keeper. He contributed vital ideas to a new theory of dynamics. He used the refracting telescope to make revolutionary astronomical discoveries. He improved or invented several scientific instruments, including a thermometer and a calculating machine.

He was convinced of the heliocentric theory early in life and later wrote a book, *Dialogue Concerning the Two World Systems,* defending Copernicus. Galileo never studied Kepler's *Astronomia Nova,* which contained his first two laws, even though Kepler had sent him a copy. (It is possible that Kepler's highly complex writing style and his mysticism were responsible for Galileo's disinterest.) Galileo insisted on circular planetary orbits all his life, thereby passing up one of the strongest pieces of evidence for the heliocentric theory. On the other hand, he erroneously believed that ocean tides were caused by the motion of the earth, which in turn caused the oceans to slosh back and forth in their basins. He took this as the clearest evidence for the motion of the earth. Kepler had earlier proposed that the moon pulled up the tides, a guess later confirmed when Newton applied his theory of gravitation. Galileo explicitly rejected Kepler's hypothesis and died before Newton's time.

---

cess. Astronomically, the seventeenth century was the age of dynamics and the telescope.

6.2 GALILEO AND FALLING

The theory of dynamics found in the works of Aristotle had been the accepted theory since the time of his death in about 320 B.C. Among the ideas in this theory were these, expressed here in somewhat simplified form:

1. The natural motion of objects is straight down, toward the center of the earth. Any other motion is unnatural.
2. A *force* (a push or a pull) is required to maintain an unnatural motion. Left to itself, an object will naturally stop.
3. Heavier objects fall faster than lighter ones.

These seem to be natural conclusions. Even today, teachers of physics find that many students come to class with these conceptions.

Galileo (see Box 6.1) studied Aristotle's theory of dynamics at the

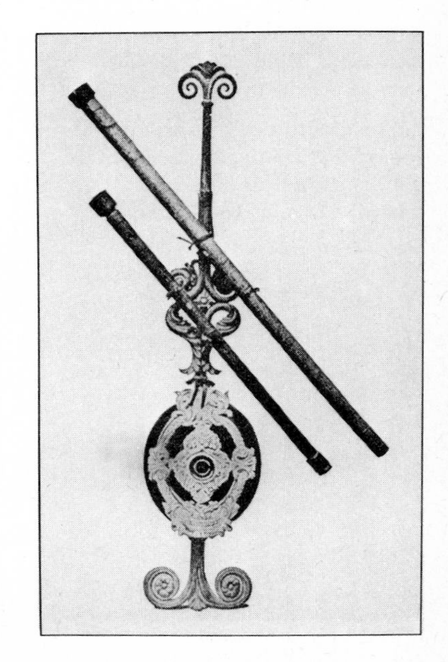

Figure 6.1 Galileo's telescopes.

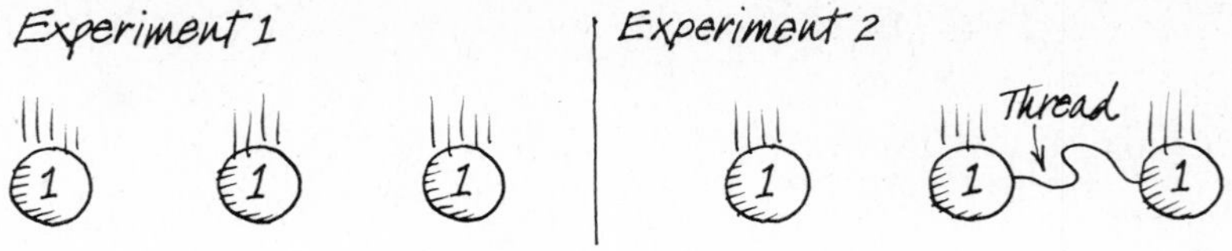

Experiment 1

Experiment 2

Thread

Figure 6.2 Galileo's thought experiment concerning falling objects. Experiment 1: Imagine three identical 1-kg balls all dropped at the same time from the same height. Anyone would agree that all three will hit the ground at the same time.

Experiment 2: Now drop a 1-kg ball and a 2-kg object, composed of two 1-kg balls connected by a thread. All objects should still hit the ground together, as in experiment 1. The result will persuade a nonbeliever that all objects fall at the same rate.

Aristotle was an intellectual giant whose ideas were accepted for thousands of years. Any new theories were tested by comparison with those of Aristotle. Some of his ideas are no longer accepted, but this does not diminish his stature or importance in the development of Western thought.

Italian University at Pisa. But instead of merely learning the theory, he also began to question the ideas he was taught. Galileo loved attacking established ideas and turning them on their heads. He eventually came to the conclusion that all three of Aristotle's ideas mentioned above are faulty.

Galileo proposed the following concept: *In the absence of air friction, all objects fall at the same rate.* Who was correct, Aristotle or Galileo? Galileo did something that seems obvious to us but which was something of an innovation at the time—he experimented. He reports that he dropped a 100-lb cannon ball and a ½-lb musket ball together from a great height. He found that both reached the ground at practically the same time. The lighter ball lagged behind the heavier one by a few centimeters but this, according to Galileo, was due to the greater effect of air friction on the lighter ball. (Study the thought experiment in Figure 6.2)

No historical records prove it, but it is said that Galileo performed this experiment in his hometown, Pisa, from the famous leaning tower.

Nonbelievers can try this experiment. Drop a piece of paper and a book. The book falls faster. Now crumple the paper into a small ball. The piece of paper still has the same weight, but now the air friction on it is reduced. Now drop the book and the crumpled paper.

We, who live in the space age, have had many opportunities to see Galileo's idea confirmed. Several times astronauts on the way back to the earth from the moon have shown how objects in their spacecraft seem to float as if they were "weightless." The explanation is that the heavy ship and all of the lighter contents of the ship fall toward the earth at the same rate. Since they fall together, the contents seem to float. (See Figure 6.3.)

6.3 INERTIA

Aristotle had assumed that a force (a push or a pull) has to act continuously on a moving object to keep it in motion. He assumed that if no force acts on an object, it will stop. Kepler adopted this idea and applied it to the planets. He devised a solar paddlewheel model to explain the continuing motion of each planet. (See Section 5.9.)

Galileo, in his characteristic way, considered turning this idea upside-down and decided he liked it better that way. He proposed that, *if no force acts on an object, it will move forever at a constant speed in a straight line.* In other words, an object in motion will continue to move on its own, without any outside help. Another way of putting it is to say that every object is endowed with inertia. **Inertia** is the property of an object that makes it sluggish, hard to put into motion, and hard to stop once it is moving. The more matter in an object, the more inertia it has. Thus a force is required to start a stationary object moving, or to stop a moving object, or to deflect it from a straight path. Left to itself, that is, in the absence of external forces, the object will naturally follow a straight path at constant speed.

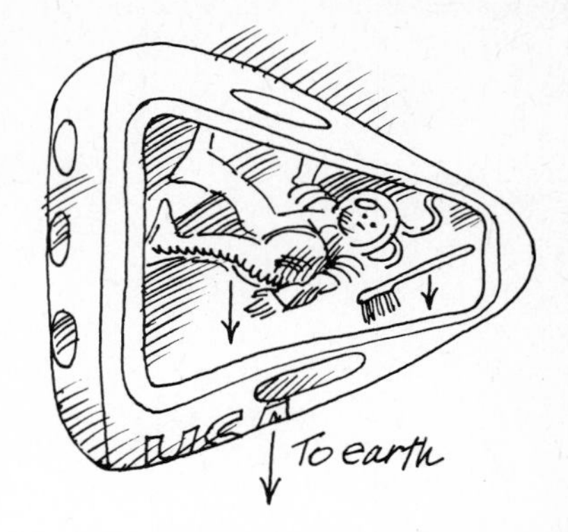

Figure 6.3 An astronaut demonstrates that all objects fall at the same rate. While returning to earth from the moon, an astronaut performed a televised physics demonstration. He demonstrated how he could float freely in the spaceship. He also released his toothbrush and it too floated. This shows Aristotle was not right in asserting that heavy objects fall faster than light ones. A person in a freely falling elevator would experience the same effects, but for a shorter time.

When people objected to his new idea, Galileo used arguments similar to this: Put a block of wood on the floor and kick it. It will soon stop, but this is because a retarding force is acting on it. The force of friction between the block and the floor brings it to rest. Now imagine putting the block on a roller skate and giving it the same kick. It will go farther than before because the wheels have reduced the force of friction. Now imagine the block in the emptiness of space. There is no friction there. Doesn't it make sense to say that, once kicked, the block will go on forever?

If the concept of inertia is adopted, no force from the sun is required to explain why a planet does not stop. It forges ahead on its own—it has inertia. We will see, in Section 6.7, how Newton found the concept of inertia to be very useful.

Other scientists, among them René Descartes (French philosopher-scientist, 1596–1650) and, late in his life, Kepler, independently conceived of inertia. The phrase "in a straight line" is not from Galileo. We add it here as a correction, anticipating the complete form of the statement.

6.4 GALILEO REINVENTS THE TELESCOPE

In early 1609, Galileo received a letter from a former pupil. The writer mentioned that he had seen an invention one looked through that made distant objects seem near and dim objects seem brighter. That such an extraordinary instrument was possible caused Galileo to "wholeheartedly inquire into the means by which I might arrive at the invention of a similar instrument. This I did shortly afterwards, my basis being the theory of refraction." (See Figure 6.4.)

The quotation is tantalizing, for Galileo did not explain the steps by which he reinvented the telescope. How did he arrive at a discovery of such importance? We will probably never know. As a playful (yet, I hope, instructive) substitute, there follows an account of the invention of the telescope by a fictitious physicist. It is meant to illustrate the steps by which someone *might* have invented the telescope but has no basis in historical fact.

Figure 6.4 Galileo demonstrating his telescope.

From the Journal of Gilbert Leo
How I Invented the Telescope

April 8: Cool and sunny. I visited a friend who wished to see the ships in the harbor more clearly from his window. I told him that, if I found time, I would invent an instrument that would make objects appear to be nearer and brighter. I did not feel that this was a boast as I am a genius, after all.

April 9: Rained all day so I spent the day at work in my laboratory. I started my invention of the new instrument by assuming that, if objects were to be made brighter, I would need something to gather light and funnel it through the viewer's eye, much as a funnel would collect rain water and funnel it through the narrow neck of a bottle.

But how to cause a beam of light to change direction? I knew of only two ways, reflection and refraction, and decided to concentrate on the latter. Thus I reviewed the definition of refraction and the law of refraction for myself: **Refraction** is the bending of a beam of light that takes place when the beam passes from one transparent medium to another. Upon passing from a less dense to a more dense medium, the beam is bent *toward* the perpendicular. [See Figure 6.5.] Then, to make sure I understood the law, I solved Puzzle 6.1. [The reader should attempt the puzzle after studying Figure 6.6.]

As I gazed at the light beam after it had traveled through the prism, I realized that the prism behaved a bit like half of a light funnel. I placed two prisms base to base [as shown in Fig. 6.7] and noted how the large light beam

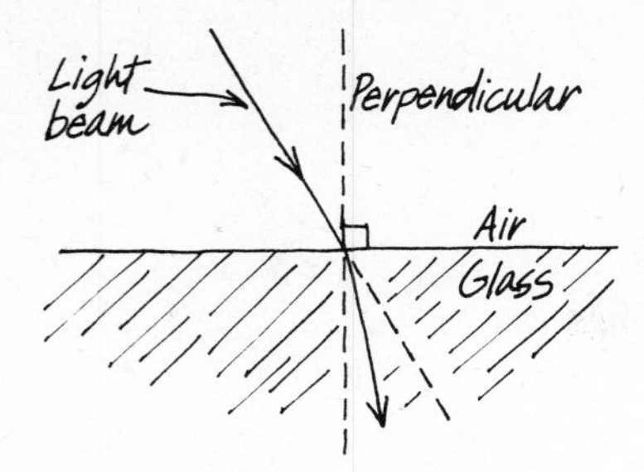

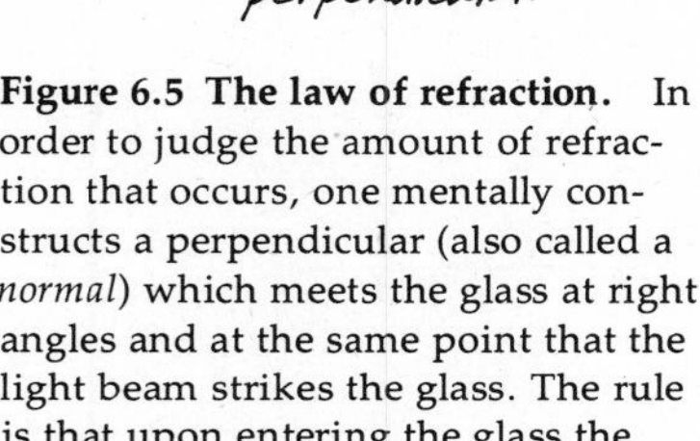

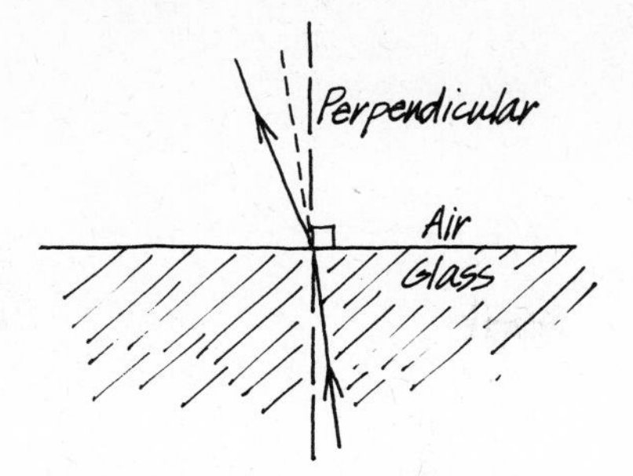

Figure 6.5 The law of refraction. In order to judge the amount of refraction that occurs, one mentally constructs a perpendicular (also called a *normal*) which meets the glass at right angles and at the same point that the light beam strikes the glass. The rule is that upon entering the glass the light beam is bent away from its straight-line path *toward* the perpendicular. Upon leaving the glass, the light beam is bent *away from* the perpendicular. A beam leaving or entering exactly along the perpendicular is unrefracted.

PUZZLE 6.1

TRACING LIGHT THROUGH A PRISM

As practice in applying the laws of refraction, try this puzzle (see Figure 6.6):

1. When the light beam enters the prism, will it be bent upward or downward?
2. Extend the beam inside the prism until it strikes the right-hand edge. Draw the perpendicular there. As the light emerges, will it be bent further downward or upward?

---

was condensed to a smaller region by the two prisms. I excitedly placed my eye in the region of condensed light but, alas, could see nothing clearly, only a bright light. It being dinner time, I stopped for the day.

April 10: Sunny. I hurried to my laboratory today, because I'd had one of my usual brilliant ideas. My prisms did not collect the light properly because, in a prism, each beam is bent by the same amount. I needed something of the same general shape as two prisms, thicker in the middle, thinner near the edges, but shaped so that rays near the top would be bent *more* than rays nearer the center. I felt sure that a piece of glass shaped like two prisms but with rounded edges would produce the desired effect. I needed a *lens*. The lens worked perfectly. It caused a beam of light to be focused to a nice, sharp point. I had my light funnel! (Study Figures 6.8 and 6.9 which define several important terms.)

I took great pleasure in toying with my lens. I aimed it at the sun and placed a piece of paper on the other side. When the paper was exactly at the focal point of the lens, a very bright, concentrated dot of light was formed. The result was so intense that the paper burst into flames!

I next noticed that, if a window were uncovered in a darkened room and a lens aimed at the window, an image of the distant scene was produced on a piece of paper held at the focal point of the lens. The image was upside down and in full color. (See Figure 6.10.) It occurred to me that the spot of light I had previously produced using the sun was a small image of the disk of the sun itself.

I then pointed the lens at the

Lenses have been known since antiquity. They were first used for eyeglasses in the late thirteenth century.

A lens used to create a fire is sometimes called a "burning glass."

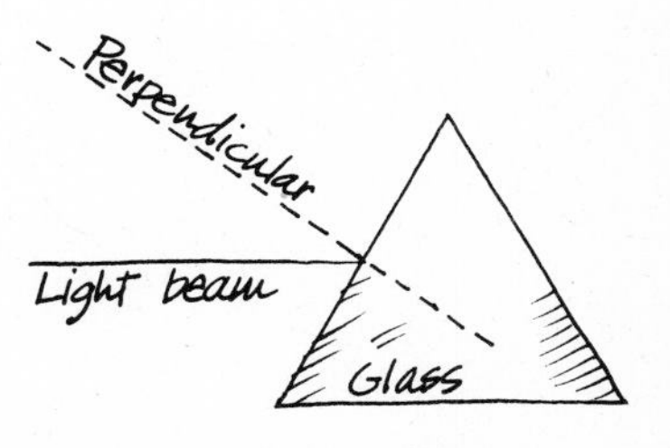

Figure 6.6 Light passing through a prism—Puzzle 6.1.

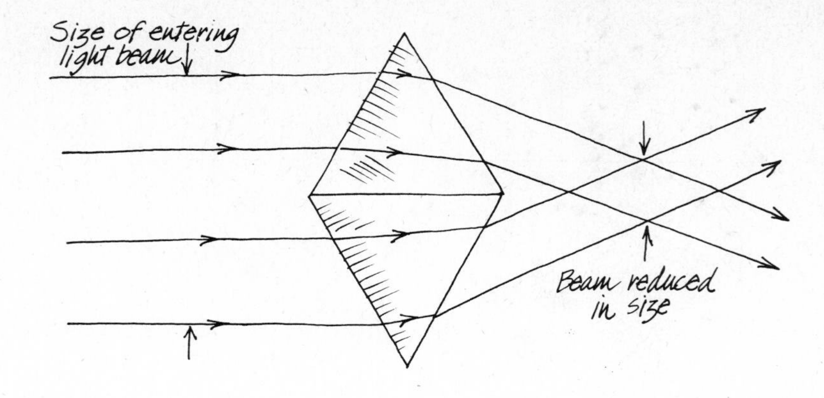

Figure 6.7 Two prisms used as a light-gathering device.

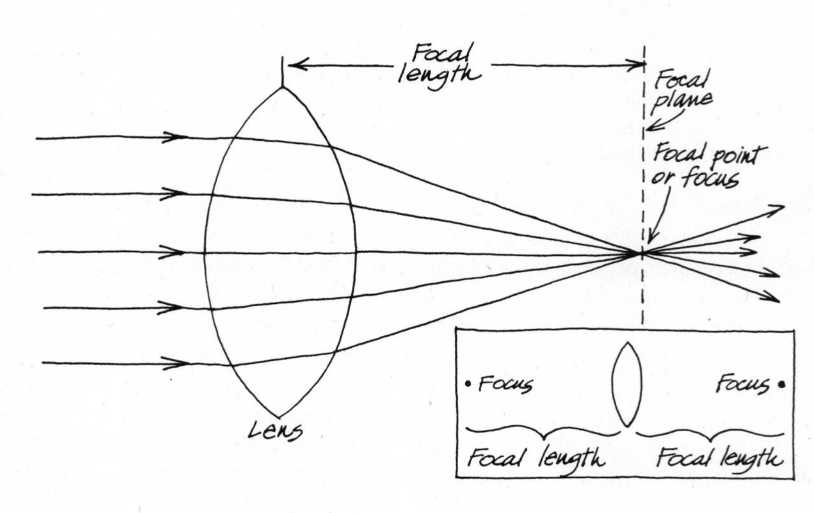

Figure 6.8 The lens, a successful collector of light. The lens succeeds totally where two prisms (Figure 6.7) fail. A beam striking the center of the lens passes through undeflected, since it is traveling along the perpendicular. A beam above (or below) the center is refracted downward (or upward). The result is that all the beams from a distant point of light are refracted by the right amount to meet at a point, called the **focal point** or **focus.** The distance from the center of the lens to the focal plane is called the **focal length** of the lens. As shown in the inset, each lens has two foci, one on each side. Each focus is the same distance from the center of the lens. See also Figure 6.9.

harbor. When I placed my eye at the focal point of the lens, it still failed to give a clear image, only a bright, confused one. Yet, I went to dinner (roast lamb and asparagus) well satisfied that I had solved half of my problem. I could collect light. Now I needed a way to make things look nearer.

April 11: Partly cloudy. I happened to pick up a lens in my laboratory and look through it by holding it up to my eye. Distant objects looked distorted and fuzzy, but I noted that, if I held my finger close to the lens, it appeared clear and larger. As I moved my finger further from the lens, the magnification increased until, when the finger went past the focal point of the lens, it became blurred. I found that, for best magnification, the finger should be placed just inside the

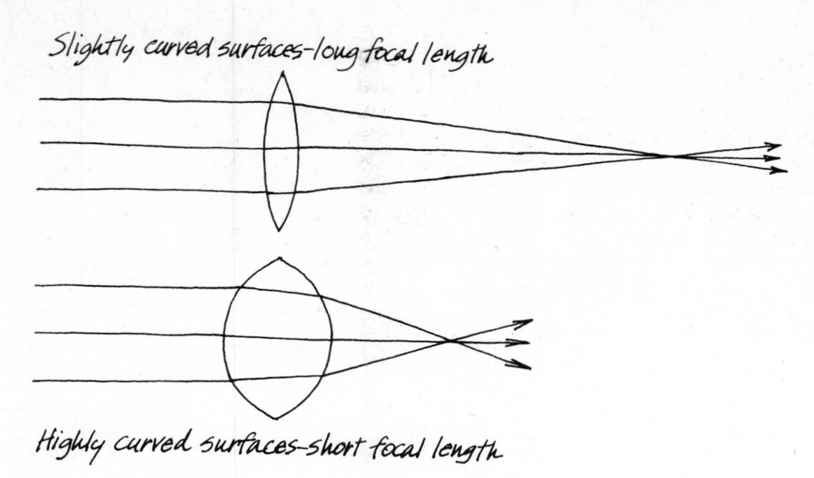

Figure 6.9 The focal length of a lens. The focal length of a lens is determined by its shape, not by its diameter. A lensmaker who wishes to grind a lens with a long focal length produces surfaces having little curvature. If a lens having a short focal length is desired, the surfaces should be highly curved. Loosely speaking, if two lenses of equal diameter are compared, the thinner lens will have the longer focal length.

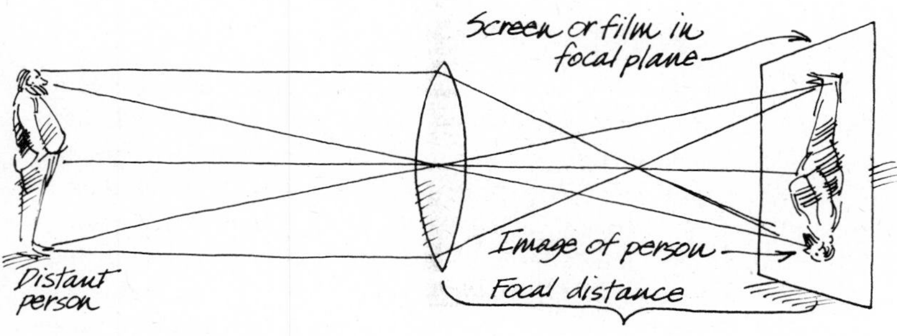

Figure 6.10 A lens forms an image. When the lens receives light from a distant scene consisting of many points each reflecting or emitting light, it focuses the light from each point at a different point. This way, a complete image of the scene is produced. The figure shows only five of the infinite number of rays that pass through the lens.

Figure 6.11 A lens used as a magnifier. The object to be viewed must be placed between the lens and its focal point. One obtains the greatest magnification if the object is placed just inside the focal point.

focal point. (See Figure 6.11.) I spent the day magnifying many objects in my laboratory and enjoying the view.

April 12: Clear and colder. The big day! Enough of playing, I said to myself. Put aside your lens and think. You need to collect light and then make the resulting image bigger. Then it hit me—aha!—I had the answer before me. My solution required two lenses! A first lens would act as a light funnel, and a second lens would help with magnification. I hurriedly put one lens at arm's length and held another up to my eye and looked through both. After some adjustment, I was delighted! Distant objects looked both nearer and brighter! My friend thanked me profusely. He needn't have. I was glad to help.

In fact, magnifying glasses were used centuries before the invention of the telescope.

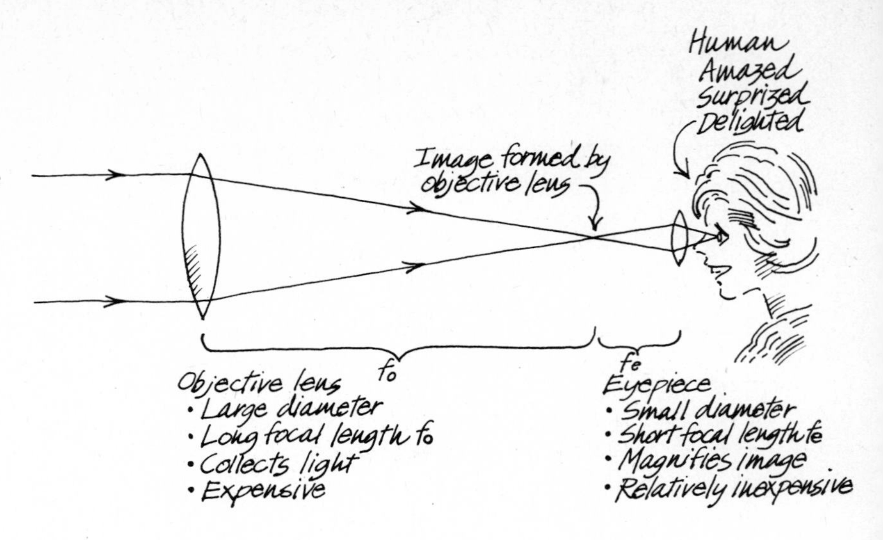

Figure 6.12 The refracting telescope. In its barest essentials, the refracting telescope consists of two lenses, an objective lens, which collects the light and forms an image, and an eyepiece, which magnifies the image.

The preceding is only an example of the way the telescope could have been invented. We do know that Galileo's telescope actually used an eyepiece lens with surfaces that curved inward rather than bulging outward. The type described in the above piece of fiction is a more common kind of astronomical telescope. It was first described in a book that originated much of the theory of optical instruments, a book written by none other than the brilliant—and busy—man, Kepler.

Figure 6.12 shows the operation of a simple refracting telescope which works in the following (somewhat simplified) way. The first, large lens is called the objective lens. It should have a diameter as large as possible for two reasons. Its job is to collect the light that falls upon it. The larger it is, the more light it collects. A telescope with an objective lens of twice the diameter will collect four times as much light, since it has four times the area. The diameter of the objective lens also determines the telescope's **resolution,** the ability to distinguish details of distant objects. For example, a telescope having high resolution can detect a pair of close stars, where a smaller telescope would show the pair as a single spot of light. A telescope having an objective lens with twice the diameter can distinguish details having half the angular separation.

The objective lens is the expensive part of the telescope. The first thing an astronomer might inquire about a refracting telescope is, What is the diameter of the objective lens? Finally, the objective lens has a long focal length compared to the eyepiece lens. The length of the refracting telescope is largely determined by the focal length of the objective lens.

The eyepiece cooperates in magnifying the small, bright image formed by the objective lens. The eyepiece need not be large, since its function is to send light into the eye. Thus it is the cheaper part of the telescope. The eyepiece has a short focal length.

The combination of the two lenses presents an inverted (upside-down) image to the eye. When used for viewing objects on earth this property can be a nuisance but, since there is no up or down to objects in space, it does not concern the astronomer.

The magnification (sometimes called the "power") produced by a telescope depends on f_o, the focal length of the objective lens, and f_e, the focal length of the eyepiece. The formula is

$$\text{Magnification} = \frac{f_o}{f_e}$$

To achieve *greater* magnification one can change either to an objective lens of *longer* focal length or to an eyepiece of *shorter* focal length. Since the eyepiece is by far the cheaper component, most good telescopes are supplied with a number of eyepieces of varying focal length. If the astronomer wishes to have a closer look at an object, he or she merely pulls out one eyepiece and replaces it with one having a shorter focal length.

One might think that an astronomer would always use an extremely high magnification, say, several thousand power. Usually, though, the telescope is hardly ever used at a magnification much over several hundred power, and much viewing is done at significantly lower powers. A magnification of 50 gives an excellent view of many objects in the sky. The problem with using very high powers is that the wavering of the atmosphere and imperfections in the telescope are magnified along with the object being viewed. Unless the air is quite steady, a magnification of, say, 700 times can produce a billowing, messy image.

The largest refracting telescope ever made is still in use at the Yerkes Observatory in Wisconsin. Completed in 1895, it has a diameter of 40 inches [102 centimeters (cm)]. It is a marvel—a carefully made lens over a yard in diameter.

(Box 6.2 gives some advice on buying one's first telescope.)

6.5 THE TELESCOPIC DISCOVERIES OF GALILEO

While using his telescope, Galileo must have felt like a traveler in a new land. He could hardly point his telescope in any direction without seeing something new and wonderful. This section relates some of his discoveries.

Each of Galileo's discoveries is treated in more detail in subsequent chapters.

"It is a most beautiful and delightful sight to behold the body of the Moon," he wrote. (See Figure 6.13.) Surely anyone seeing the moon through a telescope for the first time would agree. Galileo observed that there are basically two kinds of regions on the moon. He called them "brighter and darker" regions. Today's astronomers use the terms **highlands** and **maria.** The naked eye can distinguish these two regions on the moon. Galileo discovered that the bright highlands are rough and mountainous, while the dark maria are relatively smooth. He also discovered many smaller spots, invisible to the naked eye—"so thickly

BOX 6.2

A FIRST TELESCOPE

All too often an initial interest in astronomy has been squelched when a person buys his or her first telescope. The person reports, "I set up the telescope but had a lot of trouble pointing it at a star. I almost gave up but finally managed to get a star in view. The star looked wrong, though; it was a shapeless blob. Also, each time the breeze blew, the blob seemed to jump all over the place. I put the telescope away in the attic and left it there."

The problem was that the telescope was worthless. The attractive advertisement for the telescope may have said "300-Power Deep Space Telescope, $19.95." This telescope has an eyepiece that causes much too much magnification for a beginner to use. The eyepiece is glued in so that a more appropriate eyepiece cannot be substituted. A cheap telescope has cheap lenses that produce poor images. The tripod support will be flimsy and hard to use and will sway in the wind. And another potential amateur astronomer will be lost.

Telescopes are expensive. A good one, which will give years of pleasure, might cost several hundred dollars. The affluent enthusiast has no problem, but what about the average beginner? Fortunately, there is a solution—a pair of binoculars.

One could not make a better start than to get a decent pair of binoculars, which can be had for roughly $80.00 (at the time of writing). The author recommends a pair of 7 × 50 binoculars. (The first number gives the magnification and the second the all-important diameter of each objective lens, 50 millimeters (mm) in this case.)

Only 7 power? Definitely. The low power shows a large piece of the sky, making it easy to sight in on an object. Such a low power also minimizes the shaking of one's hands and any defects in the lenses. Yet, a 7-power pair of binoculars allows the viewer to observe every one of the discoveries made by Galileo (except the "ears" of Saturn) and much, much more. The hours of pleasure to be had viewing the sky with a pair of binoculars are literally endless. Even seasoned amateurs who own large telescopes find their binoculars an important tool.

If one wishes to invest further in viewing pleasure, one might consider purchasing a sturdy camera tripod and a clamp to attach the binoculars to the tripod. This eliminates the problem of shaky hands. The moons of Jupiter, the Milky Way, the Orion nebula, and, especially, the moon are all beautiful, interesting, and waiting for the owner of a pair of binoculars.

---

scattered that they sprinkle the whole surface of the moon." These spots are called **craters** today.

Many thinkers had asserted that the moon's surface must be smooth, since they believed that all heavenly bodies are perfect. However, Galileo asserted, "I feel certain the Moon's surface is . . . uneven, full of cavities and prominences." He went on to compare the appearance of the moon's surface to the mountains and valleys on earth. One interesting piece of evidence for this roughness was the appearance of the moon near the **terminator,** which is the dividing line on the moon that separates the dark side of the moon (night on the moon) from the bright (day) side. He could see bright spots near the terminator, but on the dark side of the terminator. He deduced that these islands of light were mountaintops catching the early rays of the morning sun. "Is it not the same on Earth before sunrise that while the plain is still in

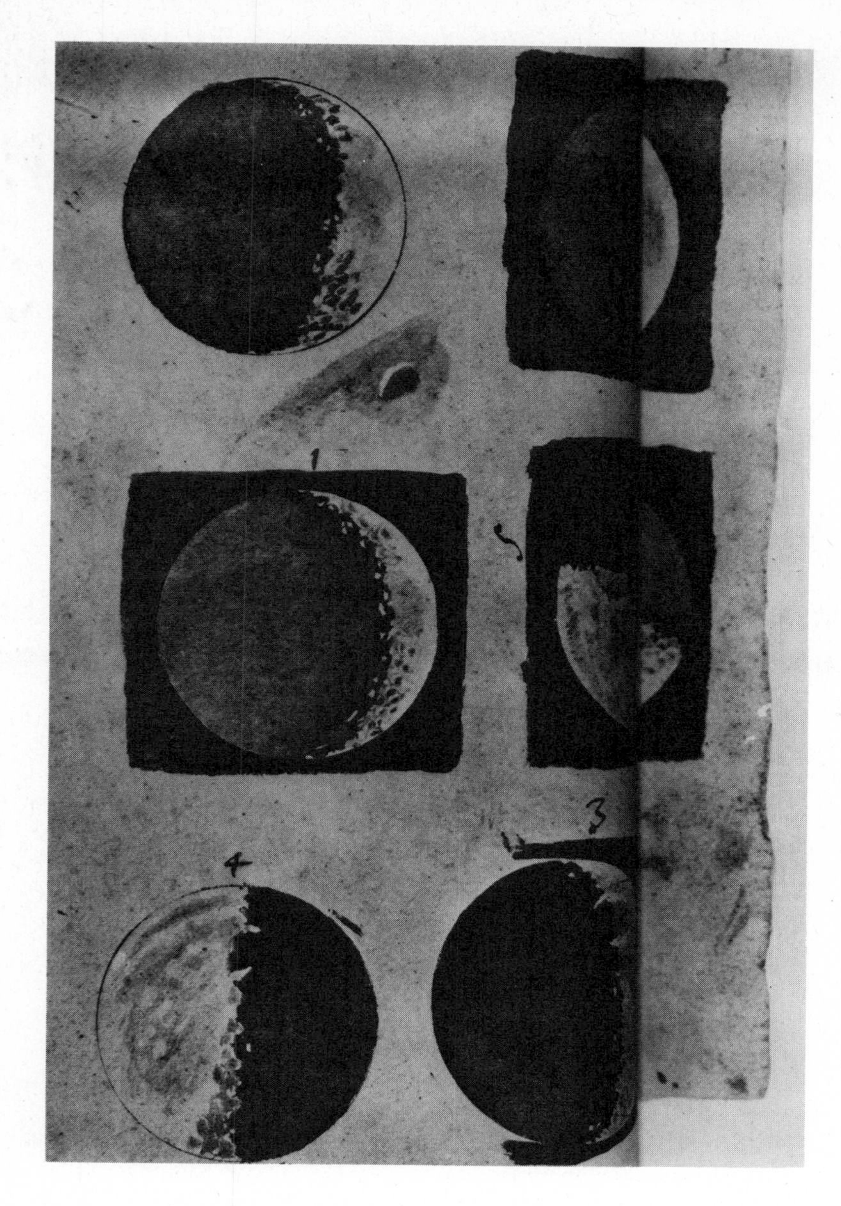

Figure 6.13 Galileo's drawings of the moon.

shadow, the highest mountain peaks are illuminated by the Sun's rays?'' he asked. By measuring the lengths of the mountains' shadows, he estimated that some of the higher mountains on the moon are about 6.5 km (4 mi) high, which roughly agrees with modern measurements. (Mount Everest, by comparison, stands about 8.9 km (5.5 mi) above sea level.)

Then Galileo looked at the stars; wherever he looked, he saw new stars. These stars were too dim to be seen by the naked eye, but his telescope collected sufficient light to show stars ''so numerous as to be

Figure 6.14 The phases of Venus as viewed from the earth. Venus usually does not appear to pass directly behind the sun and may often be observed by experienced observers at such times. Others should not try to view it near conjunction because of the danger of exposing the eye to direct sunlight.

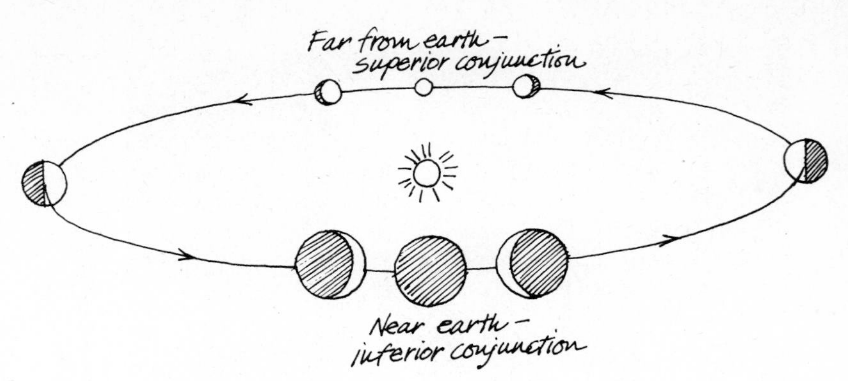

beyond belief." Both of these discoveries impressed Galileo's contemporaries greatly. Knowing that there are countless stars at unthinkable distances led to wonder at the vastness of the universe. Kepler's reaction was a startled cry, "The infinite is unthinkable!"

Earliest humans had noticed an irregular band of faint white light stretching across the night sky. It is called the **Milky Way.** (One must observe the Milky Way from a location far from city lights.) Galileo settled an old question with one glance through his telescope. Is the Milky Way a continuous band of light? No, Galileo saw that the Milky Way is "nothing else but a mass of uncountable stars grouped together in clusters."

On January, 7, 1610, Galileo chanced to look at the planet Jupiter. He noted that it looked like a small disk, as all the planets did. But he also noticed what seemed to be three stars near the position of Jupiter, like this (* * J *). It seemed peculiar to him that "they seemed to lie exactly on a straight line, parallel to the ecliptic." The next night he looked at Jupiter again. This time the arrangement was (J * * *). This excited him. Jupiter could not have moved so far with respect to the fixed stars in one day. Eventually he noticed a fourth "star" which also swung from side to side of Jupiter like the other three. He announced to the world his discovery that there were four moons which orbited Jupiter, much as our moon orbits the earth.

This discovery probably interested Galileo's contemporaries more than any other. One person exclaimed that Columbus had only discovered a new continent, but Galileo had found four new worlds. The discovery led some to claim that here was proof that Copernicus was right. Here were four small objects orbiting a large central object. It was like a small model of the Copernican solar system. Others discounted such a proof. After all, neither Ptolemy nor Copernicus had made any claim about moons orbiting Jupiter.

Galileo's discoveries seemed to go on without end. When he observed Venus, he found another surprise. Venus goes through phases and apparent changes in size. See Figure 6.14. The implications of this discovery were very important. First, it settled the argument as to whether the planets shine by reflecting sunlight or by emitting light on their own. The side of Venus that is turned away from the sun is dark. Venus has night and day, as do the earth and the moon.

Second, a study of Figure 6.14 will reveal that the most natural interpretation of the phases of Venus is that Venus revolves around the sun. Venus appears full when at superior conjunction (on the far side of the sun) and dark when at inferior conjunction (nearest the earth). This observation, that Venus passes behind the sun, was the *very first* piece of data that definitely *contradicted* the geocentric theory. The reader will recall Ptolemy's idea that Venus travels on a circle entirely in the region *between* the earth and the sun. Until the discovery of the phases of Venus, there had never been any disproof of Ptolemy's theory. But now this single observation cast the theory into doubt.

Scientists who still wished to deny the heliocentric theory and maintain a theory that allowed the earth to be at rest had two choices. They could try to modify the geocentric theory, at least as it relates to Venus, or they could accept Tycho's theory that all the planets revolve around the sun, but that the sun revolves around the stationary earth. At any rate, from this time forward more and more thinkers began to accept Copernicus' heliocentric theory.

When Galileo decided to observe the sun, he made a dangerous mistake. He held the telescope to his eye and pointed it at the sun. The concentrated sunlight blinded his eye for a week and left permanent damage. Later, he hit upon a safe manner of observing the sun. He used the telescope to project the sun's image onto a piece of paper (See Box 8.1.) He then confirmed the discovery, already made by others, that the sun has spots from time to time. The spots appeared to be on the surface of the sun, and their motion indicated that the sun rotates on an axis. Once again, a heavenly object was seen to be "imperfect," a shocking idea to many. (See Figure 6.15.)

How could so brilliant a man have made such a mistake? It is a valuable warning to telescope and binocular owners.

Perhaps the observation that puzzled Galileo most was that Saturn appeared to have what were called "handles" or "ears" on either side. These appendages grew in size and then gradually disappeared over the years. Galileo was baffled by this. It was only later that better telescopes showed that the planet has a flat ring around it. The ring system is a source of amazement to this day.

Galileo gathered his earliest discoveries together in a book called *Sidereus Nuncius*, or *Star Messenger*. It was so clearly written and so intriguing that it was widely read. The concept of the smallness of the earth and the size and diversity of the heavens accelerated the crumbling of old ideas about the makeup of the universe.

The discoveries made by the users of telescopes prompted three interesting prophecies. Galileo wrote, "Perhaps other discoveries still more excellent will be made from time to time by me or by others with the assistance of a similar instrument." Much of the rest of this text is testimony that bears him out. Two other statements show how the view provided by the telescope brought humans closer to a feeling of oneness with the universe. Kepler, one of the most farsighted, was

Figure 6.15 **Galileo's drawings of sunspots.**

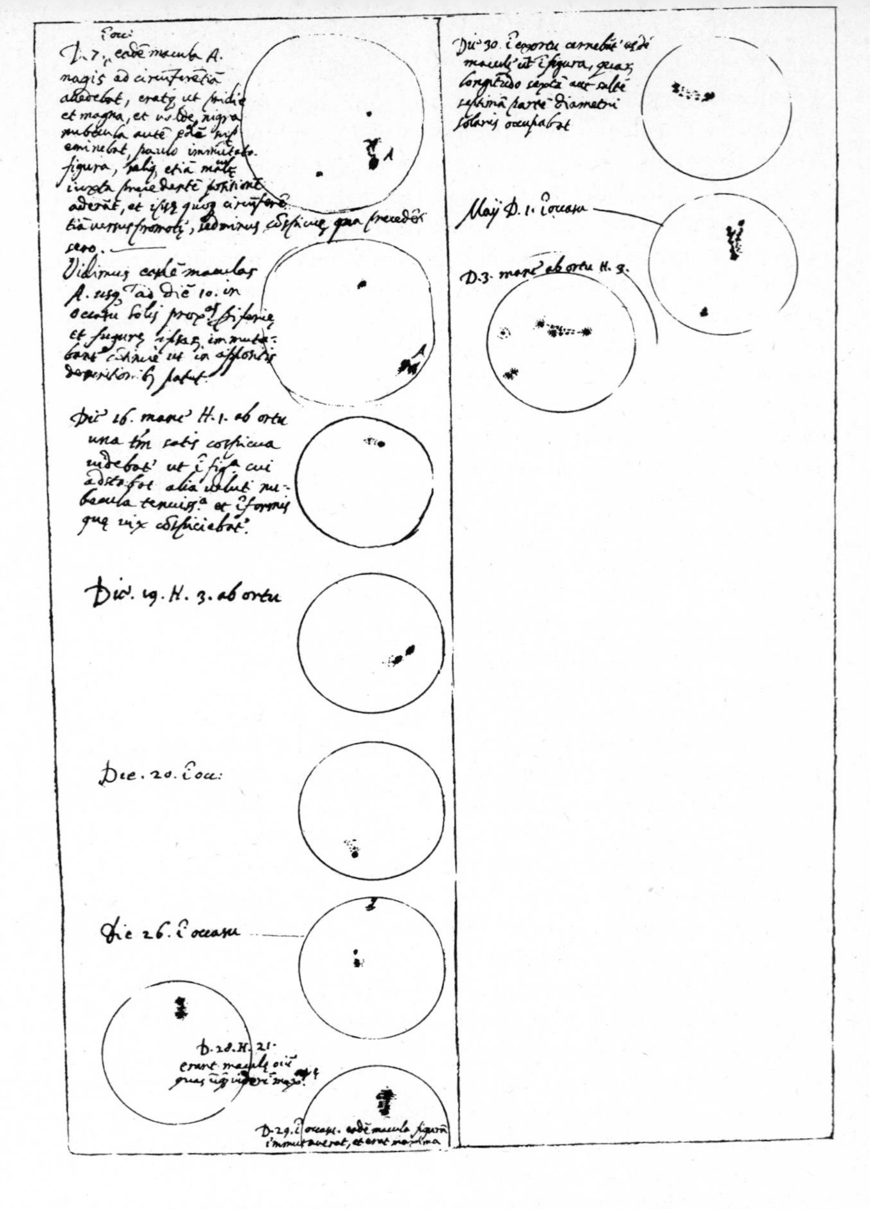

inspired to write to Galileo about space travel! He wrote, "There will certainly be no lack of human pioneers when we have mastered the art of flight. . . . Let us create vessels and sails adjusted to the heavenly ether and there will be plenty of people unafraid of the empty wastes. In the meantime, we shall use the telescope to prepare, for the brave sky travelers, maps of the celestial bodies." Finally, John Donne wrote:

Man has weav'd out a net, and this net throwne
Upon the heavens, and now they are his owne.

Another reaction to Galileo's work is discussed in Box 6.3.

BOX 6.3

SCIENCE AND RELIGION: GALILEO'S DILEMMA

The interaction of science and religion is an interesting and often emotional topic. Extreme positions can be found on both sides. Some scientists take the position that "science proves that there is no God." Some believers feel the need to attack and repress scientific theories if they seem to conflict with their religious beliefs.

One of the most famous episodes of religious oppression of a scientific theory involved Galileo and the Roman Catholic church. In 1615, a debate was raging in Rome as to whether the heliocentric theory was a threat to the church. The 41-year-old Galileo went to Rome to argue for Copernicus. Galileo was a deeply religious Catholic and feared that the church might commit itself on the wrong side. Galileo lost. He was ordered not to hold or defend the heliocentric theory, and the Decree of 1616 placed Copernicus' book in The Index, the list of forbidden books.

In 1623, Galileo thought he saw a chance to strike a blow against the decree. In that year Pope Urban VIII was installed in Rome. The pope was a lover of art and science and a friend of Galileo's. He had also opposed the Decree of 1616. Galileo met with Pope Urban and discussed the writing of a book on the heliocentric theory. The pope gave Galileo permission to write the book, as long as he gave even-handed treatment to both sides and made it clear that, in the end, heliocentrism had not won.

Encouraged, Galileo wrote his *Dialogue Concerning the Two World Systems.* It was cast in the form of a conversation between three philosophers, one obviously representing Galileo, one representing a geocentric theorist, and the third an open-minded commentator. The book came out in 1630 and was quickly sold out. It was written in such an interesting, amusing style that it was the talk of the scientific and nonscientific world.

When the pope learned of the contents of the book, he was interested but definitely not amused. It seemed to the pope that Galileo had not kept his word. The bulk of the book was a strong piece of writing in favor of Copernicus, which belittled anyone who thought otherwise. Galileo showed the open-minded character consistently being won over to Copernicus' side.

Near the end, the geocentrist claims that the arguments for the motion of the earth do not seem "true and conclusive." Furthermore, he says that all surely agree that God could have brought about the motions we observe in many different ways. He finishes with, "It would be excessive boldness for anyone to admit a limitation on the Divine power and wisdom to insist on some particular fancy of his own." The Galileo character agrees with this, and the book ends.

The pope felt he had been made to look like a fool by Galileo. The final page of the book seemed merely tossed in to placate him. It was so out of character with the rest of the book that most people read the final disclaimer with tongue in cheek. Surely, the pope was not amused to find his own sentiments mouthed by a character who had just lost 450 pages of argument. Besides, the character was named Simplicio, which can be rendered "simple one."

Galileo seems to have been aware that he had not really kept his agreement with the pope but, it appears, he expected the force of his arguments to cause the authorities to keep quiet or, possibly, even admit acceptance. Galileo was surprised to find that he had guessed quite wrongly. In spite of Galileo's 70 some years and ill health, the pope ordered him hauled before the Roman Inquisition for punishment. (At the time, the pope had vast powers in both the governmental as well as the spiritual realm.) In a shocking display, Galileo was forced to kneel and sign a statement saying, "I renounce, curse, and detest the . . . errors and heresies." He was sentenced to life in prison but, at least in this, he was treated somewhat more humanely and considerately. He was confined to his own home and allowed to continue his nonastronomical research with his pupils.

Galileo was sure that his viewpoint would eventually triumph. He saw no challenge to religion in heliocentrism and felt that the strong evidence in its favor would gradually win over the church authorities. In this he was right. In 1744, his *Dialogue* was published with only trivial changes. It was released from the Index in 1822. The 1910 edition of the *Catholic Encyclopedia* calls the Decree of 1616 a "grave and deplorable error."

Yet, the tension between science and religion continues to trouble many to the present day. Some have tried to resolve the conflict, but no one has succeeded to the satisfaction of all. The reader is invited to compare his or her thoughts on the subject with those of a hypothetical philosopher. (The viewpoints that follow do not represent anyone in particular but are offered merely as a point of departure for discussion.)

The conflict between science and religion is unfortunate and unnecessary. Look, for example, at the similarities between the two fields. Finding oneself alive and conscious in an amazing, puzzling world, every person asks, "Why am I here? What is the meaning of it all?" Science and religion are two ways of seeking answers to such questions. The religious person uses the approach called faith. Each religion also has certain rituals and practices designed to clarify and strengthen the believer's faith. The scientist also seeks to understand the mystery of nature and humanity. The scientist's practices involve measuring and observing natural phenomena. Scientists are also guided by a faith, namely, that nature is ultimately understandable and that a good theory will be simple and beautiful. The scientist also can experience a mystical inspiration (the aha)—the moment when a theory is born. At that moment, the scientist has the feeling of being a bit closer to understanding the meaning of things.

Several modern historians of science have pointed out an important feature of science bearing on this question. Each age tends to feel that the science taught in their schools is the proven, ultimate answer, carved forever in stone. From our perspective, looking back over the thousands of years, however, we can see that the theories of each age were temporary, sooner or later to be replaced by other ideas, often by ideas that completely contradicted the former ones. But, each time new theories won out, the tendency was to adopt the new theories as if they, in turn, were frozen, permanent results. Today, it is reasonable to suppose that our theories are themselves temporary and that, before too long, other theories will take their places.

With this point in mind, we should not be too concerned whether any theory agrees with or contradicts our personal beliefs. A religion that represses a scientific theory and a scientist who dogmatically claims that science shows there is no God have not yet learned this lesson.

The human mind at its most profound, whether working with science, philosophy, religion, literature, art, or music, is wrestling with our complex world and the fact of our mysterious existence. It is not strange that the products of such complex human activities will, at times, be at odds. But, after all, part of the joy of living and thinking is having a rich mixture of many points of view present in the world, each expressing in its own way humanity's quest for understanding.

BOX 6.4

ISAAC NEWTON
1642–1727

Newton was born on Christmas Day in 1642, the year Galileo died. He was born so prematurely that his mother said he would have fit into a quart mug. Those who assisted at his birth assumed he would soon die, but he lived to become one of the glories of England and of science.

As a youth, he showed a gift for tinkering with gadgets. This gift was later put to use in his many laboratory experiments. An uncle recognized his promise and prompted the country boy to attend college at Cambridge. Newton earned his B.A. there from 1661 to 1664, quietly learning the mathematics and science of his day and earning his way with odd jobs. It appears that he made no great impression on anyone in those years.

Then the plague closed Cambridge in 1664, and the 22-year-old Newton went home. During the plague years he invented the branch of mathematics having to do with rates of change, the all-important field called calculus. Today, most mathematicians rate Newton, along with Archimedes and Gauss, as the greatest in their profession. During the years of quiet at home, the idea of universal gravitation also came to him. Furthermore, he began his important studies of light at that time, studies that later led to the all-important device called the spectroscope (discussed in Chapter 7). He returned to Cambridge in 1667 and invented the reflecting telescope in 1668.

Years passed until 1684, when Edmond Halley visited Newton and found that he had worked out the basis of a new theory of dynamics. In a burst of devotion to science, Halley encouraged Newton to complete his theory and write it out. From 1684 to 1686, Newton did so, in one of the most strenuous bouts of concentration known to humanity. Halley prepared the manuscript for publication, read and corrected the proofs (a difficult and thankless task), and even paid for publication of the book out of his none-too-ample funds. The book stands today as one of the supreme classics of science.

In spite of his overwhelming importance in physics, mathematics, and astronomy, Newton did not confine his interests to these fields. He spent a great deal of time working on alchemy, the chemistry of his day. He delved into the study of religion and considered his achievements in this field to be his most important.

6.6 NEWTON AND THE REFLECTING TELESCOPE

Galileo died in 1642, the same year Isaac Newton was born, a man destined to become one of the greatest geniuses the world has seen. Like Galileo, Newton was a well-rounded genius, skilled in working with his hands in experiments and brilliant when composing theories. (See Box 6.4.)

During his early experiments with light, Newton studied the refraction of white light passing through a prism. He noticed that a white beam of light becomes **dispersed** upon refraction; that is, it is broken up into its component colors. See Figure 6.16. The resulting spread of colors is called a **spectrum.** The spectrum is produced because red light is refracted least of all the colors and violet light is refracted the most. All the other colors of the spectrum fall in between red and violet.

Newton had been trying to obtain accurate observations of the mo-

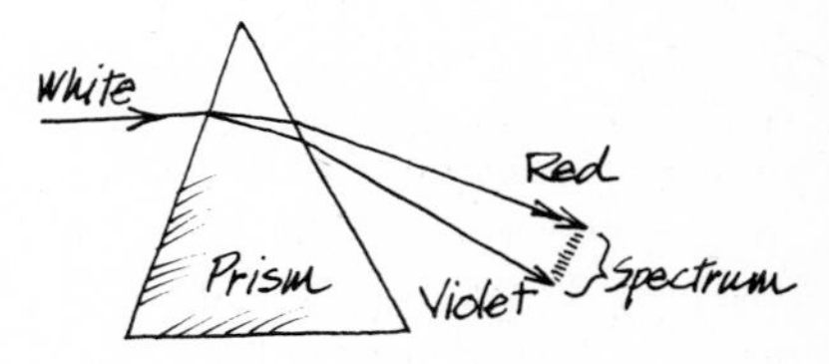

Figure 6.16 Dispersion of white light by a prism. In one of his most famous experiments, Newton passed a beam of white light from the sun through a prism. The emerging beam was spread out into a wider beam having different colors at different angles. Newton concluded that white was not, strictly speaking, a color, but rather a combination of all the colors in the spectrum.

Figure 6.17 **Chromatic aberration.** As this figure shows, dispersion causes red to be focused farther from the lens than violet. Other colors are focused at places in between these two extremes. The effect is called *chromatic aberration* or color disturbance.

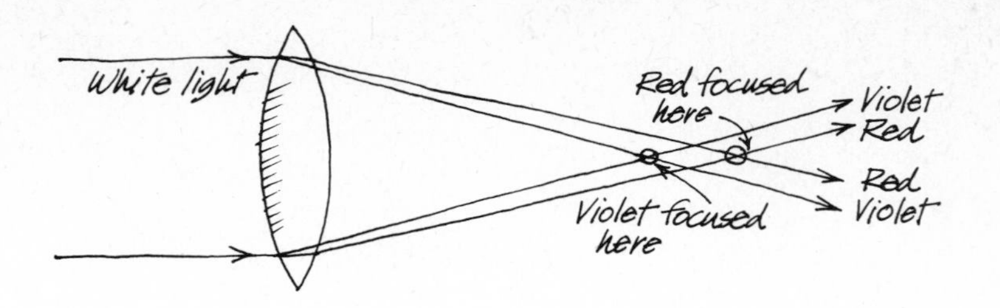

Figure 6.18 The Newtonian reflecting telescope. If an eyepiece were brought near the focal point in (a), the observer's head would block all light from the objective mirror. Newton solved this problem (b) by introducting a diagonal, flat mirror which reflected the light to the side, where the observer could then view the image with an eyepiece. The formula for computing magnification is $M = f_o/f_e$ where f_o is the focal length of the objective mirror and f_e is the focal length of the eyepiece. The flat mirror has no effect on the magnification.

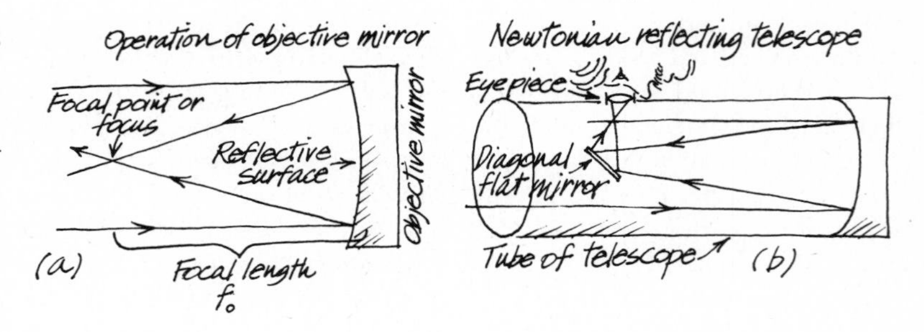

In 1757, John Dolland discovered a method of reducing chromatic aberration in refracting telescopes. He combined lenses made of different kinds of glass.

tions of the satellites of Jupiter to use in testing his new theory of dynamics. He was annoyed that all refracting telescopes showed colored edges on objects where no such colors existed. It was thought that this effect, called **chromatic aberration** (color disturbance), was due to imperfectly ground lenses.

The correct explanation occurred to Newton after his work with prisms. The lens acted like a prism and refracted different-colored light by different amounts. Each color was focused at a slightly different place, resulting in the observed chromatic aberration. (See Figure 6.17.)

Newton became convinced that the problem of chromatic aberration is unavoidable when a lens is used to gather light. He knew of only one other way, besides refraction, to change the direction of a beam of light, and that was by **reflection** (the rebounding of light from a smooth surface). And (aha!) the individual colors of a light beam are all reflected alike. Newton had the key to the solution: replace the objective *lens* with an objective *mirror.*

Newton obtained a mirror that had been ground and polished so that it was no longer flat but concave and with a shape called a *paraboloid.* A mirror of this shape accepts parallel rays of light from a distant object and reflects each ray at the proper angle so that it meets all the other rays at the focal point. See Figure 6.18. He realized that he could not place an eyepiece near the focal point to view and magnify the image, because his head would block the incoming light. To solve this problem he added a small, flat mirror which reflected the converging light rays to the side where an eyepiece could be used. In this design, the flat mirror blocks a small amount of the incoming light, but the mirror is so small that little light is lost. See Figure 6.18b.

Newton sent a working example of his design, called a **Newtonian reflecting telescope,** to the British Royal Society in 1671. Although it was only 6 inches long and had a mirror of only 1 inch diameter, it aroused great interest. Over the years, the advantages of the reflecting telescope over the refracting telescope have become more and more evident. The objective mirror causes no chromatic aberration. Since one looks into the side of the tube, one's body position is often more comfortable while viewing.

Advantages of a Reflecting Telescope As very large telescopes were developed and, once the technology of mirror construction had matured, even more of the advantages of reflecting telescopes became clear. As mentioned earlier, the largest refracting telescope ever made, the Yerkes 40 inch, was completed in 1895. Since then, each record-breaking telescope has been a reflector. The reasons are many. To reduce chromatic aberration to a minimum a refracting telescope must have a long focal length. The resulting elongated telescope requires a huge, expensive dome to protect it. Reflecting telescopes can be made with shorter focal lengths, thus allowing smaller domes. A mirror is cheaper to make than a lens of the same diameter, because less perfect glass can be used (the light will not pass through the mirror) and because only *one* side of the mirror glass need be ground and polished. The 40-inch lens for the Yerkes Observatory is the largest in use. A lens significantly larger would absorb too much light as the light passed through its thickness. Furthermore, a larger lens might sag markedly under its own immense weight. A mirror, on the other hand, can be made very large because light does not pass through it, and so the mirror may be rigidly supported at its edge *and* back, assuring that no sag will occur.

That a lens larger than 40 inches would absorb too much light and sag seems to be the majority opinion. However, Alvan Clark, who made the 40-inch lens, claimed that such problems seemed trivial to him. Clark died not long after the 40-inch lens reached Yerkes, and no one since has tried to top his achievement.

Because of these many advantages, all the world's largest telescopes are reflecting telescopes. (See Figure 6.19.) Until the mid-1970s the world's record was held by the 200-inch (5.1-meter)-diameter mirror at Mount Palomar, California. (See Figure 6.20.) Since then, the 236-inch (6.0-meter) mirror located near Zelenchukskaya, Russia, has taken the lead.

6.7 NEWTON'S LAWS OF DYNAMICS

From 1684 until 1686, Newton concentrated on writing his book of dynamics. *The Principia,* published in 1687, stands as one of the greatest achievements of the human mind. In this book, Newton described a theory of dynamics still used to this day to explain nearly all motion. The book is laid out in a **deductive** fashion, that is, it states Newton's three laws of motion and then shows how they are used to explain the motions of objects. (See Box 6.5.) Since there is no room to do more than mention the laws and a few of their applications here, perhaps the interested reader will look at one or another of the many

Figure 6.19 A large, modern telescope. This is the 153-inch-diameter telescope at Siding Spring, Australia. It went into regular operation in 1975. It is one of four large, modern telescopes operating in the southern hemisphere. The other three are in Chile.

Figure 6.20 The 200-inch mirror of the Mount Palomar telescope being polished.

BOX 6.5

EMPIRICAL KEPLER AND DEDUCTIVE NEWTON

Both Kepler and Newton were concerned, in part, with Kepler's laws of planetary motion. Both men fitted these laws into their respective schemes in very different ways. The difference is a good illustration of the two levels of science, the empirical method and the deductive method.

Kepler, the pattern finder, had nothing to go on but a mass of numbers—Tycho's observations of the positions of the planets. To most of us, the numbers would seem a meaningless jumble, but Kepler was sure there was an underlying pattern. Guided by this faith and by his genius, he spent many hard years of trial and error refining his three laws of planetary motion from the data. The result is a triumph of human ingenuity, but the laws carry little obvious meaning beyond themselves. The way Kepler discovered laws is called the *empirical* method, finding a law by looking for patterns in the observations. Many laws have been found by this method, especially in the early stages of a branch of science.

Newton's goal was very different. He was delighted, as so many scientists have been, with the way Euclid wrote his geometry books. Euclid, an ancient Greek mathematician, began his books with a set of assumptions (or axioms). Then, from this small set of assumptions, he proved all his further results by applying logic. This is called the method of *deduction.* It was Newton's aim, in *The Principia,* to cast his theory of dynamics in the same mold. He thought about the motions he was aware of (Kepler's three laws of planetary motion were especially important guides) and boiled all of them down to his three laws of motion, laws that applied to *any* motion. Near the beginning of *The Principia,* Newton states these three laws. The rest of the book uses logic to deduce the motions of a huge number of specific cases—balls, bullets, satellites, and so on. The result was an impressive new kind of theory, one that used a few simple laws as the basis to explain all known motions. Newton showed how, for example, Kepler's laws of planetary motion follow logically from Newton's laws of motion (it takes only a few pages).

Kepler had used the many specific positions of the planets at specific times to derive three laws for the planets. Newton proposed three laws of motion that applied generally and showed how to use them to explain any specific motion.

---

books explaining Newton's theory. But let us at least introduce each of Newton's three laws of motion (which should not be confused with Kepler's three laws of *planetary* motion).

Newton's First Law of Motion This law is basically a restatement of Galileo's idea of inertia.

> *If no force acts on an object, it will continue its state of motion; that is, if an object is at rest, it will remain at rest and, if the object is moving, it will continue to move in a straight line at a constant speed.*

Newton's Second Law of Motion This law answers the question, What if a force *is* acting on an object? Newton's brilliant insight was that the concepts of mass and acceleration are crucial. The **mass** of an object is related to the amount of matter in the object. The more mass an object has, the more inertia it has, that is, the harder it is to get it going or to deflect it out of a straight-line path. The concept of mass should not be confused with the concept of **volume,** which is the amount of space an object occupies. A styrofoam ball and a lead ball of the same size have the same volume, but the lead ball has much more mass.

The other concept that is crucial to Newton's dynamics is acceleration. Other theories had been concerned with the *position* of an object or with the **speed** of an object (the rate at which the position of the object changes). Newton declared that the fundamental consideration is the object's **acceleration,** the rate of change of speed and/or direction. If the speed of an object changes (it slows down or speeds up) or if the direction in which the object is traveling changes (the object follows a curved path), acceleration is taking place. The realization that acceleration is the key was not an easy one to arrive at. Perhaps that was because most people had never experienced large accelerations in Newton's day. Drivers who make jackrabbit starts or who squeal the tires in a turn know about acceleration and find it easier to think along Newton's lines.

Given these ideas, *Newton's second law of motion* may be paraphrased in this way:

In mathematical form, Newton's second law reads $A = F/M$, where A is the object's acceleration, F is the force, and M is the object's mass. The formula states that A is directly proportional to F and indirectly proportional to M.

The acceleration of an object is caused by a force acting on the object. The larger the force, the greater the acceleration. Also, the greater the mass of the object, the less the acceleration.

Thus, the harder you push on a car, the faster it will *gain* speed. If you push equally hard on a car and a bicycle, the bicycle, having the lower mass, will gain speed more rapidly.

Newton's Third Law of Motion This law concerns the interaction of two objects. Call the objects A and B. Then the third law states:

If object A exerts a force on object B, then object B will exert a force of the same size on object A but in the opposite direction.

This law is often less exactly stated: For every action there is an equal but opposite reaction. The law states that forces always act in pairs. Examples are everywhere. If you push on an open door to close it, the door pushes back just as hard on you. (If you were on roller skates, you would roll backward because of the door pushing on you.) If you push on a 100-lb bag of cement to throw it, the bag pushes back just as hard on you. (If you were on roller skates, you'd roll backward.) A rocket ship pushes on its fuel to throw it out the exhaust port. The fuel pushes back on the rocket, causing it to move.

These three simply stated laws of motion were the basis of Newton's theory of motion. He achieved one of the goals of all science—to reduce complicated observations to underlying simple laws. Of course, merely stating the laws does not make clear their power to explain motion.

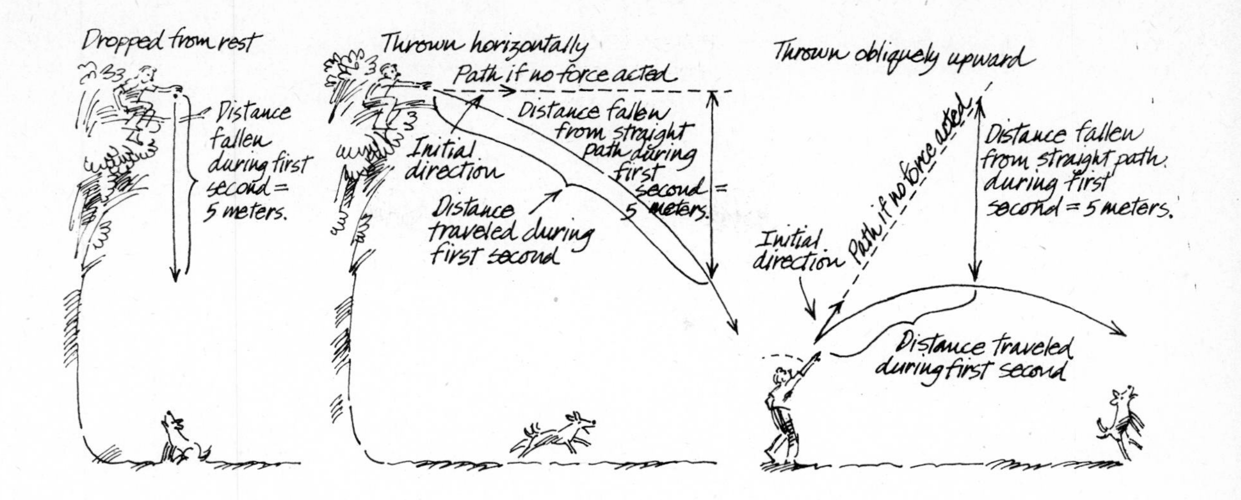

Figure 6.21 Falling objects. Near the earth's surface and in the absence of air friction, an object falls about 5 meters during the first second of flight.

One needs to see the laws used. The following sections only give a glimpse of the power of these three laws. Nevertheless, experiment upon experiment has found that Newton's laws explain all motions except those occurring at speeds close to the speed of light or in a region of very strong gravitation. In such cases, one must resort to Einstein's theory of relativity.

6.8 THE LAW OF UNIVERSAL GRAVITATION

Seeing an apple fall from a tree in his yard one day started a chain of thought in Newton's mind, which resulted in one of his most important contributions to astronomy.

Things accelerate (speed up or follow a curve) when a force is applied to them. Many forces are due to contact—one pushes on a door or pulls a bell rope. The fall of the apple is an example of an acceleration due to a force of another kind—gravitation. The earth seems to pull on objects near its surface. Measurements show that objects falling from rest fall about 5 meters (16 ft) during the first second of fall. A thrown object also falls 5 meters below its original path during the first second of flight. See Figure 6.21.

This last fact is the basis of earth satellites. The round earth curves downward 5 meters in a distance of about 8 km (about 5 mi). This means that an object shot horizontally at a speed of 8 km/sec will, in 1 sec, fall away from the horizontal the same distance the curve of the earth falls away (Figure 6.22). The result is that (ignoring the fact that air resistance slows it down) the object falls away from the horizontal but gets no closer to the surface of the earth. The object will follow a circle completely around the earth and return to the starting point (in about 1.4 hr). It is remarkable that Newton was contemplating artificial earth satellites over 250 years before they became a reality.

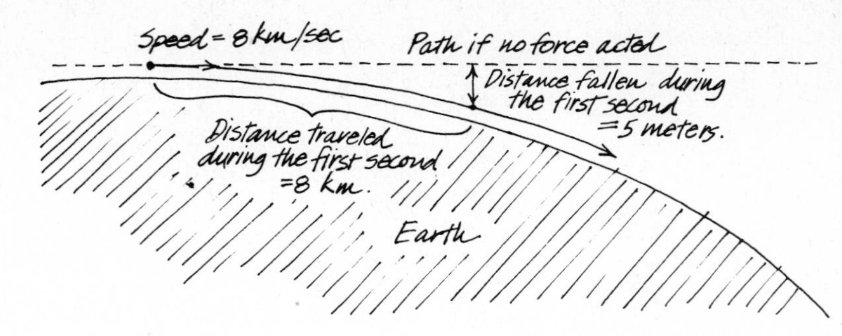

Figure 6.22 An earth satellite at ground level (distances not drawn to scale).

Figure 6.23 Evidence that the earth pulls on the moon (distances wildly not to scale).

Newton also contemplated the moon, our only natural satellite. Why does it follow a curved path around the earth? According to Newton's first law of motion, if no force acted on the moon, it would travel in a straight line and soon disappear into the distance. Some sideways force had to be causing the moon to fall away from its natural straight path. (Figure 6.23) Newton assumed that this force was the same force that made the apple fall—the gravitational pull of the earth.

Newton, a firm believer in the heliocentric theory, also studied Copernicus' assumption that the planets orbit the sun. What sideways force acts on the planets to cause them to curve away from their natural straight paths? Surely it is the gravitational pull of the sun acting on them. Notice the contrast between the approach of Kepler and that of Newton in treating this question. Both could see that the sun affected planetary motions. Kepler looked for a way for the sun to push the planets along. Newton claimed that the planets' own inertia kept them going. To him, it was the inward pull of the sun on the planets that caused them to follow curved orbits.

Newton enlarged these ideas, finally formulating the following law.

The Law of Universal Gravitation

Every object in the universe attracts every other object with a force directly proportional to the product of the masses of the two objects and inversely proportional to the square of the distance between them. (See Figure 6.24.)

The law may be expressed mathematically by $F = G\,(M_1M_2)/R^2$, where F is the force of attraction on one object due to the other, M_1 and M_2 are the masses of the respective objects, and R is the distance between the objects. G is a numerical constant, a number that is the same no matter which two masses in the universe the formula is applied to.

The law of gravitation is one of Newton's greatest contributions to astronomy. It will be used again and again throughout this book. A few of its applications can be mentioned at this point.

Applications of the Law of Gravitation **Weight** is defined as the force on an object due to gravitational attraction. When I say that my weight is 845 newtons (190 lb), I mean that, near the earth's surface, the earth pulls on me with that force. The astronauts who went to the moon had the interesting experience of walking on an object that exerts only one-sixth as much force on objects at its surface as the earth does. That is, a person weighs one-sixth as much on the moon as on the earth. I

A newton is a metric unit of force equivalent to 0.225 lb.

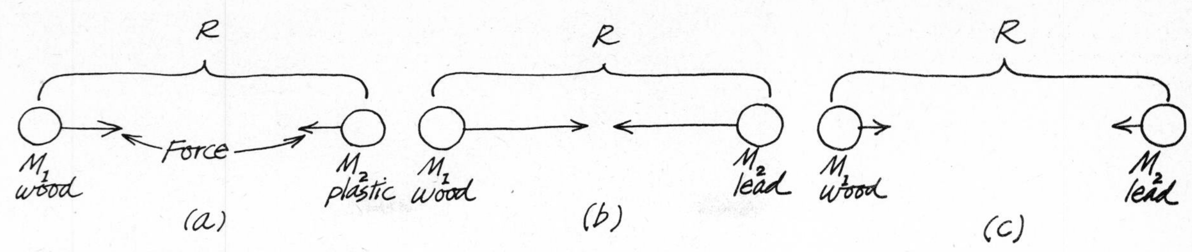

Figure 6.24 Illustrations of Newton's law of universal gravitation. (a) Newton's law of universal gravitation states that any two objects attract each other. The force is the same no matter what other masses lie near or even between the two masses in question. That is, one cannot shield gravitation. (b) The plastic ball in (a) has been replaced by a lead ball having the same size. Thus the mass of the second ball has been increased. As shown by the length of the arrows, the force of gravitation has increased, since the numerical value of M_1M_2 has increased. (The force of gravitation between ordinary objects is so small that it can be measured only with difficulty.) (c) The two balls have been separated so that they are now twice as far apart. The force between them has decreased. By Newton's formula, the force is $1/R^2 = 1/2^2 = 1/4$ as large as previously. Notice that, in each illustration, each of the pair experiences the same force as its partner, but in the opposite direction, as required by Newton's third law.

would weigh 141 newtons (32 lb) on the moon. My weight has been reduced. Note, however, that my mass, the amount of matter in me, is unchanged no matter where I am. In the depths of outer space, in a region where gravitation is small, my weight would be nearly zero, but my mass would be the same as ever [86 kilograms (kg), if anyone cares].

To compute your mass, divide your weight (as measured at the earth's surface) in pounds by 2.2. The result is your mass in kilograms. For example, a person having a sea-level weight of 110 lb has a mass of 50 kg.

Newton's theory of dynamics also gives an explanation for Galileo's statement that, in the absence of air friction, all objects fall at the same rate. Suppose this book is released from rest. It accelerates (speeds up) because the earth exerts a force on it. According to Newton's second law of motion, the acceleration is

$$A = \frac{F}{M}$$

M is the mass of the book and F is its weight, the force due to gravity. Now suppose another identical book is tied to this one. When the combination is dropped, it accelerates exactly as the single book did. The reason is that the force of gravity on the books is doubled, but so is the mass of the combination (and therefore its inertia). The equation is

$$A = \frac{2F}{2M} = \frac{F}{M}$$

The combination is pulled twice as hard, but it is twice as resistant to an increase in speed. (See Figure 6.25.)

If gravitation is universal, why don't we notice it between objects in our daily experience? Why don't two marbles on a table roll together

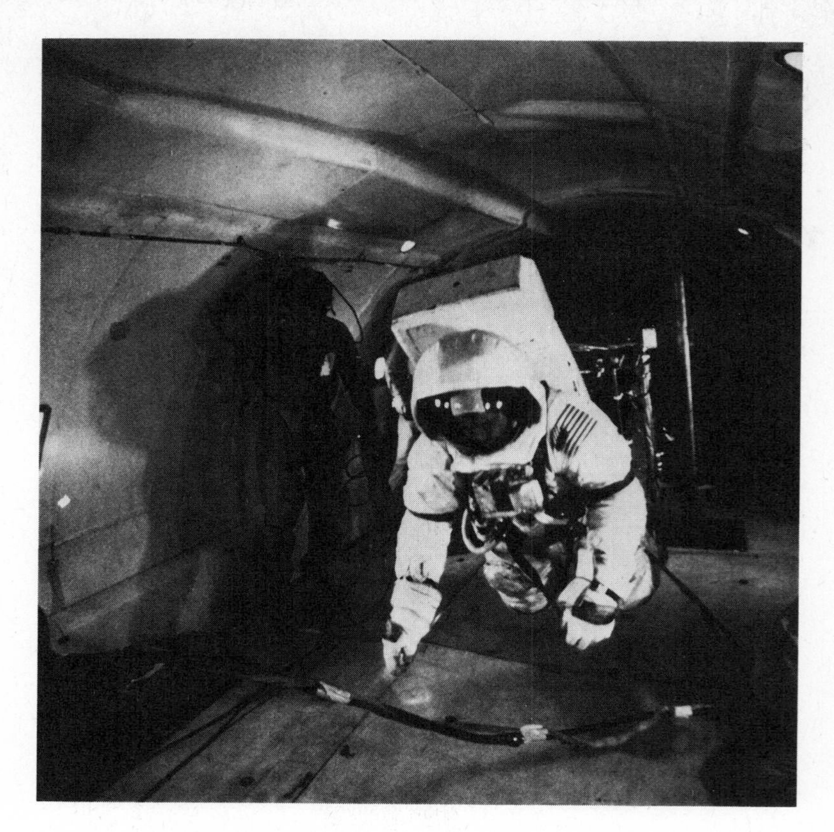

Figure 6.25 Fallinggggggggg! In practice for activities in outer space, astronauts were placed in airplanes and made to experience freefall, (sometimes called weightlessness). The pilot of the plane would go into a dive, following the same path a stone would follow in the absence of air friction. Here we see Alan B. Shepard floating, while the man in the background hangs on so that he will not float away.

because of their mutual gravitational attraction? The reason is that the gravitational force between two objects depends on the product of their masses and the masses of marbles are very small. However, a sensitive plumb bob placed near the base of a mountain hangs slightly away from the vertical and toward the mountain because of the mountain's gravitational attraction.

The sun is a huge ball of gas under pressure. What prevents the gas from rushing out into space the way the gas in a punctured balloon does? The answer is that gravitation holds the sun together. Every particle of the sun attracts every other particle gravitationally. The result is an inward force on each portion of the sun, a force sufficient to confine the gas of the sun under enormous pressure.

What an awesome vision Newton had. The same force that we fight when we climb stairs is acting to hold the planets in their orbits and to hold the sun together. Some stars, called binary or double stars (see

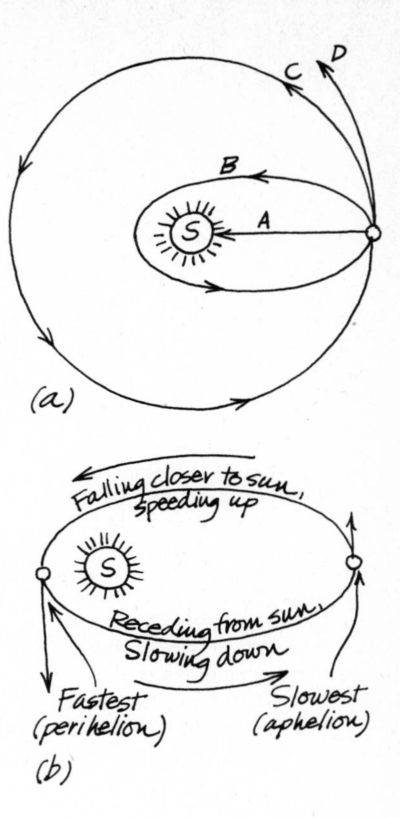

Figure 6.26 Newton's laws applied to planetary motion. After Newton had formulated his theory of dynamics he found that, when it was applied to the motions of planets, Kepler's laws followed. The reader may not have the background necessary to follow Newton's mathematical derivation, but it is not hard to grasp at least some of the ideas intuitively. If the planet in (a) is released from rest, it will be drawn into the sun in a straight line, path A. If it is given a small sideways speed, it will fall toward the sun along path B. The sideways motion causes it to miss the sun. The planet's path is curved by the sun. When the planet is close to the sun, its speed increases sufficiently because of its falling, so that the sun cannot curve it into the sun. However, the planet's inhabitants will be roasted once each year. Newton showed mathematically that his theory predicted that path B would be an ellipse—Kepler's first law. If the planet is given the proper speed, faster than path B, the orbit will be a circle, ideal for sustaining life. Speeds slightly greater or less than this result in nearly circular elliptical orbits. The earth's average orbital speed, 30 km/sec (18 mi/sec), results in a nearly circular orbit. If the planet were given a somewhat greater sideways speed (42 km/sec for the earth), the planet would be going so fast that it would escape the sun along path D, never to return. It was, in part, the fact that all the planets have nearly circular orbits that led Newton to feel that the solar system must have been designed by an intelligence. Why a planet in an elliptical orbit speeds up as it goes from aphelion to perihelion is illustrated in (b). The planet is falling toward the sun, and the gravitational pull of the sun is pulling it forward, increasing its speed. After perihelion, the planet is coasting away from the sun. The pull of the sun now is "backward," retarding its speed. Analogously to an upward thrown ball, the planet is eventually slowed enough so that it "turns the corner" and begins its long fall again. Again, Newton was easily able to derive Kepler's second law as a mathematical deduction from his theory of dynamics.

Section 12.6), are a system of two stars. If watched carefully in a telescope for long enough, they can be observed performing orbital motion about each other. The shapes of their orbits confirm that, here too, Newton's laws apply. The gravitational attraction of one star for the other holds the system together.

6.9 NEWTON'S DYNAMICS AND ORBITAL MOTION

Newton was in part guided by Kepler's three laws of planetary motion as he formulated his theory of dynamics. His completed theory can be looked on as an explanation of all motion, including the motions of the planets. Thus Newton explained why Kepler's laws took the form they did—not only in a qualitative way but also mathematically. See Figures 6.26 and 6.27.

As discussed earlier, an object dropped from rest accelerates downward at the same rate as any other object, independent of its mass (ignoring air friction). An object in orbit also is accelerating, because it constantly falls away from its natural straight path. Newton's laws say that, here too, the mass of the object does not matter. This means that, for example, all objects moving on the same path orbit the earth to-

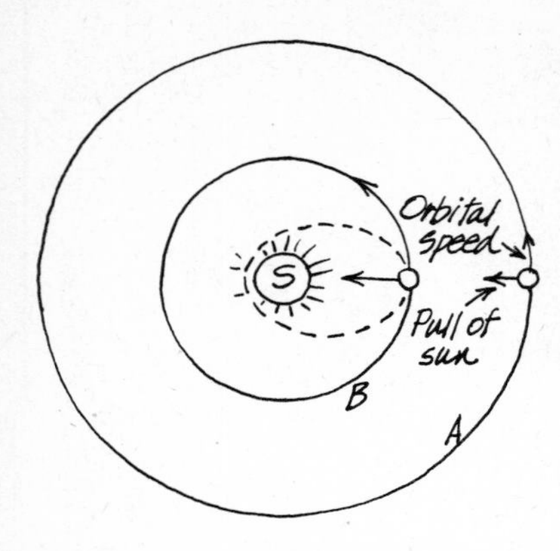

Figure 6.27 The speed of a planet in its orbit. Why does Mercury move more rapidly in its orbit than, say, earth? A represents the earth in its present position. It travels in a nearly circular orbit at a certain speed and feels a certain pull because of the sun's gravitation. What if the earth were placed in Mercury's orbit (B) but without the benefit of any additional orbital speed? Since it is nearer the sun, it would experience a stronger inward pull. Its speed would be insufficient to allow it to maintain a circular orbit. It would be pulled into a highly eccentric orbit (dotted ellipse). Mercury follows a nearly circular orbit (B) because of its higher speed which prevents the powerful pull of the sun from curving its orbit too drastically. Similarly, if the earth were placed in Jupiter's orbit but with the earth's present speed, the weaker pull of the sun would not be sufficient to retain the earth and it would escape from the solar system. When Newton put these ideas into mathematical form he was able to derive Kepler's third law, $P^2 = R^3$.

gether at the same speed. Figure 6.28 shows an astronaut during a space walk outside his spaceship. He orbits the earth at the same rate as his more massive ship. He need not fear that the ship will race on ahead of him.

Similarly, inside the ship, all objects move at the same rate as the ship. Objects seem to float. Gravity does not press objects to the floor of the ship. The astronaut floats freely inside the ship, feeling no sense of up or down. He constantly feels as if he were falling, a sensation that takes some getting used to. Several astronauts have experienced nausea and disorientation before they adjusted to this state called **weightlessness.** The term is a good description of the feeling but is also misleading. The earth still exerts a gravitational pull on the astronaut. He still has weight in that sense. But, since he falls with the rocket ship there is no feeling of the floor holding him up, he cannot *feel* his weight. A better term for weightlessness is **"freefall."**

6.10 NEWTON'S DYNAMICS: THE ULTIMATE TRIUMPH OF COPERNICUS

Newton's theory is used by all physicists throughout the world. It has been a total success wherever it has been applied (except at very high speeds or very high gravitation, where Einstein's theory of relativity must be used). The National Aeronautics and Space Administration (NASA) uses Newton's three laws of motion and his law of gravitation during all space shots to predict where the vehicle will be at each point in time. It is interesting that Newton provided the dynamic theory and the mathematics for space flight some 250 years before technological expertise (rocket construction, radio communication, electronic computers, etc.) made space flight possible.

For all of its success, Newton's dynamics did come under attack. Gottfried Wilhelm von Leibnitz (1646–1716), the well-known philosopher who had originated calculus independently of Newton, complained that *The Principia* was "built upon miracles, and . . . occult qualities." Leibnitz could not see how the sun, separated from the earth by millions of kilometers of empty space, could have any effect on the earth's motion. This was "action at a distance," and it was very hard to understand. Newton's response was, "I don't make hypotheses," as if to say, "The theory works, it explains motion, whether or not we can figure out the inner mechanism of gravitation." Newton need not have been upset by such criticism. The history of science shows that, each time a successful theory is proposed to explain something, it always leads to a search for an explanation of the explanation. We shall discuss an advance in the understanding of gravity in Section 17.4.

And Newton's theory was a triumphant success. Those who learned the theory solved one problem concerning motion after another. No other theory of dynamics was one-hundredth as useful.

The widespread acceptance and success of Newton's theory was also

Figure 6.28 Astronaut Edward White on a space walk.

a climax to the thread of astronomical history we have been following. In Chapter 2 we related some of the observations of apparent motions that had been recorded. In Chapter 3, we saw how Ptolemy explained these motions assuming an earth at rest and planets moving in epicycles as they traveled around the earth. Chapter 4 showed how Copernicus explained the same observations based on a rotating, revolving earth and planets that travel around the sun in circles. Chapter 5 told how Tycho observed the heavens, obtaining excellent data which Kepler then used to correct Copernicus' theory. Kepler showed that the

BOX 6.6

THE SPINNING EARTH: THE VIEWS FROM THE SEVENTEENTH AND THE TWENTIETH CENTURIES

In Section 4.5, we described a debate concerning the rotation of the earth as it might have been contested in the sixteenth century, just after the publication of Copernicus' book. This question was seen as a crucial confrontation between the geocentric theory and the heliocentric theory. We now discuss how that debate would have concluded just after Newton's work and then just after Einstein's.

In Section 4.5, one of the challenges the defender of the geocentric theory (earth at rest) put to the defender of the heliocentric theory (earth spins) was to explain why, if the earth rotates, it does not fly apart.

Newton himself gave a satisfying answer, one based on his theory of dynamics. He calculated what would happen to an earth that rotated once in 24 hr. He found that it would bulge at its equator and be flattened at its poles. When a sufficiently accurate survey of the earth had been completed, it was found that, in fact, the earth did bulge as Newton had predicted. Other tests were performed which, in the light of Newton's dynamics, revealed the rotation of the earth. Each test came out positive. Even school children came to know that the earth really did spin.

This was the situation until the early part of this century. At that time, Einstein became dissatisfied with one of the basic assumptions in Newton's theory and began a search for a new theory of dynamics. He was eventually successful; the result is known as the general theory of relativity. (See also Section 17.4.) In some ways, this theory is even more beautiful that Newton's, but it has not been accepted on that basis alone. Einstein proposed certain tests in which Newton's dynamics gave predictions measurably different from those of relativity. Because relativity triumphed in every case, it has come to replace Newton's ideas about the nature of the laws of physics.

One should hasten to add that, in most cases, Newton's and Einstein's theories give the same answers when applied to ordinary events. Since Newton's theory is mathematically simpler, it is used whenever it is known that no measurable error will result. Nevertheless, Einstein's dynamics is the theory that is more

---

planets follow elliptical orbits. At this stage, Copernicus' heliocentric theory became a success, able to predict planetary positions to a high degree of accuracy.

Still, the acceptance of Copernicus' idea in preference to Ptolemy's (or Tycho's) was slow. The clincher came from Newton. His theory of dynamics explained nearly all motion, and so it became firmly entrenched. But Newton's theory assumed the correctness of the heliocentric theory. It could explain planet orbits to a mathematical exactness only if the planets were orbiting the sun. No other dynamical theory existed that approached the success of Newton's. Thus it was Newton's work that forced eventual acceptance of the idea of heliocentrism as part of a unified, successful theory. (See also Box 6.6.)

SUMMING UP

Two frontiers were crossed in the seventeenth century. On the one hand, Galileo reinvented the refracting telescope and Newton invented satisfying both intellectually and aesthetically and is the only theory that gives the right answers in all situations.

The general theory of relativity has revolutionized many areas of thought, including the debate over the earth's rotation. As long as Newton's theory was dominant, one could hold that the earth must rotate. Now, general relativity has brought a new perspective to the debate. According to Einstein, all motion is relative. One cannot say that any object is definitely at rest. Before beginning a calculation of some motion, one may *choose* any object in the universe to be at rest and then use Einstein's theory to compute the motions of the other objects relative to it. Two people may choose different objects to be at rest and the calculations one performs may be more complex than the other's, but the answers they obtain will be identical as far as relative motion is concerned. In Einstein's theory, there is no test one can perform to determine whether an object is really at rest or really moving.

Up until the early 1900s, one could use Newton's theory to demonstrate the earth was not at rest. After that time, it became possible to appeal to general relativity and to think of the earth spinning or at rest as one wished. For example, if one chooses to assume that the earth is at rest, one may use general relativity to compute the resulting effects. In this case, the rest of the universe whirls around the earth. This produces new gravitational effects, effects undreamed of by Newton, which pull the earth's equator out into a bulge. Indeed, any of the tests formerly used to prove the rotation of the earth on the basis of Newton's theory are now seen to be inconclusive. The tests come out the same way whether one initially assumes that the earth rotates or not.

Thus, from Newton's perspective, Copernicus was right and Ptolemy wrong. If a debate between them were held *today*, however, a judge who knew of general relativity would rule the debate a draw. For example, Fred Hoyle writes, "Today we cannot say that the Copernican theory is 'right' and the Ptolemaic theory is 'wrong' in any meaningful sense. The two theories . . . are physically equivalent to each other." Max Born agrees: "Thus, from Einstein's point of view Ptolemy and Copernicus are equally right. What point of view is chosen is a matter of expediency."

the reflecting telescope, inventions that led to a large number of impressive discoveries.

On the other hand, the science of dynamics was greatly advanced by the work of these same two men. Building upon Galileo's innovations, Newton produced a theory of dynamics still widely used today. Newton's three laws of motion and his law of gravitation led to greatly increased understanding not only of astronomical motions but also of the motions encountered on earth. The widespread acceptance of Newton's dynamics produced an automatic victory for the heliocentric theory because the two theories go hand in hand.

The next chapter discusses light and another instrument, the spectroscope. With the use of the telescope and spectroscope, astronomers have learned most of what we know about the universe.

EXERCISES

1. Define the following terms: refraction, dispersion, chromatic aberration, focal point (or focus) of a lens, focal point (or focus) of a mirror, focal length, magnification, resolution.

2. Define the following terms: dynamics, inertia, mass, volume, weight, weightlessness, speed, acceleration.

3. Define the following terms: highlands of the moon, maria, terminator, Milky Way, satellite.

4. What were Galileo's innovations concerning (a) falling objects and (b) inertia? In each case, give evidence to support Galileo's innovation.

5. A stone on the end of a string is whirled in a circle over someone's head. Then the string breaks. Describe the subsequent path of the stone according to the law of inertia, that is, according to Newton's first law of motion. (Ignore the effects of the earth's gravity.)

6. How can a convex lens (one that bulges outward on both sides) be used as a "burning glass"? As a magnifying glass?

7. Describe the operation of a simple refracting telescope. Why should the diameter of the objective lens be large (give two reasons)? Why can the diameter of the eyepiece be small? Why should the focal length of the objective lens be long and that of the eyepiece short?

8. Suppose a telescope has an objective mirror with a diameter of 12 cm and a focal length of 160 cm. If it is used with an eyepiece having a focal length of 4 cm, what magnification will be obtained? Now, suppose a magnification of 80 times is desired. What should the focal length of the new eyepiece be? (Answers: 40 times, 2 cm)

9. Describe Galileo's discoveries when he viewed each of these objects in his telescope: the moon, the stars, the Milky Way, Jupiter, Venus, the sun, Saturn.

10. Draw a labeled sketch showing the parts and operation of a Newtonian reflecting telescope.

11. Give at least four reasons why all the large telescopes made in this century have been reflecting telescopes and not refracting telescopes.

12. State Newton's three laws of motion.

13. State Newton's law of universal gravitation. Use it and Newton's second law to explain Galileo's law of falling objects.

14. Explain the difference between these two methods of discovering a law of science: the empirical method and the deductive method.

15. Why did acceptance of Newton's dynamics lead to nearly universal acceptance of the heliocentric theory in the seventeenth century?

READINGS

Among the many sources of information on Galileo's and Newton's life work are

"Galileo," by I. B. Cohen, *Scientific American*, August 1949, p. 40.

Galileo, by Colin A. Ronan, Putnam, New York, 1974, a recent, wonderfully illustrated biography, and "Isaac Newton," by I. B. Cohen, *Scientific American*, December 1955, p. 73.

Dictionary of Scientific Biography, cited in Chapter 3 and useful throughout this book.

A wonderful book, one which helped hook me on astronomy and physics and which details in a highly readable way the early interactions between these two fields, is

The Birth of a New Physics, by I. Bernard Cohen, Doubleday (Anchor Books), New York, 1960.

Telescope owners will want to own

Amateur Astronomy, by Patrick Moore, Norton, New York, 1968, by the most famous amateur astronomer-author.

Exploring the Moon Through

Binoculars, by Ernest H. Cherrington, Jr., McGraw-Hill, New York, 1969, which will give a lifetime of pleasure.

Norton's Star Atlas and Reference Handbook, Sky Publishing Corporation, Cambridge, Mass., 1973, for its star charts.

Handypersons may want to build their own telescope and may consult

Amateur Telescope Making, edited by Albert G. Ingalls, Scientific American, New York, 1970.

Standard Handbook for Telescope Making, by N. E. Howard, Harper & Row, New York, 1959.

The story of telescopes of the past is told in

Eyes on the Universe, by Isaac Asimov, Houghton Mifflin, Boston, 1975, which also covers much of the astronomy discussed in this text.

The Giant Glass of Palomar, by David O. Woodbury, Dodd, Mead, New York, 1970, concentrates on the 200-inch reflector but also discusses earlier instruments.

Finally, travelers should plan to visit some of the large telescopes, guided by

U.S. Observatories: A Directory and Travel Guide, by H. T. Kirby-Smith, Van Nostrand, New York, 1976.

LIGHT AND THE SPECTROSCOPE: ACCOMPLISHING THE IMPOSSIBLE

Prophesying what scientists cannot do is a tricky business. In 1850 C. H. Davis stated that astronomers would never know much about the physical nature of planets. In 1930, Forest Ray Moulton, an outstanding astronomer, wrote that travel between the planets would forever be impossible. Chapter 10 of this book illustrates how unreliable these prophecies were.

Another interesting prediction was made by the French philosopher Auguste Comte. As an example of what is forever unknowable, he stated in 1835 that humans would never discover the chemical makeup of the stars. It is easy to understand why this seemed like a safe assertion. The telescope had revealed many details about objects in the solar system but little about the stars. Even using today's largest telescopes, the stars appear as minute points of light. The distances to the stars are vast. How could one ever know anything about their size, temperature, or chemical makeup? The stars seemed to represent the ultimate unknown.

In this chapter, after a brief discussion of the nature of light, we relate how a new instrument was invented, an instrument that brought the stars into the realm of the knowable.

7.1 LIGHT: PARTICLES OR WAVES?

And God said, Let there be light; and there was light.

Genesis 1:3

The ability to see is one of the most important features of human life, making possible countless avenues of knowledge, communication, and pleasure. Without light and the ability to see it, astronomy would be unthinkable. Humans would not even have known that stars and planets exist were it not for the light we receive from them. Small wonder, then, that we have always been fascinated by light and have inquired about its nature and properties.

One of the conclusions Newton came to during his work with light was that it consists of small particles. He imagined that a candle flame emits tiny bullets of light at an enormous speed. If the light particles enter the eye, they stimulate the eye and we see the candle. If the particles strike a piece of paper, they bounce off like billiard balls (they are reflected) and we see the paper. Newton was able to account for many of his experiments using his particle theory of light.

At about the same time, a rival theory of light was proposed by Huygens. He thought of light as a wave phenomenon. Imagine, as an analogy, the quiet surface of the water in a swimming pool. Suppose a duck is floating at one end and that you bob up and down at the other end. You will create waves which spread over the surface of the water. When the waves reach the duck, they will make it bob up and down. Your motion has been communicated to the duck by means of waves. Similarly, Huygens supposed that a candle sent out light waves in all directions. It is the action of the waves on the eye that allows us to see the candle. Huygens found that his theory could explain many of the phenomena of light. For example, water waves approaching the side of a swimming pool are reflected back into the pool. Huygens could account for the reflection of light in a similar manner.

Christiaan Huygens (hi'-genz, 1629-1695) was a Dutch mathematician, physicist, and astronomer. Among his many accomplishments, he first perceived Saturn's "ears" as a ring, discovered Saturn's largest satellite, made several contributions to dynamics which were incorporated by Newton, and invented one of the earliest pendulum clocks. Newton declared that, among living scientists, he regarded Huygens most highly.

Who was right? Both theories accounted for many of the known properties of light. At first it was fashionable to choose the particle theory, probably because of Newton's towering reputation. But gradually, more and more observations were made that could only be explained by the wave theory. For example, Newton deduced that when light enters a transparent medium it speeds up. The wave theory predicts that the light will slow down. When an experiment was eventually performed to test the predictions, it was found that light travels more slowly in a medium such as water or glass than it does in air. By about 1900 the wave theory was assumed by all to be the total victor.

Then, in the early twentieth century, Einstein interpreted certain other experiments by assuming that light is absorbed and emitted only in discrete packets. These packets of light are called **photons,** particles of light. When quantum mechanics, a new theory of dynamics concerning objects the size of atoms or smaller, was developed, it was based on the assumption that all subatomic particles, such as electrons, protons, and photons, exhibit *both* wave and particle properties. This may seem a strange state of affairs, but the theory works splendidly.

The study of light is another illustration of the way science progresses. A theory may seem whole and complete—proven—only to be turned upside-down by the next generation. In science, truth is a somewhat temporary thing, subject to modification and even revolution in times to come. This is one of the profoundly fascinating features of the pursuit called science.

7.2 LIGHT AS A WAVE

When a choice is available, a scientist uses whichever theory is most convenient and simplest for the purpose at hand. Surveyors mapping the earth and astronomers mapping the sky tend to use a geocentric approach, thinking of the earth at rest, surrounded by a rotating celes-

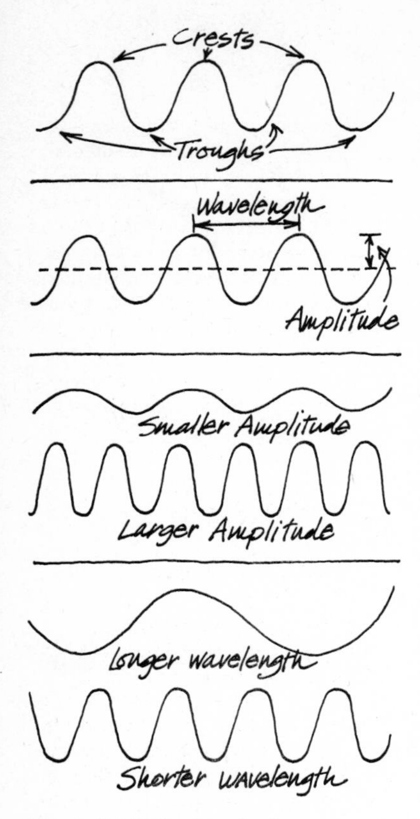

Figure 7.1 Wave terminology.

The spectrum displays an infinite number of colors, one shading into the next imperceptibly, yet it does not display all possible colors. Most colors, such as puce or mauve, are combinations of two or more pure colors. Only pure colors, those having a single wavelength, are found in the spectrum.

We use the convenient units known as angstroms (Å) to describe the wavelength of light. 10,000,000,000 Å = 1×10^{10} Å = 1 meter. The micrometer (μm) is also in use. 1,000,000 μm = 1×10^6 μm = 1 meter, and 1000 Å = 0.1 μm.

tial sphere. When predicting planetary positions, the heliocentric approach is much more convenient. Similarly, when thinking about light, one uses the most convenient theory, particles or waves, depending on the circumstances. We use the wave theory a great deal in this book. In this section we describe a few of the concepts involved in this theory.

As illustrated in Figure 7.1, a simple wave can be described by two quantities. The **amplitude** of the wave is one-half the distance between the top of a crest and the bottom of a trough. In the case of a light wave, the amplitude determines the brightness of the light; the larger the amplitude, the brighter the light.

The distance between the crests is called the **wavelength.** The interpretation of the wavelength of light comes from Huygens' wave theory. The theory predicts that light of different wavelengths will be refracted by different amounts in a prism. (See Figure 7.2.) Although light of any wavelength slows down upon entering glass, longer-wavelength light travels faster in the glass than shorter-wavelength light. As a result, longer-wavelength light is refracted less. Shorter-wavelength light is slowed down more and is refracted more. The natural conclusion is that the colors of the spectrum are different because their wavelengths differ. Red is the longest wavelength of light, and violet the shortest. Each of the other places in the spectrum is occupied by light of a wavelength that differs slightly from that of its neighbors.

Other conclusions may be drawn. In Figure 7.2, the light entering the prism is white light. No white is seen in the spectrum. Rather, white must be a combination of all the wavelengths in the spectrum. Similarly, black has no wavelength; it is the absence of light. Newton named seven *regions* of the spectrum. In order of *decreasing* wavelength they are red, orange, yellow, green, blue, indigo, and violet (ROY G. BIV is an easy way to remember the order).

After methods of measuring the wavelength of light were discovered, it was found that light has an unimaginably short wavelength. For example, the wavelength of violet light is about 4000 Å (0.4 μm). In 1 cm (0.39 inch) of a beam of violet light, there are 25,000 crests. The wavelengths of visible light extend up to red light having a wavelength of about 7000 Å (0.7 μm). (See Figure 7.2.)

7.3 LIGHT'S COUSINS: THE ELECTROMAGNETIC SPECTRUM

The eye sees the spectrum gradually fade out beyond the red end of the spectrum. In 1800, William Herschel, of whom we will be hearing much more in chapters to come, was experimenting with sunlight that had passed through a prism. He found that a thermometer placed in the spectrum absorbed light and converted it to heat. It was surprising to find that, if the thermometer were placed beyond the red end of the spectrum, where the eye could see nothing, the thermometer still regis-

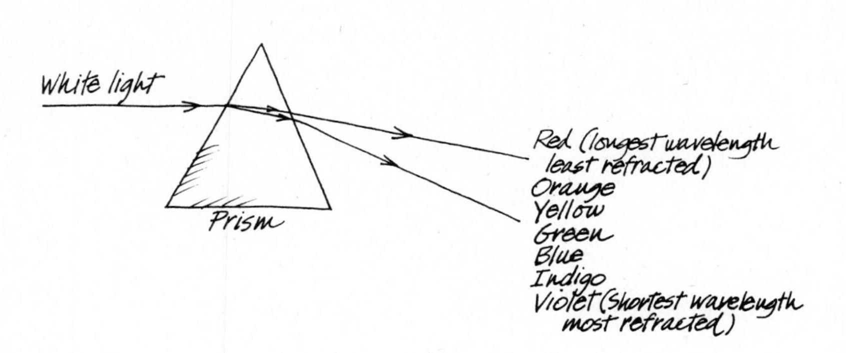

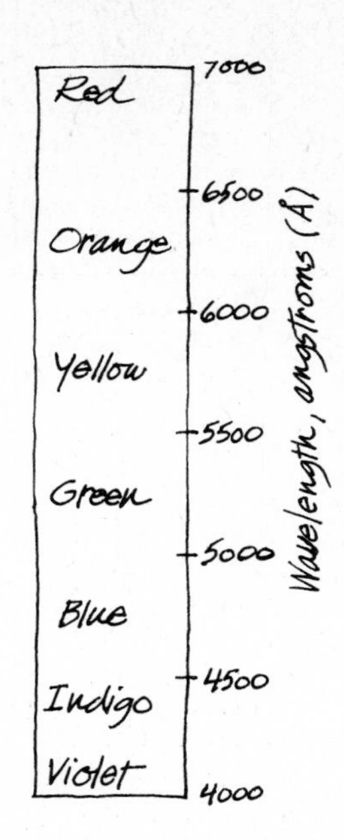

Figure 7.2 The visible spectrum. When white light is passed through a spectroscope, it is dispersed into the colors of the rainbow, technically called a spectrum. Each position in the spectrum is occupied by a different wavelength. There are an infinite number of colors in the spectrum, each of a slightly different shade. The names of seven regions of the spectrum, selected by Newton, are labeled.

tered. Herschel's interpretation was that sunlight contains light having a wavelength longer than that of visible light. Since this radiation lies beyond the red end of the spectrum, it is called **infrared radiation.** The eye cannot see it, but many ways have been found to detect infrared radiation. One good detector is infrared-sensitive photographic film. Such devices show that all objects emit infrared radiation, warmer objects emitting more than cooler ones.

One year after Herschel's discovery, it was found that visible light made the white chemical silver chloride turn black. When the chemical was placed in the region beyond the violet end of the visible spectrum, darkening occurred there too. The radiation responsible for this result has a wavelength shorter than violet light and is called **ultraviolet radiation.** Ultraviolet radiation is sometimes called "black light." Black light displays are produced by shining ultraviolet radiation on certain paints that **fluoresce,** that is, convert ultraviolet radiation into visible light. It has been discovered that it is the ultraviolet radiation from the sun that causes sunburn.

Relatively inexpensive bulbs that emit ultraviolet radiation are available. The reader might enjoy discovering which household objects fluoresce.

During this century, advancing technology has provided ways of detecting radiation of ever longer and ever shorter wavelengths. Radiation having wavelengths of about 1 cm is called **microwaves** and has been put to use in microwave ovens. Radiation having still longer wavelengths is called **radio waves** and is used in the field of communications.

Radiation having wavelengths shorter than that of ultraviolet radiation has been discovered also. In the wavelength range of roughly 10 to 1 Å, the radiation is called *x-rays.* X-rays penetrate matter quite well and are a familiar medical tool. Radiation having an even shorter wavelength is known as **gamma radiation.** Gamma rays are found to be

Figure 7.3 The electromagnetic spectrum. Various regions of the spectrum have been named because the radiation in that region interacts with matter in a different way. The boundaries shown are only conventional. There is no distinct difference between short-wavelength ultraviolet and long-wavelength x-rays, for example.

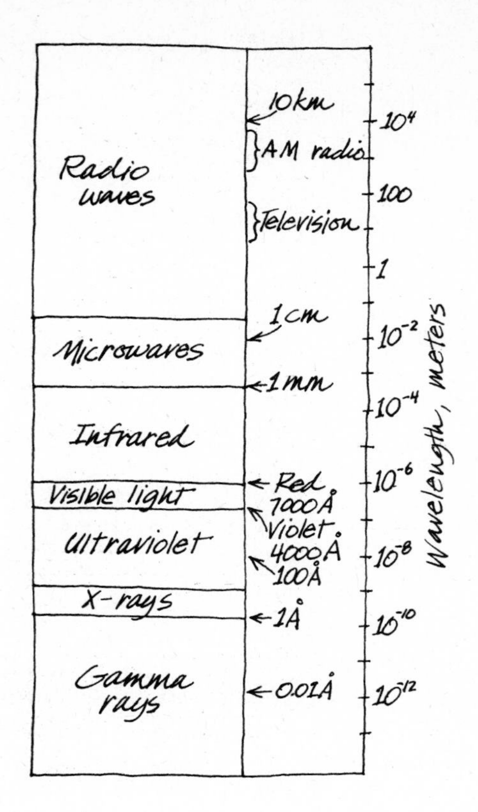

emitted during the breakup of nuclei; such breaking up is known as **radioactivity.**

One of the great accomplishments of physics is the recognition that each of these kinds of radiation, gamma rays, x-rays, ultraviolet, visible light, infrared, microwaves, and radio waves are all the same phenomenon. The only essential difference between them is in their wavelength. All such radiation has been given the name **electromagnetic radiation.** Figure 7.3 shows the entire electromagnetic spectrum, of which the visible spectrum is only a tiny part.

Every known type of radiation has been found coming from a variety of sources in outer space. Some of the astronomical discoveries made using these waves, short and long, are described in the chapters to come.

7.4 THE SPEED OF LIGHT, *c*

Light travels much faster than sound. A flash of lightning often precedes a crack of thunder by a number of seconds. How fast does light travel? It travels too rapidly to be measured easily. Perhaps light travels at an infinite speed?

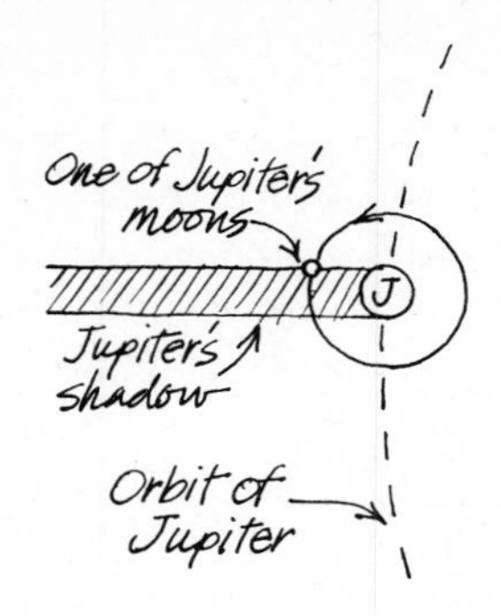

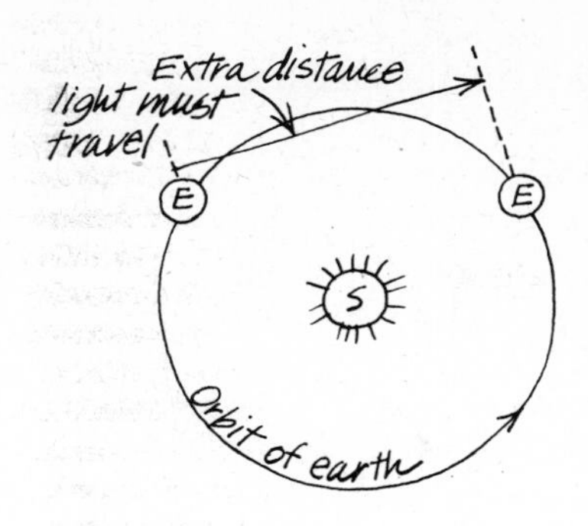

Figure 7.4 The first estimate of the speed of light. When Roemer observed eclipses of the moons of Jupiter, he concluded that he was seeing the eclipses late because the reflected sunlight from the moons of Jupiter had to travel longer to reach the earth when the earth was farther away.

In order to make a measurement of the speed of light, large distances were needed so that a time lag, if any, would be noticeable. A Danish astronomer named Olaus Roemer found that the enormous distances in the solar system provided such a test. Roemer was compiling data on the times when each of the moons of Jupiter were eclipsed by Jupiter's huge shadow. In effect, each moon was like a clock, repeating the same motion over and over. But, Roemer found, the clock did not seem very regular. When the earth was near Jupiter, the clock ran at a certain rate. But the farther the earth got from Jupiter, the more the clock lost time. When Jupiter was near conjunction, the eclipses were more than 15 min late. Roemer deduced what was happening. The eclipses seemed to occur late when the earth was far from Jupiter, because it took the light extra time to travel across a portion of the earth's orbit. See Figure 7.4. Using the size of the earth's orbit as it was then reckoned (a figure somewhat too small), Roemer calculated in 1676 that the speed of light is about 230,000 km/sec.

Ever more accurate determinations of the speed of light have been made since. The most accurate come from ingenious laboratory experiments. The modern value is close to 299,793 km/sec. Most scientists remember a handy rounded-off value for this important number:

$$c = 300{,}000 \text{ km/sec } (186{,}000 \text{ miles/sec})$$

(c is the customary symbol used for the speed of light.) Strictly speaking, this is the speed of light in a vacuum. When traveling through matter, light is slowed down. It is slowed down very little by air but more by glass and water. It has been found that c is the speed in a vacuum of *any* electromagnetic radiation, whether x-rays, visible light, or radio waves.

Light travels fast; by comparison, almost any other known speed is nearly negligible. A plane traveling at the speed of sound (331 meters per second = 1090 ft/sec) would take about 10.5 days to go as far as light travels in 1 sec. A beam of light (bounced off a series of mirrors) could go around the earth's equator 7.5 times in 1 sec. The fastest

BOX 7.1

THE UNIVERSAL SPEED LIMIT, c

When Einstein was a teenager he was already thinking about light. Light is peculiar in that it disappears when it is stopped. When light is captured, the energy in the light is always converted to some other form, often into heat. You can't hold a pool of light in the palm of your hand. But what if you were in a rocket ship moving at a speed of $c = 300,000$ km/sec? Couldn't you merely look out the window and, if a beam of light were traveling beside the ship, see it at rest relative to you? Einstein's intuition told him that something was wrong with this thought experiment. But what?

Einstein pursued this and other thoughts about light and was led to a new theory of dynamics called the theory of relativity. According to relativity, the error in the above thought experiment was the assumption that the rocket ship was moving at the speed of light. One of the many surprising conclusions drawn from the theory of relativity is that no object can go as fast as the speed of light. A very expensive rocket engine operating for a very long time might reach a speed of 99.999... percent of the speed of light, but that is the limit.

This conclusion, like so many in the theory of relativity, seems to contradict common sense. Yet, it has been experimentally verified. Particle accelerators have been used to verify Einstein's claim. One type of particle accelerator consists of a hollow, evacuated ring. An electron is placed in the ring and given a magnetic impulse. It is guided around the ring and given repeated impulses to make it travel in a circle at an ever-increasing speed. But no matter how strong the impulses or how long the experiment runs, the electron will not break the speed barrier c.

This creates a problem for humans if they ever wish to travel among the stars. Suppose light takes 50 years to go from our sun to a certain star. The rocket takes off from earth heading for the star. The people on earth know that they will not see the rocket again for an absolute minimum of 100 years. This limitation has led to some interesting science fiction stories. Are the astronauts placed in suspended animation for the trip? Does a whole colony travel in the ship, knowing that only their grandchildren will reach the destination? Does someone invent a "hyperwarp drive" that goes faster than c? Perhaps humans will never travel to the distant stars. But, on the other hand, prophesying what humans cannot do is a tricky business.

Table 7.1 The Time Required for Light to Travel From the Sun to Each Planet

Planet	*Average distance from sun (AU)*	*Light travel time*
Mercury	0.39	3.2 min
Venus	0.72	6.0 min
Earth	1.00	8.3 min
Mars	1.52	12.6 min
Jupiter	5.20	43.2 min
Saturn	9.54	1.3 hr
Uranus	19.20	2.7 hr
Neptune	30.10	4.2 hr
Pluto	39.40	5.5 hr

experimental jet planes travel at only about 2 km/sec. When the *Pioneer 11* spacecraft flew past Jupiter, it was moving at a record breaking 48 km/sec, but this was still only 0.01 percent of the speed of light.

If any government ever approves enough money (the technology already exists) to send humans to Mars, the time delay in a radio conversation with earth will become a source of annoyance. A radio wave will take between 9 and 42 min to make a round trip depending on the distance between earth and Mars at the time. (The round-trip time for a radio wave to Pluto is about 11 hr.)

Table 7.1 lists the time required for sunlight to reach each planet. We see the sun's surface as it was 8.3 min ago. If Saturn were to explode, we would not know it until about an hour later. When we look at astronomical objects, we are, in effect, looking back in time. (See Box 7.1.)

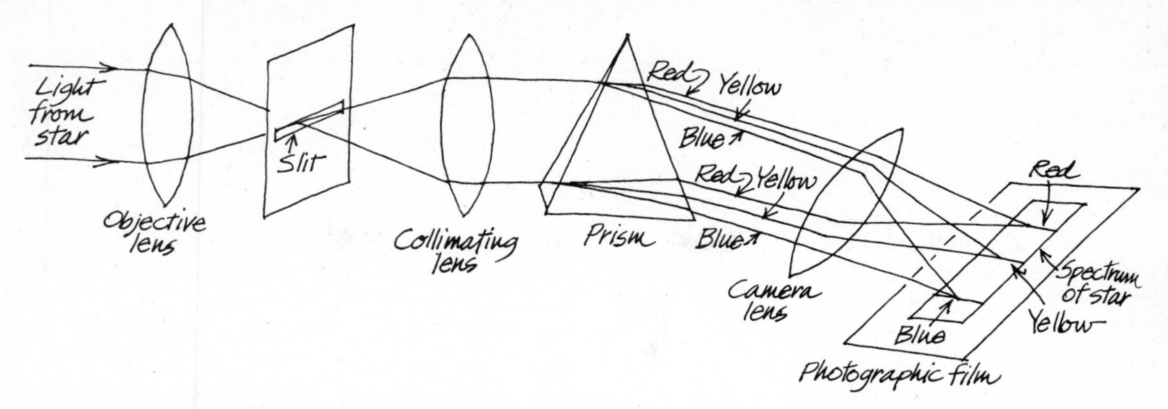

Figure 7.5 Principal components of a prism spectroscope. The objective lens first collects the light from a faint object. Next, the slit forms the light into a narrow beam. The second lens causes a parallel beam which is dispersed by the prism. The third lens focuses each separate color into its proper place in the spectrum. As an example, three rays, each of a different wavelength, are traced out to show how the third lens guides each to its proper place.

7.5 THE MIRACULOUS SPECTROSCOPE

Telescopes, as effective as they are, are unable to bring the stars close enough for a good, detailed look. Will we never know what the stars are? Are they similar to the sun? How big and how hot are they? What are they made of? These intriguing questions, which have been asked for thousands of years, have finally been answered by another invention, the spectroscope, which has enabled astronomers to see stars in a new way.

The key to the working of the spectroscope is **dispersion,** the breaking up of light into its separate wavelengths. The ability of a prism to disperse light, formerly a nuisance to makers of refracting telescopes, turned out to be a priceless gift to astronomy.

Look again at Figure 7.2. Suppose that the light beam is coming from a star. Before the light reaches the prism, it is a jumble of wavelengths mixed together. But, after being dispersed, the resulting spectrum is an orderly array. Light of each particular wavelength has been separated from its neighbors. The differing wavelengths have been placed side by side in order of wavelength. The shorter the wavelength, the further down the spectrum it appears. The light beam has been analyzed, that is, resolved into its constituent parts.

A sketch of the basic parts of a prism spectroscope is shown in Figure 7.5. The objective lens collects light from a star, making the star seem brighter. The slit forms the beam into a shaft of light. The collimating lens forms a beam of parallel rays. The prism then disperses the light which in turn is focused onto a photographic film by a third lens. The reason for the slit is explained in Figure 7.6. Each separate wavelength of the entering light appears as a colored image of the slit.

Devices known as gratings are also used to produce dispersion. They have largely replaced prisms in modern, professional-quality spectroscopes. Gratings are discussed in many physics texts.

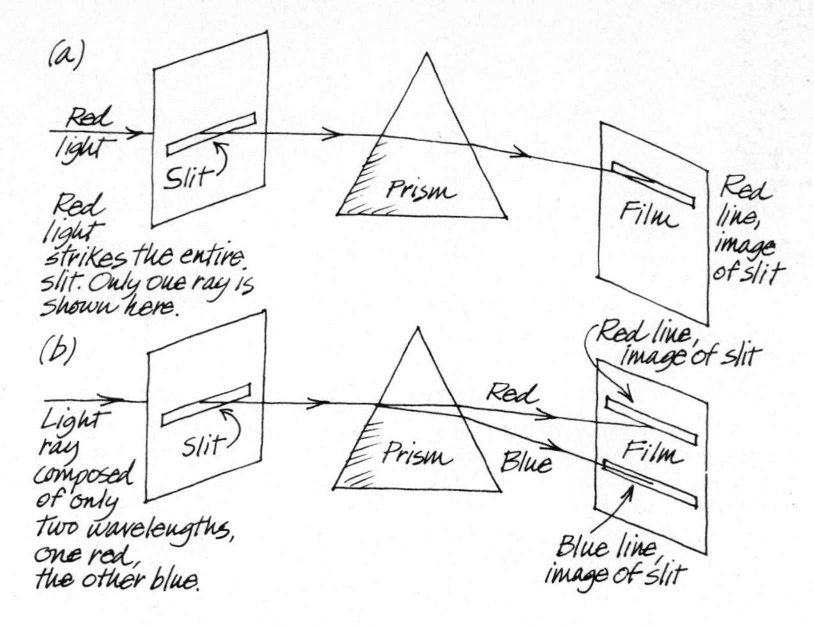

Figure 7.6 The spectroscope slit. The slit forms the incoming light into a narrow beam. (The lenses have been omitted here for simplicity.) After dispersion by the prism, each wavelength of light results in a narrow beam which is an image of the slit at that wavelength. The result of sending in light of only one wavelength is shown in (a). The result is a spectrum that is dark except at the position corresponding to that wavelength. A red image of the slit, a *line*, as it is called, occurs at the appropriate place. How two incoming wavelengths cause two separate lines is shown in (b).

Once the spectrum of the star is photographed (see Figure 7.7), a great amount of detail is apparent, detail that was not apparent in the original, scrambled beam of light. One notices the usual colors stretching from red to violet. Not all colors are equally bright, because the star broadcasts more powerfully at some wavelengths than at others. Dark lines are also noticeable at certain wavelengths where the star is not broadcasting any significant radiation.

The spectra of stars can be bewilderingly complex, and they vary from star to star. Once spectroscopes were invented, the next task was to learn to decipher the information provided. This difficult research is still going on, but much has been learned. The following chapters will relate how a wealth of data about stars and other objects has been obtained using the spectroscope. Let's briefly mention a few of the discoveries. The wavelength of the brightest color in a star's spectrum reveals a star's surface temperature. (See Section 9.4.) The kinds of atoms the star is made of, its chemical composition, are revealed by a study of the dark lines in the star's spectrum. (See Section 9.5 and Plate 2.)

7.6 THE DOPPLER EFFECT

The effect is named for Christian Doppler (1803–1853), an Austrian physicist.

There is another way in which the spectroscope provides valuable information. Called the Doppler effect, it is a feature of any type of wave phenomenon. Take an example involving sound waves. Suppose you are driving on a two-lane highway. An approaching driver wishes

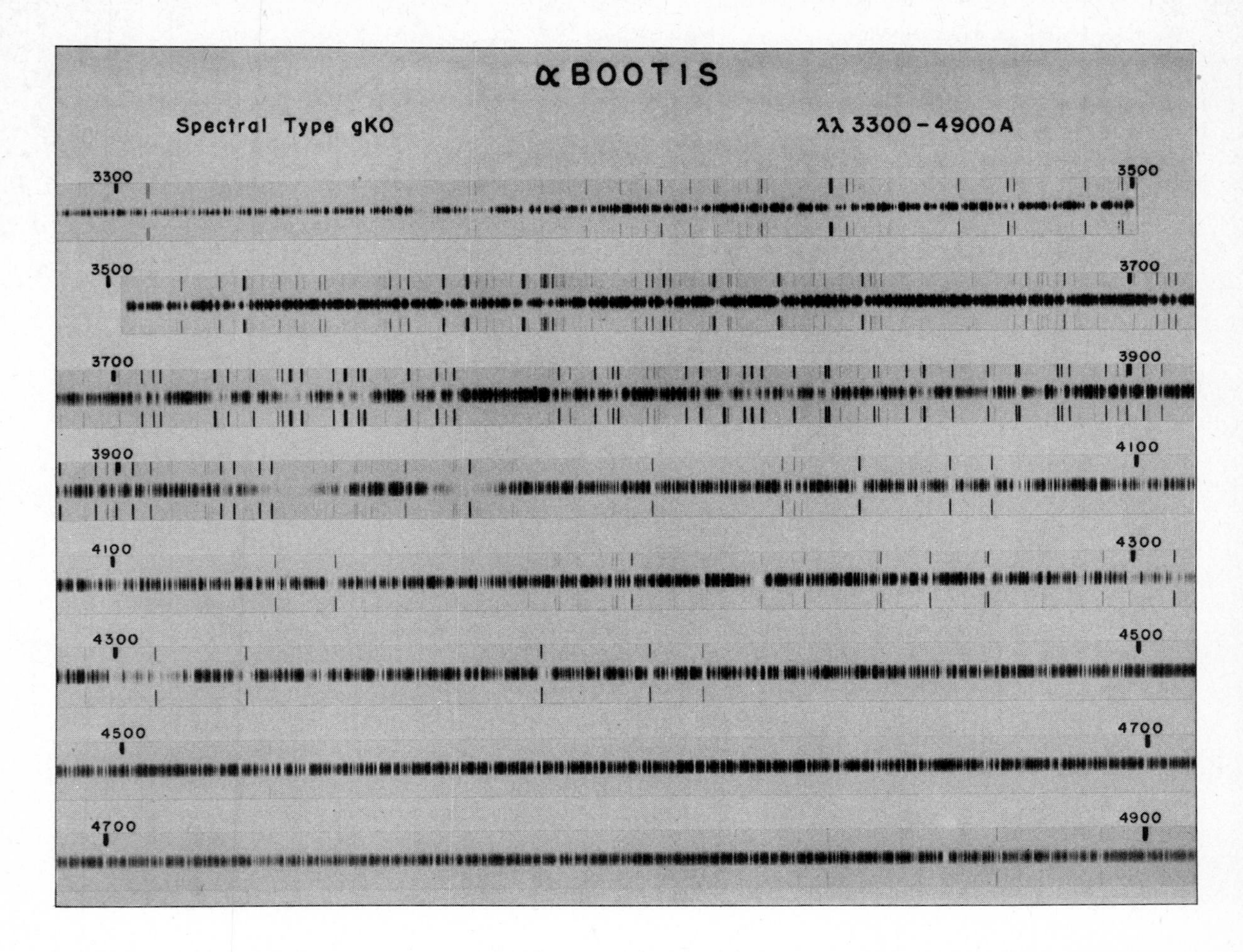

Figure 7.7 Portion of the spectrum of the star Alpha Bootis (Arcturus). The spectrum has been cut into eight pieces. The star's spectrum may be seen running through the center of each strip. Above and below the star's spectrum are reference lines, used for making measurements. The wavelength near the end of each strip is marked in angstroms. A color photograph of this portion of the spectrum would be violet at the upper left and shade to blue-green at the lower right.

to attract your attention and blows her horn as she passes. You will notice that, as the car passes, the pitch of the horn drops from high to low. As the car is approaching, you hear a higher pitch. As the car is receding, you hear a lower pitch. If you and the car are standing still, you hear a pitch intermediate between the two.

The Doppler effect occurs whenever a source of waves and an observer are moving toward or away from each other. Suppose a source of waves is emitting a certain wavelength. If the source of waves and the observer are approaching each other (either or both may be moving), the observer perceives a shorter wavelength. Or, if the source and the observer are receding from each other, the observer perceives a longer

The *pitch* of a musical note indicates how high the note is. Sopranos sing high notes; basses sing low notes. A high-pitched note has a short wavelength, and a low-pitched note a long wavelength. The concept of pitch should not be confused with loudness. A high-pitched note can be sung loudly or softly.

Figure 7.8 Definition of radial velocity and tangential velocity. Radial velocity (a and b) is the speed of an object either directly toward or directly away from an observer. (It is the speed along a *radius* drawn from the observer to the object, hence the name.) Sideways speed is called **tangential velocity** (c), motion at right angles to the observer's line of sight. Most motions are composed of both radial and tangential parts as illustrated in (d). The Doppler effect reveals only the radial part of any motion.

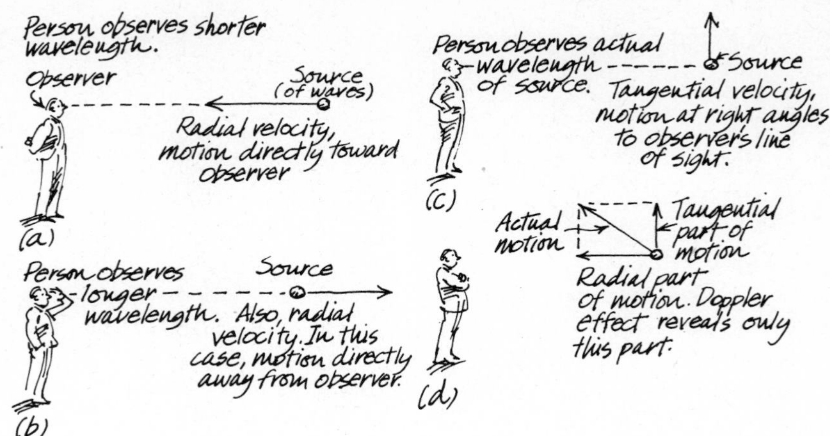

wavelength. The effect is appreciable only if the source and the observer are moving toward or away from each other at a speed that is appreciable compared to the speed of the waves.

How does the Doppler effect work? Here is an illustration using water waves. Imagine that you are in a motor boat which is anchored and that waves are moving past you at a rate of, say, two crests per second. Next, you pull up the anchor and start the motor. If you steer the boat so that it travels into the waves, you will meet crests more frequently than before, perhaps at a rate of three crests per second. Because the crests are hitting you at a higher rate, you can say that, in effect, the wavelength has decreased; that is, the time between crests has decreased, giving the same effect that would occur if the boat were at rest and the length of the waves had decreased. Suppose you now turn the boat around and travel in the same direction as, but slower than, the waves. You now find that crests move past you at a slower rate than when you were at rest, perhaps at a rate of one crest per second. The time between crests has lengthened, giving the same effect that would occur if the boat were at rest and the length of the waves had increased.

The effect is similar if you, the observer, are at rest in the water and the source of the waves moves. If the source moves toward you, it chases, as it were, a wave crest after producing it and, as a result, produces the next crest closer to the previous one than if the source were at rest. Conversely, if the source moves away from you, the crests are produced farther apart, resulting in a longer wavelength.

Doppler proposed that the same effect should be possible for light. Suppose a yellow light bulb were somehow caused to move toward an observer at a very high speed, at an appreciable fraction of the speed of light. The waves of the light would be crowded together, so to speak, by the motion, and the observer would see a shorter-wavelength light, perhaps blue. If the bulb were rushing away at the same speed, the bulb might appear red because the waves would be stretched apart by

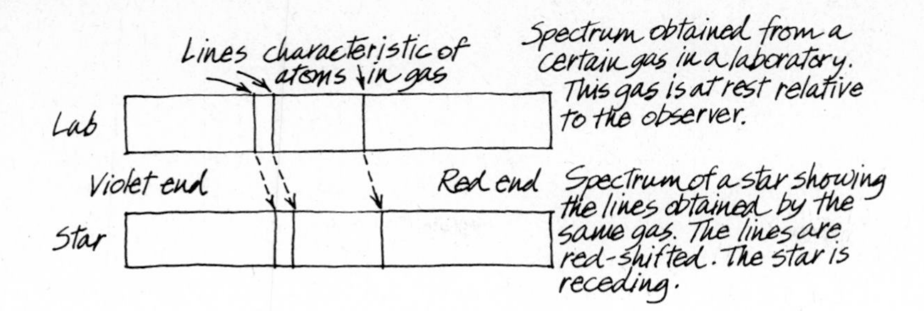

Figure 7.9 Using the Doppler effect to determine a star's radial velocity.

the motion. For this reason, a shift in wavelength toward a shorter wavelength is commonly called a **blue shift.** A shift toward a longer wavelength is called a **red shift.**

Now, how is the Doppler effect used to determine the **radial velocity** (the speed of approach or recession; see Figure 7.8) of a star? Each kind of atom in the atmosphere of a star causes a particular pattern of dark lines in the star's spectrum. These dark lines are used as markers to detect the Doppler effect caused by the radial motion of the star. Suppose an astronomer wishes to measure the radial velocity of a particular star. He or she photographs the star's spectrum and examines the dark lines, choosing any particularly clear line. Perhaps the astronomer recognizes, by knowing the pattern of the lines, that the chosen line is caused by hydrogen atoms. One finds by the placement of the line in the spectrum that the line lies at a wavelength of, say, 4345 Å. One next consults a physics book to obtain data on hydrogen spectra produced in a laboratory. In the laboratory, spectral lines have been measured that were produced by hydrogen at rest. Let's say that the chosen hydrogen line occurs at a wavelength of 4340 Å in the laboratory; we see that the wavelength of the line in the star's spectrum is longer. The astronomer may immediately conclude that the star is receding from us. It is this motion that caused the red shift of the line, a shift toward a longer wavelength. Then, using a mathematical formula, one can calculate the star's radial velocity. The larger the shift, the larger the radial velocity. The astronomer will probably repeat the process for other dark lines as a check of the results. (See Figure 7.9.) To obtain an accurate result, one must allow for the motion of the earth in its orbit around the sun, which also causes a Doppler shift of the lines.

This experiment is not practicable. A yellow light bulb would have to travel toward an observer at about 10 percent of the speed of light in order to appear green. This is a speed of 30,000 km/sec (18,600 mi/sec). At that speed, the bulb would be able to travel from New York to Los Angeles in about $\frac{1}{7}$ sec.

Doppler hypothesized that star colors could be explained using the Doppler effect. Might not red stars appear red because they are receding rapidly from us? Are all blue stars rapidly rushing toward us? One way to check is to look at the dark lines in the spectra of red stars. Unfortunately for Doppler's hypothesis, the dark lines of all red stars show only a very small shift which indicates a speed much too slow, compared to the speed of light, to cause a color change. Furthermore, some red stars show lines that are slightly blue-shifted. No stars of any color have been found that are moving rapidly enough to affect their color noticeably. The colors of stars must be caused by some other factor. (See Section 9.4.)

SUMMING UP

The many types of electromagnetic radiation may be listed in order, beginning with those of shortest wavelength: gamma rays, x-rays, ultraviolet radiation, visible light, infrared radiation, microwaves, and radio waves. All electromagnetic radiation travels at the same speed in a vacuum, about 300,000 km/sec.

The spectroscope has proven invaluable to astronomers in their research, as the following chapters will show. It uses the phenomenon known as dispersion to analyze the radiation from an object. One of the important applications of the spectroscope is to determine an object's radial velocity by means of the Doppler effect.

The next chapter begins a description of our principal source of light, the sun.

EXERCISES

1. Define the following terms: amplitude, wavelength, spectrum, fluorescence, line (in a spectrum), radial velocity, tangential velocity.

2. Describe briefly the changes in scientific opinion over the ages concerning the nature of light.

3. Name the seven regions of color in the spectrum of visible light as chosen by Newton in order of *decreasing* wavelength. Give the approximate wavelengths of red and violet light in angstroms.

4. Name the seven general types of radiation in the electromagnetic spectum in order of *increasing* wavelength. Give a fact about each type which characterizes it.

5. What is the approximate speed of light expressed in kilometers per second? How many times can light travel around the earth in 1 sec? How long does light from the sun take to reach the earth? In a movie, people supposedly on Mars talked to their spouses on earth and got instantaneous replies. What would be the *minimum* time necessary for an answer? (Use data from Table 7.1. Answer: round-trip time, about 8.6 min.)

6. How does a spectroscope work? What does it do to the incoming wavelengths?

7. (a) Describe the Doppler effect for light. (b) What is meant by each of these terms and what do they imply about the motion of the light source? (i) Red shift (ii) Blue shift. (c) How do we know that red stars do not look red because of a large speed of recession?

8. A motorist appeared before a judge. The charge was running a red light. The motorist pleaded, "I am innocent. Due to the Doppler effect, my motion toward the red light made it look green!" The judge replied, "Fine, we'll drop the charge of running the light. *However*, the speed limit on that street is 90 km/hr (about 55 mi/hr) and, since you saw the red light as green, you must have been. . . ." Supply the punch line for this joke. No numerical answer is required.

9. List the steps involved in determining the radial velocity of a star.

READINGS

The following two references are valuable for this chapter as well as all following chapters. They are highly recommended sources of information at a somewhat higher level than this text.

Principles of Astronomy, by Stanley P. Wyatt, Allyn and Bacon, Boston, 1977.

University Astronomy, by Jay M. Pasachoff and Marc L. Kutner, Saunders, Philadelphia, 1978.

Also valuable for its discussion of spectroscopy is

Atoms and Astronomy, NASA, Washington, D.C., 1976, NAS 1.19:128.

Many NASA publications are quite instructive and relatively inexpensive. Write for a publication listing to: NASA, Washington D.C. 20546.

THE SUN, THE DAZZLING MYSTERY: A PRELIMINARY LOOK

The sun does much more than supply us with illumination. Its gravity holds the earth in orbit, and its light, when absorbed by the earth, changes to heat. This heat keeps the oceans from freezing and the air from condensing into a liquid. Unequal heating of the earth produces thunderstorms, tornadoes, hurricanes, and other weather systems. And more and more solar influences on the earth are being discovered as research continues.

The sun also has a strong effect on the earth's biosphere, the thin region of living matter on the surface of our planet. Plants require sunlight, and many other forms of life, in turn, require plants for food. The sun regulates the activity of living things; we wake, sleep, and base our time-keeping system on the motion of the sun. In late winter it is the lengthening of the day that signals tree sap to flow and birds to migrate, and, in general, urges the biosphere to reactivate.

Further effects of the sun on the earth are discussed in Sections 9.6 and 9.7.

We begin our study of the sun with a description of some of the observations made by early astronomers. We first describe two sun-related activities which early astronomers pursued and which amateur astronomers still enjoy: observing the sun with binoculars or a telescope, and watching a total eclipse of the sun. We conclude this chapter by describing how early astronomers measured the distance to and mass of the sun. Chapter 9 is concerned with more recent solar research.

8.1 THE SUN IN AN AMATEUR'S TELESCOPE

It is natural to think of getting out your telescope only after sunset, but the instrument can be put to splendid use on any clear day.

We now discuss the view of the sun obtainable using optical instruments. You will need a telescope or mounted binoculars. With such an instrument you can observe many fascinating solar phenomena. Before you begin, though, please read Box 8.1 so that you will not endanger your vision.

"Faculae" is the plural of "facula" and is pronounced "fack'-yuh-lee." It means "small torches" in Latin.

Faculae or Plages Let us assume that an instrument has been properly set up to observe the sun following the instructions and cautions of Box 8.1. You may wish to start observing by searching the rim of the sun's disk for **faculae** or **plages.** These are bright patches on the sun. Because they are only slightly brighter than the rest of the surface of the

BOX 8.1

VIEWING THE SUN SAFELY

Neither the Sun nor Death can be looked at with a steady eye.

François Duc de La Rochefoucauld

It is *extremely* dangerous to look at the sun! The sensitive retina of the eye is easily damaged by direct sunlight. The damage is insidious in that the retina has no pain sensors and the injury may not become evident for many years afterward. To observe the sun properly, you must have proper instruction. In 1963 a partial eclipse of the sun was visible from many locations in the United States. Over 240 cases of eye damage were recorded after that event. Do not ever look directly at the sun when it is high in the sky, even for an instant.

Never use binoculars or a telescope to view the sun except by using the **method of projection,** which is relatively safe. This method requires a telescope or binoculars. Mount the instrument on a sturdy support; a tripod is often used for this purpose. When binoculars are employed, a binoculars clamp is used to attach the instrument to the tripod. For casual observations, the binoculars may be hand-held.

Without looking through the instrument, aim it at the sun by minimizing the instrument's shadow. Then hold a screen (white paper mounted on cardboard works fine) so that the light falls upon it. Arrange things so that a shadow falls on the rest of the card. Focus the image by moving the paper and/or adjusting the instrument's focus. The solar disk will be displayed for all to see. An advantage of this arrangement is that several observers can view simultaneously.

Cautionary Note: I do not recommend that beginners use solar filters for direct viewing. They can be dangerous if used incorrectly. Also, cap your finder telescope, if you have one, to prevent anyone from looking through it and to protect the cross-hairs from melting. Let the sun drift out of view from time to time to allow the eyepiece to cool off. Never leave the instrument unattended and make sure that bystanders do not endanger themselves. Above all, *do not look through the instrument at the sun.* Instantaneous blindness can result.

---

sun, they are best seen when near the sun's rim where the surface does not appear to be as bright (see Figure 8.1.) A facula precedes a sunspot by a few days and usually last days longer than the spot. An expensive instrument known as a spectrohelioscope can reveal faculae more easily than can a telescope.

Sunspots When they are present and large enough, sunspots (Figure 8.1) easily catch your eye when you are observing with a telescope. They come in many sizes, individually, in groups, and in greater or lesser numbers. It is fascinating to watch them from day to day and from year to year.

The discovery of sunspots allowed the first determination of the sense and period of rotation of the sun. The spots on the sun's surface are carried along as the sun rotates. It was revealed that the sun rotates in the same direction (or sense) in which the nine planets revolve around the sun. See Figure 8.2. Furthermore, the motions of the spots reveal that the sun's axis is tilted only 7° relative to the earth's orbit—more properly, it is 7° away from being perpendicular to the plane of the ecliptic.

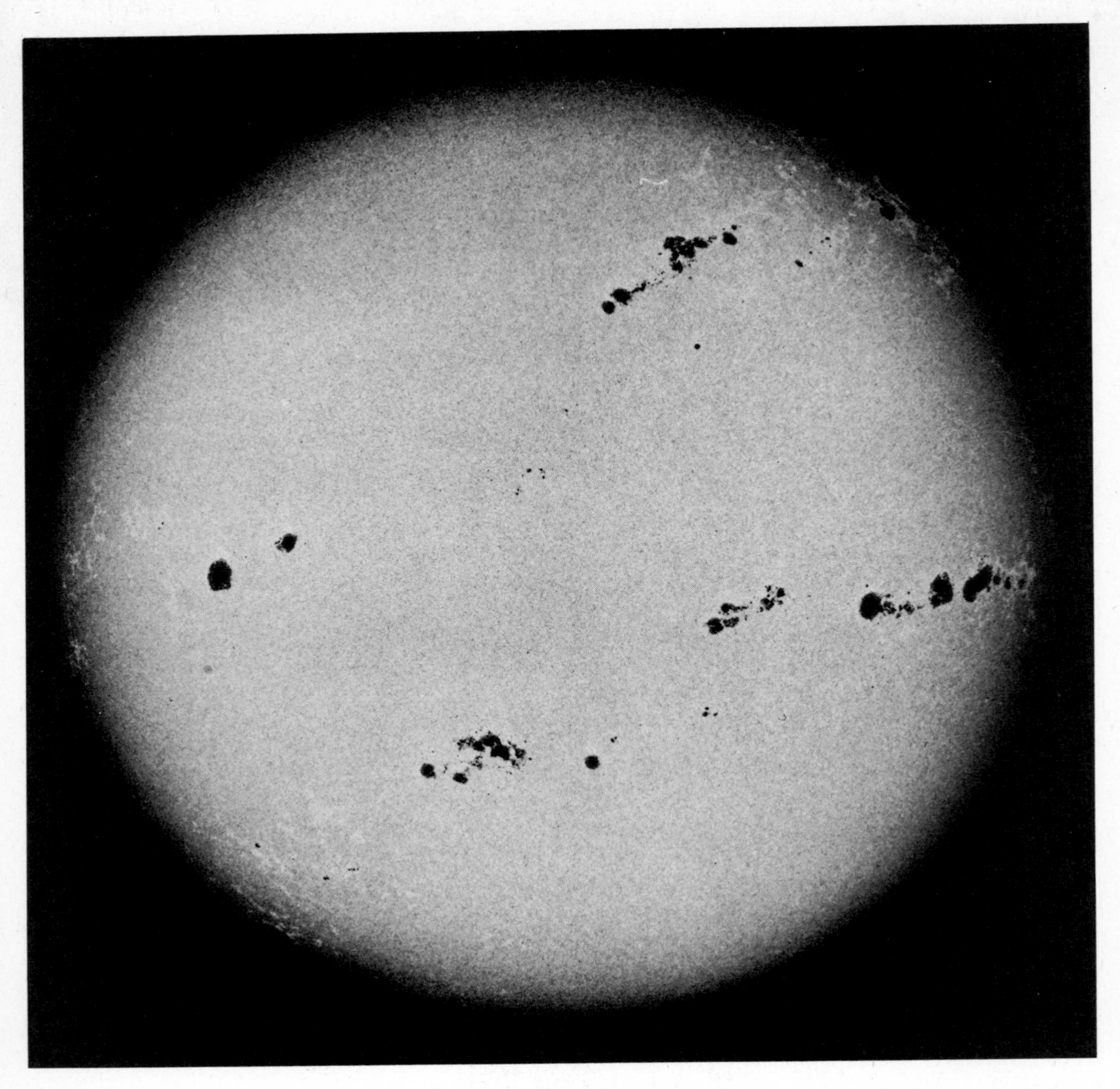

Figure 8.1 A very active sun. This photograph was taken on December 21, 1957, during cycle 19 (explained in a following subsection) of the solar cycle. Cycle 19 broke the record cycle 18 had set for the number of sunspots recorded. Besides sunspots, faculae (white regions) may be seen near the rim of the sun's disk.

When a large spot lasts sufficiently long for the sun to make more than one complete rotation, you can determine the rotational period of the sun. It is remarkable that spots make one rotation in differing times depending on their distance from the sun's equator. See Figure 8.3. Regions near the sun's equator make one rotation in about 25 days. At

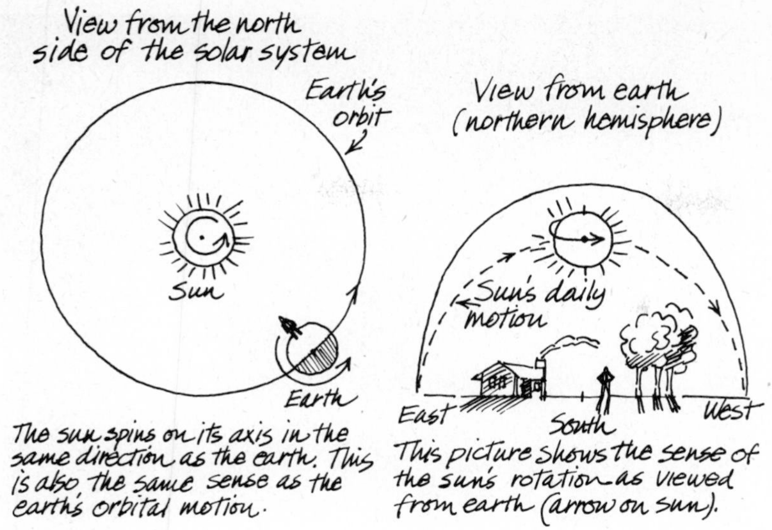

View from the north side of the solar system

Earth's orbit

Sun

Earth

The sun spins on its axis in the same direction as the earth. This is also the same sense as the earth's orbital motion.

View from earth (northern hemisphere)

Sun's daily motion

East

South

West

This picture shows the sense of the sun's rotation as viewed from earth (arrow on sun).

Figure 8.2 The sense of the sun's rotation. These two sketches indicate the sense of the sun's rotation from two points of view. Are they consistent?

latitude 35° the time is about 27 days. At 45°, it is 28 days. Sunspots never form near the sun's poles, but measurements using the Doppler effect indicate that the regions near the poles take more than 30 days for one rotation. An object that rotates with differing periods at differing latitudes is said to exhibit **differential rotation.** This is evidence that the surface of the sun is not solid. There is no universally accepted theory to explain why differential rotation occurs.

You may have to observe the sun repeatedly, but eventually a spot will appear that is large enough for one to discern its structure. A well-developed spot has a dark central region called the **umbra.** Surrounding the umbra is an irregular, lighter, gray region called the **penumbra.**

An ordinary, small sunspot lasts no more than a week. It develops from a small black dot, 1000 km (600 mi) in diameter, called a **pore.** In a few hours to a few days the spot matures and then gradually fades away. Only gigantic spots last much longer.

A large, long-lasting spot is usually a member of a **sunspot group.** (See Figure 8.4.) A typical group's structure is quite fascinating. One spot is usually larger than the other members of the group and typically precedes them in their motion around the sun. Thus it is called the **leader** or **preceding spot.** The leader is usually the first spot of any group to appear and the last to fade. (A large, lone spot is often a very new or a very old leader.) Next in size may be a second, sizable spot which trails the leader. This **follower** or **following spot** almost always lies a bit farther from the sun's equator than the leader. The remaining spots in the group are small.. They tend to cluster around the two major spots. Thus, at its fullest development, the sunspot group is assembled into two subgroups. Many variations on the above pattern will be noticed by the patient sunspot observer.

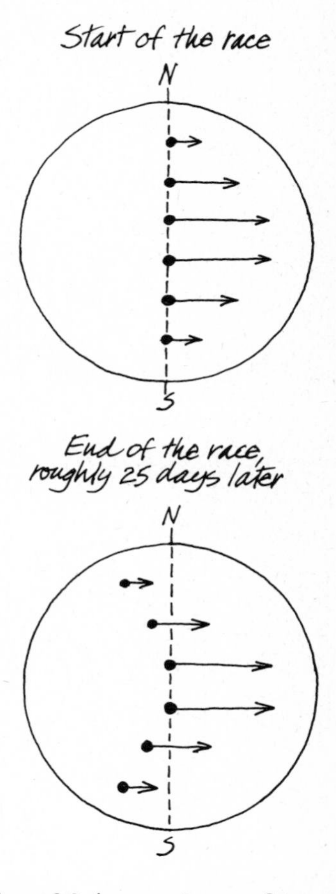

Figure 8.3 A sunspot race. Suppose a number of sunspots were lined up as shown at the top. Which spot or spots would return to the finish line first? On the bottom it can be seen that differential rotation carries the spots nearest the sun's equator around the sun most rapidly.

Figure 8.4 A large sunspot group. This enormous sunspot group was photographed April 7, 1947, near the peak of cycle 18. It is shown on the entire solar disk and also in an enlargement. It displays many of the features of a typical sunspot group. Another, smaller sunspot group may be seen near the left edge of the sun.

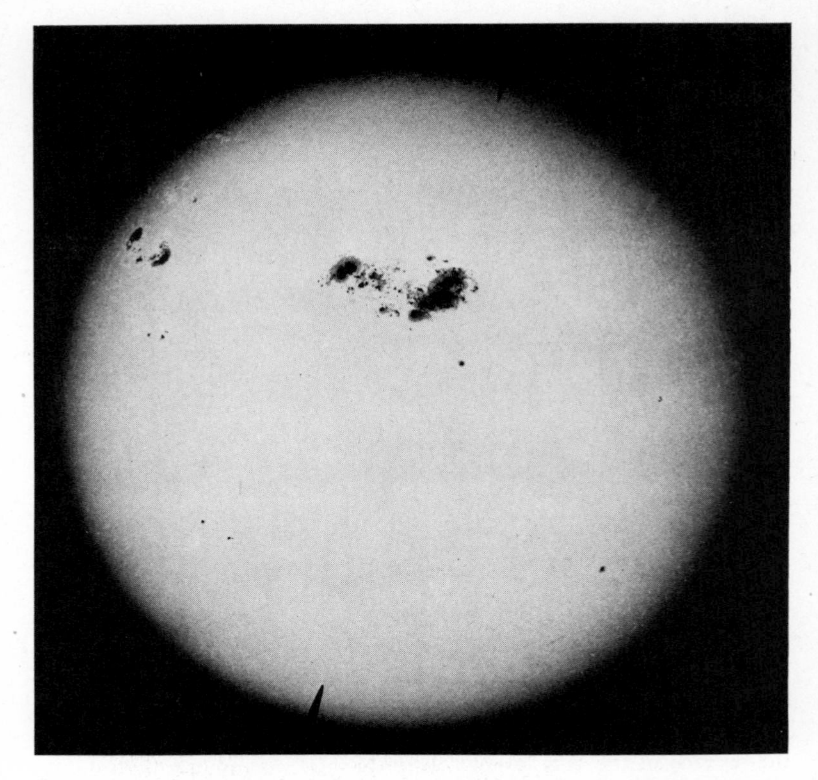

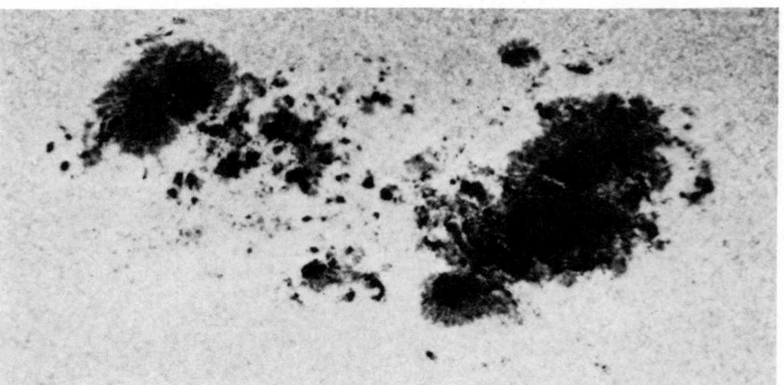

The Solar Cycle If you continue to watch the sun over the years, you will find that the number of spots varies. For a while there may be fewer and fewer spots until the sun becomes completely free of spots. Then the number of spots will increase year by year until the sun's surface is quite blotched. This pattern, called the **solar cycle,** was discovered in the 1840s. The average time from one sunspot minimum to the next is about 11 years, but this interval has ranged from roughly 8 up to 14 years.

Near a solar minimum, sunspots that form are commonly single. Sunspot groups tend to form near a solar maximum.

Figure 8.5 shows the number of sunspots graphed versus time for some recent cycles. (Actually, the vertical axis is the **mean annual sunspot number.** This number takes into account both groups and individual spots, using a rather complicated formula.) It is a fascinating record. Each cycle from 1744 on has been numbered. Cycle number 20 ended in 1975. The spots signaling cycle 21 began to appear in mid-1977. It is expected to peak in 1980–1981.

There is more to the pattern of the sunspot cycle than the number of spots. Near the end of an old cycle, the last spots form at roughly 5° on both sides of the sun's equator. At about the same time, the first spots associated with the new cycle appear at roughly 35° on both sides of the equator. As the new cycle wears on, the regions of sunspot formation tend to move toward the equatorial regions. Spots are rarely found outside the bands between 5° and 35° from the sun's equator.

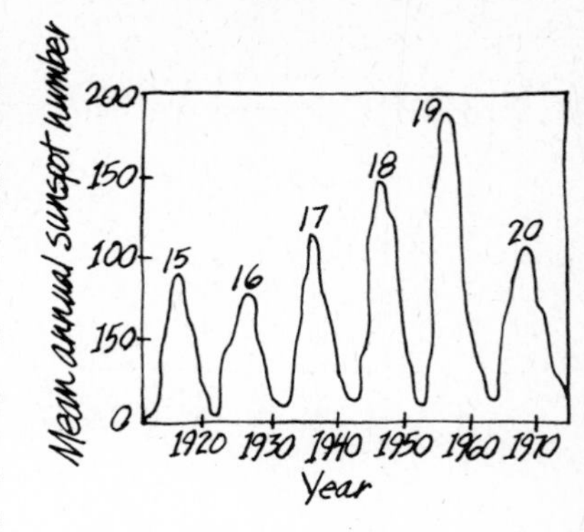

Figure 8.5 Sunspot numbers for recent years. The sun exhibits cycles of greater and lesser activity. One index of solar activity is the mean annual sunspot number, which shows the overall trend. From 1923, each cycle was stronger than the last through cycle 19, which broke the record for solar activity. After the solar minimum of 1964, cycle 20 was watched with great interest but did not continue the trend.

The Maunder Minimum Many theories of geology use a concept known as **uniformitarianism,** which assumes that, unless evidence to the contrary is available, geological processes have proceeded in the past at the same rate at which they are now proceeding. It is a strategy used when an educated guess must take the place of more reliable information.

Astronomers use a concept closely related to uniformitarianism in much of their work. For example, a profoundly important assumption is that the physical laws discovered on earth hold, in the absence of evidence to the contrary, elsewhere in the universe. For example, when analyzing distant star clusters, one customarily assumes that the law of gravitation is the same as that which governs motions in the solar system. Similarly, it is customary to assume that the solar system as we observe it now has been relatively unchanged for the previous thousands, if not millions, of years. Naturally, observations are sometimes made showing that this assumption, and others like it, have led to error in one way or another. In such cases, earlier workers need not be embarrassed; they were making an educated guess.

After the acceptance of the discovery of the solar cycle, it became an unspoken assumption that the sun had always experienced an approximately 11-year solar cycle. When, in 1894, E. Walter Maunder published a paper claiming that the sun had exhibited very few spots in the period from roughly 1645 to 1715, his assertion was ignored. Those who read his paper may have assumed that the astronomers of that era, which began only about 35 years after Galileo first saw sunspots, simply weren't paying much attention to the sun or that they failed to record the sun's activity.

Then, in the mid-1970s, solar astronomer John A. Eddy reexamined historical records for the time period Maunder had pointed out and came to agree with him: During that period of about 70 years the solar cycle was not in operation, and the sun was relatively free of spots. (See Figure 8.6.) Laboriously checking journal articles of the period, Eddy found statements such as that of G. D. Cassini in 1671: "It is now about

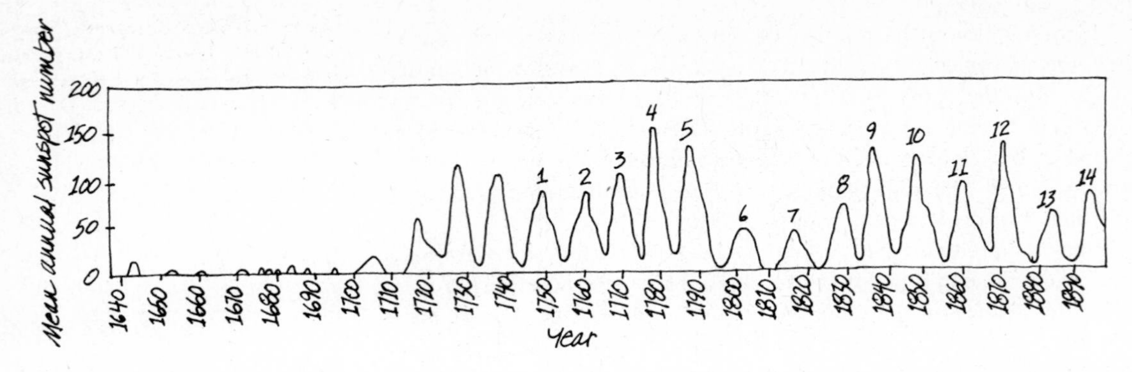

Figure 8.6 Earlier solar cycles and the Maunder minimum. This graph shows the sunspot number for most of the three centuries preceding ours. See also Figure 8.5. The data for the earlier portion of the graph are not from systematic observations of the sun but only from occasional observations reported in journals, diaries, and the like. As a result, they are not as reliable as the data in the latter portion.

twenty years since astronomers have seen any considerable spots on the Sun. . . ." Becoming increasingly convinced of the validity of Maunder's claim, Eddy proposed calling the period the "Maunder minimum."

Eddy has assembled an impressive array of independent evidence that the sun was indeed much less active during the Maunder minimum. Observations of sunspots large enough to be seen by the naked eye are found in the records of oriental astronomers. No such spots were reported during the Maunder minimum. The aurora borealis, also known as the northern lights (see also Section 9.6), occurs more frequently during the maxima of the solar cycle. Eddy points out that there were markedly fewer aurora reports during the Maunder minimum. In 1716, near the end of the Maunder minimum, Edmund Halley (see Section 6.7), an active, persistent observer of the sky, saw the first aurora of his life. Halley, 60 years old at the time, wrote a paper explaining the phenomenon, an indication of the novelty of an aurora in his era. Furthermore, Eddy has studied the solar observations of the era in question and has discovered that the pattern of differential rotation of the sun at the beginning of the Maunder minimum was different from today's pattern. In fact, in the early 1640s, the equatorial region of the sun rotated significantly faster than it does today.

Another line of evidence comes from the ^{14}C (read "carbon fourteen") in tree rings. Cosmic rays (discussed in Section 14.6) from outside the solar system constantly strike the earth's atmosphere. One of their effects is to produce the radioactive atom ^{14}C. This atom is ingested by trees as they grow, and this process leaves a record of the number of cosmic rays striking the earth during that period. It is known that, when the sun is active, some of the cosmic rays are prevented from reaching the earth, causing a reduction in the abundance of ^{14}C. Tree rings formed during the Maunder minimum show a pronounced increase in ^{14}C, another confirmation of the sun's low level of activity during that period.

Figure 8.7 Chinese representation of an eclipse. Looking at this picture, one can feel something of the emotional impact the eclipse had on the artist.

Eddy considers the evidence compelling. The sun does not always exhibit the solar cycle, but rather the phenomenon switches on and off at times. Not only is the sun not perfectly spotless, as western pretelescope astronomers presumed, it is not even regular in the rhythm of the coming and going of spots. This is one more example of the complexity of the sun's surface phenomena, which some future, successful solar theory will need to explain. It is just a bit unsettling to contemplate the idea that our sun, upon which we depend so abjectly, can seemingly change so much in such a short period of time.

Further implications of Eddy's research into historical solar activity are explored in Sections 8.2 and 9.1.

8.2 HOW TO ENJOY A SOLAR ECLIPSE

One of the most awesome events a human can witness is a total eclipse of the sun (Figure 8.7). Such events are rare in any one location but are of such scientific worth that astronomers are willing to travel almost anywhere in the world to view one. If you have a chance to travel to the location of a total solar eclipse, you are urged to do so. In this section, we describe what such a fortunate individual will see and discuss some interesting solar phenomena along the way.

Conditions Necessary for a Total Solar Eclipse If a total solar eclipse happened everyday, few people would pay much attention. Since they are rare, such eclipses are much more impressive. A total solar eclipse is visible from *somewhere* on earth roughly once every 1.5 years. Yet, if you remained at *any one place* on earth, you would see a total solar eclipse once roughly each 400 years, on the average.

To see a total solar eclipse, you must be carried by the earth into the part of the moon's shadow called the **umbra.** (See Box 8.2 and Figures

BOX 8.2

A QUICK LESSON ON SHADOWS

A tiny source of light (a **point source**) causes simple black shadows as shown in Figure 8.8. Such a shadow becomes wider the further it is from the object.

A large source of light (an **extended source**), like the sun, produces a more interesting shadow. In Figure 8.9, lines connect the edges of the sun and the moon. These lines divide the region behind the moon into three subregions. To see a total solar eclipse, you must be in ship A, which is in the **umbra.** In any other region, at least some of the sun's disk is visible. The umbra narrows down to a point, marked E. Viewed from a distance from the moon farther than this, the moon's angular size is smaller than the sun's, and so no total eclipse is possible.

Figure 8.9 also shows the view obtained at various positions in and around the moon's shadow.

Figure 8.8 The shadow caused by a point source.

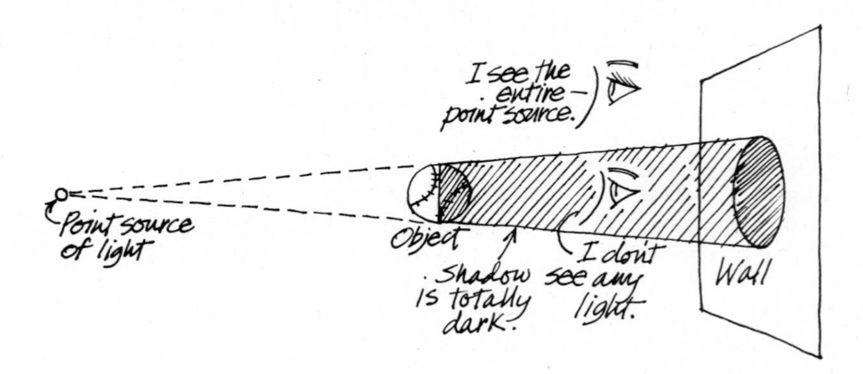

8.8 and 8.9.) ("Umbra" is also the word used for the central, dark region of a sunspot, you will recall.) The moon's umbra has an average length of 376,000 km (234,000 mi). The average distance from the center of the moon to the surface of the earth is 378,000 km (235,000 mi). These numbers indicate that, if the moon followed a circular path around the earth, the earth would never quite enter the moon's umbra. However, the moon's orbit is elliptical. The moon, at closest approach, is only 350,000 km from the earth's surface. This makes total solar eclipses possible, but just barely. The earth can travel only a short distance into the moon's umbra. The size of the moon's umbra on the earth is roughly 300 km (190 mi) wide during most of an eclipse. A very narrow band, called the **path of totality,** falls on the earth. It is defined by the path of the umbra during the eclipse, and only those in this path see a total eclipse. See Figure 8.10.

Other kinds of solar eclipses, partial and annular (see Figure 8.9) are well worth observing, but they are much less impressive than a *total* solar eclipse. As long as direct light can reach you from any portion of the sun's brilliant disk, it produces glare that drowns out the rest of the show.

A number of circumstances have to coincide to produce a total solar eclipse. There must be a new moon, so that the moon's shadow is

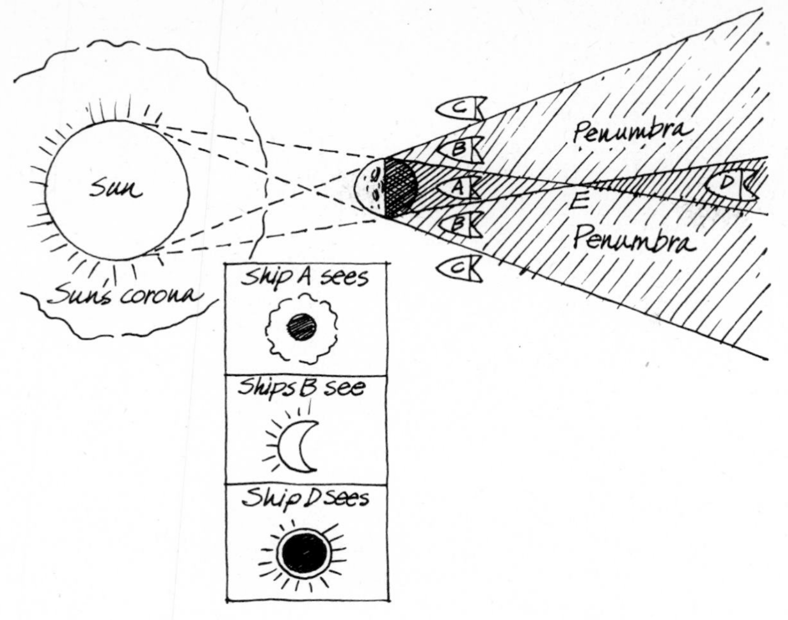

Figure 8.9 The moon's shadow. The rocketships shown obtain different views as they travel in or near the moon's shadow. Ship A reports a total eclipse of the sun. The ships marked B report a partial eclipse. The ships marked C can view the entire disk of the sun, and ship D is in a position from which the sun looks larger than the moon. A ship at point E sees the moon's disk exactly cover the sun.

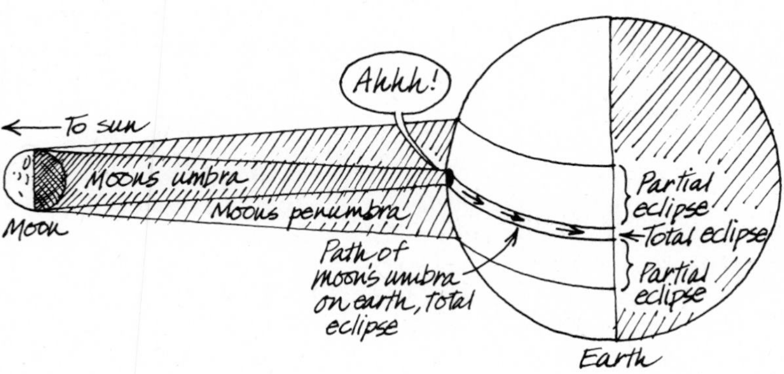

Figure 8.10 The moon's shadow strikes the earth: The crowd is awed. The resulting total solar eclipse is visible at only a very small region on the earth at any one time. As in most sketches of astronomical phenomena, this sketch is necessarily not drawn to scale. The moon should be about 28 earth diameters from the earth, which would place it far off to the left of the page.

pointed at the earth. The moon must pass through the line from the sun to the region of the earth where the observer is stationed. The moon must be nearer the earth than about 376,000 km, so that the umbra will reach the earth's surface. One must also be in the path of totality at the right time, of course. Finally, the sky must be clear.

One member of an amateur astronomical society traveled to five consecutive solar eclipses, all of which were clouded out.

Of these conditions, the only ones you can control are (1) being in the right place and (2) being there at the right time. Affluent readers may wish to consult the list of total solar eclipses in Table 8.1 to find an eclipse to observe. The total solar eclipse of February 1979 was the last in the contiguous United States in this century. If you do intend to travel to an eclipse, begin watching an astronomy magazine (such as *Sky and Telescope* or *Astronomy*) about a year in advance for detailed descriptions of the eclipse path and timings. Those of us who are less mobile can at least experience an eclipse vicariously by reading the rest of this section.

Table 8.1 Some Total Eclipses of the Sun—Past and Future

Date	*Place*
October 23, 1976	Africa, Indian Ocean, Australia
October 12, 1977	Northern South America
February 26, 1979	Northwest United States, Canada
February 16, 1980	Central Africa, India, China
July 31, 1981	Siberia
June 11, 1983	Indian Ocean, Indonesia
November 22, 1984	Indonesia, South America
November 12, 1985	South Pacific Ocean
March 29, 1987	Central Africa
March 18, 1988	Philippines, Indonesia
July 22, 1990	Finland, Arctic, Alaska
July 11, 1991	Hawaii, Central America, Brazil

The Day of the Eclipse For billions of years the force of gravitation has been producing an intricate maneuver. The earth has been orbiting the sun while the moon orbits the earth. The motions are such that today the moon will pass between your eye and the sun. Get up early (set *two* alarm clocks). Check the sky. (It's clear!) Leave for the observation site with all your equipment early enough to allow for unforeseen delays. Set up your equipment, check your accurately set watch, and then observe the following.

First Contact The instant the moon begins to cover the first bit of the sun is called **first contact.** Time and record this instant. (Later you can compare this and the other contacts with the predictions, which are sometimes very slightly in error.)

The best way to watch this part of the eclipse, the partial phase, is by projection using binoculars or a telescope. (Beware: Read Box 8.1 first!) If you cannot beg, buy, or borrow such an instrument, at least make a hole telescope as described in Box 8.3. Never look at the sun directly, either with the naked eye or especially with a telescope.

Crescents in Tree Shadows As the eclipse proceeds, look at the shadow under a tree. Often you can see small crescents formed on the ground. This surprising effect is the same as that obtained with a hole telescope. The holes are provided by spaces between the leaves and branches of the tree. When an eclipse is not in progress, these images of the sun are round and don't catch your attention.

Shadow Bands Before the eclipse starts, spread a white sheet on the ground. This will act as a screen for **shadow bands,** faint patterns of light and dark lines which are in agitated motion. When only a thin crescent of the sun remains, look away from the telescope screen long enough to see if shadow bands are rippling across the sheet. This effect is caused by the usually undetectable rippling of the earth's atmo-

BOX 8.3

A HOLE TELESCOPE

If you cannot obtain a telescope to view the partial phase of a solar eclipse, here is another safe method. Obtain some pieces of cardboard (paper plates work well) and punch a hole in one. Let sunlight pass through the hole and catch the light on a screen (another piece of cardboard). Before the eclipse, prepare a variety of holes of different sizes and then during the eclipse try them at varying distances from the screen until you find the one that works best. Such a telescope does not reveal fine detail, but it is cheap and safe and shows the shape of the partially eclipsed sun. *Do not* use smoked glass, layers of negatives, or any other homemade filter to view the sun. These filters may pass invisible radiation which can cause serious damage to the eye.

---

sphere. Stars twinkle because of the same motion of our atmosphere. The shadow bands are apparent only when there is a thin source of light. The usual disk of the sun is too thick. (Shadow bands have also been seen on east-facing walls when just part of the sun has risen.)

The Approach of the Moon's Umbra During totality you will be in the moon's umbra. As the sun thins down to a sliver, the umbra approaches, in most eclipses, more or less from the west. Can you see an eerie darkening in the western sky? If the sky is cloudy, you may be able to detect the shadow on the clouds.

Bailey's Beads and the Diamond Ring Effect In 1836 Francis Bailey reported on an eclipse, and his report sparked renewed interest in total solar eclipses. One of the intriguing things Bailey reported was that, just as the sun disappears, the last rays of the sun shine through the valleys at the edge of the moon. The effect resembles a string of brilliant beads. Often the last bead seems especially brilliant and dazzling. This is called the "diamond ring effect."

Second Contact and the Chromosphere Things are happening fast now. Watch your telescope screen carefully. Roughly one hour after first contact, the visible surface of the sun, the **photosphere** (ball of light) finally disappears. This is **second contact.** At the time of second contact you may see some reddish color for a few seconds at the place where the last of the photosphere was last seen. This is a glimpse of the **chromosphere** (ball of color). You must be alert, since it is easy to miss this important apparition. The chromosphere is a pink layer of gas lying above the photosphere. (The photosphere and chromosphere are discussed further in Section 9.6.)

See three views of an eclipse in Plate 13.

The Corona Totality! Congratulations: You have joined the select group composed of those who have witnessed a total solar eclipse.

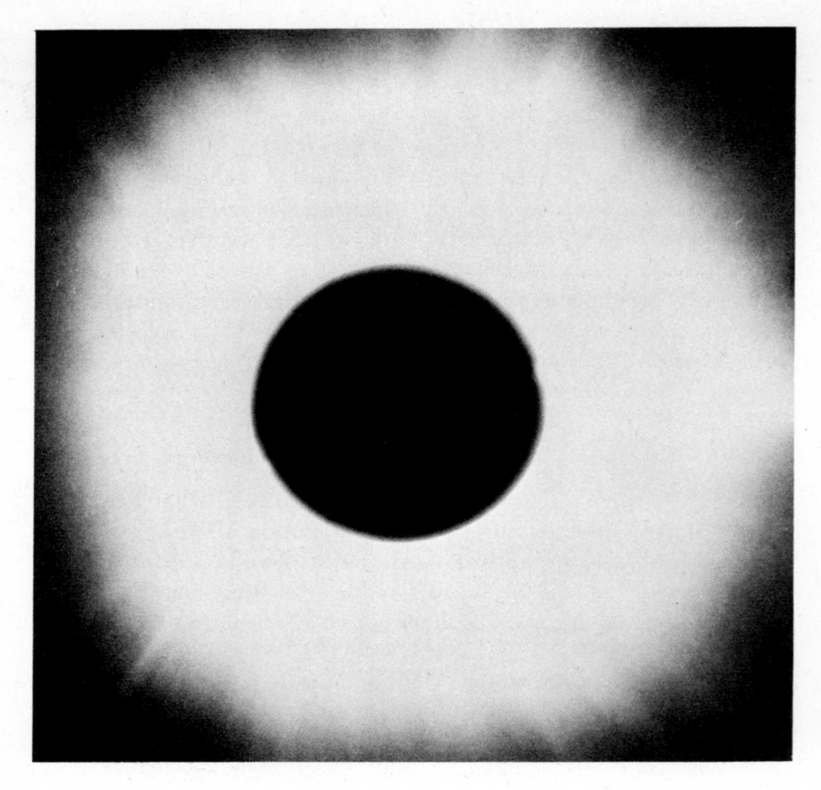

Figure 8.11 Total eclipse of the sun, Mexico, March 7, 1970. People traveled from the far corners of the earth to view this awesome phenomenon.

Once you are *sure* totality has begun, you may at last lift your eyes and watch the spectacular event directly. As you stand in awe in the moon's umbra, you will see a dark circle where the sun was. But more than that, you can now see one of nature's most wonderful sights, the sun's **corona** (crown) (Figure 8.11). The corona is the next layer of the sun above the chromosphere. It is, in effect, the sun's atmosphere. It looks like a beautiful, delicate halo of white light surrounding the sun.

The corona is always present, but it is visible only during a total eclipse because when you are not in the moon's shadow, scattered sunlight masks the corona. Now, however, the sky is sufficiently darkened so that the spectacular corona can shine through. It is quite fortunate that the moon is just the right size and distance away to hide the sun's photosphere, yet reveal its corona. Those who have seen the corona report that the sight is unforgettable.

Early on, astronomers agreed that the corona was caused by glare produced by the sun's radiation shining through an atmosphere. But where was the atmosphere that caused the effect? Was it the atmosphere of the sun, the moon, or the earth? Today's astronomers know that the moon has no atmosphere and because, during an eclipse, the atmosphere of the earth near the observer is in the moon's shadow, we may conclude that the corona is the sun's atmosphere. The corona glows because of energy received from the sun's photosphere.

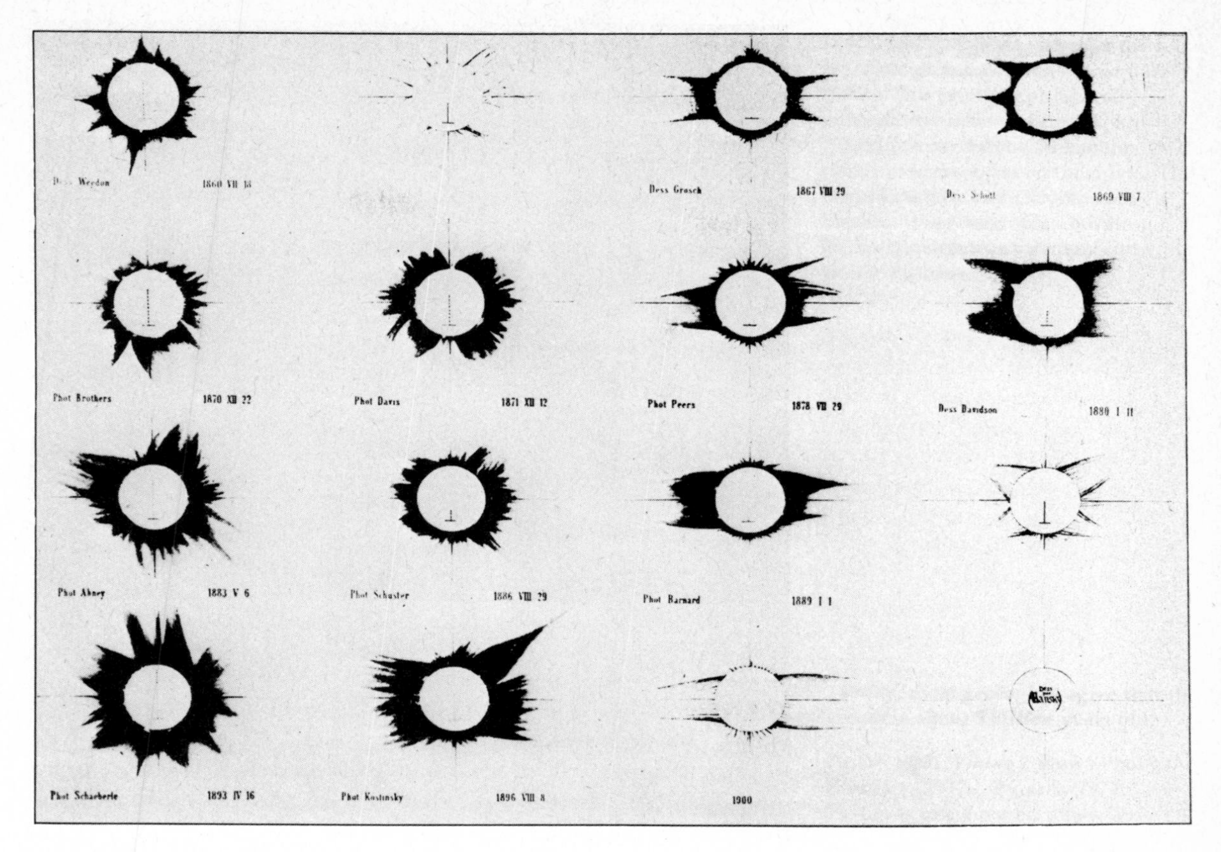

Figure 8.12 Drawings of the solar corona: 1860–1900. You may find it interesting to compare this figure with Figure 8.6 in order to compare the shape of the corona with the activity of the sun in the corresponding year. For example, the eclipse of 1871 occurred near the peak of cycle 12, while that of 1878 was near a solar minimum. Note that many of the drawings indicate, by small white protrusions, the presence of prominences.

The appearance of the corona varies dramatically from one eclipse to the next. The general appearance seems to change depending on the current stage of the solar cycle. Near sunspot maximum, when the sun's surface exhibits many spots, the corona has many streamers projecting in many directions from the sun and resembling petals of a flower. Near sunspot minimum, the corona is smaller, less circular, and devoid of much detail except for long extensions above the sun's equator and shorter plumes near the poles. (See Figure 8.12.) The streamers are difficult to photograph accurately, but make a strong visual impact.

John Eddy's historical research (see Section 8.1) led him to speculate on another astounding proposal. The earliest report he can find of prominent streamers in the sun's corona dates from 1715. Earlier reports mention only a dull, narrow ring of light around the sun. Could it

Figure 8.13 Solar prominences, August 31, 1932.

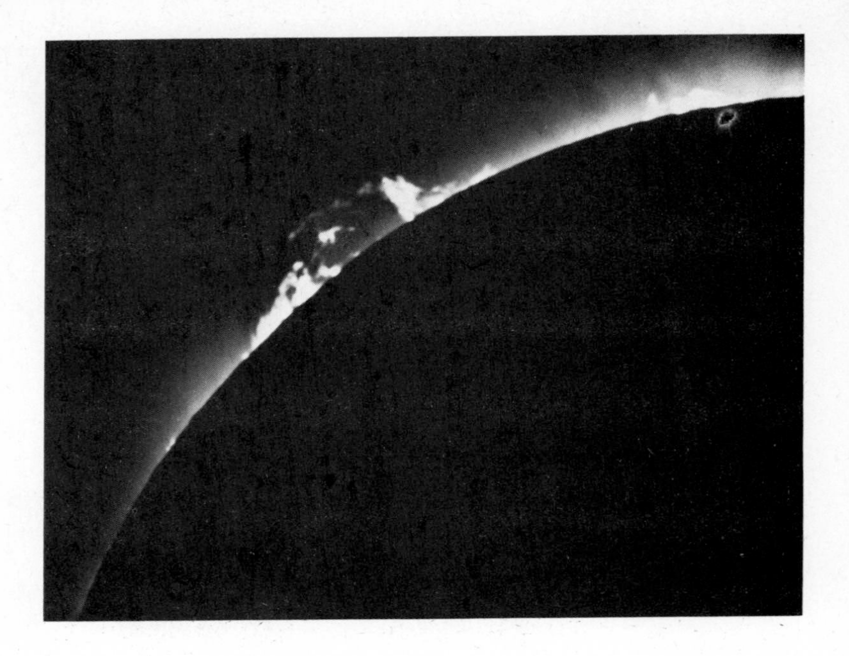

be that earlier astronomers did not notice the corona or failed to report it? Eddy feels that this is doubtful. Who could not have reported "a thing so breath-taking and beautiful?" he asks. Is it possible that the spectacular appearance of the corona is a relatively new feature of the sun? If so, it is another indication that we may live in a period of abnormal activity on the sun.

Prominences During totality, the naked eye may see tiny pink tongues of light protruding from the sun's disk. These are called **prominences** (Figures 8.12 and 8.13). The number of prominences on the sun increases as the maximum of the solar cycle approaches. They are among the most long-lasting features of the sun. Prominences at the edge of the sun's disk are visible to the naked eye only during an eclipse. They are revealed in some photographs as long, dark, irregular lines on the sun's face. (See Figure 9.6.) When so revealed, prominences are also known as **filaments.** Most prominences are so immense that they could swallow up many earths. We discuss prominences further in Section 9.6.

Nature During an Eclipse Quickly now, not much totality is left. Take your eyes away from the corona long enough to observe conditions on earth. How dark is it? What color is the sky? Often planets and the brighter stars become visible. Has the temperature dropped? How do living things react? According to some reports, flowers close up, cattle stand still, bees return to the hive, and chickens frantically scramble back into the coop. During the 1973 eclipse in Africa, a nasty kind of gnat that ordinarily swarms at night attacked observers. How do other people react? Do they cheer or stand in quiet awe? What are your emotions? Do you feel amazement, awe, or fear? How does this display compare with other natural phenomena you have seen?

The End of the Eclipse A solar eclipse can be total for no more than seven minutes. Most are considerably shorter. The eclipse of February 1979 lasted only three minutes. You may have just settled down to serious observing when again the chromosphere becomes visible briefly and sunlight spills around the edge of the moon. Bailey's beads flare out again at this, the **third contact.** Time it.

This is the most hazardous moment of the eclipse. The viewer's pupils are open wide because of the relative darkness and therefore most susceptible to damage from direct sunlight. You must be prepared to look away from the sun the instant the slightest hint is given that totality is over.

You may now look again for shadow bands and the receding umbra in the east. Finally, roughly one hour after third contact, you may use the telescope to time the **fourth contact** as the moon steps aside.

We next discuss two of the earliest and most basic measurements of the sun made by astronomers: the sun's distance and mass.

8.3 DISTANCES IN THE SOLAR SYSTEM AND THE SIZE OF THE SUN

Measuring Distances in the Solar System One of the attractive features of the heliocentric theory is that it can be used to construct a scale model of the solar system. For example, as discussed in Chapter 4, measuring the changing position of Venus in the sky relative to the sun allows us to calculate that its orbit is 72 percent as large as the earth's orbit. Thus, on a scale model of the solar system that places the earth 1 meter (3.3 ft) from the sun, Venus should be placed 0.72 meter (2.36 ft) from the sun. Similarly, Saturn should lie 9.54 meters (10.4 yd) from the sun. Astronomers employ a standard scale model of the solar system based on the average distance from the sun to the earth, the **astronomical unit.** Other distances in the solar system may be expressed in terms of the astronomical unit: "Venus lies 0.72 AU from the sun." "The distance from the sun to Saturn is 9.54 AU."

On a map of Wisconsin, Green Bay may be 8 cm from Milwaukee. To calculate how many kilometers this represents, we must look for the scale used on the map. If it reads "1 cm equals 20.4 km," we can compute that the distance from Green Bay to Milwaukee is $8 \times 20.4 = 163$ km (about 101 mi).

Now suppose the map maker has failed to reveal the scale upon which the map is based. The map would be useless unless we knew from some other source the distance between any two points shown on the map. From this information and the measured distance on the map, we could compute the scale of the map and, from this, any other required distances.

The astronomers of Kepler's day (about 1600) were in this position: The heliocentric theory had provided a map (scale model) of the solar system, but the map did not have an accompanying scale. To provide the scale the astronomers needed to measure the actual distance, in (say) *kilometers,* from the earth to the sun or from the earth to any object orbiting the sun. Then the scale of the solar system, "x kilometers equals 1 AU," could be computed.

Several ancient astronomers attempted to measure the distance to the sun. Their methods were correct, in principle. However, the measurements on which the methods were based were very difficult to perform accurately. Hipparchus estimated that the sun is about 15 million km from us. Why not use the method of parallax? The problem is that no one has thought of an accurate way to measure the sun's parallax directly, that is, by using the method by which the moon's distance was obtained. (See Section 5.3.) The trouble is that stars near the sun's position in the sky are invisible by day so that no distant background is available. The scale of the solar system remained in doubt until the invention of the telescope. Then, G. D. Cassini greatly improved upon the ancient measurement.

Cassini was the first director of the Paris Observatory. In 1671 he sent one of his assistants to a distant location across the Atlantic Ocean, instructing him to note carefully the angular position of Mars with respect to the fixed stars at preselected times. At the same time Cassini was similarly measuring Mars' angular position as viewed from Paris. Later, Cassini compared the two sets of observations and found, as he had hoped, that each man had seen Mars at slightly different angular positions in the sky. Using the method of parallax, he calculated the distance to Mars and used this information to compute the scale of the solar system. His result was about nine times larger than Hipparchus'. He found 1 AU = 135 million km (84 million mi).

Asteroids are tiny objects that orbit the sun. They are found chiefly between the orbits of Mars and Jupiter. (See Section 11.7.) Eros is an exceptional asteroid in that it occasionally passes relatively near the earth, *only* 22 million km away, which is about 60 times the distance of the moon. At such a time it is a good candidate for an accurate parallax measurement, because its small angular diameter permits a pinpoint determination of its angular position in the sky. Furthermore, its nearness to the earth gives it a relatively large parallax angle, one that is easier to measure accurately. As Eros passed by in 1900, the parallax angle was measured and its distance determined. The resulting scale indicated that 1 AU = 1.492×10^8 km = 149.2 million km (92.71 million mi), about 10 percent larger than Cassini's determination.

Two sophisticated contemporary methods have yielded highly accurate results. One method is to reflect radar pulses off Venus. Since the speed of radar waves is known, the length of time required for a pulse of radar (a radio wave) to reach Venus, reflect from its surface, and return to Earth gives the distance to Venus and thus the scale of the solar system. When Venus is at inferior conjunction, the pulse takes about 4.6 min round trip.

Another method uses space probes. The position of such a probe in orbit around the sun is very well known. The probe emits its own radio signal by which it can be tracked accurately. Thus its position and distance from the earth can be very accurately determined.

The Distance to the Sun The measurements discused above have given modern astronomers an accurate determination of the average distance to the sun: 1 AU = 1.495985×10^8 km. This number is so important that every educated person should be familiar with this rounded value:

The average distance from the sun to the earth, 1 AU, is about 150 million km.

This is an enormous distance—about 93 million mi. At a speed of 88 km/hr (55 mi/hr) it would take about 194 *years* to travel to the sun. This distance is equivalent to driving around the earth's equator about 3740 times.

This, then, sets the scale of the solar system. It means that we can now compute that Venus is 0.72 AU = 108 million km from the sun and Saturn is 9.54 AU = 1427 million km from the sun.

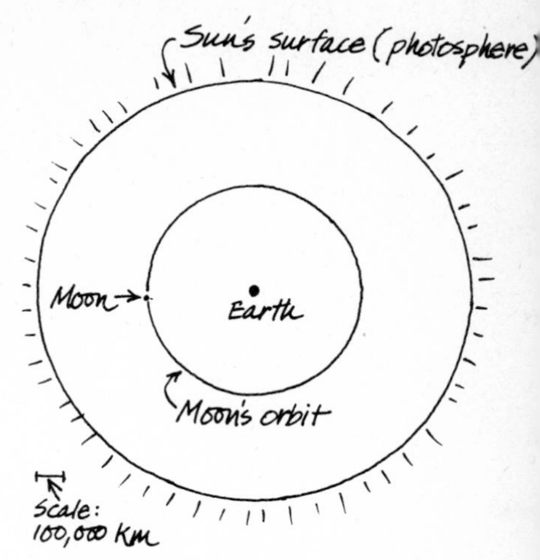

Figure 8.14 A comparison of sizes: The earth, the sun, and the moon's orbit. This figure is drawn to scale. It shows that the diameter of the sun is roughly twice the diameter of the moon's orbit. (Note: The earth is not actually inside the sun!)

The Size of the Sun The frequently used method of determining the actual diameter D of a distant object is to measure its angular diameter A and its distance R. Then, a formula from geometry may be used. The sun's average angular diameter is $A = 0.533°$, and its distance is R = 150 million km. Thus the sun's actual diameter is D = about 1.39×10^6 km = 1,390,000 km, or about 864,000 mi.

For those who are interested, the formula is $D = A \times R/57.296$, where A is in degrees.

An object this large is very hard to imagine. (See Figure 8.14.) The sun's diameter is 109 times the earth's diameter. More than 1 million earths would fit inside the sun. Driving at a speed of 88 km/hr (55mi/hr), you would need 19 days to drive around the earth's equator, but 5 years, 8 months to drive around the sun's. For all its bulk, the sun occupies relatively little space in the solar system. For example, it would take about 107 objects the size of the sun placed in a row to cover a distance of 1 AU.

8.4 THE MASS OF THE SUN

The enormous size of the sun naturally raises the question, How much matter is in the sun? or, more technically, What is the mass of the sun? You might guess that anything so large should contain a great deal of matter, but this is not necessarily so. A relatively large box of feathers might contain less mass than a much smaller car battery.

The concept of mass is discussed in Section 6.3.

The straightforward method of determining the mass of an object, balancing it on a scale against a standard mass, is clearly not applicable to the sun. Once again, astronomers have had to resort to a clever, indirect method. Astronomers were highly motivated to find the sun's mass. Aside from pure curiosity, they wanted to be able to answer such questions as, How much matter does the sun have to use as fuel? But how are we to make the mass determination?

It was Newton's theory of dynamics that solved the problem. If a massive object having a mass of M kilograms is being orbited by a body having a mass that is insignificant compared to M, and if the average distance between the objects is R meters and the period of the orbit is P seconds, then Newton's laws give the formula

$$M = \frac{4\pi^2}{G} \times \frac{R^3}{P^2}$$

In this formula, $\pi = 3.14\ldots$ and G is the constant of gravitation.

Data concerning the earth may be used to calculate the mass of the sun using this formula. However, the calculation was not carried out by Newton. The orbital period of the earth, $P = 1$ year $= 3.156 \times 10^7$ sec, was well known and, after Cassini's work (see Section 8.3), done when Newton was about 30 years old, the approximate distance to the sun was available for R. The problem was the G, the constant in Newton's law of gravitation (See section 6.8), was not known. Once again a scale was unknown.

To determine G and thus the scale of gravitational forces, it was necessary to measure the gravitational force between a pair of objects of known mass. Such an experiment is difficult to perform. The gravitational attraction between ordinary objects on earth is so tiny that it escapes casual notice and is not even detectable using ordinary student laboratory apparatus. Finally, in 1798, 71 years after Newton's death, the physicist Cavendish designed a device sufficiently sensitive that it could accurately measure the gravitational force between two lead balls. The modern value is

$$G = 6.668 \times 10^{-11} \frac{\text{meters}^3}{\text{kilogram-second}^2}$$

As expected, G is a very small number. Using it, along with Newton's law of gravitation, we can compute the gravitational force between two 1-ton cars separated by a distance of 10 meters (33 ft). The resulting force would only be able to suspend one forty-thousandth of a dime.

With G known, we may then proceed to compute the sun's mass (see also Box 8.4) from the known data concerning the earth's orbital motion. The result is $M = 1.99 \times 10^{30}$ kg. Can the imagination grasp an amount of matter which, expressed in kilograms and rounded slightly, has a mass of 2 followed by 30 zeros? Suppose we were to line up

BOX 8.4

DETERMINING THE MASS OF A PLANET

Newton's method for determining the mass of the sun may be applied in a straightforward manner to determine the mass of any planet with a satellite (moon). For example, our moon has a sidereal period of P = 27.321 days = 2.361×10^6 sec. Its average distance from the earth is $R = 3.844 \times 10^8$ meters. Therefore, by means of the formula in Section 8.5, we may calculate the earth's mass, $M = 6.03 \times 10^{24}$ kg or about 6000 billion billion metric tons. On a cosmic pan balance, about 330,000 earths would be required to balance the sun's mass. The masses of other planets with one or more satellites are found similarly.

It is considerably more difficult to determine the mass of a moonless planet such as Venus or Mercury. Astronomers have developed the **method of perturbations,** which also makes use of Newton's law of gravitation. Each planet exerts a gravitational attraction on every other planet, causing it to deviate slightly from its exact elliptical path. For example, the gravitational attraction of earth acting on Venus causes a measurable deviation of the path of Venus. The amount of deviation depends on the mass of Earth, the Earth-Venus distance, and the mass of Venus. Because only the mass of Venus is unknown, it can be calculated using Newton's laws.

After all the mass in the solar system is determined and added up, the sun still dominates. It contains 99.9 percent of the solar system's mass.

1-metric-ton cars each 5 meters (16.4 ft) long until the line had the same mass as the sun. The line would be about 67,000 million million AU long. This is many times larger than the diameter of the known universe. Evidently, the sun's mass is not easily comprehended in earthly terms.

This example is not to be confused with the one-liner: "If all the cars in the nation were lined up end-to-end, it would be Chicago."

SUMMING UP

Of all astronomical objects, the sun influences us the most significantly. Early telescope users discovered that the surface of the sun exhibits faculae and sunspots. The latter in turn led to the discovery of the sun's differential rotation and the sunspot cycle, a rhythm that may have been interrupted during the Maunder minimum.

When the sun is studied during a total eclipse, the chromosphere and the corona, the layers of the sun lying above its photosphere, are revealed. Often, prominences are visible as well. A total eclipse is one of the most exotic and impressive of nature's displays.

The scale of the solar system has been determined with ever-increasing accuracy. The distance to the sun is now known to be about 150 million km and its diameter to be about 1.4 million km. The sun's mass was determined using Newton's theory of dynamics and is 2×10^{30} kg.

In the next chapter, we discuss some of the more recent investigations of the sun.

EXERCISES

1. Describe the following features and regions found on and above the surface of the sun: photosphere, chromosphere, corona, prominence.

2. Describe the appearance and typical life of a sunspot. How does its period of rotation around the sun depend on its location relative to the sun's equator? Describe the solar cycle. Where are we in the solar cycle at the present time?

3. What is meant by the Maunder minimum? What is the evidence that it actually occurred?

4. Describe the principal events one should observe before and during a total solar eclipse.

5. How was the sun's distance determined? What is 1 AU equivalent to in kilometers (approximately)? What is the average distance from the earth to the sun?

6. How many times larger is the sun's diameter than the earth's?

7. How was the sun's mass determined?

READINGS

A delightful book concerning the sun and solar observing activities is

The Sun and the Amateur Astronomer, by W. M. Baxter, David and Charles, South Pomfret, Vt., 1973.

See also the references at the end of Chapter 9.

COLOR PLATES

Plate 1 An absorption spectrum and an emission spectrum. The top spectrum consists of the continuous background plus the major absorption lines in the sun's spectrum. The second spectrum is the emission spectrum of sodium. Observe sodium's twin yellow lines. Notice that longer wavelengths (red) are at the right in these spectra.

Plate 2 The H-R diagram and star sizes and colors. This H-R diagram displays surface temperature (horizontal axis, increasing to the left) and luminosity (vertical axis). Also indicated are the sizes of the stars, shown to scale within a luminosity class. Between classes (white dwarf, main sequence, giant, supergiant) only absolute differences are relevant. Given the size of the sun here, a white dwarf should be an insignificant speck and Betelgeuse should be larger than the page.

Plate 3 The geologist on the moon. NASA sent a scientist-geologist (now senator), Harrison Schmitt, to the moon on the last moon mission, *Apollo 17*. This huge fractured rock rolled down a mountain nearly a mile away before stopping here. Moon dust has accumulated on it.

Plate 4 The moon, far side on the right. The *Apollo 17* astronauts, the last to visit the moon, took this picture after their rocket engines had blasted them back toward the earth. It is a view that we cannot get from earth—the near side is on the left, the far side on the right. See also Figure 10.13.

Plate 5 Spaceship earth. On a particularly clear day on earth, astronauts took this stunning shot. Africa is clearly visible, as is snow-covered Antarctica at the South Pole.

Plate 6 Mars from *Viking 1* orbiter. North on this photograph is to the upper left. The large impact basin at right bottom center is named Argyre. Valles Marineris is to the left. The "mitten" of Sabaeus Sinus is at the upper right. Find these same features on Figure 10.43.

Plate 7 The top of Olympus Mons on Mars. This artist's conception, drawn using *Viking 1* orbiter photographs, shows the caldera (volcanic crater) of the largest volcano known in the solar system, Olympus Mons, on a day when the clouds were below the mountain top. The caldera is about 80 km (50 mi) across.

Plate 8 The surface of Mars at sunset. The *Viking 1* lander sent back this view of the surface of Mars about 15 minutes before local sunset. The low altitude of the sun emphasizes surface features. Note the depression at near center. *Viking 1* verified that Mars definitely has red soil.

Plate 9 Valles Marineris from orbit around Mars. *Viking 1* orbiter took this splended shot looking down on the end of Valles Marineris, the gigantic canyon on Mars. The area shown in this photograph is about

1800 × 2000 km, over five times the area of Texas.

Plate 10 Jupiter from *Voyager 1*. Two *Voyager 1* space probes, bearing cameras much better than the *Pioneer* probes, flew by Jupiter in early 1979. When 33 million km (20 million mi) from Jupiter, *Voyager 1* took this extremely detailed view of the planet. The details of the red spot, the other spots, and the many plumes and swirls will keep researchers busy pondering them for years to come. (A bonus came when a photograph was made to the side of Jupiter showing a very thin ring not visible here.)

Plate 11 The four largest moons of Jupiter: Ganymede (top), Callisto, Io, and Europa. For me, seeing these four closeups of Jupiter's Galilean satellites (those discovered by Galileo) taken by *Voyager 1* was the most moving experience of the space age. Formerly four tiny disks in the world's largest telescopes, these satellites are now worlds to us, each with its own unique character. Each is shown in its correct relative size. Starting at the top and working down, Ganymede superficially looks most like our moon, having rayed craters and dark, maria-like regions. Closeups reveal mysterious grooves. Callisto, the darkest of the four, has craters "standing shoulder to shoulder," but, surprisingly, no very large craters. Io, the most sensational of the four, resembles a pizza. The total lack of craters was puzzling until eight active volcanoes were spotted, spewing sulfur far out into space. When this material falls back onto Io, it accumulates planet-wide at a rate of several centimeters per year. Europa at first seemed to resemble Mars, Lowell's canals and all, but later was seen to be the solar system's smoothest object, having both dark and bright streaks on a surface of frozen ice. The question for planetologists: How can four bodies that formed in the same region have turned out so differently?

Plate 12 Jupiter's Great Red Spot. The Great Red Spot, photographed by *Voyager 1*, shows here much more detail than ever seen before. A spiral structure is revealed, indicating rotations. Streams of clouds swirl around the vicinity in a riot of color. The large white spot appeared in 1940 and has persisted since then.

Plate 13 Total solar eclipse: three views. Taken through a telephoto lens during the total eclipse of February 26, 1979, from Washington State, these photographs reveal three aspects of this awesome phenomenon. The left-hand photograph shows inner detail near the covered photosphere. Notice especially the pink prominences that were even visible to the naked eye. The middle photograph shows roughly the appearance of the corona to the naked eye. The third photograph is of the diamond ring effect that occurs when the first bit of the photosphere reappears from behind the moon.

Plate 14 An explosion on the sun. On December 19, 1973, a huge outburst on the sun was captured by Skylab equipment, which recorded light emitted by ionized helium. The resulting burst of gas spans 590,000 km on the sun's surface, 1.7 times the distance from the earth to the moon. See also Figure 9.12.

Plate 15 An x-ray view of the sun. If our eyes could see x-rays, the sun might look like this computer-processed image. The computer uses colors to represent intensity: white (strongest x-rays), followed by yellow, red, blue, and black (no x-rays). Coronal holes (dark) are seen at the sun's poles and to the side of the central portion of the disk. See also Figure 9.14.

Plate 16 A dark nebula, the "Horsehead" nebula in Orion. Located in the sky just south of the lowest star in Orion's belt (seen shining through the cloud at lower left), this dark blob was at first thought to be a gap in the bright nebulosity. Now it is recognized as a protruding part of the lower dark cloud of dust that blocks the more distant bright nebula. Notice that more stars are visible above the region of the dark nebula than below. See also Figure 15.16.

Plate 17 The "Lagoon" nebula, M8. This diffuse nebula, located in Sagittarius, toward the central region of our galaxy, is visible to the naked eye on a dark night as a faint patch. Illuminated by bright stars within it, it looks greenish in a telescope. The photographic film used here emphasizes the red light the nebula emits. See also Figure 13.6.

Plate 18 The Great nebula in Orion, M42. This Great nebula is considered the finest diffuse nebula in our skies. Color photography reveals it in all its glory. William Herschel correctly described it as "the chaotic material of future suns." See also Figure 13.5.

Plate 19 The Dumbell nebula, M27. This planetary nebula is expanding at a rate of about 25 km per sec (about 56,000 mi per hr). It is illuminated by a very hot central star. One of my students described this photograph as "the ghost of the planet earth, fading from view."

Plate 20 The Crab nebula, M1. This supernova remnant is the result of a star that was seen exploding in 1054 A.D. and is the leading character in Chapter 14. See also Figure 14.2.

Plate 21 The Pleiades, M45, an open star cluster. This cluster of stars, only a portion of which is shown here, is easily visible as a group of six or seven naked-eye stars on winter evenings. (See Appendix 1.) The stars seem to make a small dipper and this group is often mistaken for the actual "Little Dipper," of which Polaris is a member. The nebulosity wreathing the stars is gas and dust left over from when the cluster was formed. See also Figure 13.12.

Plate 22 The Andromeda galaxy, M31. This enormous disk of stars was photographed using light that left the galaxy about 2 million years ago. The color photograph shows that the galaxy's outer regions look blue and the inner nucleus reddish. See also Figure 16.7.

INVESTIGATING THE SUN IN THE TWENTIETH CENTURY

The McMath solar telescope at Kitt Peak National Observatory has been designed specifically for solar research. In addition to being a strikingly beautiful structure over ten stories tall (see Figure 9.1), it is a technical marvel as well. Atop the vertical tower is a tracking mirror which pivots so as to reflect sunlight down the slanting shaft to an objective mirror about 90 meters (300 ft) underground. The light is then reflected upward again to an observing room where an image of the sun 76 cm (30 in) in diameter is formed.

Astronomers have used such instruments to make a large number of discoveries about the sun. We discuss some of these discoveries in this chapter.

9.1 THE LUMINOSITY OF THE SUN

Although the sun is 150 million km away, it is much too bright to look at. How bright is it?

Apparent Brightness and Luminosity We first discuss the distinction between two terms relating to brightness. The **apparent brightness** of an object describes how bright it looks—its appearance. The apparent brightness of an object depends both on the object's distance from the eye (if it is closer, it looks brighter) and on its actual brightness (if it actually increases in brightness, it will look brighter). The other term used to discuss brightness is **luminosity.** This is the object's actual brightness. It does *not* depend on distance from the eye. Everyone is familiar with one of the units used to measure luminosity: watts. The wattage of a light bulb is stamped on it. It is correct to speak of a bulb with a luminosity of 100 watts. Note that a 100-watt bulb is a 100-watt bulb no matter how distant it is. A 5-watt bulb may have a greater *apparent* brightness than a 100-watt bulb if the 100-watt bulb is much farther away. Nevertheless, the 100-watt bulb still has a *luminosity* 20 times greater than that of the 5-watt bulb.

A precise way to ask, How bright is the sun? is to say, What is the sun's luminosity? We wish to know, in effect, how many 100-watt bulbs would emit the same quantity of electromagnetic radiation as the sun.

9.1 The McMath solar telescope at Kitt Peak National Observatory, Arizona.

Measuring the Sun's Luminosity Serious attempts to measure the luminosity of the sun began in the late 1800s. The procedure requires the following steps.

1. Measure the number of watts of sunlight reaching a square meter of area on the surface of the earth.

2. Modify this result to allow for the absorption of sunlight by the earth's atmosphere. The number of watts of sunlight striking a square meter of area just above the atmosphere is known as the **solar constant.**

3. Find the distance to the sun.

4. Imagine a sphere with a radius of 1 AU which is centered on the sun. Such a sphere would collect all the sun's radiation. Calculate the number of square meters of area on the surface of such a sphere.

5. The sun's luminosity is equal to the area calculated in step 4 times the solar constant obtained in step 2.

Only steps 4 and 5 are easy to accomplish; they are mere calculations. Step 3 was largely accomplished by the late 1800s, and steps 1 and 2 remained.

In 1881 Samuel Langley invented an instrument that accomplished step 1. Called a **bolometer** (accent on the second syllable), it changes electromagnetic radiation into an electric current. Even tiny amounts of

Samuel Pierpont Langley (1834–1906), American astronomer and inventor, was one of the nineteenth century's outstanding solar observers. Besides inventing the bolometer, he observed the sun and drew excellent representations of its surface. He was a master of solar spectroscopy (see Section 9.5). He also attempted powered flight three times but failed for the last time shortly before the Wright brothers' success. Langley Air Field in Virginia and NASA's Langley Research Center are named in his honor.

radiation produce a measurable current. Langley said that his bolometer could detect the infrared radiation emitted by a cow 10 mi away.

Since the age of rocketry, astronomers have sent bolometers above the atmosphere to make direct measurements unimpeded by absorption of light by the atmosphere. The modern value for the solar constant is 1400 watts per square meter. That is, the amount of radiation from the sun striking a square 1 meter on each side and above the atmosphere is equivalent to that emitted by fourteen 100-watt light bulbs.

For the record, the solar constant may also be expressed as 1.99 calories per centimeter squared per minute or 1.99 langley per minute.

Now we can calculate the sun's luminosity. At a distance of 150 million km (1 AU), it would take 2.8×10^{23} square meters to block all the sun's rays. The sun's luminosity is therefore 2.8×10^{23} square meters times 1400 watts per square meter or about 4×10^{24} (4 million million million million) 100-watt light bulbs.

Suppose the United States lit enough light bulbs to produce the equivalent radiation it absorbs from the sun. At a cost of 4 cents per kilowatt-hour, this would use up the annual federal budget in about 3 hr. Small wonder that investigations are underway to find a way to use solar energy more efficiently.

The very term "solar constant" emphasizes that astronomers tend to think of the light output of the sun as constant. Is this more than wishful thinking? Data seem to show that, from 1920 to 1952, the solar constant may have increased by about 0.2 percent. More dramatic, but less direct, evidence comes from John Eddy's study of the Maunder minimum (see Section 8.1). The minimum occurred during the worst period of the "little ice age," a period of unusually severe winters in Europe. Could it be that, when solar activity (sunspots, prominences, etc.) dies away, the solar constant also decreases slightly? It makes us ponder how precariously our existence may depend on the continued good behavior of the sun.

9.2 WHAT IS HEAT?

We all have experienced a fire so hot that we could not bear to venture within several meters of it. The sun is so hot that it can sometimes make us very uncomfortable at the vast distance of 150 million km. But how hot is it?

Section 9.4 describes the ingenious method devised to measure the surface temperature of the sun. In order to understand this method, we must know a little about heat.

Before the atomic theory became popular in scientific circles, heat was imagined to be a type of fluid which flowed out of objects as they cooled. During the 1800s, the atomic theory provided a much better understanding of heat. The theory states that all matter is composed of atoms. The atoms in a piece of solid matter are vibrating rapidly in a random way, undergoing thermal agitation. **Heat** may be thought of as the random motion of the atoms of a material.

Suppose you could watch the atoms in a piece of iron as the iron is heated in a furnace. The atoms vibrate more and more rapidly as the iron grows hotter. Eventually, the structure of the iron breaks down, the atoms tumbling over each other. The solid iron has melted and become a liquid. Even more intense temperatures will cause the atoms to fly free of the liquid, producing a gas, in this case iron vapor. In a

Table 9.1 The Kelvin Temperature Scale

Temperature in kelvins	*Example of that temperature*	*Celsius equivalent*	*Fahrenheit equivalent*
0	Atoms stop vibrating	−273°	−460°
154	Oxygen is a liquid at or below this temperature	−119°	−182°
273	Water is a solid (ice) below this temperature	0°	32.0°
310	A hot day on earth or average body temperature	37°	98.6°
600	Lead melts	327°	620°
6,000	Surface of sun	5,700°	10,300°
15,000,000	Center of the sun	Close to 15,000,000°	27,000,000°

gas, the atoms are separated from each other and rush around at high speed, colliding with the walls of the container. If you increase the temperature of the gas, the atoms rush around more rapidly, striking the container with more force; we say the pressure exerted by the gas has increased. Some atoms move more rapidly than others at any temperature, but in any case, *the higher the temperature of a material, the greater the average speed of the atoms of the material.*

Absolute Zero and the Kelvin Scale of Temperature The United States is gradually changing over to the metric system. We are getting used to hearing temperatures given in Celsius degrees. But neither the Fahrenheit nor the Celsius scale is satisfying to physicists. For example, suppose the temperature is 0°C and then doubles. What is the new temperature? Also, is there any real meaning to a negative temperature such as "10 below zero"?

Such considerations and the new theory of heat encouraged Lord Kelvin to introduce a scale for measuring temperature. The underlying idea of the Kelvin scale is this. As an object cools, its atoms vibrate more and more slowly. In principle, a temperature will eventually be reached at which the atoms stop. No further cooling will be possible; the object will have reached **absolute zero.** Kelvin defined 0° Kelvin (now called 0 kelvins, 0 K) on his scale as absolute zero. Although absolute zero has never been reached in the laboratory, in 1969 a group of physicists produced a temperature of 0.0000000006 K. Since the Kelvin scale is widely used, you may wish to become familiar with the comparisons in Table 9.1.

9.3 THERMAL RADIATION

In this section we discuss another concept essential to understanding the method used to take the sun's temperature: thermal radiation. We shall not discuss the most sophisticated modern theory explaining thermal radiation but will be content with a simplified but useful approach.

Each atom has an outer cloud of negative charge consisting of a number of orbiting electrons, subatomic particles of negative charge. *In nearly all situations in which electromagnetic radiation is produced, electrons are accelerated.* (An object is **accelerated** if it is sped up, slowed down, or has its direction changed.)

If you repeatedly accelerate the end of a rope by shaking it up and down, waves will travel down the rope. If an electron is accelerated similarly, it will cause electromagnetic waves to travel out in all directions. Radio and television broadcasting works this way, for example. At the top of a tall tower is a wire connected to the broadcast studio. The broadcaster causes some of the electrons to move up and down the stationary wire. The electrons vibrate at just the right rate to produce a radio wave of the wavelength assigned to that station. Because the electrons are vibrated, they emit a series of electromagnetic waves. When these waves strike your radio antenna, some of the electrons in the antenna are caused to move up and down in time with the waves. This constitutes a signal in the antenna, a signal which is sent into the radio and electronically changed into sound.

This concept of radiation production can be used to explain **thermal radiation,** which is the electromagnetic radiation emitted by any matter as a result of thermal agitation of its atoms. The atoms in any object are vibrating, which is to say, the matter has a certain temperature. The electrons associated with the atom are therefore being vibrated, and these accelerated electrons produce electromagnetic radiation. This idea explains why any sample of matter above the temperature of absolute zero emits radiation.

To reiterate, our explanation has been highly simplified. Many physics texts give the details of a more sophisticated theory.

The act of seeing a candle can now be compared to the act of listening to a radio. The atoms in the hot candle flame are agitated so that their electrons emit radiation of many wavelengths, including short, visible-wavelength radiation. The resulting waves enter the eye and cause the electrons on the back of the eyeball to vibrate in time with the waves. The resulting electric signal is sent along the optic nerve to the brain, and the brain changes the signal into a picture of a candle flame.

The simple concept of radiation production we have been using can also help us understand why objects at different temperatures emit thermal radiation of different wavelengths. Observation shows that the hotter an object, the shorter the wavelength of the thermal radiation it emits. For example, the heating element of an electric range is not visible when it is at room temperature if the room is dark. The element and all other objects in the room are emitting invisible infrared radiation; this radiation can be detected using the proper instruments, but the wavelength of infrared radiation is too long to be detected by the eye. When current is applied to the element, it heats up and eventually begins to glow red; that is, it begins to emit shorter wavelengths.

You may wish to review Section 7.3 before going on.

An analogy will help. Imagine causing waves in a rope by shaking the end of the rope. If you shake the end slowly, the crests in the wave will be far apart; the wavelength will be long. If you increase the rate of vibration, the crests will be closer together; the wavelength will be smaller. The analogy with thermal radiation is this. The hotter the object, the more rapidly its electrons are vibrating and the shorter the wavelength of the thermal radiation they emit. If an object glows red and is heated further, it will begin to emit orange light, then yellow light, and so on up the spectrum.

Not all of the atoms in an object at any particular temperature vibrate at the same rate. At any temperature, there is a range of accelerations. As a result, no (highly compacted) object emits one exclusive wavelength but rather a range of wavelengths. An object said to be "white hot" is emitting radiation over a large range of wavelengths, a range that spans the visible spectrum. The resulting radiation looks rather white because white light is a combination of all visible wavelengths.

You must be warned not to confuse the color an object appears to have when viewed by reflected light with the color it appears to have by virtue of its thermal radiation. A piece of blue steel and a piece of yellow china have different colors when viewed in a lit room because their surfaces absorb all but the wavelengths associated with their color. An object that reflects all colors equally well appears white by reflection. However, if the steel and the china are in a dark room and are heated to equal temperatures, both will emit the same wavelengths by the process of thermal radiation. Both will glow red if heated to the appropriate temperature. The planet Mars looks red, but you must not conclude that it is red hot. The red soil of Mars reflects the red wavelengths in sunlight well. The *thermal* radiation emitted by Mars is infrared radiation, and the surface of Mars is quite cold. On the other hand, stars do not shine by reflected radiation. They are so hot that much of their thermal radiation is visible. Red stars look red because they are sufficiently hot to emit long-wavelength visible light.

9.4 WIEN'S LAW AND THE SURFACE TEMPERATURE OF THE SUN

An astronomy student jokingly suggested using a thermometer on the sun at night when the sun was not so hot.

How, then, are we to measure the sun's surface temperature? We cannot simply plunge a thermometer into the sun. It would vaporize. Many attempts to use an indirect method were made last century. The reported surface temperatures varied from 2000 to 1 million K depending on the method used.

Wilhelm Wien ("Veen," German physicist, 1864–1928) studied thermal radiation. His work led to puzzles which when solved by others gave birth to quantum mechanics. Wien received the Nobel Prize for his work in 1911.

By 1895, Wilhelm Wien had sufficiently clarified the ideas we have been discussing so that the sun's surface temperature could be measured with confidence. Every object emits thermal radiation over a range of wavelengths. But, at any given temperature, there is always one wavelength at which the thermal radiation is strongest, that is, the most intense. This wavelength is called the **principal wavelength.** If the principal wavelength happens to be in the visible range of wavelengths, we can say that the principal wavelength emitted by the object is the brightest wavelength, the color that comes through the most intensely. An object so hot that it emits an orange glow also emits some violet light, more blue light, even more yellow, a lot of orange, less red, and even less infrared radiation. The object's principal wavelength is in the orange region of the visible spectrum.

Wien's law states the connection between principal wavelength and surface temperature: *The higher the surface temperature of an object, the shorter the principal wavelength of its thermal radiation.* Observe that it is the temperature of the object's *surface* that counts.

λ is the Greek letter lambda. It is the standard symbol for wavelength.

Wien expressed his law even more exactly as a mathematical formula. Let λ_p be an object's principal wavelength and T be its surface temperature. Then Wien's law states that K is a constant (pretty closely) applicable to any object. If λ_p is in angstroms and T is in kelvins, then $K = 2.9 \times 10^7$.

$$T = \frac{K}{\lambda_p}$$

In Section 7.6, we discussed Doppler's hypothesis that star colors differ because some stars are rushing away from us and others rushing toward us. We discarded that idea. No stars are moving fast enough to cause a noticeable color change. Today it is believed that Wien's law is the proper explanation for star colors. Red stars (longer principal wavelength) are cooler than blue stars (shorter principal wavelength).

Here, then, is how Wien's law can be used to measure the sun's surface temperature: (1) Take a photograph of the sun's spectrum using a spectroscope. (See Section 7.5.) (2) Determine at which of the wavelengths in the spectrum sunlight is most intense. This is λ_p, the sun's principal wavelength. (3) Use Wien's law to compute the sun's surface temperature. Measurements indicate that λ_p for the sun is about 5000 Å, which implies that *the sun's surface has a temperature of 5800 K.* (In conversation, this figure is usually rounded up to 6000 K.)

By terrestrial standards this is an extremely high temperature. A raging forest fire may reach a temperature of 650 K (920°C or 1700°F). At a temperature well below 6000 K, all known materials become gases; that is, their atoms fly off individually in all directions. This is one reason for believing that the sun is a ball of gas.

Max Planck ("Plonk," German physicist, 1858–1947) studied Wien's law and the other laws of radiation known at the time and explained them in a theory. Einstein interpreted Planck's work and proposed that radiation is absorbed or radiated as if it were composed of particles—photons. (See Section 7.1.) Planck received the Nobel Prize in 1918 for his work.

Planck's Law and the Surface Temperature of the Sun In the early 1900s Max Planck invented a theory explaining the varying amounts of thermal radiation emitted by an object at different wavelengths. Planck's mathematical formula is rather complex, but the results are fairly easy to comprehend when presented as a graph. See Figure 9.2.

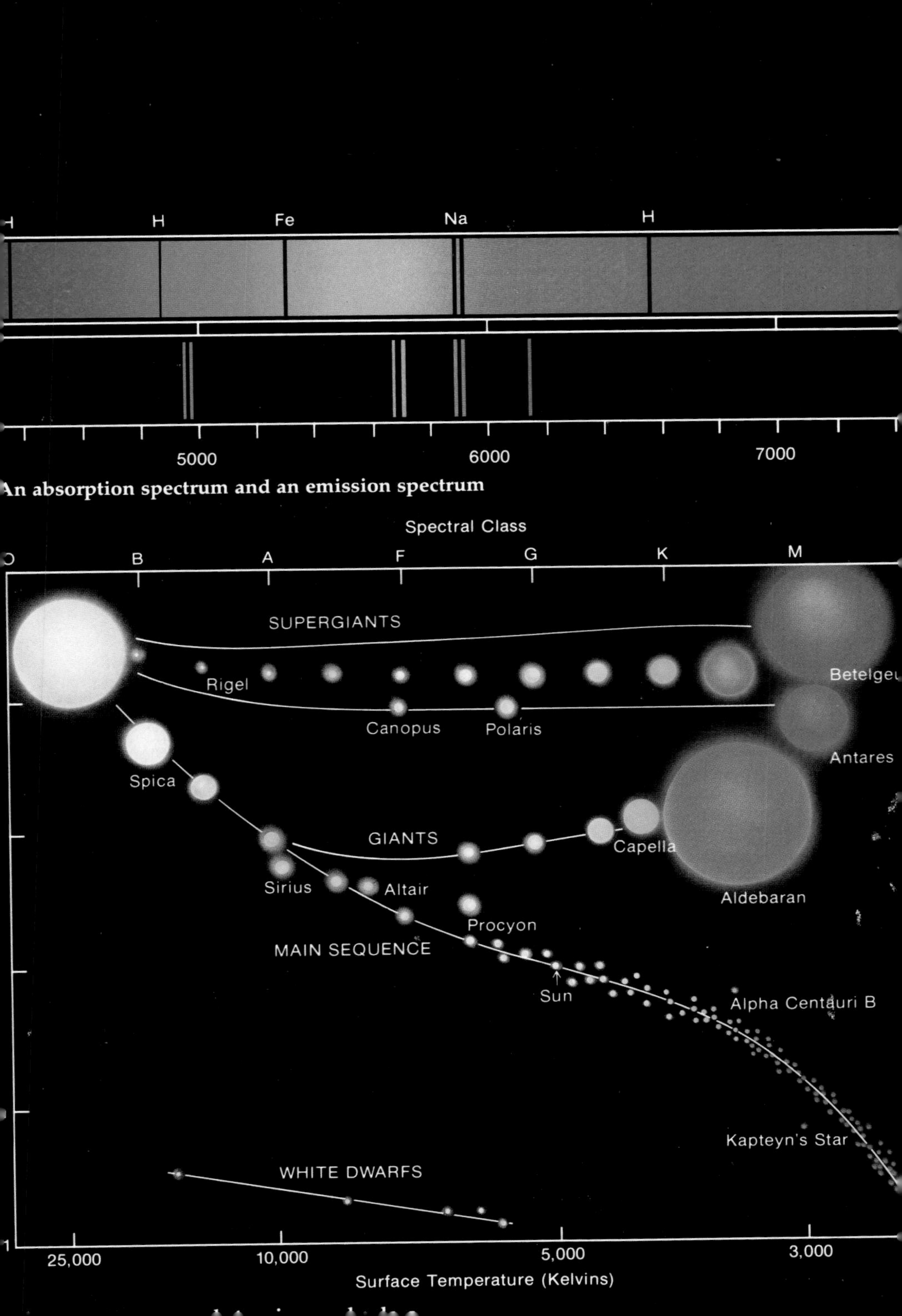

An absorption spectrum and an emission spectrum

Plate 3 The geologist on the moon

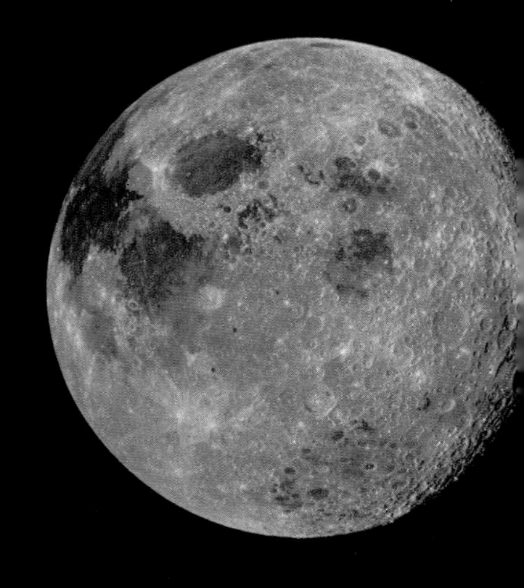

Plate 4 The moon, far side on the right

Plate 5 Spaceship earth

Plate 6 Mars from *Viking 1* orbiter

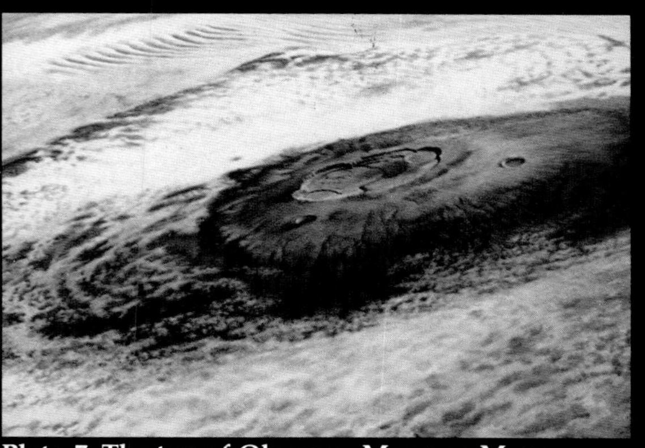

Plate 7 The top of Olympus Mons on Mars

Plate 8 The surface of Mars at sunset

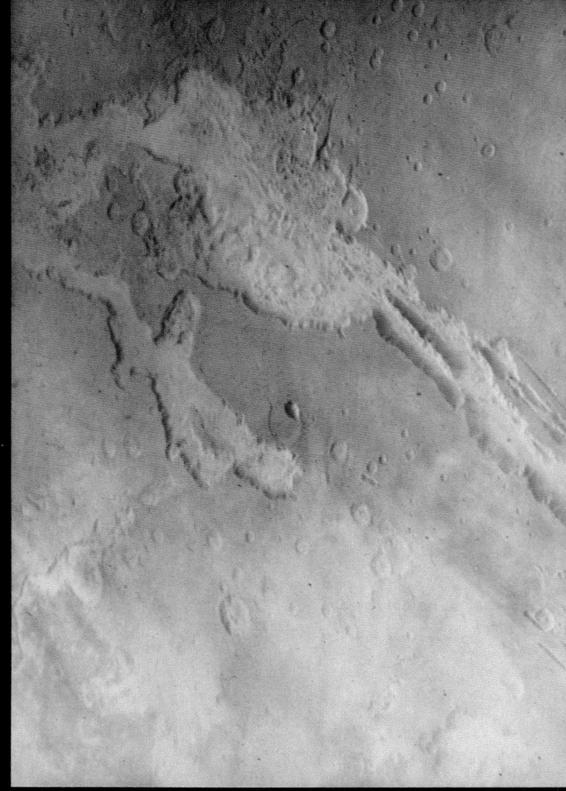

Plate 9 Valles Marineris from orbit around Mars

Plate 10 Jupiter from *Voyager 1*

Plate 11 The four largest moons of Jupiter: Ganymede (top), Callisto, Io, and Europa

Plate 12 Jupiter's Great Red Spot

Plate 13 Total solar eclipse: Three views

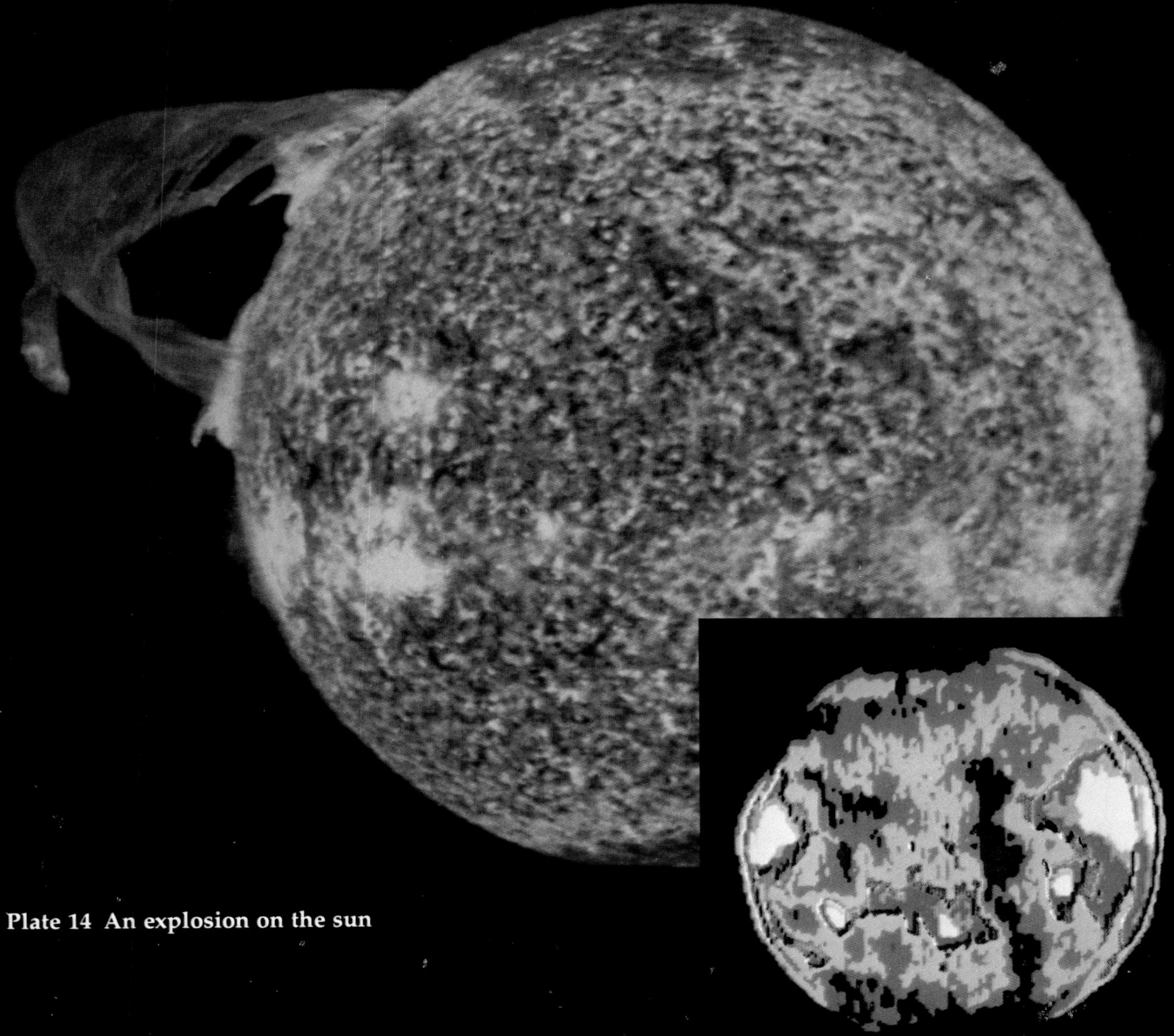

Plate 14 An explosion on the sun

Plate 15 An x-ray view of the sun

Plate 16 A dark nebula, the "Horsehead" nebula in Orion

Plate 17 The "Lagoon" nebula, M8

Plate 18 The Great nebula in Orion, M42

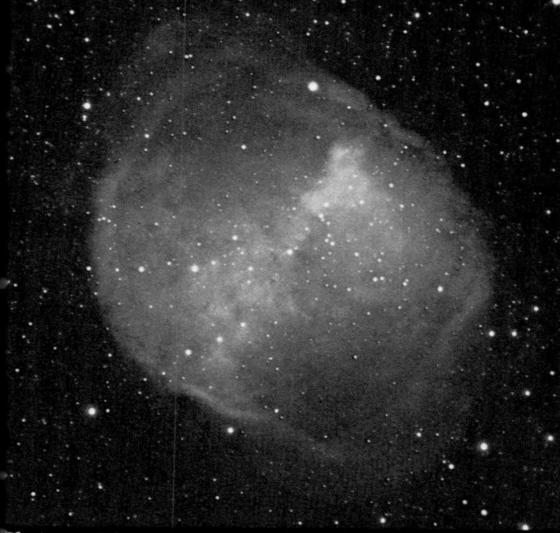

Plate 19 The Dumbell nebula, M27

Plate 20 The Crab nebula, M1

Plate 21 The Pleiades, M45, an open star cluster

Plate 22 The Andromeda galaxy, M31

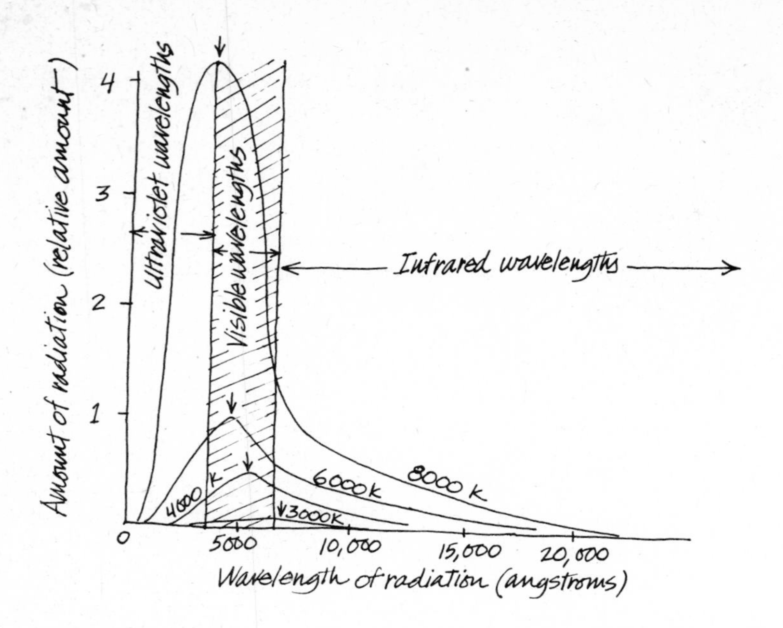

9.2 Planck's law of thermal radiation. An object's radiation is graphed versus wavelength for four representative temperatures. The vertical axis indicates the relative amount of radiation an object emits at each particular wavelength. The graph illustrates Wien's law. As an object's temperature increases, the peak of the curve (arrow), which marks the principal wavelength, shifts toward shorter wavelengths, that is, to the left.

Such a graph shows how an object's thermal radiation changes as its temperature changes. Planck's law can also be used to deduce the sun's surface temperature. We can measure the amount of solar thermal radiation received at many wavelengths and graph the results. Then, by trial and error, we can determine which form of the curves representing Planck's law looks most like the solar data. The resulting temperature determination agrees more or less with the result from Wien's law; $T = 6000$ K.

9.5 SPECTROSCOPIC CHEMICAL ANALYSIS AND THE CHEMICAL COMPOSITION OF THE SUN

The results so far in this chapter have indicated several great differences between the earth and the sun. Do these differences extend to the type of matter of which the earth and the sun are composed? We cannot perform a chemical analysis of a solar sample as we do of the earth's soil and (more recently) the moon's rocks. Beginning in the mid-1800s an ingenious indirect method was gradually worked out. The spectroscope eventually scored one of its most triumphant breakthroughs here.

The Discovery and Interpretation of Spectral Lines Progress toward the goal began in 1802 when W. H. Wollaston improved the spectroscope. He placed a slit between the source of light and the prism. If the source of light emitted only one wavelength, say in the red, then Wollaston saw the slit as a red line in the spectroscope. But when the source emitted several wavelengths, he saw a nicely defined separate line for each wavelength present. When Wollaston viewed

William H. Wollaston (English chemist and physicist, 1766–1828) worked in many fields. He discovered two chemical elements, palladium and rhodium, invented the goniometer for geologists, and worked with the spectroscope, for example. In 1819, he was influential in keeping Britain, and thus America, on the English system of measurement, a move from which America has not yet recovered.

The *continuous spectrum* consists of all the colors of light gradually shading from one wavelength to the next. No wavelengths are omitted. A rainbow is a continuous spectrum.

sunlight in his spectroscope, he saw the usual continuous spectrum, shading from violet to red. But he also noticed three dark lines in the spectrum. These were wavelengths missing from the sun's spectrum. (Wollaston's incorrect interpretation was that these lines marked boundaries between four colors in sunlight.)

By 1817, Joseph Fraunhofer, an instrument maker, had made further improvements in the spectroscope. With his improved instrument he saw not three but about 700 dark lines in the solar spectrum. He studied the lines and gave names to nine of the darkest ones. He called them A (in the red) through I (in the violet), and these names are still in use. This kind of a spectrum, dark lines on a colored background, is now called an **absorption spectrum.** Figure 9.3 shows the sun's absorption spectrum. (See also Plate 1.)

As the century progressed, more discoveries were made. In 1822, John Herschel (son of William Herschel; see Box 15.1) found that if materials in a flame are heated until light is given off, the light produces bright lines on a dark background when viewed in a spectroscope. Today we call such a result an **emission spectrum.**

Gustav Kirchhoff ("Kirkhh'-uff," German physicist, 1824–1887) is also well known for his laws of electric circuits. Robert Bunsen (German chemist, 1811–1899) made many discoveries in chemistry but is best known for the burner that bears his name.

A chemical *element* is a material composed of only one kind of atom. Ninety-two different elements occur in nature. Oxygen, nitrogen, lead, and gold are familiar examples.

A breakthrough in understanding the lines in spectra occurred when, in 1859, two German physicists used the earlier findings as a guide for a series of experiments. Gustav Kirchhoff and Robert Bunsen worked together to find the key to the puzzle. Their approach was to heat various chemical elements with a Bunsen burner until they emitted light. Viewing this light in a spectroscope revealed an emission spectrum. They found that every chemical element produces an emission spectrum having a different set of lines. In other words, each element has its own unique emission spectrum by which it can be identified, just as each human has identifying fingerprints. This discovery is used in industry. For example, a steel company may heat a sample of steel until it yields an emission spectrum. The lines in the spectrum reveal which impurities are in the steel. This analysis can be done in much less time than the usual chemical analysis would require.

Kirchhoff summarized his findings in what are today called *Kirchhoff's laws of spectroscopy:*

1. *An object will emit a continuous spectrum if it is a solid, a liquid, or a gas at high pressure.* For example, the hot filament in a light bulb is a glowing solid. It emits a continuous spectrum.

2. *A mass of glowing gas at low pressure will, if properly stimulated, emit an emission spectrum* (bright lines on a dark background). For example, in a neon light, neon gas at low pressure is stimulated by a flow of electrons. The gas emits red-orange light. Many of the bright lines in the emission spectrum of neon are in the red-orange region. In space, there are huge clouds of gas at low pressure which emit emission spectra. These emission nebulas, Figure 13.5, are stimulated by radiation from hot stars.

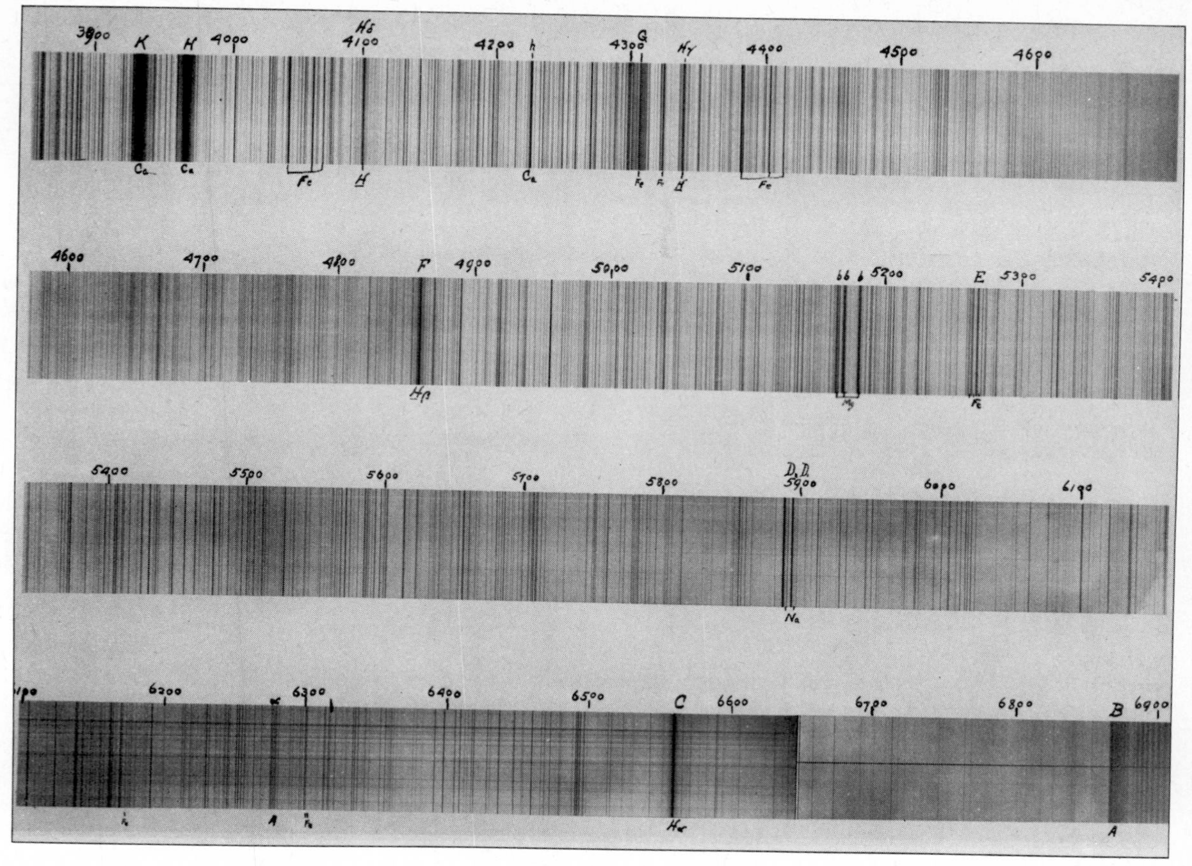

Figure 9.3 The sun's absorption spectrum. In a color photograph, this spectrum would appear violet at the upper left end and shade to red at the lower right. Many of the thousands of dark lines in the sun's spectrum are visible. The numbers indicate the wavelengths, expressed in angstroms. Below selected lines are the chemical symbols indicating the type of atom causing the absorption at that wavelength. Some of the more prominent lines are labeled with the capital letter symbols chosen by Fraunhofer. The twin lines produced by sodium mentioned in the text are located just short of 5900 Å and are marked "Na," the chemical symbol of sodium. You can begin to understand the complexity of the task of unraveling the meaning of these dark lines in the sun's spectrum as you study this photograph.

3. *If light passes through a low-pressure gas, the gas will absorb certain wavelengths. Dark lines appear in the spectrum at these wavelengths, producing an absorption spectrum. The light absorbed by the gas is reemitted in all directions.* (See Figure 9.4.) The research of Kirchhoff and Bunsen revealed that, when a chemical element produces black *absorption* lines in a spectrum, these lines appear at the same positions in the spectroscope as the bright lines in the element's *emission* spectrum. In other words, the wavelengths absorbed in the absorption spectrum of a gas are the same as those emitted in an emission spectrum. An important conclusion is that *the dark lines in the absorption spectrum can be used to determine the composition of the absorbing gas.*

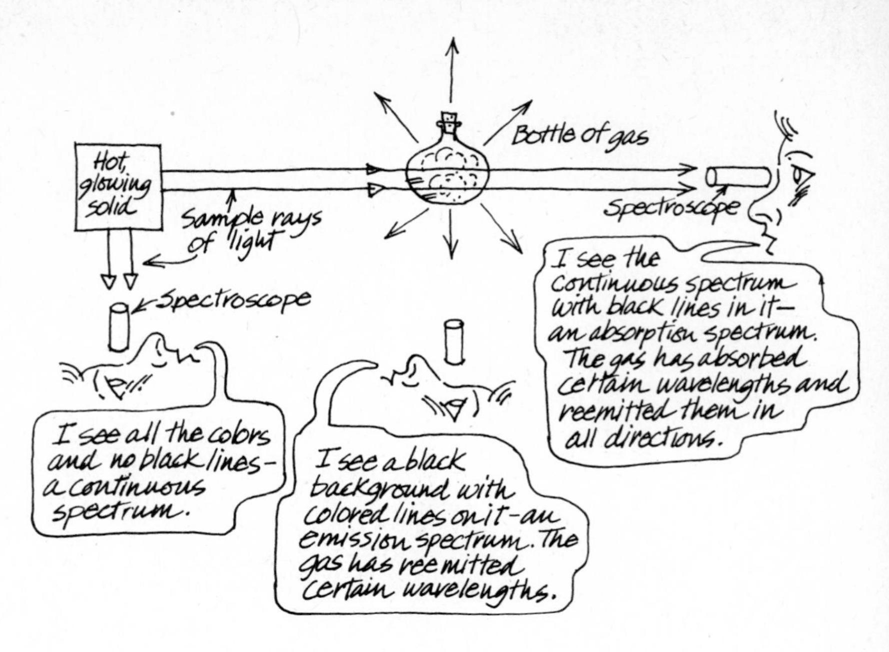

Figure 9.4 A cool, low-pressure gas creates spectra. On the sun, an analogous phenomenon occurs, except that the photosphere, a compressed gas, emits the continuous spectrum, and cooler layers in the sun's atmosphere act like the bottle of absorbing gas. The flash spectrum observed by Young (see Box 9.1) is analogous to that described here by the observer at the bottom center.

Kirchhoff and Bunsen went on to suggest that the sun's absorption spectrum is produced in the following manner. The visible surface of the sun, the photosphere, is a hot sphere of gas under pressure. Thus, by law 1, it should emit a continuous spectrum. Above the photosphere is a layer of cooler gas under lower pressure. This layer causes the sun's absorption spectrum. If this hypothesis is right, we can determine the composition of the sun. For example, if sodium is heated, it gives off a yellow light. In a good spectroscope we see two yellow lines close together at the wavelengths 5890 and 5896 Å. These are known as "sodium D lines." The sun's spectrum has two black lines in the yellow region at exactly the same two wavelengths. (See Figure 9.3.) Kirchhoff and Bunsen proposed that this indicates the presence of sodium in the sun. The other lines in the spectrum indicate other elements.

During the early part of this century, the new theory of quantum mechanics was used to explain the empirical findings of Kirchhoff and Bunsen. Many physics texts describe their ideas, and the interested reader may wish to investigate them. More on spectral analysis is discussed in Section 12.3.

This suggestion was gradually accepted, particularly after the work of C. A. Young (see Box 9.1). Since then, a great deal of time and energy has been spent deciphering the sun's spectrum to determine which chemical elements are present.

The Interpretation of the Solar Spectrum The spectral analysis of the sun is one of the most difficult tasks in all of science. In spite of the many difficulties, interesting results have been achieved. The overall finding is that the sun is composed of the same chemical elements as the earth. Nothing resembling "celestial matter" has been found yet. Of the 92 natural elements, one for each of the 92 kinds of atoms found in nature, over 60 have been identified in the sun's spectrum and more are likely to be found as research continues. The sun is chiefly made up

BOX 9.1

YOUNG'S EXPERIMENTUM CRUCIS

Kirchhoff and Bunsen proposed that the sun's atmosphere absorbs light from the photosphere and reemits it in all directions. How could this hypothesis be checked? A crucial experiment was needed (an *experimentum crucis* as they used to call it).

In 1870, an astronomer from Princeton University, C. A. Young (1834–1908), performed such an experiment. Kirchhoff had suggested watching the sun in a spectroscope during a total solar eclipse just after the last bit of the photosphere disappeared. If Kirchhoff's explanation of the absorption spectrum were right, the light coming to earth just after the photosphere was covered by the moon would be light absorbed by the sun's atmosphere and then reemitted in another direction. This light, since it was emitted by a gas under low pressure, should look like an emission spectrum in a spectroscope.

Young's account of the eclipse of 1870 conveys some of his excitement. "As the crescent [of the uneclipsed sun] grew narrower, . . . the dark lines of the spectrum and the [continuous] spectrum itself gradually faded away until all at once, as suddenly as a bursting rocket shoots out its stars, the whole field of view was filled with bright lines more numerous than one could count. The phenomenon was so sudden, so unexpected, and so wonderfully beautiful, as to force an involuntary exclamation."

The beautiful bright lines Young saw are now known as the **flash spectrum.**

---

of hydrogen atoms, the simplest of atoms. The next most abundant atom in the sun is helium. The percentages are approximately 90 percent hydrogen, 9.9 percent helium, and 0.1 percent traces of other atoms. The other elements found in the sun include many familiar substances, aluminum, calcium, iron, and sulfur, for example. Of course, in the sun each of these substances is in the gaseous state.

Each atom consists of an outer cloud of electrons and a tiny central nucleus. The nucleus is composed of protons and neutrons. Each type of atom may be distinguished by the number of protons in its nucleus. Hydrogen has one proton and helium has two.

9.6 CONTEMPORARY IDEAS ABOUT THE SUN: A JOURNEY OUTWARD FROM THE CENTER OF THE SUN

We now discuss some of the features of the sun as reported by professional astronomers, touching upon only a few of the highlights.

In an attempt to grasp the sizes of the regions under discussion, let's suppose that we are riding a highly advanced (most likely, impossible) space vehicle outward from the center of the sun. The windows automatically let in just enough light for comfortable viewing and somehow protect us from dangerous heat and radiation. We set the ship to travel at 160 km/hr (about 100 mi/hr).

The Photosphere The initial stage of our journey is relatively uninteresting. The light at the windows is brilliant, but we cannot see very far beyond the windows because the sun's gas is too dense to allow light to reach us directly from distant regions. Our thermometer tells us that the temperature at the center is about 15 million K. The first real event we notice occurs after we have traveled about 5 months. We are now about 580,000 km from the center (roughly 80 percent of the way to the surface), and we can feel our ship being jostled about. We are entering the convective zone of the sun's interior.

Theoretical calculations of the sun's interior indicate that the energy produced near the center of the sun flows outward by means of radiation. That is, the energy is passed outward as atoms of the sun's interior emit electromagnetic radiation. This radiation is absorbed by other atoms and then reemitted; thus the energy is handed on. The inner region where this occurs is called the **radiative zone.**

Calculations indicate further that, in about the top 20 percent of the sun's interior, the process of energy transport changes to one of **convection.** In this region, hot globules of gas rise toward the surface, lose their energy at the surface, and then, being cooler and so denser, fall back down again. As the cooler globules fall back, other, hotter globules take their place. This region, in which energy is transported by the bulk motions of gas rather than by radiation, is called the **convective zone.** The motions of the gas are turbulent, and so our ride has become a rough one.

We require about one month to travel through the approximately 120,000-km-thick convective zone. The temperature outside our vehicle drops to 6000 K and then, above us, we begin to see the vast edge of the sun when we are within about 400 km of it. We cover the remaining distance in a few hours and then begin to rise above the **photosphere,** the visible surface of the sun. Below us it seems to stretch on forever to a vaguely defined horizon.

As we rise higher, we see that the photosphere has an appearance resembling that of rice soup. The brighter "grains of rice," known as granules, are the regions where the globules of hot gas mentioned before are rising to the surface. We also see dark lanes between the granules; these are regions where the cooled gas is descending. The Doppler effect has been used by astronomers on earth to verify that the granules are rising and the gas in the dark lanes is descending.

Sunspots We have now risen high enough to notice a dark region of the photosphere in the distance—a sunspot. Sunspots are not really dark, astronomers now realize. If isolated, a large sunspot would provide the earth with about the same illumination that a full moon does. The spectrum of a sunspot shows that its temperature is about 4500 K (in the umbra). Spots look dark because they are only 40 percent as bright as the contrasting photosphere. (In general, the hotter an object, the more radiation it produces. See Section 12.5.)

Many modern findings link sunspots to magnetism. C. A. Young, the first astronomer to see the flash spectrum, noticed in 1892 that some of the lines in the spectrum of a sunspot are double. A few years later, it was found that the same phenomenon could be produced on earth. When a glowing gas is placed in a strong magnetic field, the lines it produces become multiple. Early this century George E. Hale (see Chapter 16) pointed out that this means that the sunspots are centers of intense magnetic activity. The strength of a magnetic field is measured in gauss. At the surface of the earth, the magnetic field, which causes magnetic compass needles to point north, is comparatively weak, about ½ gauss. Magnetic fields as large as 10,000 gauss have been observed in sunspots.

Hale made many observations of magnetism in sunspot groups and reported several interesting discoveries. Just as a magnet has a north pole and a south pole, the leading and following spots in a group are also opposite magnetic poles. Furthermore, all the leading spots on the northern half of the sun are the same type of pole. The leading spots on the southern half of the sun are the opposite kind of pole. Finally, and most interestingly, this whole pattern reverses each solar cycle. (See Figure 9.5.)

The question, What is a sunspot? is still largely unanswered. John Herschel rejected his father's idea that the spots were holes revealing inner layers of the sun. He speculated that a sunspot is caused when a meteor (a rock from outer space; see Section 11.9) hits the sun. He suggested that meteors travel in a swarm with a period of 11 years. This was supposed to account for the sunspot cycle.

John Herschel's idea is no longer accepted, but no contemporary theory has been able to explain all the details of sunspots either. Are the leading and following spots linked by a magnetic tube under the sun? Are sunspots some kind of magnetic thunderstorm? There is plenty of room for research in this field.

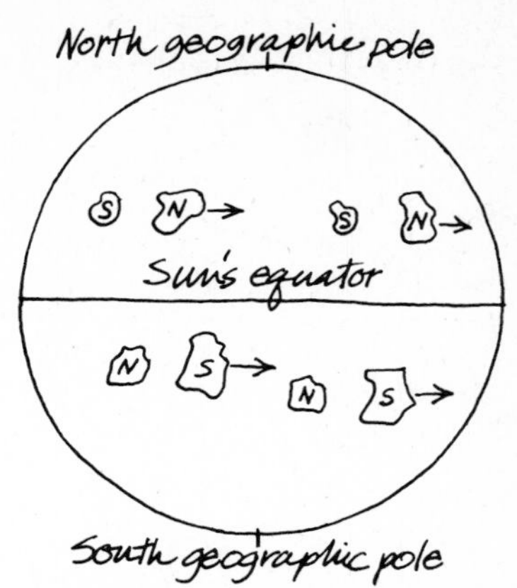

The sun near a time of high sunspot activity

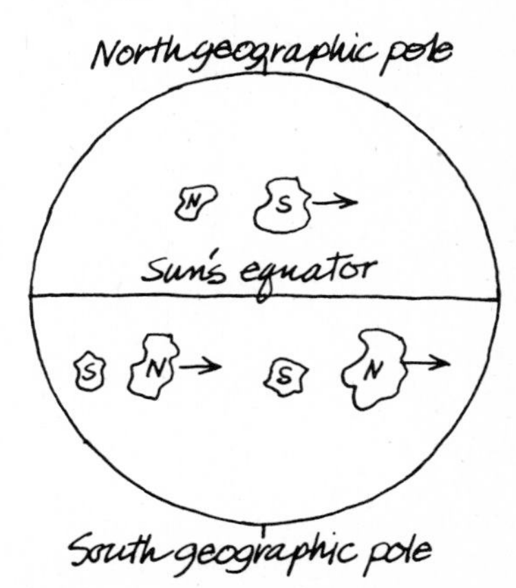

The sun about 11 years later, during the next interval of sunspot activity

Figure 9.5 Sunspots and their magnetic polarity. On the top, leading sunspots in the sun's northern hemisphere are north magnetic poles, and the following spots are south magnetic poles. The situation is reversed in the southern hemisphere. During the next solar cycle, on the bottom, the entire pattern is reversed.

The Chromosphere We will spend 38 hr traveling in the sun's 6000-km-thick lower atmosphere, the **chromosphere.** The light outside has become pink, just as the chromosphere looks from the earth. The temperature outside is rising again and, near the top of the chromosphere, is about 1 million K. This fact is revealed in measurements made of the flash spectrum during an eclipse. As we near the top of the chromosphere we see that its surface seems to resemble a burning prairie. Jets of gas about 800 km (500 mi) wide surge up to an average height of 3000 km (1900 mi), lasting about 10 min. These **spicules,** seen from earth on the edge of the sun's disk, are caused by bursts of energy from unknown causes in the sun's interior.

Solar Flares Off in the distance we see a brilliant white region of light. A violent explosion is taking place above a sunspot. This is called a **solar flare.** (See Figure 9.6.) Flares are most frequent during the active part of the solar cycle. On the average there are only a few large ones each year, and they usually fade in about an hour. Hundreds of smaller flares occur each year. At the maximum of the solar cycle, there may be hundreds of small flares per day. Large flares are the most violent of events occurring on the sun's surface. The flare emits strong radio, x-ray, ultraviolet, and visible radiation. The temperature may go as high as 5 million K in a flare.

Large flares throw a huge stream of plasma into space. (As used by physicists, the term **plasma** refers to a form of matter consisting of

Figure 9.6 A solar flare: July 16, 1959. This photograph was made using light from the hydrogen atoms on the sun's surface. A large, violent flare is occurring. Also visible are a number of dark, snakelike streaks on the sun's surface. These are called filaments and are prominences viewed against the disk of the sun rather than at its edge.

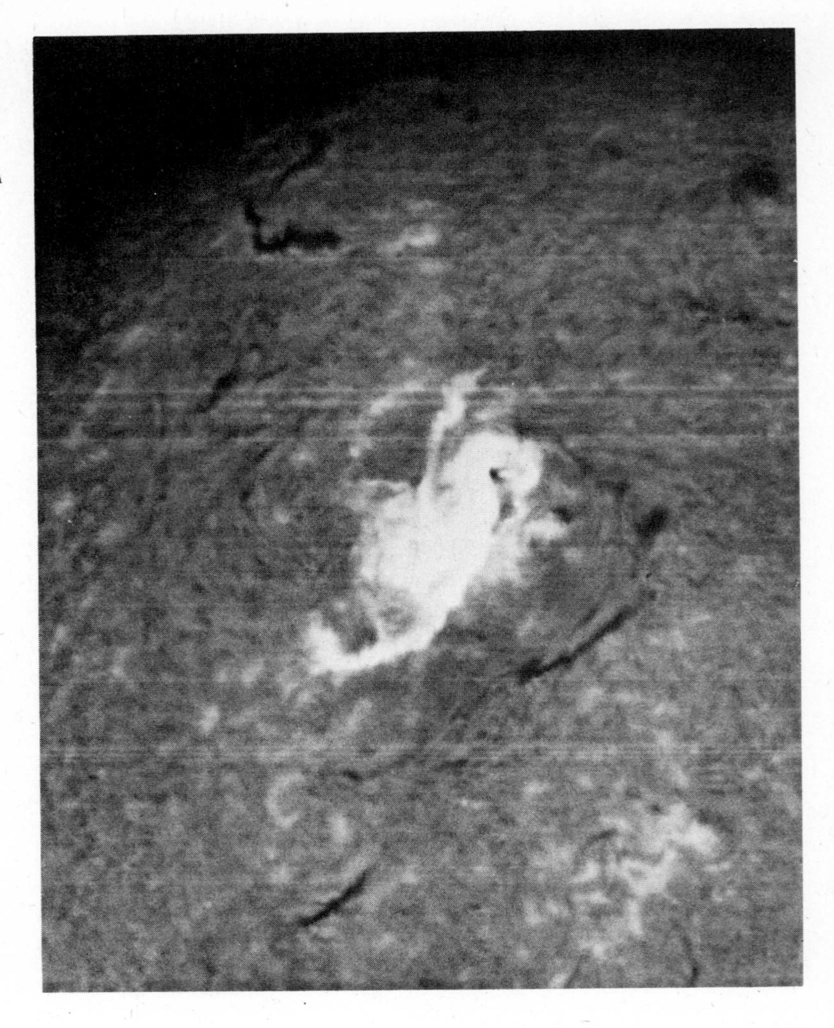

electrons and the partial atoms from which they have been torn. Plasma from the sun consists chiefly of protons and electrons, constituents of the hydrogen atom.) If the plasma is projected in the proper direction, it will eventually envelop the earth. An intense burst of radio, visible, ultraviolet, and x-ray radiation precedes the plasma, often interfering with world radio communications. The plasma itself induces the beautiful display known as the **aurora.** (See Figure 9.7.) By a mechanism that is not fully understood, the plasma causes electrons to shower down into the atmosphere. These electrons stimulate atoms of the atmosphere to produce visible light. The sun is monitored from a number of places on the earth to give warning of solar flares because of, among other things, the radiation danger to astronauts in space and to passengers in very high-altitude aircraft. As with most phenomena on the solar surface, little is known about the causes of solar flares.

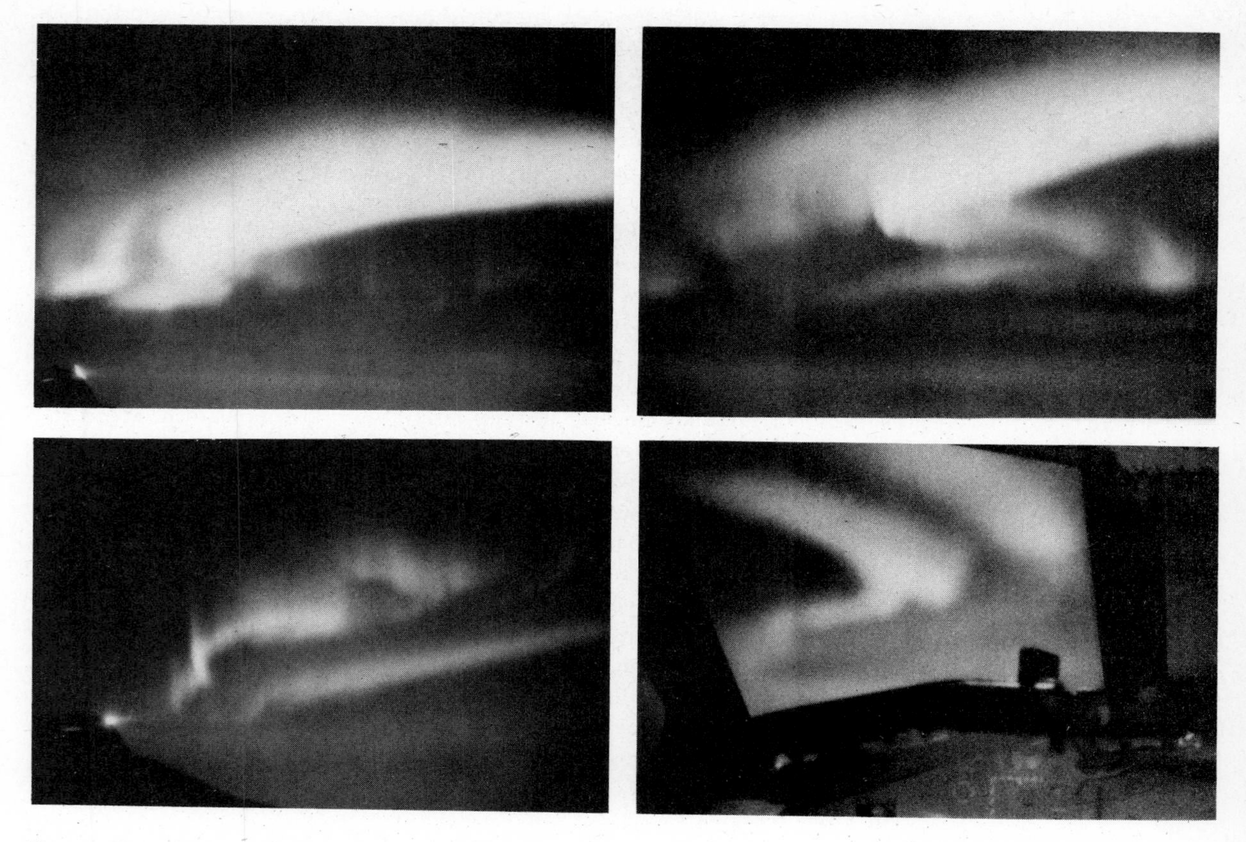

Figure 9.7 Four views of the magnificent aurora. Few people are ever privileged to view a total eclipse of the sun. Many people, however, especially those in northern climes, can, if persistent, see the aurora. A good display can be awesome indeed. These photographs were made from a high-altitude aircraft, NASA's *Galileo,* during a study of the aurora.

The Corona We are now about 40 hr and about 6400 km (4000 mi) above the photosphere. We gradually leave the chromosphere and enter the second layer of the sun's atmosphere, the **corona,** which is visible to the naked eye from earth only during a total solar eclipse. The temperature here is about 2 million K. This high temperature indicates that the particles of matter making up the corona are moving at very high speeds. However, the gas of the corona is very tenuous, or rarefied. If you thrust a bare hand into the corona, the few particles striking it would not be sufficient to prevent your hand from rapidly freezing. Why is the corona so hot? Perhaps powerful turbulence in the photosphere causes shock waves that roar up into the corona, thus maintaining its high temperature.

These shock waves apparently propel some of the upper reaches of the corona completely away from the sun. This results in the **solar**

wind, which has been measured near earth and even beyond the orbit of Jupiter by space probes. It too consists of a plasma. The solar wind leaves the sun at a speed of about 400 km/sec (900,000 mi/hr). It is estimated that the solar wind carries 1 million tons of gas from the sun each second.

There is no definite top to the corona. Some authorities indicate that the corona may end at a height of about 1.5 million km (930,000 mi), so we spend about a year in the corona as we travel outward.

Prominences At this great height we can make out the form of an incredibly large arch of glowing matter spanning the vast space between two sunspots. It is a **prominence.** Prominences are often associated with sunspots but can outlast them for months. They can reach heights of 1 million km (600,000 mi), and some move upward at about 1500 km/sec (3 million mi/hr). (See Figure 9.8.) Other prominences are relatively stationary. No good detailed explanation for prominences exists. Figure 9.6 shows some **filaments,** prominences seen contrasted against the sun's surface.

If we were to continue our trip at the same speed, 160 km/hr, it would take 41 years to reach the orbit of Mercury, 36 *more* years to reach the orbit of Venus, and an additional 30 years (107 years in all) to reach the earth's orbit.

Having studied the wonders on the surface of the sun, we turn our attention to the sun's mysterious interior. What is the source of the sun's huge supply of energy?

9.7 THE SOURCE OF THE SUN'S ENERGY

The Sun's Abundant Energy Humans depend on energy, and our energy demands increase as time goes on. It is interesting to realize that almost all of our energy is provided, directly or indirectly, by the sun. The sun supplies the energy to heat the earth and drive the weather systems. Most likely you are reading this book by means of light provided by the sun, either by direct sunlight or by a lamp. The lamp's energy is supplied by a power station which, unless it is a nuclear reactor, uses stored solar energy. For example, hydroelectric stations depend on falling water, water that was evaporated from oceans and lakes by sunlight and later fell on high ground. A coal- or oil-burning station burns fossil plants which, in the act of combustion, release energy stored by plants in prehistoric times. It is only with the advent of nuclear stations that the first trickle of energy not derived from the sun is beginning to meet human needs.

The earth as a whole receives only about one unit of every billion units of the energy the sun produces. The rest streams out into space. Ths sun has been producing an enormous quantity of energy for a very long time.

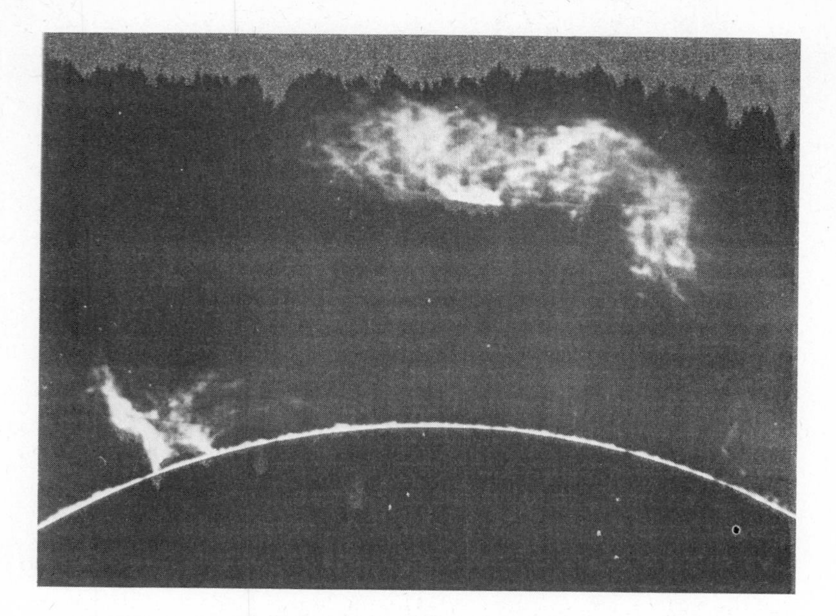

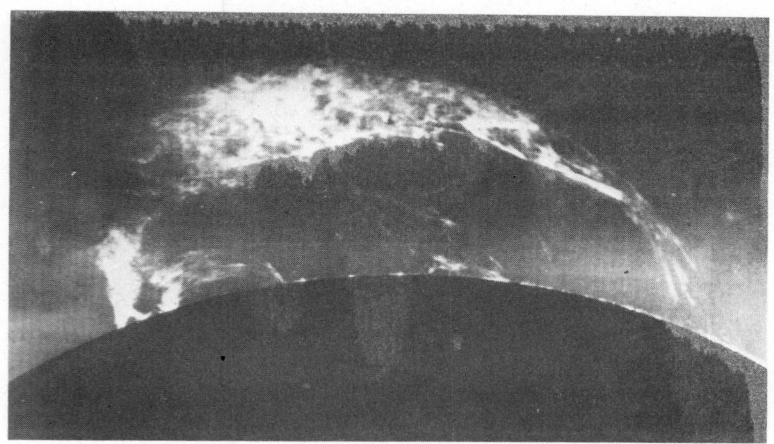

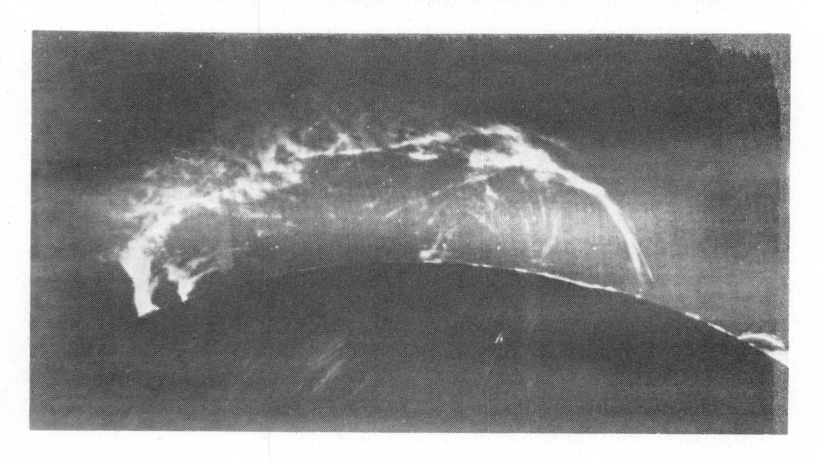

Figure 9.8 The great prominence of May 29, 1919. This mighty arch of relatively cool gas erupted from the sun and rushed outward into the corona near the end of solar cycle 15. This is an extraordinary example of a prominence, one of the most mystifying phenomena on the surface of the sun.

Such considerations lead to a difficult question. According to the conservation of energy principle, discussed later, the energy produced by the sun must originate from the release of some other form of stored energy. How can the sun produce such a lavish flow of energy for so long? What is the source of this energy? In what form is it stored, to be later released as light? We outline three early theories proposed to answer this question.

Attempt One: The Combustion Theory Anyone casting about for an explanation of the sun's output naturally considers fire. Is the sun burning, that is, undergoing combustion, like a piece of burning coal?

A chemist would quickly spot two serious flaws in this theory. Fire results when a fuel combines with oxygen. When an atom of carbon (the main constituent of coal) combines with two atoms of oxygen, the result is a molecule of carbon dioxide. Energy, stored in the electron clouds of the atoms is released as light and heat when the electron clouds interact. One says chemical energy has been released.

The chemist would criticize the combustion theory by pointing out that the surface temperature is 6000 K, and higher yet in the interior. At such extreme temperatures, the atoms of the sun are speeding about at a terrific rate. So fast, in fact, that the formation of molecules is impossible. If a carbon atom joined with two oxygen atoms for an instant, the resulting molecule would be rapidly battered to pieces again by collisions with nearby atoms. It sounds odd, but the sun is too hot to burn.

Sunspots are cool enough so that a few types of molecules, combinations of atoms, can be observed in their spectra.

The chemist would not stop here, but would attack the combustion theory from another angle. Using the known mass of the sun, she could calculate how much fuel, say carbon and oxygen, the sun must have had to start with. Then, knowing at what rate combustion produces energy, she could calculate how long the sun could burn until the fuel was used up. She would find that 2×10^{30} kg of fuel (the mass of the sun) would be used up in roughly 2000 years. We have ample evidence that the sun has been shining a good deal longer than that.

Attempt Two: The Meteor Infall Theory If the sun is not on fire, how else can its enormous output of energy be explained? In the mid-1800s, a scientist named Julius Mayer proposed that huge numbers of meteors may be striking the sun and that this might be the source of the sun's energy.

We shall meet Mayer again in the next subsection.

There are large numbers of meteoroids in orbit around the sun. (See also Section 11.9.) They are mostly tiny specks of material, although there are some larger pieces as well. When one of these meteoroids plunges into the atmosphere of the earth, it is known as a meteor. The friction between the meteor and the atmosphere heats up the air and the meteor and usually vaporizes the meteor completely. The process produces a momentary streak of light in the sky which is sometimes called a "falling star." If meteors in large enough numbers were to strike the sun, the friction would add energy to the sun in sufficient quantities to maintain its outflow of energy.

The meteor infall theory has been discarded. One reason involves the motion of the earth and the theory of gravitation. If meteors in such large numbers were to strike the sun, it would increase the sun's mass significantly. According to Newton's law of gravitation, such a mass increase would increase the gravitational attraction of the sun on each object in the solar system. In particular, the increased pull on the earth would cause it to move gradually closer to the sun. This would have a measurable effect on the length of our year: According to Kepler's third law, the more closely an object orbits the sun, the shorter its year. This would have the result that each successive year would be about 2 sec shorter than the last. Because this would be an easily detectable change and because no such change was observed, the meteor infall theory was not the end of the search for the sun's energy source.

Interlude: The Conservation of Energy Energy, a topic of much contemporary interest in our personal lives, not to mention national and even international politics, is a relatively new concept to physics. It began to emerge in the mid-1800s as Julius Mayer meditated about the sun. He was one of the first to propose that most of the energy on earth could be traced back to the sun. Then he began to wonder how the sun derived its energy. In one series of experiments, energy was expended by a horse and Mayer measured the heat energy created in its place. Involved in Mayer's thinking was the recurrent question, Where did this energy come from?

Julius R. Mayer ("My'-er," German physicist, 1814–1878) was a medical doctor but enjoyed physics more than his practice. When others received credit for his idea about the conservation of energy, the lack of recognition and several personal problems drove him to attempt suicide. About ten years later his work began to receive notice, and he eventually received a medal in recognition of his contributions.

Mayer's work led up to one of the greatest principles of physics. We now recognize many forms of energy—energy of motion, potential energy (two examples are an object raised above the ground and a compressed spring), heat energy, sound energy, electromagnetic energy, electric energy, and chemical energy, to name just a few. It is recognized that energy can change forms. Examples of energy transformations are plentiful. A block slides across a table and stops; energy of motion has been converted to sound and heat energy (the table and block are warmed by friction). In an automobile, the chemical energy of gasoline is converted into energy of motion. The energy of motion of a generator in a power plant is converted into electric energy, sent to our homes, and converted into heat energy and electromagnetic energy by a lamp.

Mayer's work, as well as that of a number of others, finally led to what is now known as the principle of **conservation of energy.** It can be stated in a number of *equivalent* ways:

In its ordinary sense, "conservation" implies "saving, preserving, or using wisely." We see that, in physics, the meaning is somewhat different.

1. In any transformation of energy, the total energy before the transformation equals the total energy after the transformation is completed.

2. Energy cannot be created or destroyed, only transformed. When energy appears, it must come from some other form (forms) of energy.

3. Imagine any region in space. If no energy crosses the borders of the region, the total energy in the region will remain constant for all time.

This principle may be likened to an article of faith. It is such a beautiful and useful idea that physicists the world over accept it as true even though no one can show that it is exactly true in all places at all times. Think what a surprising statement the principle is. Suppose there were a large room with a wall impervious to any form of energy. People could be born, live, and die in the room. Clocks could run down and stop. Food could be consumed. Bombs could go off. But no matter what transformations of energy occurred, the total quantity of energy would remain the same forever. Some physicists put it this way: "The total quantity of energy in the universe is constant."

When applied to the sun, conservation of energy demands an explanation of the electromagnetic radiation the sun spews forth. One must ask, "Which form of energy is used up in the sun to produce the sun's radiation?" We have seen that combustion and meteor infall (proposed by Mayer) will not do.

Attempt Three: The Contraction Theory Mayer proposed another possible source of energy for the sun. This idea was also propounded by another German physicist, Hermann Helmholtz, with whom the theory is most often associated. Helmholtz, like Mayer, independently conceived of the conservation of energy. He discussed the principle in highly mathematical terms, which brought the idea to prominence.

Hermann Helmholtz (1821–1894) was a man for all scientific seasons. He was interested in, did research in, and published works in most branches of physics, a feat that has seldom been equaled. His investigations ranged from electricity to the physics of seeing, and from the physics of music to abstruse topics in mathematics.

Also, like Mayer, Helmholtz sought a possible source of energy for the sun and hit upon the same solution as one proposed by Mayer. "What large store of energy does the sun have?" Helmholtz asked. "Gravitational potential energy," was his answer.

In order to understand his answer, consider the following example. A ball of putty is raised to a height of 1 meter and then released. It falls and sticks to the floor. A measurable amount of heat is produced by the collision. Since heat energy appears, it has to have come from somewhere. One says that the ball had gravitational potential energy before it was released, energy due to its height above the floor. This potential energy was converted to heat when the ball collided with the floor.

The sun is a large ball of gas. Every atom of the sun attracts every other atom of the sun because of mutual gravitational force. Thus the sun experiences a strong gravitational force tending to squeeze it down to a smaller size. The rapid motions of the atoms produce an outward pressure which tends to counteract the gravitational force so that the sun does not collapse at once. The large mass and size of the sun imply that it has a large store of gravitational potential energy. If the sun were to contract—if the atoms were allowed to fall inward—they would collide and release the gravitational potential energy as heat energy and electromagnetic radiation, also a form of energy.

Helmholtz's suggestion, known as the **contraction theory,** was that the sun was contracting slowly and that the transformation of gravitational potential energy to heat and radiation was the process by which it gained energy. The theory was open to the objection that the sun has been observed for ages and has not been observed to contract. Helmholtz answered the objection by computing that in order to maintain its temperature and its output of radiation, the sun's radius need shrink by only 40 meters (about 2.5 percent of a mile) each year. But the sun's radius is about 700 million meters. Such a small change would not be measurable, even over many years of observation. At that rate, it would take 175,000 years for the sun to shrink by even 1 percent.

The contraction theory met with other objections, but not from astronomers and physicists, most of whom found the idea quite satisfying. Whereas the combustion theory allowed the sun a mere few thousands of years to shine, Helmholtz computed that the slow conversion of the sun's vast reserve of gravitational potential energy would have allowed it to shine for about 20 million years in the past. He also gave the sun another 17 million years in the future. To many astronomers, this seemed an almost endless length of time, very comfortable and reasonable.

However, strong objections came from other quarters. The new field of geology was opening up. Geologists, by studying rocks, rivers, and oceans, found that these objects seemed to be at least 100 million years old and some demanded upward of 300 million years of sunshine in the past. In addition, those who advanced the theory of evolution could not see how the evolution of simple, one-celled creatures to the animals we see today could have occurred in such a short time as a mere 20 million years. The debate was vociferous in some quarters. Kelvin, originator of the Kelvin temperature scale, was a prominent proponent of the contraction theory. Charles Darwin, who proposed the theory of evolution, called Kelvin an "odious spectre," strong language for those days, no doubt, and then revised his book, trying to find ways to speed up the process of evolution. The sciences were at loggerheads. Something had to give.

Today, most geologists agree that the earth is about 5 *billion* years old.

Interlude: Einstein and Mass-Energy In the end it was the astronomers who changed their opinion. This change was only one of the numerous by-products of the ideas conceived by the man whom future centuries may well consider the most important of our century, Albert Einstein. (See Figure 9.9.)

Einstein, one of the few ranked with the likes of Newton, was in some ways an indifferent student. A high school teacher made the famous prediction that he would "never amount to anything." In college he studied hard only topics that interested him, skipping most of the lectures and even flunking an important comprehensive exam. Nevertheless, his mind was always extremely active. He could not find an academic position upon graduating and wound up working in a patent office. Then, in 1905, he published the first of a series of papers, many of which were concerned with what we call the theory of relativity. These ideas were destined to affect the course of history and physics and to touch the lives of all who followed him.

We mention a few of the ideas contained in the theory of relativity from time to time in this book and we will discuss more of them later.

Figure 9.9 **Albert Einstein (1879–1955): the man who shook the foundations of physics.**

In this section, we describe Einstein's radical modification of the principle of conservation of energy. About a century before the formulation of this principle, another, similar scientific law had come into prominence. The principle of the **conservation of mass** states that mass cannot be created or destroyed. When a piece of paper is burned, the ashes that remain have less mass than the original paper. Has mass disappeared? Carefully controlled measurements show that it has not. The total mass of the paper plus the oxygen in the air before the burning equals the total mass of the ashes, remaining unused oxygen, and gasses given off by the burning. Mass has been conserved.

Einstein thought long and hard about the basic ideas of physics and, in some cases, found them unsatisfying. This led him to modify and revitalize these foundations. One of his most striking proposals was that mass and energy are equivalent. He proposed, in other words, that one should consider mass a form of energy. This idea altered the outlook of all physicists. He was saying that the two great conservation principles are not true; at best they are incomplete. Einstein raised the possibility that mass might be made to disappear; the result would be an equivalent quantity of energy. Similarly, if energy were to be destroyed, an equivalent quantity of mass would appear in its place. Einstein proposed a new principle to replace the other two, the **conservation of mass-energy.**

In the course of Einstein's research, he derived an equation expressing the equivalence of mass and energy in quantitative terms. This equation, probably the most famous equation in all of physics, allows one to calculate how much energy E is derived when a given amount of mass m is transformed into energy:

Anyone who knows only one equation of physics most likely knows that $E = mc^2$.

$$E = mc^2$$

In this equation c is a constant, a number which, when squared, is to be multiplied by the amount of mass that disappears in order to compute the amount of derived energy. The equation may also be used to compute the amount of resulting mass derived when a given amount of energy disappears.

The constant c has some interesting aspects. Let us express energy in joules. One **joule** ("jool") is a unit of energy equal to that gained by a 1-kg mass (which weighs about 2.2 lb on earth) when it is raised a distance of 10.2 cm (4.02 inches). Equivalently, if a 1-kg mass is dropped from rest from a height of 10.2 cm, it produces 1 joule of heat and sound energy when it strikes the ground. A 100-watt light bulb consumes 100 joules of energy per second. Now, if we express mass in kilograms, then, Einstein demonstrated, $c^2 = 9 \times 10^{16}$ joules per kilogram. The striking implication of the enormous size of c^2 is that a tiny amount of mass is equivalent to a large amount of energy. We may calculate that, if we somehow transformed all of the mass in a dime (2.5 grams) completely into energy, we could light 71,000 100-watt light bulbs for one year. We would have to burn roughly $1\frac{1}{2}$ million gallons (gal) of gasoline to produce the same quantity of energy.

The constant c in Einstein's equation can also be expressed in the form $c = 300{,}000$ km/sec, the speed of light. The speed of light, so fundamental in Einstein's theory, appears here in an unexpected place—as the constant used to convert mass to energy. In this context, the appearance of c has *nothing* to do with an object moving at the speed of light. We must not interpret $E = mc^2$ to mean, for example, that when an object moves at the speed of light it is changed into energy.

After the initial shock and disbelief wore off, scientists began to realize that here was a theory offering a potentially awesome supply of energy, if one could only find a way to convert mass into energy. The first notable application of $E = mc^2$ was in the construction of bombs, the atomic and the hydrogen bombs. The A-bomb was first used in Japan to end World War II. Today we are only a pushbutton away from having even more powerful nuclear bombs unleashed again. A more peaceful application of the conversion of mass into energy occurs in nuclear power stations.

Attempt Four: $E = mc^2$ and the Source of the Sun's Energy $E = mc^2$ also intrigued a number of astronomers. It might explain how the sun has created so much energy for so long a time, namely, by changing a small amount of its matter into a large amount of energy. Going on this assumption, one can use the known rate at which the sun emits energy to calculate the rate at which it is using up its mass. The answer is staggering at first glance. The calculation using $E = mc^2$ shows that the sun converts about 5 billion kg (or about 5 million metric tons) of itself into energy every *second!* "Surely," one might gasp, "at that rate, the sun would not last long!" This initial reaction, however, is wrong: The sun has a great deal of mass to splurge.

To convince yourself of this, write down a 2 followed by 30 zeros. This is the mass of the sun in kilograms. Now subtract the number 5 billion. Do it again. Continue subtracting 5 billion repeatedly until you get tired. You will notice that the original number has not changed significantly. In fact, it would take the sun roughly 130 billion years to use up even 1 percent of its present mass at the rate of 5×10^9 kg/sec. (The sun is not expected to remain unchanged nearly that long. This topic is discussed in Chapter 13.)

Thus Einstein's equation broke the deadlock between the sciences. Here was a new source of solar energy to be considered. If the concept proved feasible, the sun might be found to be old enough to please nearly everyone.

Sir Arthur Eddington is further discussed in Chapter 13.

Eddington's Fusion Theory It seems that the sun *might* be making energy by converting mass to energy. But *how?* The first glimmering of what the actual process might entail was conceived by Arthur Eddington. He appears to have been the first to suggest that the sun is converting its supply of hydrogen into helium, losing mass and gaining energy in the process.

The discussion here is, in the main, true to Eddington's thoughts. For convenience, it has been modified slightly in the light of some later discoveries. For example, at the time, neutrons had not been discovered.

Eddington's concept flowed from considerations such as the following. The sun is composed chiefly of hydrogen, the simplest atom. The nucleus of a hydrogen atom was what interested Eddington. The nucleus of any atom contains more than 99.99 percent of the mass of the atom. (Electrons have little mass.) A hydrogen nucleus consists of one proton, a particle having a positive electric charge. Helium is the next atom in nature in order of complexity. Its nucleus is composed of two protons and two neutrons. (A neutron has a mass roughly the same as that of a proton, but no electric charge.) Now, it had been found not long before that a hydrogen nucleus has a mass of 1.008 units and that a helium nucleus has a mass of 4.003. Eddington noticed that *four* hydrogens have a mass just slightly *more* than that of one helium ($4 \times 1.008 = 4.032$, which is greater than 4.003.)

Eddington's proposal, then, was this. Deep inside the sun where, as

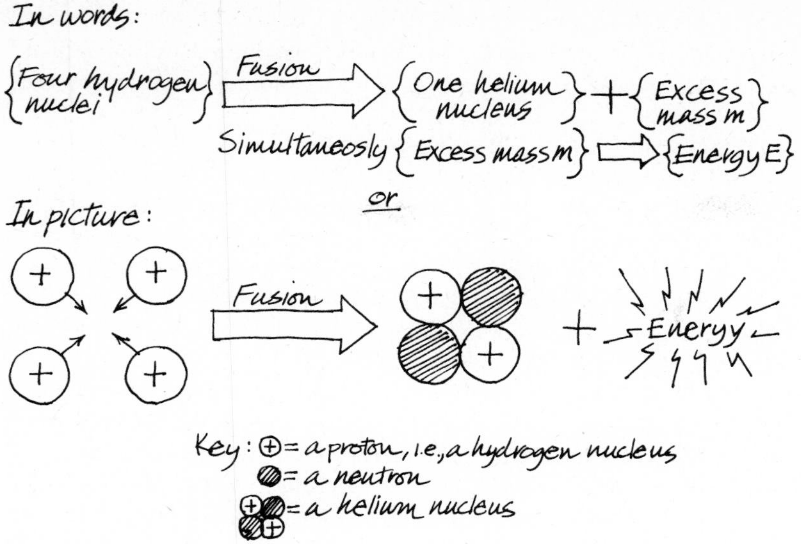

Figure 9.10 Fusion. Two simple ways of imagining the process of fusion, thought to produce the sun's energy, are shown. As discussed in the text, this picture of fusion is incomplete. How can four protons colliding simultaneously occur with any reasonable frequency? How are two protons converted into neutrons?

his studies indicated (see Section 13.3), the temperature and pressure are enormous, four hydrogen nuclei come together and fuse (combine). Two of the protons somehow turn into neutrons. The result is a helium nucleus. But that cannot be the sole result, for some of the mass from the four protons is left over, as noted above. The excess mass, according to the proposal, is changed to energy according to the formula $E = mc^2$. The reaction is shown schematically in Figure 9.10. The process is called **fusion.** Later, in the hydrogen bomb, fusion was employed on earth. Research is underway to harness fusion for the peaceful production of energy.

Everyone realized that Eddington's insight was only the germ of an idea. The calculated probability of four protons happening to collide simultaneously was so small as to be insignificant. Furthermore, how could two protons be converted into neutrons? These and many other, more technical problems were solved by Hans Bethe in 1939. He found a multistep process with a high probability of occurrence in which protons are, in effect, added one at a time. In this process, neutrons are produced as required. The overall result of Bethe's process proceeds as outlined above, but with many intermediate steps. Bethe found a way to make everything work out correctly and plausibly by the end of the cycle. Bethe's proposal, known as the **carbon cycle,** is a complex, monumental achievement for which Bethe received the Nobel Prize in 1967.

Hans Bethe (German-American physicist, born 1906) contributed important work in a number of fields. He worked on the theory of the emission of electromagnetic radiation by accelerating charged particles and helped the United States develop the atomic bomb during World War II.

The carbon cycle begins when a proton is added to one of the relatively rare carbon atoms in the sun. A number of other reactions ensue. The upshot of the cycle is that four protons are consumed and one helium nucleus plus energy are produced. In the last step of the process, a carbon nucleus is also produced, so that the carbon is not used up; it merely serves as a vehicle for the process. Calculations indicate that about 8 percent of the sun's energy is produced by the carbon cycle. The remaining 92 percent, it is now believed, is created by the **proton-proton reaction.**

The proton-proton reaction is a process which was discovered after the carbon cycle. In it, no carbon nuclei are needed. Two protons combine at first and then a third proton is added, forming a three-particle nucleus. Two such nuclei then combine. The resulting products of this last reaction are a helium nucleus, two leftover protons, and excess mass which is converted into energy. In stars hotter than the sun, the carbon cycle is more significant than it is in the sun, and in very hot stars it is dominant.

A high temperature is required to initiate the process of fusion. Modern calculations indicate that it can occur only in the central region of the sun, where the temperature reaches 15 million K. At this extreme temperature, Wien's law implies that very short-wavelength radiation, x-ray radiation, is produced. One astronomer asserted that a pinhead of matter at a temperature of 15 million K could cause death from a distance of about 100 km (62 mi). The liberated x-rays in the sun's central region are greatly hampered in their motion by the immense quantity of matter lying between them and the surface. Roughly 5 million years are required for the radiation to work its way out to the surface. As it reaches successively cooler layers of the sun, the wavelength of the radiation lengthens and it emerges from the surface as light.

In the sun the process of fusion yields surprisingly little energy per unit of mass. One kilogram of the sun produces about 18 joules of energy per day. For comparison, 1 kg of a human produces about 92 joules of energy per day. The reason for the low output of fusion is that very little of the mass of the initial protons is converted to energy in each fusion reaction. It is only the fact that the sun has such a large mass that enables it to produce such enormous quantities of energy.

We can see that the process of fusion is very different from a chemical reaction such as occurs in combustion. In a chemical reaction, energy is released by the interaction of the outer electron clouds of whole atoms. In fusion, the atoms, stripped of their electrons by the high temperature, interact as nuclei. Much more energy is available in the massive nuclei. Fusion is said to be a **nuclear reaction.**

Given the success of the fusion theory, the long history of the search for the source of the sun's energy seemed to have reached a satisfying conclusion. Based on this theory, the sun's age was estimated to be about 5 billion years, in satisfying conformity with geological estimates of the age of the earth. Textbook authors wrote that the fusion theory had been proven. But then, Raymond Davis bought 100,000 gal of cleaning fluid.

The Missing Neutrinos One of the most unusual subatomic particles discovered in modern times is the **neutrino.** First detected in 1956, it is a particle of matter with no charge and no mass when at rest. It can only exist when traveling at the speed of light. Perhaps its most interesting property is its ability to penetrate matter. If 100 neutrinos were to strike a lead wall 6 million AU thick, about 60 of them would be stopped, but the other 40 would emerge from the other side 100 years later.

In the mid-1960s, Raymond Davis, Jr., of the Brookhaven National Laboratory assembled a neutrino detector about 1.6 km (1 mi) underground in the Homestake Gold Mine in Lead, South Dakota. He intended to observe neutrinos from the sun. The detailed theory of fusion predicts that the sun produces large numbers of neutrinos. Unlike the radiation produced by fusion, most of the neutrinos escape immediately from the sun. Davis calculated that, on the average, out of the billions of solar neutrinos passing through his detector, one per day would be stopped and detected. The detector consisted of 400,000 liters (about 100,000 gal) of a common dry-cleaning fluid. The idea was that a captured neutrino would cause a nuclear reaction in the fluid, the results of which should be detectable.

The neutrino detector was placed deep underground to shield it from interference from cosmic rays (see Section 14.6) which are unable to penetrate so far.

This experiment, one of the most difficult ever attempted, was considered worthwhile for many reasons. The light we see from the sun was produced near its center about 5 million years ago. The neutrinos, on the other hand, would be only 8 min old and would yield information on the present state of the sun's interior. Success would also confirm the fusion theory—not that confirmation was really felt necessary.

The results of the experiment rocked the world of astronomy. Davis detected considerably less than a quarter of the number of neutrinos required by the theory of fusion. His experimental apparatus has been subjected to the most critical scrutinizing and constant refinement, and yet the discrepancy with the theory of fusion remains and may remain for some time.

What does this mean? Speculation is still going on, and no really satisfying resolution has been found. Does the experiment, which is much more complex than the foregoing description can convey, have an undiscovered flaw? Do some neutrinos disintegrate before they can reach the earth? Has the process of fusion in the sun slowed down for some reason? If so, will it speed up again?

The discrepancy between observation and theory would be lessened if the sun's central temperature were lower than current theory calls for. A lower temperature would result in a slower rate of fusion in the sun and thus a slower rate of production of neutrinos. Research is going on to determine whether the theory might be flexible enough to allow such modifications.

The most likely outcome of this conflict is that the theory of fusion will have to be patched up somehow. Yet, although the possibility is thought to be very remote, we may be witnessing what has happened so often in the past, the decay of an established theory and, perhaps someday, the rise of a new one in its place.

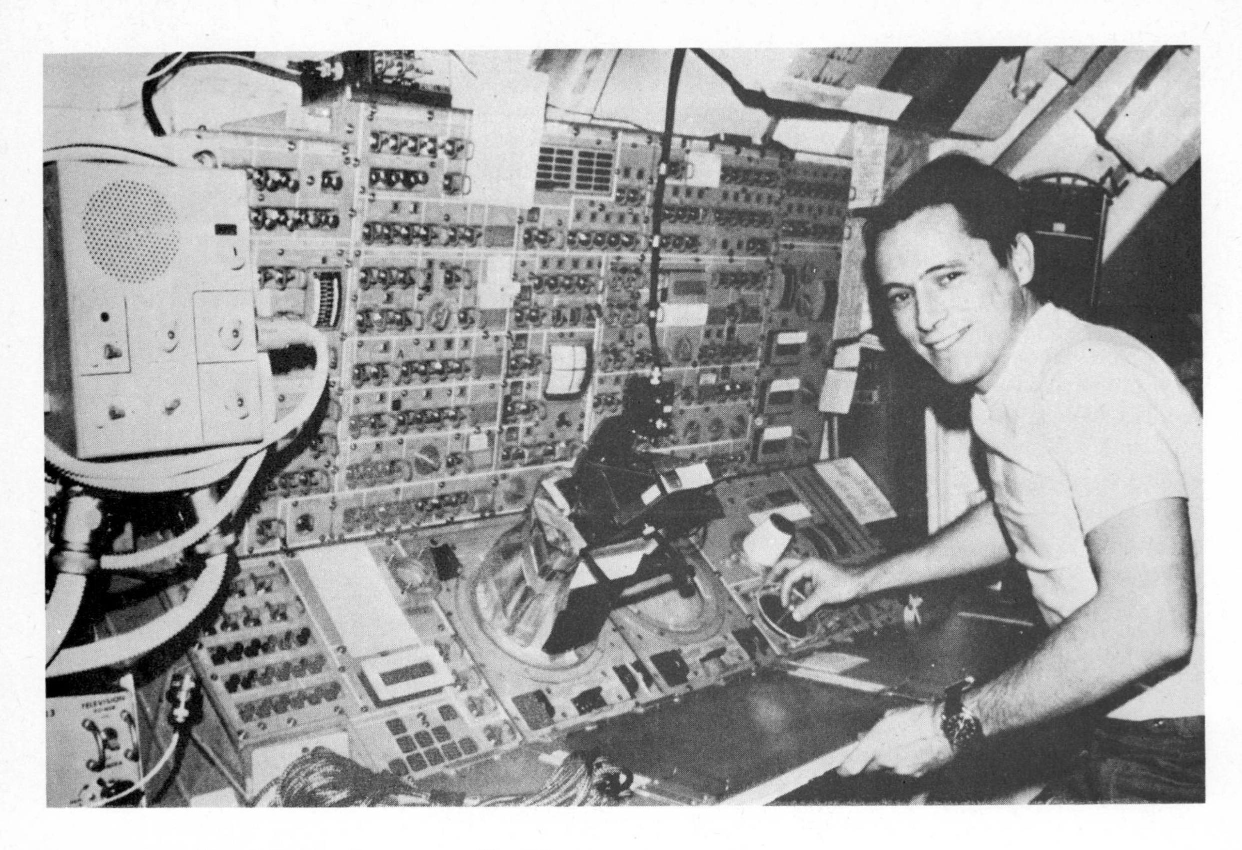

Figure 9.11 On board *Skylab:* the telescope mount. Astronaut Edward Gibson demonstrates the control board used to study the sun and other objects such as Comet Kohoutek during his 84-day stay aloft. Apparently Gibson is holding onto the desk so that he will not float away.

9.8 SKYLAB AND THE SUN

Imagine that the chair you are sitting on is suddenly jerked out from under you. The physical sensation you would have at that moment is known as "freefall," or, somewhat inaccurately, "weightlessness," or "zero gravity." (See Section 6.9.) Space travelers may be subjected to long periods of freefall. How will it affect their health?

An experiment was begun in 1973 which had, in part, the purpose of investigating this question. *Skylab* was an earth satellite as large as a house. It was designed to support three human lives for extended periods. Three crews of three men each spent, respectively, about one, two, and finally three months aboard *Skylab.*

The results were encouraging for the prospects of extended space travel. All astronauts adjusted quickly to the feeling of constantly floating, performed their tasks well, remained healthy, and resumed their normal lives on earth after a short period of readjustment. One of the astronauts, Owen Garritt, reported that his two-month experience of freefall was "quite pleasant, enjoyable, and absolutely fascinating."

The two other major objectives of *Skylab* were to study the earth from space and to study the sun for extended periods of time as viewed from above the earth's atmosphere.

The experiments were carried out using the *Apollo* telescope mount, an extensive array of cameras, telescopes, and other equipment. (See Figure 9.11.) The atmosphere of the earth absorbs much of the sun's radiation that does not have wavelengths in the visible region. The sun's ultraviolet and x-ray radiation was investigated by the astronauts.

For the first time the ultraviolet and x-ray radiation from the intensely hot corona could be studied for extensive periods. It was found that the corona exhibits much more activity than had previously been known. Figures 9.12 and 9.13 show the appearance the sun would have if we could see it with eyes sensitive to two different wavelengths of ultraviolet radiation. In each photograph, the image at the right was produced by radiation having a wavelength of 304 Å caused by helium atoms which lack one electron. The vividly contrasting image at the left, which partially overlaps the helium image, was produced by iron atoms which lack 14 of their normal 26 electrons and which produce radiation having a wavelength of 284 Å. The contrast is striking. The iron atoms emit radiation strongly in regions of the lower corona, which are regions of intense magnetic activity. The helium atoms radiate from much of the corona and reveal enormous prominences as well. It had been hoped that a few eruptions of the corona might be observed during the mission. Actually, more than 60 violent eruptions were observed, indicating its newly revealed dynamic nature.

As viewed at even shorter (x-ray) wavelengths, 3 to 60 Å, the radiation from the over 1-million-K corona revealed several surprises (Figure 9.14). Solar x-rays are emitted primarily from its hottest regions, and complex loops and arches are revealed, many of which seem to connect separate active regions.

Figure 9.14 also shows examples of **bright points.** Not apparent in visible light, they show clearly in x-ray photographs. They are tiny, intense regions of activity which can vary in brightness from one minute to the next and which last an average of 8 hr. It is estimated that about 1500 bright points are formed on the sun each day.

A major *Skylab* program involved the observation of **coronal holes,** regions of the corona in which x-ray emission is weak. In Figure 9.14, examples of coronal holes may be seen at the sun's north pole (just to the left of the top of the sun's disk) and near the center of the disk. It is speculated that these are the principal regions from which the solar wind pours out into space.

See Plates 14 and 15 for color photographs of the sun.

SUMMING UP

The luminosity of the sun (4×10^{26} watts) is derived from a measurement of the solar constant. An analysis of the sun's thermal radiation permits a determination of the sun's surface temperature (6000 K). Study of the sun's absorption spectrum leads to a chemical analysis of the sun; it is composed chiefly of hydrogen, with some helium and traces of other atoms.

Figure 9.12 The sun at two wavelengths in the ultraviolet. In addition to the features discussed in the text, the helium image of the sun shows a huge "spider" prominence which the sun expelled. Consisting of cooler helium gas, its upward motion appears to have been partially blocked. This photograph was the first indication that helium gas can be spewed from the sun and hold together over such enormous distances. (See the dot representing the size of the earth to give one the scale).

Figure 9.13 The sun at two wavelengths in the ultraviolet. Near the bottom of the image, one can see a helium prominence. When viewed in white light, this eruption was seen to produce a huge coronal bubble of luminous gas much larger than the sun, which expanded at a rate of about 1.6 million km/hr (1 million mi/hr).

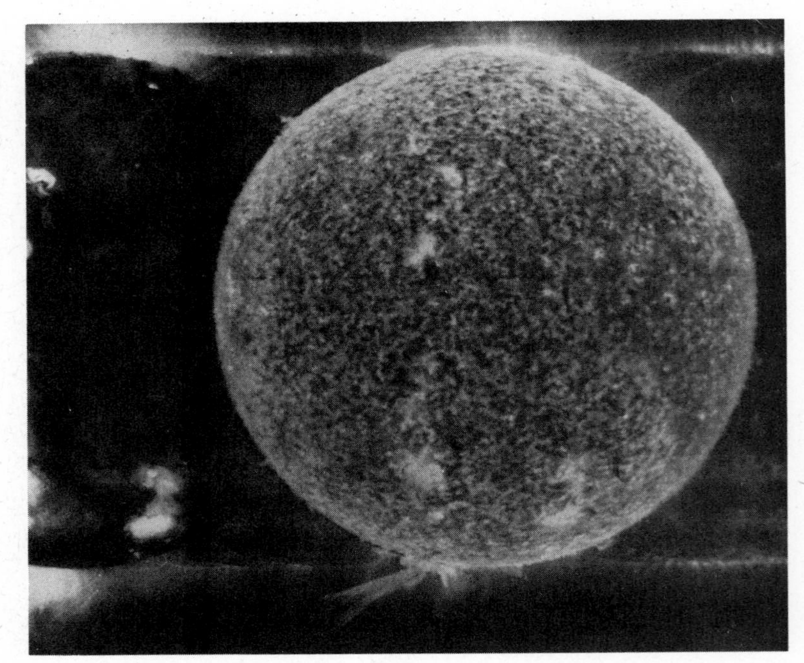

The sun exhibits a large number of mysterious surface phenomena. On or near the photosphere are found granules, sunspots, and solar flares. Higher up is the chromosphere, topped by spicules. Higher yet is the corona into which prominences project and from which the solar wind emanates. Bright points and coronal holes have recently been discovered.

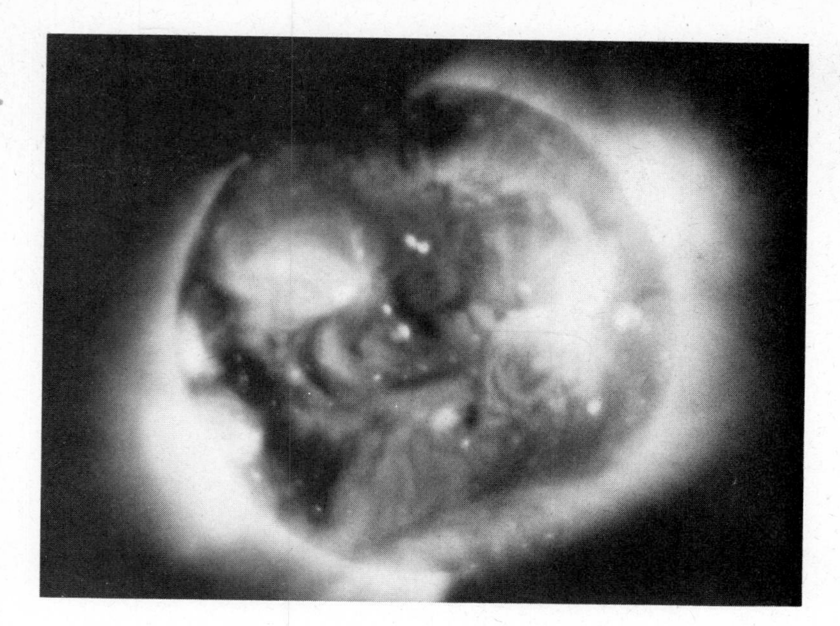

Figure 9.14 The sun as viewed using x-rays. The hottest regions of the corona are revealed by photographing the x-ray emissions from the sun.

The sun is currently thought to produce energy by a process known as fusion. This theory has failed to receive a clear confirmation from the attempt to detect solar neutrinos.

In the next chapter we investigate the nearest members of the solar system, the moon and the planets Mercury, Venus, Earth, and Mars.

EXERCISES

1. Describe these solar phenomena: solar wind, flare, bright point, coronal hole.

2. Describe the magnetic phenomena associated with sunspots and the solar cycle.

3. Define the following terms: bolometer, heat, absolute zero, thermal radiation, principal wavelength, (chemical) element, continuous spectrum, absorption spectrum, emission spectrum, plasma.

4. What is the conceptual difference between the apparent brightness of an object and its luminosity?

5. Why is the Kelvin scale of temperature an appropriate scale to use in the light of our understanding of the nature of heat? Should Lord Kelvin wear a coat outside if his outdoor thermometer reads 273 K? What is normal body temperature in kelvins? What is the approximate surface temperature of the sun in kelvins?

6. What is Wien's law? How does it explain why different stars have different colors? What steps does one follow to measure the surface temperature of the sun or a star?

7. Sketch a rough graph of the amount of radiation versus the wavelength of radiation for two objects having differing surface temperatures. What is the connection between the peak of a curve on this graph and Wien's law?

8. Give Kirchhoff's laws of spectroscopy and an example of a situation in which each law is applicable. What is the importance of the third law to astronomy? What is the chemical composition of the sun?

9. Name six examples of the

sun's influence on the earth and its inhabitants.

10. What is the principle of the conservation of energy? Give some examples in which one sort of energy is converted to another. What sort of energy conversion is envisioned by the contraction theory? By the fusion theory?

11. Explain why each of the following theories fails to explain the source of the sun's energy satisfactorily: (a) combustion, (b) meteor infall, (c) contraction.

12. Explain the symbols in the formula $E = mc^2$. What is the significance to solar astronomy that c is a large number?

13. Describe the process of fusion as it is presently thought to occur in the sun's central regions. What experiment has cast some doubt upon the fusion theory?

READINGS

Valuable and authoritative is

The Quiet Sun, by Edward G. Gibson, NASA, Washington, D.C., 1973, NAS 1.21:303.

Although later chapters are more technical, Chapter 2 is appropriate for readers of this text. Gibson studied the sun as an astronaut on *Skylab*.

Atoms and Astronomy, NASA, Washington, D.C., 1976, NAS 1.19:128

discusses the production of spectral lines by atoms.

Two good biographies of Einstein are

Albert Einstein, Creator and Rebel, by Banesh Hoffman, Viking Press, New York, 1972.

Einstein, the Life and Times, by Ronald W. Clark, Avon Books, New York, 1971.

THE INNER SOLAR SYSTEM: A VIEW OF OUR NEIGHBORHOOD

Study Figure 10.1, imagining that you are an astronaut passing close to this planet and searching for signs of life. What would you conclude? This photograph was taken by NASA from an orbit around one of the planets in the solar system.

Would you believe that this is one of the more densely populated regions on planet earth? Long Island, New York, stretches from east to west. A ship in the Atlantic could sail eastward along Long Island's southern coast into Lower New York Bay, turn north, and pass through the Narrows into Upper New York Bay. The southern tip of Manhattan is the place where the bay is divided into the East River and the Hudson on the west. Life is not easy to recognize from orbit.

In this and the next chapter, we discuss the objects in the solar system that orbit the sun. We will take three points of view: how each looks to the naked eye, through a telescope, and from a space probe.

10.1 THE BLUE PLANET

The Earth As Observed from Space A visitor to our solar system might find each of the planets rather fascinating. The third planet from the sun (Figure 10.2), might catch the being's eye (if any) and perhaps arouse enough curiosity to encourage a closer look. What would the being find?

The earth is relatively close to the sun, only 2.5 percent as far from the sun as the most distant known planet. Its orbit is nearly circular, and its distance from the sun only varies by about 3 percent. A small planet, although not the smallest, it has a diameter of about 12,700 km (7900 mi). This planet spins somewhat slowly on its axis, once every 24 hr. The rotation produces a slight bulge in the planet; the equatorial diameter is about 42 km (26 mi) longer than the polar diameter. The axis of spin of the planet is tilted by an angle of about $23\frac{1}{2}°$ from the perpendicular to its orbit. This tilt produces a yearly cycle of mild weather changes known as seasons.

The visitor can observe the moon in orbit around the earth and, using Newton's laws, compute the earth's mass (see Box 8.4): It is roughly 6×10^{24} kg (6000 billion billion tons). This makes it the most massive of the smaller planets. Next the visitor may wish to compute

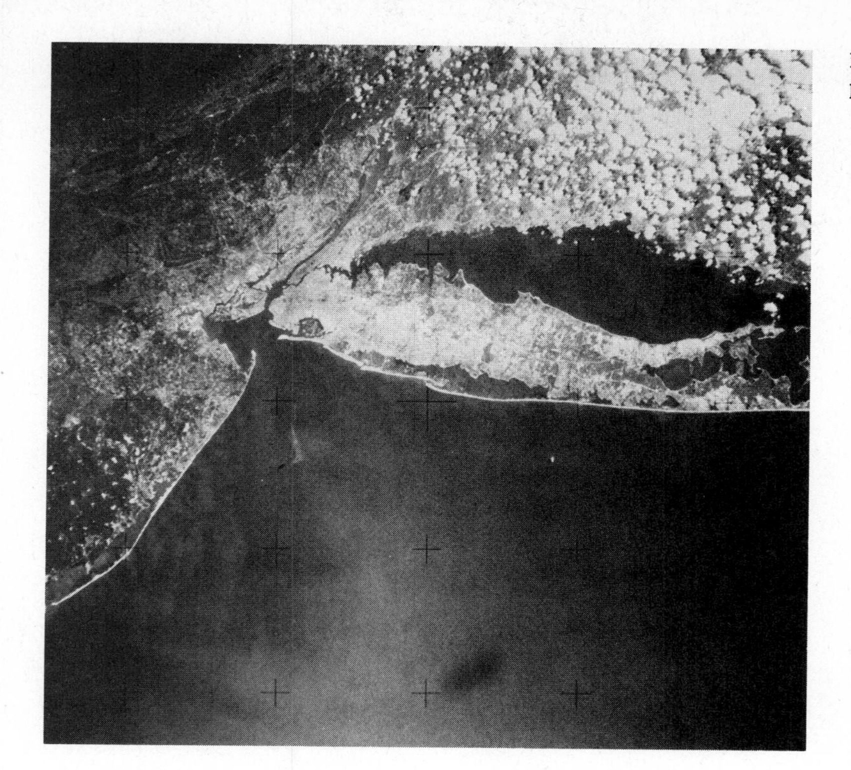

Figure 10.1 Does life exist on this planet?

the earth's density, for the necessary measurements have already been made.

The mass of an object is the amount of matter in it, and the **volume** of an object is the amount of space it occupies. An important concept in physics is the **density** of an object, which is its mass divided by its volume. You may express the mass of an object in grams and the volume of an object in cubic centimeters. If so, the density of the object is then expressed in grams divided by cubic centimeters or, in the more usual wording, grams per cubic centimeter. The density of liquid water is a handy (and easy) piece of data to remember; it is 1 gram per cubic centimeter.

A note for those who may be unfamiliar with grams and cubic centimeters. On earth, a pound object has a mass of about 454 grams or 0.454 kg. A fluid ounce ($\frac{1}{128}$ gal) is a volume equivalent to about 29.6 cubic centimeters.

The density of an object gives information about the type of substance and the state of the substance in the object. The densities of some common substances are given in Table 10.1.

Table 10.1 The Density of Some Common Substances

Substance	*Density (grams per cubic centimeters)*
Wood	About 0.5–0.9
Alcohol	0.8
Water	1.0
Aluminum	2.7
A typical rock	About 2.7
Iron	7.6
Lead	11.3

Upon dividing the earth's mass by its volume (computed by using the known average diameter), our visitor finds that the earth's density is about 5.5 grams per cubic centimeter, a value intermediate between those of rock and iron.

The mass and the diameter of the earth may also be used in a formula derived from Newton's law of gravitation to compute the earth's surface gravity. This, combined with the average temperature on the surface of the planet, 20°C (or 68°F), indicates that the earth is able to gravitationally hold all gasses in its atmosphere

Figure 10.2 Nearly full earth. This was the view of the earth obtained by the Apollo 10 astronauts on the way back to their home planet after visiting the region of the moon. The west coast of North America can be seen through the swirling clouds.

except those composed of the lightest atoms, hydrogen and helium. The average temperature could be determined by finding the principal wavelength of the infrared radiation produced by the earth and using Wien's law (Section 9.4).

It would be obvious to the visitor that the earth has an atmosphere. The visitor would see the ever-changing cloud patterns (Figure 10.2) and the blue color caused by the scattering of sunlight in the atmosphere. A spectroscopic analysis of the earth's atmosphere would reveal that it is composed chiefly of nitrogen (78 percent) and oxygen (21 percent) with traces of other gases, water vapor being one of the most important.

Plate 5 is a color photograph of the earth.

The atmosphere of the earth presses down on the earth's surface with a force of about 1000 newtons on each square meter of surface or, equivalently, 1.1 tons per square foot. This is sufficient pressure to allow liquid water to exist on the surface of the planet. In the solar system, standing water is unique to the earth. (If the pressure were too low, the atoms in liquid water would fly off in all directions; that is, the water would boil.) Indeed, the visitor would see that about two-thirds of the planet's surface is covered with bodies of that precious substance, water, and could then deduce that the clouds consist of water.

The visitor's on-board detectors would also find that the earth has a magnetic field. The earth is a large magnet having one pole in the northern hemisphere and the other in the southern hemisphere. The earth's magnetic field produces many effects. It causes compasses to point but also traps electrons and protons in large radiation belts, the Van Allen belts, which lie high above the earth's surface. The visitor would probably not want to spend much time in these belts because they might affect its health.

The visitor would probably detect our radio, television, and radar broadcasts and come in closer to look for the civilization that produces them.

A Brief Sketch of the Earth's Early History A number of techniques have been worked out in an attempt to determine the age of the earth. The most commonly accepted methods analyze radioactive elements found in rocks. These methods indicate an age of about 4.5 billion years. Interestingly, similar analyses of certain moon rocks and meteors have arrived at approximately the same age for them. This, then, is taken to be roughly the age of the solar system as a whole.

As discussed in Section 11.10, the commonly held view is that the planets formed as small chunks of matter, which somehow became clumped together to produce bigger chunks. One of these eventually became the earth. It is thought that, early in its history, the earth was in a melted state. Suggested heat sources are the gravitational infall that produced the earth, meteor bombardment (see Section 11.9), and the heat released as radioactive elements decayed. In this liquid state, the earth **differentiated,** that is, the various substances separated. The more dense substances went to the center and the less dense floated to the surface. In this way, it is assumed, an iron core formed in the center of the earth and such elements as aluminum, silicon, sodium, and potassium floated up and formed its crust. Studies of the waves produced in the earth by earthquakes verify that there are three main layers in the earth's interior.

The pressure at the center of the earth must be enormous, so enor-

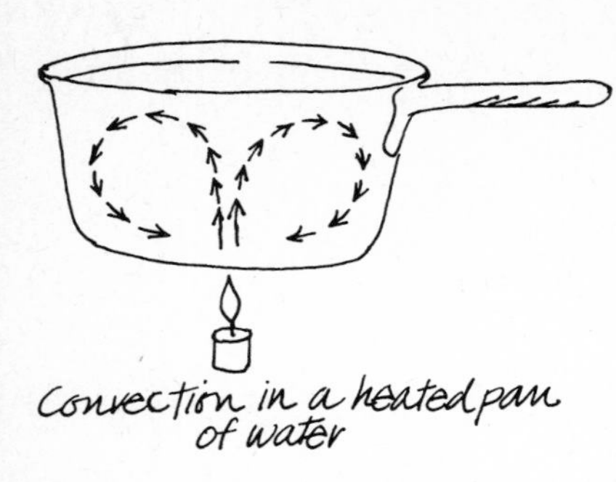

Figure 10.3 Convection. Sawdust may be added to the water to make the motion of the water visible.

mous that, even at its elevated temperature (perhaps 4000°C or 7000°F), the iron at the center is a solid. Surrounding the solid iron is thought to be a thick layer of liquid iron. Above that there seems to be a thick layer of dense rock known as the mantle. Above this is the relatively thin rocky crust upon which we live.

The layer of liquid iron inside the earth is interesting to those who study our planet. Presumably, its lower levels are considerably hotter than its upper levels. This is likely to induce **convection,** a type of vertical circulation, in the liquid. One can observe convection in a pan of water heating on a stove if a substance like sawdust is sprinkled into it to act as a tracer. The hot water near the bottom of the pan rises, and cooler water falls to take its place. See Figure 10.3. Convection occurs in our atmosphere. On a summer's day, puffy cumulus clouds are evidence of moving columns of air which have risen high enough that the water vapor in the air has cooled and condensed into water droplets.

A widely held hypothesis supposes that the convection of the liquid iron has an effect observable above the earth's surface: the magnetic field. It might be that this convection and the rotation of the earth supply the energy required to produce an electric current which, in turn, creates a magnetic field. A test of this hypothesis, based on recent observations of other planets, is mentioned in Section 11.10.

The rocky mantle and lower crust are sufficiently hot that they too behave like a sluggish fluid and undergo gradual movement. When melted rock, under high pressure and at high temperature, reaches the surface, the resulting phenomena are known as **volcanism.** The most spectacular event of this type is the eruption of a volcano. An important occurrence in the early history of the earth, it is thought, was the release to the atmosphere of large amounts of gasses through volcanism. Most likely, carbon dioxide, methane, water vapor, and sulfur were the most important contributions, particularly water which later condensed and rained onto the earth, producing the oceans. One then pictures life forming in the oceans (a topic further explored in Section 17.7) and plants developing. The plants may well have been one of the principal sources of that all-important gas in our atmosphere, oxygen.

The convection in the earth's interior has recently been incorporated into one of the most interesting theories of geology, **plate tectonics.** This theory envisions matter welling up in oceanic ridges and forcing the surrounding surface of the earth to spread apart in a slow but irresistible manner. The earth's surface is divided into slowly moving plates upon which the continents ride. Much of the interest lies in the regions of the earth's surface where the plates meet and grind against each other. The motion of these plates, also known as **continental drift,** produces effects easily noticed by surface dwellers. Where certain plates meet, the subsequent buckling of the crust raises mountains. For example, the motion of the Nazca plate (under the Pacific) toward the South American plate raised the Andes Mountains, according to this theory. Even more spectacularly, the motion of the plate we call India northward into Asia raised the Himalayas. As the North American plate rubs against the Pacific plate, tension builds up until it is suddenly released, resulting in an earthquake, an event familiar to many Californians. The places where the plates meet are regions where earthquakes are common. One fascinating feature of this theory is that the continents may all have been united in one gigantic continent roughly 200 million years ago. Looking at a globe of the earth today, one can see how some of the continents might well fit together like pieces of a jigsaw puzzle.

The earth has a sizable companion in its yearly travel around the sun. We study our only natural satellite next.

10.2 OUR COMPANION IN SPACE: THE MOON

The Moon Viewed with the Naked Eye The moon is one of the most fascinating celestial objects to observe with the naked eye. It moves with respect to the fixed stars, and it changes its phases (see Sections 2.3 and 4.7). Both phenomena are easily observed without optical aid. The moon is near enough to us (roughly 400,000 km, 250,000 mi, or 60 earth radii) that we can see some of its largest surface features without a telescope. Children in various countries are taught to see the "man in the moon," a woman's head, a rabbit, and other designs on the moon's face. The easiest features to see are dark regions, many of them circular, called **maria.** The singular form of the word "maria" is "mare," which means "sea." It is now known that there are no bodies of water on the moon, but the name has been retained. Figure 10.4 is a photograph of the full moon, which shows a bit more detail than the naked eye can see. See also Figure 10.5 and Box 10.1.

Earthshine When conditions are right, the naked eye can easily observe a phenomenon called **earthshine.** It can best be observed when the moon appears to be a thin crescent. You should look several days after a new moon and just after sunset. Earthshine is also visible several days before the new moon, just before sunrise. At such times, the night side of the Moon, the side turned away from the sun, appears to glow faintly. The first person to give the accepted explanation of earthshine was Galileo. He suggested that, when earthshine is visible, the dark side of the moon is illuminated by sunlight reflected from the earth, or, in his words, "the Earth, in fair and grateful exchange, pays back to the Moon an illumination similar to that which it receives from her." See Figure 10.6.

The effect of earthshine is sometimes poetically called "the old moon in the new moon's arms."

Earthshine is easy to photograph. Make a time exposure with a steadily mounted camera. Try an exposure time of about 1 sec. The result can be striking.

The Moon Effect Another curious naked-eye phenomenon occurs as the moon is rising or setting, especially if it is full or nearly so. The moon appears to be much larger when it is near the horizon than when it is high in the sky. This **moon effect,** or **moon illusion,** is so very obvious that it is surprising to learn that it is only an optical illusion. For reasons not yet totally understood, your brain incorrectly infers that

Figure 10.4 The face of the full moon. This photograph shows the moon's north pole at the top so that it may be compared to the naked-eye or binocular view. The reader may wish to become familiar with the easily visible maria (see Figure 10.5) and to identify them when the moon is visible. Some prominent craters are also labeled. At full moon, the location of Tycho can be inferred using its noticeable ray system.

Figure 10.5 Names of some of the major features on the moon.

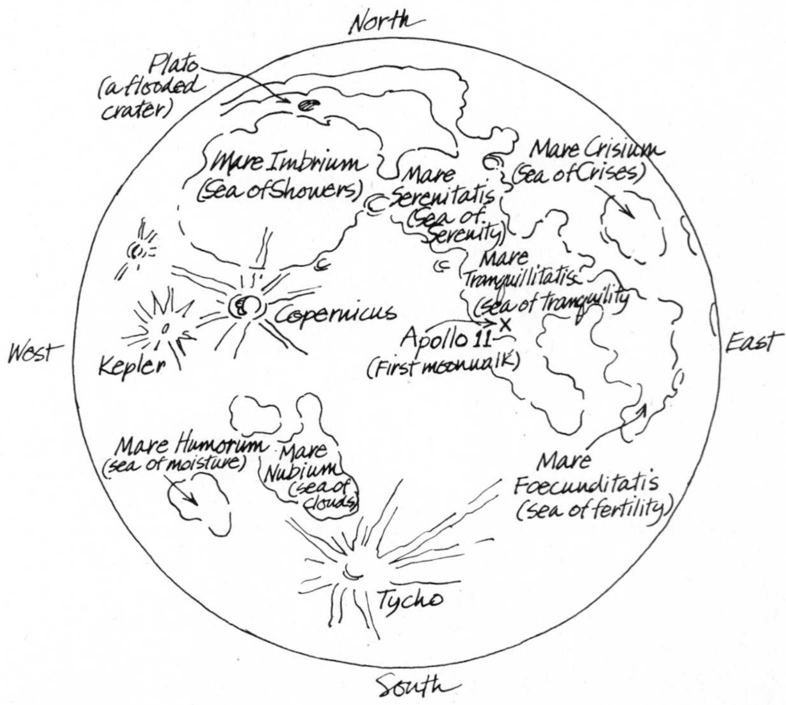

BOX 10.1

VIEWING A LUNAR ECLIPSE

When the moon is full, it is on the side of the earth opposite the sun. But, because of the 5° tilt of the moon's orbit with respect to the earth's orbit, the moon usually does not pass into the earth's shadow. On those occasions when the sun, earth, and moon are lined up correctly, the moon enters the earth's umbra and there is a total eclipse of the moon. At the distance of the moon from the earth, the earth's shadow has a diameter of about 9200 km (5700 mi). Since the moon has a diameter of about 3500 km (2200 mi), a total eclipse of the moon is not very rare. On the average, we should be able to see a total lunar eclipse once every three years or so. These eclipses are seen everywhere on the night side of earth at the same time, in strong contrast to solar eclipses.

An observer living at one place on the earth sees many more total eclipses of the moon than total eclipses of the sun. Lunar eclipses are perfectly safe to observe. The moon enters the shadow of the earth from the west. We may watch the shadow creep across the face of the moon using binoculars or a telescope. It is fascinating to see the shadow gradually engulf the craters and maria. We might expect the moon to disappear completely after the shadow has covered it, but usually it is still visible, although its appearance changes dramatically. The moon's color while eclipsed may, during different eclipses, be grayish, brown, red, or copper. When in the earth's shadow, the moon receives light which has passed through and been refracted by the earth's atmosphere.

The next total lunar eclipse visible in the United States occurs in 1982; there will be two total lunar eclipses visible that year.

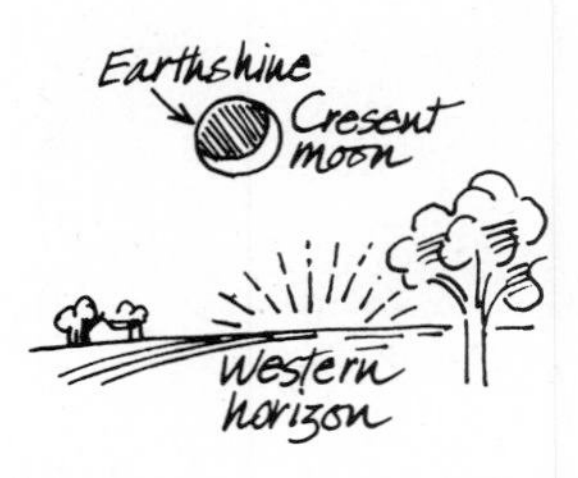

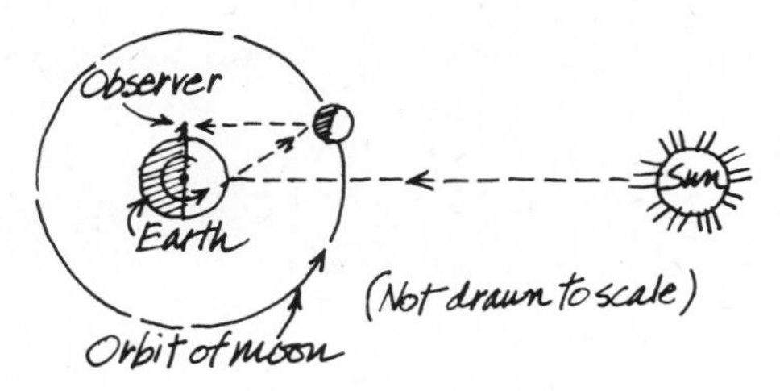

Figure 10.6 Galileo's explanation for earthshine. When the moon is a few days past new, earthshine, the faint glow from the night side of the moon, can be observed. On the left is the view from earth. The drawing on the right shows the path of a light beam (broken lines) on its roundabout travels from the sun, to the earth, to the night side of the moon, and back to the observer's eye. Can the reader make two drawings similar to these but for the case when it is several days before new moon? Earthshine is visible at that time too.

the moon is larger when low in the sky. The next time you notice the moon effect, try these experiments: Squint so that you can see only the moon; also, make a tiny hole with your fingers and then view the moon through the hole. For many people, both of these methods make the moon squeeze back down to normal size. Or, hold your little finger in front of the moon at arm's length. When I do that, the moon seems to shrink in size as my finger approaches the moon. You may be surprised

An uninhibited observer could try viewing the rising moon by turning his back on the moon, bending down, and viewing the moon through his legs. For many, the moon effect disappears.

Figure 10.7 The beauty of the moon through a telescope. This is the eastern edge of Mare Imbrium. This region can be located on Figure 10.4, in the upper left, by reference to the prominent flooded crater Plato, which has a diameter of about 100 km (62 mi). To the right of Plato, the rugged mountains known as the Alps curve downward. One can just barely see the Alpine Valley cutting through the Alps. (But see Figure 10.15.) Individual mountain peaks jut out from Mare Imbrium. The two most prominent are Pico, just below Plato, and Piton, further down and to the right. Both are roughly 2.4 km (8000 ft) tall.

to find that your little finger can easily cover the full moon. You could also photograph the moon as it rises and then, later, when it is high in the sky. The two images can then be compared. A similar effect can be noticed for the sun and familiar constellations when they are near the horizon.

The Moon's Synchronous Rotation An observer of the moon who has become familiar with the major markings shown in Figure 10.5 soon notices that they always appear in about the same place on the moon's face. In other words, the moon always keeps the same face toward us. This occurs because, on the average, its sidereal period (the length of time required for it to revolve around the earth once relative to the fixed stars) is equal to its rotational period (the length of time needed for it to turn once on its axis). We say that the moon's motion is **synchronous.** In other words, after it has gone halfway around the earth it has also turned halfway around on its axis. As a result, one side of the moon is never visible from the earth and, until the space age, this side was an intriguing subject of speculation. This side is usually called the "far side" or the "back side." It should not be called the "dark side," because all sides of the moon experience day and night.

The Moon Through Binoculars or a Telescope One look at the moon through binoculars or a telescope reveals why Galileo said, "It is a most beautiful and delightful sight to behold the body of the moon." You can see details so profuse and so clearly defined that you are tempted to spend hours viewing the moon. See Figure 10.7. (Galileo's discoveries concerning the moon are discussed in Section 6.5.) At any given time, the most interesting portion of the moon to view through a telescope is that near the *terminator,* the dividing line between night and day on the moon. For an observer on the moon at the terminator, it is either sunrise or sunset. Near the terminator, the shadows cast by mountains and crater rims are long, and detail is easiest to see for this reason. The full moon is less interesting in a telescope because much of the detail appears to be washed out because of the lack of shadows.

Through a telescope you can see that the maria are low, flat, dark regions. Many maria are roughly circular and ringed by mountains. See Figure 10.8. The maria are strongly contrasted with the **highlands,** the higher, brighter, most rugged regions on the moon. A telescope or binoculars clearly reveal that the moon's surface is pock-marked with many **craters,** more-or-less round holes in its surface, ranging in size from huge craters hundreds of kilometers across down to craters so small that the telescope can barely make them visible. Some of the largest, most famous craters are labeled in Figure 10.5. You may note that the maria have far fewer craters than the highlands.

While viewing the moon with binoculars or a telescope you may be fascinated by the many forms craters can take. See Figures 10.9 and 10.10. Some are bowl-shaped, while others, usually the largest, have one or more central mountain peaks. The larger craters have impressive dimensions. For example, Tycho (see Figures 10.11 and 10.12) is about 90 km (56 mi) in diameter. The top of its rim is about 4 km (2.5 mi) above the crater floor, and the central peak is 1.6 km (1 mi) tall.

A number of craters have an associated **ray system,** a series of bright streaks on the lunar surface that stretch out from the crater itself. The rays show up best near full moon. The most noticeable ray system is that associated with the crater Tycho in the southern region of the moon. When the moon is full, one can see some of Tycho's rays with the naked eye. The rays from a crater can extend for hundreds of kilometers.

The Moon and the Space Age The space age began October 4, 1957, when the USSR put the first successful artificial earth satellite, *Sputnik 1,* into orbit. Astronomers began to hope that they might soon be able to explore the moon by means of spacecraft.

Figure 10.8 Mare Serenitatis. This region of the moon overlaps that shown in Figure 10.7 in the upper left of this view. The Alpine Valley, extreme upper left, shows more clearly here. Like most maria, the floor of Mare Serenitatis exhibits *wrinkle ridges*, the result of lava flows which did not quite flatten out before solidfying. The rugged Caucasus Mountains above and the Apennine Mountains below nearly pinch off the connection of Mare Serenitatis' left shore with Mare Imbrium. Mare Serenitatis is roughly 650 km (400 mi) in diameter.

On October 4, 1959, exactly 2 years after Sputnik, the Soviets launched *Luna 3* which passed behind the moon and took the first pictures of the far side. For the first time since the dawn of human kind we were permitted a look at the mysterious far side. Most scientists had supposed that the far side would look much like the near side, but few had guessed correctly. *Luna 3* and succeeding space probes have revealed that the far side is quite unlike the near side in significant ways. (See Figures 10.13 and 10.14.) Most importantly, the far side consists almost entirely of highlands. There are few major dark basins. One was visible in the *Luna 3* photographs and named Mare Moscoviense (Sea of Moscow). The crater Tsiolkovsky is filled with dark material, as are some of the near-side craters. The rest of the far side is highlands, rugged and heavily cratered. There is also a significant lack of large mountain ranges on the far side. This may be evidence that whatever formed the maria also raised the mountains on the near side. Why

Figure 10.9 Crater Copernicus and surroundings. Copernicus, seen here in the upper left, is one of the most impressive craters on the moon. Roughly 100 km (62 mi) in diameter, its walls rise to a height of about 3.8 km (2.4 mi). The tallest mountain peak inside Copernicus is about 600 meters (2000 ft) tall. Copernicus has a ray system, which can be seen here as white lines spraying outward. The other end of the Apennine Mountains from the portion seen in Figure 10.8 can be seen here, stretching in from the right margin and extending to the north of the prominent crater Eratosthenes.

Figure 10.10 Julius Caeser, an old crater. Julius Caesar, the large crater near the top left-center, lies near the left margin of Mare Tranquilitatis, seen here on the right side of the photograph. Julius Caesar shows all the signs of old age. It is flooded by the lava flows from the mare, most of its southeastern wall having been obliterated. This indicates that Julius Caesar was formed before the mare. We see also that the rim mountains of Julius Caesar are worn and soft-edged, another sign of age. The Ariadaeus rille, a groove in the surface, slants through a very old, huge, mostly obliterated crater just below Julius Caesar.

Figure 10.11 Crater Tycho: telescope view. This earth-based photograph shows another of the major craters on the moon. It can easily be located on Figure 10.4 in the lower region because of its ray system, also seen in part here. Tycho lies in the lunar highlands.

Figure 10.12 Crater Tycho: view from orbit. This magnificent closeup of Tycho shows much more detail than had ever been seen from earth. It was taken from an altitude of 217 km (135 mi) by the fifth lunar orbiter. The instrument took pictures in strips, which are reassembled here to produce a mosaic.

Figure 10.13 The far side of the moon. Mare Crisium (see figure 10.4, upper right) is visible here at the upper left edge of the moon. The two other maria, less well-defined than Mare Crisium, are below Mare Crisium and just barely visible from earth (see Figure 10.4, right edge). The rest of the view shows the rugged lunar far side. The far side flooded crater Lomonsov is visible. This picture was taken during the *Apollo 10* mission.

Figure 10.14 Crater Tsiolkovsky on the lunar far side. This view of the lunar far side was taken by *Lunar Orbiter 3* from an altitude of 1450 km (900 mi). It shows the flooded crater Tsiolkovsky which is one of the few dark features on the far side.

should the far side be so different? No one is sure. Perhaps the gravitational pull of the earth somehow produced this remarkable lack of symmetry.

Starting in 1966, the United States launched a series of five satellites into orbit around the moon in order to photograph its surface. This was known as the lunar orbiter project. NASA is proud that in this series of launches, its success score was five for five. These spacecraft have radioed back nearly 2000 photographs of the lunar surface, many of them truly spectacular. See Figures 10.15 and 10.16 and others in this chapter. One of the most magnificent photographs shows a feature that can just barely be glimpsed from earth near the western edge. It is a huge, double-ringed feature variously called a crater or a mare named the Oriental Basin. If the western side of the moon faced us, we would surely talk about an "eye in the sky." See Figure 10.17.

By "the western edge of the moon" is meant the side of the moon toward the terrestrial observer's eastern horizon. Moon dwellers on the near side would see the sun set in this direction and so would call it west.

One of the most significant days in human history was July 20, 1969, when humans first walked on the moon. See Figures 10.18 through 10.20. *Apollo 11* was followed by five more manned landings. There is regret in some quarters that no more manned landings are planned, but at least the accumulated data and moon rocks still exist. A great deal of research is still going on, trying to fit all the data together into a coherent picture of the earlier states of the moon. The end of that attempt does not yet seem to be in sight.

Figure 10.15 The lunar alpine valley: *Lunar Orbiter* photograph. Compare this view with Figure 10.7. Barely glimpsed from earth, the mighty valley is revealed here in its full glory. What caused this 180-km (110-mile) long valley which is up to 21 km (13 mi) wide? Perhaps a giant projectile ripped through the mountains leaving this long, straight gash.

Figure 10.16 Crater Copernicus: *Lunar Orbiter* photograph. Compare Figure 10.9 with this slanting closeup, looking north from the southern rim. Beyond the north rim are the Carpathian Mountains. Note how rounded and smooth the mountains are, testimony to the erosion which does take place on the moon. This picture provides a truly revolutionary view, totally inconceivable until the late twentieth century.

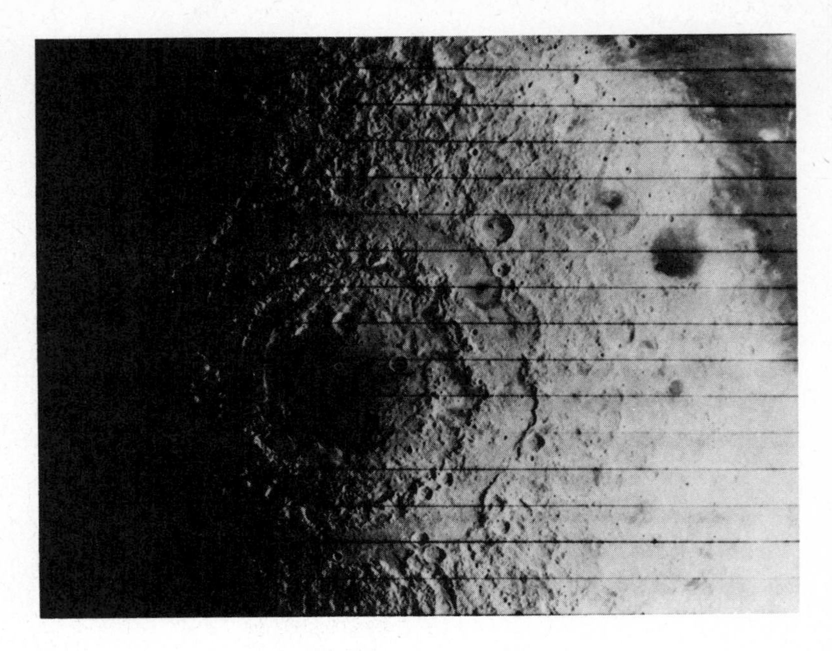

Figure 10.17 Oriental Basin: the bull's eye on the moon. The outer circular scarp is about 1000 km (about 600 mi) in diameter in this lunar orbiter photograph.

Figure 10.18 ***Apollo 11*** **launch: off to walk on the moon!** 9:32 A.M., Eastern Standard Time, July 16, 1969, Armstrong, Aldrin, and Collins are launched on a mission that will be long remembered in history. The vehicle is the mighty *Saturn 5* rocket, one of the most awe-inspiring machines ever made. As tall as a 35-story skyscraper, the rocket produced a thrust of about 3800 tons.

The Environment on the Moon Perhaps you and I will never go to the moon and walk on its surface. Yet, well before the moon walks, astronomers were able to describe what it would be like. There are two important facts about the moon that chiefly determine its environment: The moon has a low mass and practically no atmosphere.

The moon has no natural satellites, but its mass was fairly well determined by measuring such gravitational effects as the tides on the earth. These data were refined by observing the motions of the lunar orbiters. The moon's mass is about 7.35×10^{22} kg (735 followed by 20 zeros). For comparison, it would require about 81 moons to equal the mass of the earth.

The moon's diameter is roughly 3500 km (about 2200 mi), which is about one-fourth the diameter of the earth. The earth has an angular diameter of 2° as viewed from the moon. A full earth, as seen from the moon, reflecting sunlight from its white clouds, looks beautiful and very bright.

The moon's density, its mass divided by its volume, is about 3.4 grams per cubic centimeter. This is considerably less than the density of the earth, 5.5 grams per cubic centimeter. This leads us to conclude that the moon's interior is considerably more rocky than the earth's. If the moon has an iron core, it must be much less significant than the earth's.

Knowledge of the moon's mass and diameter also allows us to calculate that its surface gravity is one-sixth that of the earth. We would weigh one-sixth as much on the moon. This low gravitational pull of the moon implies that it is unable to retain an atmosphere. If the moon were given an atmosphere, the rapidly moving atoms of that atmosphere would gradually escape into space.

Figure 10.19 Buzz Aldrin descends the lunar lander: July 20, 1969. There is no air on the moon. A moonwalker must wear a spacesuit and carry a bulky air supply and air conditioner on his or her back. Do you remember where you were at this moment?

This lack of an atmosphere is the second important fact that determines the moon's environment. There is nothing to breathe. We must bring along our own atmosphere from earth and be extremely careful that our spacesuit is not punctured. Because there is no atmosphere or accompanying clouds to reflect sunlight back into space, the surface of the moon becomes extremely hot. All of the sun's fierce radiation

PUZZLE 10.1

THE VIEW FROM THE MOON

Draw a diagram showing the earth, the sun, and the moon's orbit. Draw the moon at four places in its orbit, namely, those places earthlings call new, first quarter, full, and last (or third) quarter moon. Draw an observer on the last quarter moon in such a position that the earth appears to be directly overhead. Now think about these problems.

1. Assuming that the observer stays at the same location on the moon's surface, draw the observer on the other three positions of the Moon. (Hint: The Moon rotates synchronously.)
2. Next to each sketch of the observer, write in the time of day. The four labels will be "sunrise," "noon," "sunset," and "midnight."
3. How long is a day on the moon if by "day" we mean the number of earth days from one sunrise to the next?
4. Describe the rising and setting of the earth as viewed from the moon. (Warning: trick question.)
5. Will the observer eventually see the entire surface of the earth?
6. What is the relationship between the phases of the moon as viewed from earth and the phases of the earth as viewed from the moon? For example, when earthlings see a full moon, moon dwellers see which phase of the earth?
7. Is midnight darkest on the side of the moon nearest the earth or on the side farthest from the Earth?

Figure 10.20 Neil Armstrong: The new Columbus. This man's name will live on in history as the first human to walk on the moon. The second was Buzz Aldrin. Who was the third? History's memory is very selective.

reaches the surface. The noontime temperature at the surface can reach about 370 K (about 100°C or 210°F). The astronauts usually landed at a location where the sun was low on the horizon (as we can verify by the long shadows in the photographs taken on the moon; see Figure 10.21) in order to avoid the hottest time of day. After the sun sets, there is no atmosphere to act as a blanket and hold the heat in. The nighttime temperature can drop to 120 K (about −150°C or −240°F).

The environment on the moon seems even more hostile when we realize that, because of the lack of atmosphere, bodies of water cannot exist there. A liquid is composed of molecules that are squeezed close together. On earth, the molecules of water in the oceans are forced together by the enormous weight of the overlying atmosphere. If the moon were given a body of water, the water molecules would rush apart and gradually escape into space.

The sky would be interesting as viewed from the moon. Even by day it would look black. Without an atmosphere to scatter sunlight and cause the familiar blue sky we see on earth, the blackness of space would always be visible. The stars would be visible by day, assuming that the brilliant glare from the moon's sunbathed surface did not make them to hard to see. The earth would also present a fascinating view. Solve Puzzle 10.1.

There is no weather on the moon. The only "rain" is a more-or-less steady influx of tiny micrometeors from space. This influx of micrometeors and occasional larger meteors has pulverized the surface of the moon over the millenia until it is covered with a layer of moondust. On earth most meteors burn up in the earth's atmosphere, heated by

Figure 10.21 Aldrin unpacks the scientific equipment. The lunar lander looks very little like a classical, streamlined science fiction spaceship. It did not need to be streamlined, since there is no air on the moon. The sky is dark by day. No stars are visible in this photograph, and the sun is near the horizon, as evidenced by the very long shadows.

friction with the air. A meteor of any size can reach the moon's surface, unimpeded by atmosphere. Most of the moon's craters are thought to be a result of the impact of meteors striking its surface.

Micrometeors are specks of solid matter orbiting the sun. They are smaller than about 1 mm (0.04 inch) in diameter. See also Section 11.9.

There is little erosion or weathering on the moon of the type known on earth. There is no wind or rain to carve away the lunar features. A crater or a footprint (Figure 10.22) lasts a long, long time once it has been formed on the moon's surface. There is erosion, however. It is caused by the influx of meteors and the enormous extremes of temperature which gradually cause the moon rock to crumble. Early ideas that the moon had a craggy, sharp-edged surface were wrong. Close up, the surface is smooth and rounded. See Figure 10.23. Now let us turn to the earth (Figures 10.24 and 10.25) and consider the moon's features.

Theories Concerning the Moon's Surface Features Astronomers have used the data concerning the moon to produce a subject as complex as geology, the study of the earth. We have space to mention only a few of the topics under discussion these days.

What produced the craters on the moon? For hundreds of years two theories have competed with each other. The **impact theory** proposes that meteors, chunks of rock big and small, crashed into the moon, blasting out the craters. The **volcanic activity theory** points out that volcanic action on the earth can cause craters resembling some of those

Figure 10.22 A footprint in the dust. The several-billion-year long bombardment by meteors from space has churned up the surface of the moon, leaving a layer of dust. Earlier fears that the dust might be kilometers thick proved to be unfounded, fortunately.

Figure 10.23 The lunar rover parked near Hadley rille. The *Apollo 15* mission, the fourth exploration of the surface, landed near the Hadley Mountains. The astronauts were able to cover large distances in the battery-operated rover which they brought along. The mountain Hadley Delta looms up enormously in the distant background. We see how rounded lunar mountains are, in great contrast to terrestrial mountains.

Figure 10.24 The lunar lander in orbit. Taken from the command module by Michael Collins, this photograph shows the *Apollo 11* lunar lander approaching the mother ship. After rendezvous, the moon walkers transferred to the command module, and all three headed for home, seen in the distance, 400,000 km ($\frac{1}{4}$ million mi) away.

Figure 10.25 Earthbound once more: *Apollo 11,* July 24, 1969. The return trip to earth from the moon is every bit as complex and hazardous as the outward leg. Perhaps the most difficult maneuver is the reentry into the atmosphere. The astronauts experience their largest g forces at this time, not at takeoff. The friction from the rapid descent through the atmosphere heats and tears at the craft. Some superficial damage may be seen in this photograph.

Figure 10.26 Crater chains. Crater chains are considered to be evidence of volcanic activity on the moon. The large crater is named Hyginus. Crater chains are also visible in Figure 10.9.

on the moon. It proposes that the moon's craters were formed in a similar way.

Over the years, scientific opinion has swung from one theory to the other. There was at times an odd division of opinion in that many geologists believed the impact theory, while astronomers favored the volcanic activity theory. In recent decades, the impact theory has been gaining strength. Careful comparisons of lunar craters with impact and volcanic craters on earth have produced evidence of a greater similarity to the impact craters. Yet, there is still room for disagreement among the experts, particularly on the origin of certain individual craters. There is abundant evidence that volcanism has occurred on the moon. The maria and flooded craters are evidence that lava (molten rock) has flowed on its surface. Indeed, crater chains, strings of small craters lined up edge to edge, are probably the result of underground movement of molten rock. See Figure 10.26. Still, most authorities today agree that about 99 percent of the craters on the moon were created by impacts. The rays emanating from many craters on the moon's surface are probably composed of fine, white grains produced by the impact of the meteor that formed the crater. Older, more worn craters do not have rays, an indication that the material of the moon darkens when exposed at the surface. Perhaps the solar wind is the darkening agent. Many of the maria are huge, circular basins often ringed by mountains. It is presumed that these basins were blasted out of the moon's surface by the impact of a very large meteor. The impact also raised the mountains and cliffs ringing the maria. Later, flows of dark lava, melted rock from beneath the lunar surface, filled in the basins. The lava must have been very fluid to have flowed over such large distances and solidified with such a flat surface. The maria must have been flooded well after the bulk of the meteor impacts occurred, because they are much less heavily cratered than the highlands; the maria are therefore younger than the highlands. As mentioned earlier, it remains unclear why the basins on the moon's far side did not, for the most part, fill up with lava.

The Origin of the Moon The moon is an important member of the inner solar system. Increasingly, it is being regarded as a planet along with Mercury, Venus, Earth, and Mars. The moon's diameter is about 70 percent of Mercury's, for example. The earth and the moon are sometimes thought of as a double planet, a unique situation in the solar system. This point of view is reasonable because the moon is unusually large as compared to its parent body, the earth. The moon's diameter is 27 percent of that of the earth's. Triton, the largest satellite in the solar system has a diameter 1.7 times that of the moon, yet its diameter is only 12 percent as large as that of its parent, Neptune.

How did the double planet come about? The search for a satisfactory answer to this question has proved to be one of the most difficult in all of astronomy. Three hypotheses have been under consideration for some time.

The first may be called the **fission hypothesis.** It is proposed that in the early stages of the formation of the earth, when the earth was still a hot liquid, the moon fissioned, split off, from the earth. Perhaps the earth was spinning rapidly and this caused the fission. If so, it is hard to see why the earth and moon are spinning so relatively slowly now. What slowed them down? Furthermore, how could the liquid moon avoid being torn apart by the large gravitational pull of the earth when the two bodies were close together?

A second hypothesis, the **capture hypothesis,** envisions the moon being formed somewhere else in the solar system and then being captured whole by the earth's gravitational field. A difficulty with this idea is that, in order for the moon to be captured, it would have to be slowed down. How would the large amount of energy of the moon have been lost during this braking of its motion? Energy cannot simply disappear.

A third approach is to assume the **accretion hypothesis,** which proposes that the earth captured small objects which went into orbit around it. These eventually collided and stuck together (accreted) to form the moon. An advantage of this idea is that small pieces have less energy and it is easier to dissipate this energy a little at a time as the pieces are captured over a lengthy period of time.

It was hoped that information obtained from the moon by the manned landings would help choose from among these possibilities. In the opinion of many experts, it didn't. One problem is that the moon's surface has been so churned up and mixed by the impacts of meteors that information about its early history has been largely obliterated. Some of the information recovered is revealing; some is puzzling.

The oxygen in the moon rocks was tested to discover the makeup of the nuclei of the oxygen atoms. Every oxygen nucleus has eight protons, and most oxygen nuclei have eight neutrons as well. However, it has been found that earth's rocks contain some oxygen atoms with more than eight neutrons. Two atoms having the same number of protons but differing numbers of neutrons are said to be **isotopes.** Earth's rocks have oxygen isotopes which are present in certain relative percentages. Meteorites, rocks from outer space which have landed on the earth, have been found to have the same oxygen isotopes, but in different relative amounts. This suggests that the meteors were formed at a different distance from the sun than the earth was formed. The moon rocks have the same relative abundance of oxygen istotopes as the earth's rocks, which many take to mean that the moon must have been formed at about the same distance from the sun as the earth. These oxygen data rule out variations of the capture hypothesis proposing that the moon was formed elsewhere, say, near the orbit of Mercury. Yet, the moon still may have been formed in an orbit similar to the earth's and later captured by the earth, judging by the oxygen data.

In contrast to this fairly clear data, it was also found that the outer layer of the moon has in some cases higher and in other cases lower relative amounts of certain other elements as compared to the earth. For example, when rocks known as basalts are compared, moon basalts have much lower amounts of certain elements such as bismuth, gold, and silver. How can this be? Does it rule out any of the hypotheses concerning the moon's origin?

It seems to rule out the fission hypothesis. If the moon split off from the earth, one would expect it to have the same relative amounts of elements. But what of the capture hypothesis? If the moon was formed in about the same region of the solar system as the earth, why should the two objects differ? Similar difficulties are found with the accretion hypothesis. How could the earth have captured pieces with similar oxygen isotopes and differing numbers of other elements?

Without describing the details of this continuing debate any further, one can say that the majority of astronomers have not settled with any confidence on any of the hypotheses. Nor have any been ruled out for certain. One thing, however, is sure, we do have a moon. Furthermore, this moon is enjoyable to watch and still mysterious to meditate upon.

10.3 MERCURY: THE SUN'S MOON

There is a legend that Copernicus never saw the planet Mercury. Modern scholars tend to doubt this, but it points to an important fact about this planet; Mercury is close to the sun. At times, as seen from the surface of Mercury, the sun appears to have an angular diameter of 1.7°, 3.3 times larger than when it is viewed from earth. At such times, the blazing sun pours light onto Mercury which is 10.5 times as intense as the light we receive. Let us investigate this infernal world.

Mercury As Viewed by the Naked Eye Mercury travels in a relatively small orbit. On the average, it is 0.39 AU or 58 million km (36 million mi) from the sun. Because of this, it is never seen at a large angle from the sun in the sky. For this reason, of all the planets known to the ancients, Mercury is the most elusive to the naked eye. It is never seen against a dark sky.

Mercury's orbit is fairly eccentric (flattened), more so than any planetary orbit except that of Pluto. As a result, Mercury's distance from the sun varies widely, from a minimum of 0.31 AU = 46 million km to a maximum of 0.47 AU = 70 million km. It is best to search for Mercury when it is near maximum elongation, that is, when it appears to have swung out farthest from the sun.

When all factors are favorable, Mercury is not difficult to see with the naked eye. Astronomers calculate these factors, and astronomy magazines publish in advance the times when it is best to look for Mercury. With this guidance, the reader will have no trouble finding Mercury, but much trouble believing the legend about Copernicus.

Mercury As Viewed in a Telescope Mercury shows phases similar to the moon's, but a fairly good telescope is required to see them. The planet is small and relatively distant. Professional astronomers often observe Mercury by day when it is high in the sky, out of the haze near the horizon. Very experienced observers working with professional telescopes can see faint markings on Mercury.

Only very experienced observers should try viewing Mercury through a telescope during the day. The danger of accidentally pointing the telescope at the sun is too great.

The Rotation of Mercury Most observers of the markings on Mercury came to the conclusion that it rotated in a fashion similar to the moon. That is, they declared, Mercury rotated synchronously, keeping one face toward the sun. This would mean that it had permanent day on one side and permanent night on the other. Fred Hoyle commented that this would make Mercury unique; it would at once be the hottest planet and the coldest planet in the solar system. Some astronomers calculated that the permanent night side, always turned toward the cold dark of space, would have temperatures as low as 30 K (about −240°C).

A puzzle developed when long-wavelength radiation detectors were attached to telescopes and pointed at Mercury. Such detectors, sensing the amount of thermal radiation emitted by the planet, revealed the surface temperatures on Mercury. As expected, the daytime temperature was frightfully high, about 700 K (about 430°C or 800°F), hot enough to melt lead. But the nighttime temperature, while very cold, 95 K (about −180°C or −290°F), was warmer than expected. Some supposed that Mercury might have a substantial atmosphere which carried heat to the dark side.

Then, in 1965, the puzzle was resolved in a surprising way. Radar signals were reflected off of Mercury. (See Box 10.2.) This procedure revealed that Mercury has no permanent day side. What had been taken to be a hard fact for so long was simply false.

Mercury's rotation period is now thought to be about 58.7 days. Since it orbits the sun once in 88 days, the formerly accepted rotation

When referring to other planets, the terms, "hours, days," and the like mean "earth hours, earth days," and so on.

BOX 10.2

RADAR, THE DOPPLER EFFECT, AND PLANETARY ROTATION

The easiest method of determining the length of time a planet takes to rotate once on its axis is simply to watch until a marking on the planet appears to revolve once. A new method has recently been used which is an interesting application of radar, high technology, and the Doppler effect. The large dish of a radio telescope is used to beam a train of radio waves of an accurately known wavelength toward the planet in question. This is done by placing a radio transmitter at the focus of the telescope and using the telescope in reverse. The radio waves reflect from the planet, and a small part of the original signal returns to the earth and is collected by the radio telescope. The detection of the very weak return signal is an electronic triumph in itself.

The idea behind the analysis of the return signal is this. Part of the beam reflects from one edge of the rotating planet and another from the other edge. When the beam returns from the edge moving away from us, the returning radio waves have a longer wavelength after reflection because of the Doppler effect. Similarly, the signal from the approaching edge of the planet returns with a shorter wavelength. The amount of shifting gives the speed of each edge, from which the rotational period can be calculated. Experimenters must be careful to subtract the known orbital motions of the earth and the planet, as well as the rotational motion of the earth itself. In actuality, the method is even more complex than can be described here. An involved method of analyzing the radar returns over a period of time is required. The results of this work have been worth the effort, however, for the rotational periods of Mercury and Venus have at last been measured. As explained in the text, both results came as a surprise to most astronomers.

---

rate of 88 days would indeed have produced synchronous motion. It did not take an Italian astronomer, Giuseppe Colombo, long to notice that the newly discovered rotation rate also implied a surprising variation on synchronous rotation. Mercury spins on its axis three times relative to the fixed stars in (3×58.7) 176 days. How many times does it orbit the sun in that time? The answer is "twice," because 2×88 also equals 176 days. This produces a curious result. Imagine someone living on Mercury. The time from noon to midnight for this person is one Mercury year. One complete Mercury day, from noon to noon, takes two Mercury years. Presumably Mercury is slightly elongated, and the sun's gravitational pull on this elongation keeps it locked in this peculiar rhythm.

We can see now why the night temperature on Mercury is not as low as had been anticipated. Each side of the planet is periodically heated by sunlight. No atmosphere is necessary to produce the effect. The long night on Mercury allows the planet to become very cold. The temperature range on Mercury of about 600 K is the largest of any planet.

Mercury in the Space Age The above is essentially what was known about Mercury until the magnificent flight of *Mariner 10*, launched November 3, 1973. (See Box 10.3 and Figure 10.27.) By March 1974, when *Mariner 10* arrived near Mercury, a new age in the study of Mercury had begun.

BOX 10.3

MARINER 10: A TOUR OF THE INFERIOR PLANETS

Mariner 10 had been designed and its flight path was being chosen at a meeting. Guiseppe Columbo, the Italian astronomer who had previously pointed out the peculiar nature of Mercury's rotation period, told the group that, if precisely the right path were chosen, *Mariner 10* would be able to pass Mercury and then orbit the sun once while Mercury orbited the sun twice. The result would be that *Mariner* could return to Mercury more often than the single fly-by originally planned. The necessary changes in the spacecraft could be made, it turned out, because $750,000 was still left in the billion dollar budget.

Another feature of the flight path involved an ingenious use of gravitation. *Mariner* was to be aimed at Venus, initially, in such a way that it would pass near that planet and use the gravitational field of Venus to swing it into the proper direction so that it could visit Mercury one month later. This reaiming procedure is known as a "gravity assist."

Figure 10.27 The *Mariner 10* craft. The large rectangles are solar collectors which converted the abundant sunlight near Venus and Mercury into electricity. The dish on the upper left is the radio antenna. The telescope-cameras may be seen on the right.

Figure 10.28 (facing page) Mercury, at last. The *Mariner 10* space probe was still about 6 hr and 200,000 km (120,000 mi) from its initial closest approach to Mercury when it took this photograph. This portion of Mercury is heavily cratered, giving it an appearance similar to the moon's highlands. The largest craters are about 200 km (120 mi) in diameter.

What magnificent pictures! Details stand out with impressive clarity. (See Figure 10.28.) The surface bears a striking resemblance to the surface of the moon because of the large number of craters and basins. As on the moon, sharp-rimmed young craters, many with a ray system, and worn-down older craters are in abundance. The differences between Mercury and the moon are also interesting. There are no regions precisely similar to the moon's maria. In Mercury's heavily cratered regions there are, however, unexpected plains, relatively smooth areas, totally unlike the moon's highlands. The planet has a large number of **lobate scarps,** scalloped cliffs extending for hundreds of kilometers. See Figure 10.29. It is thought that these were formed when the inner core of Mercury cooled and shrank. The crust, the outer surface, was compressed until it cracked, producing the scarps. Mercury has a bull's-eye-shaped basin somewhat similar to the moon's Oriental Basin. Named Caloris Basin, this feature is presumably due to the impact of a large piece of debris from outer space. See Figure 10.30.

The View from Mercury's Surface What would it be like to land on this small planet? We would need to bring a robust life-support system with us. *Mariner 10* detected a very insignificant atmosphere, so we would need to bring our own air. We would need a powerful heat supply for the nights and a powerful air conditioner for the days. The horizon would appear relatively close to us as we stood on the surface, for Mercury is only 38 percent as large as the earth.

Mercury has a density about the same as that of Earth, indicating a fairly sizable iron core; this is in strong contrast with the less dense moon. The moon is nearly as large as Mercury, only 29 percent smaller. Although Mercury has the same density as Earth, its smaller size results in a smaller surface gravity. On Mercury, we would weigh 37 percent as much as we do on Earth. Mercury has a magnetic field much like Earth's, but it is only about 1 percent as strong; if we bring along a sensitive compass, it will work correctly on Mercury, pointing north as it does on Earth.

Suppose we arrive when Mercury is nearing its perihelion. We would probably not choose to land near the noon location because, as stated earlier, at perihelion the nearby sun floods the landscape with 10.5 times more energy than the earth receives. (Imagine the effect of 10 suns in our sky.) Suppose we choose to land at a location on Mercury where the sun is about to rise. We will be treated to one of the most interesting sights in the solar system. First we see the sun's corona rise over the horizon and then the blazing sun itself. At this distance the sun appears to have an angular diameter of 1.7°, over three times larger

Figure 10.29 One of the many scarps (cliffs) on Mercury. Taken an hour after *Mariner 10*'s second approach to Mercury, this picture features a 300-km (480 mi)-long scarp which curves in from the upper left. The lobate (curved) shape of these scarps leads experts to believe that they were formed when Mercury's crust shrank. This region is near Mercury's south pole.

Figure 10.30 Caloris Basin. Astronomers were delighted to find a "bull's eye" on Mercury, somewhat similar to the Oriental Basin on the moon (Figure 10.17). We see here nearly half of the basin. It is surrounded by a ring roughly 1300 km (800 mi) in diameter. The hurtling object that blasted this basin into existence is estimated to have been at least tens of thousands of kilometers in diameter.

than when viewed from the earth. The sun rises for a while and then stops and sets again! The reason for this is that, near perihelion, Mercury is traveling so rapidly in its orbit that the orbital motion temporarily overwhelms the effect of its rotation. After perihelion passage, the sun rises again and this time slowly travels toward the meridian.

One-half Mercury year after sunrise (44 earth days), the sun is at our meridian. We notice that it looks smaller now (about 1° in angular diameter) because Mercury is at its aphelion. We now receive only 4.5 times the energy from the sun that the earth does. (Only?) Then, during the next ½ Mercury year the sun heads toward our western horizon, growing larger in appearance as it does so. It sets and, as we round perihelion once more, it rises again briefly in the west only to set a second time.

10.4 VENUS: THE VEILED PLANET

Venus As Viewed by the Naked Eye While it requires some alertness to see Mercury with the naked eye, Venus, the other inferior planet, has probably been seen by most people. When it is east of the sun, it sets after the sun does and is easily visible every cloudless evening in the western sky for well over nine months at a time. When Venus is brightest, only the sun and the moon appear brighter. At its brightest, Venus can deceive people. It has been shot at and has often been misidentified as an unidentified flying object.

Venus Through a Telescope The phases of Venus (see Section 6.5) are easy to follow using a telescope. (They can be seen with high-quality 7-power binoculars, at least when Venus is a thin crescent.) When Venus looks brightest to the naked eye, it is a rather slim crescent in a telescope. See Figure 10.31. There are two effects that determine Venus' apparent brightness. The nearer Venus comes to us, the bigger and brighter it seems. Opposing this is the fact that, as Venus nears us, we can see less and less of its sunlit side; that is, it is approaching the crescent phase. After its greatest brightness it becomes a very thin crescent, dims down, and enters inferior conjunction.

Considering how bright and interesting Venus is to the naked eye and how close it comes to us, at best only about 0.28 AU (41 million km or 26 million mi), it is a disappointing object when viewed through a telescope. Apart from the phases little more can be seen. Astronomers have concluded that the surface of Venus is perpetually covered with a blanket of clouds. The clouds make Venus a splendidly bright object, but they hide its surface from our view. Even its rate of rotation on its axis was unknown until recently.

That question was settled in the 1960s when radar signals were reflected off Venus (Box 10.2). Radar signals are short-wavelength radio waves which can penetrate Venus' cloud cover and detect the planet's surface. An interpretation of the returning radar beam revealed the

Figure 10.31 Venus in its crescent phase. When viewed in visible light, Venus is featureless. The telescope user must be satisfied with following the phases of Venus. Venus looks brightest when near the phase shown here, because of its nearness to the earth.

period of rotation of Venus for the first time. It turned out to be as surprising as the result for Mercury, though in a different way.

Suppose we are viewing the solar system from high above the sun's north pole. We see that each planet revolves around the sun in the same sense, counterclockwise from our point of view. Now, before the radar experiments, it was known that the sun and the planets (except for Pluto and Uranus) also rotate on their axes in the same sense. (We still aren't certain of Pluto's sense of rotation. Uranus rotates in a unique fashion; see Section 11.4.) The radar measurements determined that Venus rotates backward, that is, clockwise as viewed from the north side of the solar system, a totally unexpected result. As viewed from the surface of Venus, the sun would rise in the west. Venus rotates slowly. Were it visible, a star would appear to return to a venusian's meridian once every 243 earth days.

The Atmosphere of Venus As Studied from Earth In 1932, sunlight which had passed through the upper layers of Venus' atmosphere and been reflected toward Earth was studied in a spectroscope. An analysis of the dark lines in the spectrum revealed that the atmosphere of Venus consists mostly of carbon dioxide (CO_2), the same compound we exhale and which, when frozen, is called "dry ice." This led to speculation that Venus might be very hot.

The temperature of a planet is regulated by two opposing processes. Sunlight strikes the planet, and the fraction of sunlight absorbed is changed into heat. A planet loses heat by means of thermal radiation (see Section 9.3). The hotter the planet, the more thermal radiation, in the form of infrared radiation, it emits. The average temperature of a planet is achieved when a balance occurs. When the incoming energy in absorbed sunlight is balanced by the outgoing energy lost by thermal radiation, the planet's temperature no longer changes. If, for example, the sun were to get brighter and increase the energy the planet receives, the planet's temperature would increase until the thermal radiation equaled the new input of energy.

A key factor in this balance is the amount of heat energy the planet can trap. It is known that sunlight, radiation having wavelengths in the visible region, passes freely through carbon dioxide. On the other hand, infrared radiation, produced by the heated surface of Venus, does not pass easily through carbon dioxide. In effect, carbon dioxide acts like a heat trap; energy passes inward freely but has difficulty getting out again. This phenomenon is known as the **greenhouse effect.** It was expected that this effect produces a high temperature on Venus.

Venus As Viewed from Spacecraft The Soviet Union has had notable success exploring Venus by means of space probes. A number of the Venera series of probes have reached the surface of Venus and radioed back information concerning the environment. The surface is indeed hot, about 750 K (480°C or 890°F), comparable to the temperature on Mercury, an indication of the effectiveness of the greenhouse effect. After all, Venus is nearly twice as far from the sun as Mercury is.

The instruments reported a very dense atmosphere. The atmospheric pressure on the surface of Venus is about 90 times greater than the pressure on earth. Small wonder that the best the Venera probes could do was to function for about an hour under these unbearable conditions.

Venera 9 and *Venera 10* resembled the *Viking* missions to Mars launched by the United States. Each mission involved an orbiter which released a probe which landed on the surface of Venus, took pictures, and made measurements. The photographs revealed a smooth terrain in one case and, at a location 2000 km (1240 mi) away, a rocky hillside. Many astronomers had assumed that the surface of Venus would be dark because of the thick clouds, but it was found that a surprising amount of sunlight came through. The brightness corresponds to a cloudy day on earth. The photographs from *Venera 9* seem to show shadows, as if an observer on the surface would find the position of the sun in the sky to be rather distinct. The headlights on board *Venera 9* and *Venera 10* turned out to be unnecessary.

In 1974, *Mariner 10* (see Box 10.3) passed within 5800 km (3600 mi) of Venus on its way to Mercury. It confirmed that photographs taken with visible light show little detail. However, photographs taken with ultraviolet light show a remarkable structure of swirling clouds and high

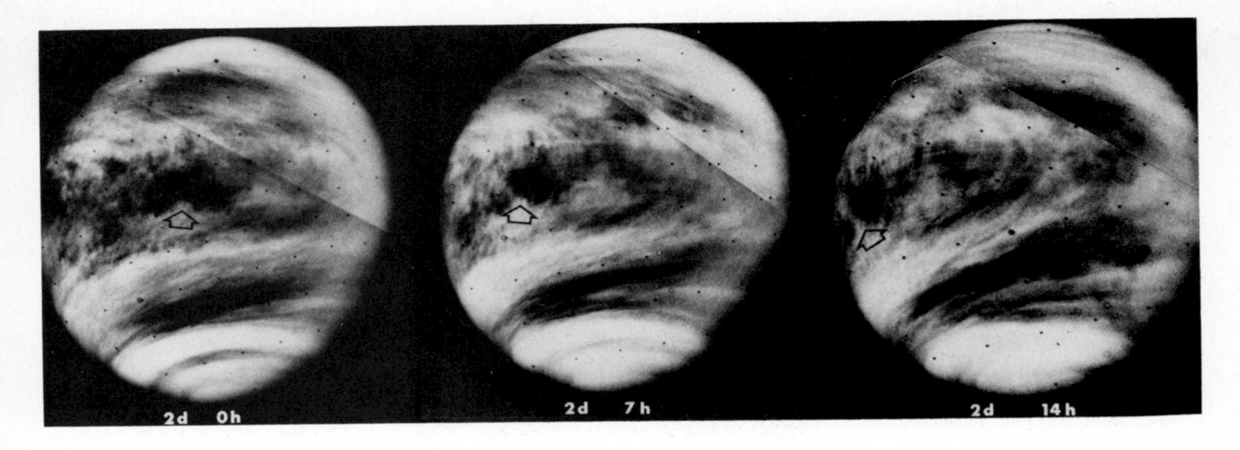

Figure 10.32 *Mariner 10* views Venus using ultraviolet radiation. When *Mariner 10* was two days past Venus and on its way to Mercury, it looked backward and took this series of three photographs once every 7 hr. Streaks and patchy regions were observed, remarkable in themselves. As the arrows show, the atmosphere rotates rapidly, making one rotation in four earth days. This too is remarkable, because the planet's surface rotates very slowly.

winds in the upper atmosphere. See Figure 10.32. These bands are not yet fully understood. Some suspect that they may indicate an uneven distribution of deadly carbon monoxide.

The combined data from space probes and Earth-based spectroscopy indicate that Venus' atmosphere is awesomely inhospitable. By volume it is 97 percent carbon dioxide and has almost no water. The clouds are most likely formed of droplets of sulfuric acid (also known as "battery acid"). The yellowish tint in the clouds may be due to particles of sulfur. The planet is close to some conceptions of Hell. (see Box 10.4.)

10.5 MARS: THE RED PLANET

***The War of the Worlds,* by H. G. Wells, is one of the earliest and best stories dealing with an invasion from Mars. Although riddled with scientific errors, the vivid narrative prose makes it exciting reading.**

For centuries the planet Mars has held a special fascination for humans. Named for the god of war, it has been associated with violence, blood, and menace. Invasions of "men from Mars" were the theme of much early science fiction. Mars continues to fascinate us in an age in which space probes have landed on the planet, the first foothold of the invasion of the earthlings.

Mars and the Naked Eye Mars is distinctive to the naked eye; it is the only reddish-brown planet. Because it is a superior planet (Section 2.4), it exhibits motions different from those of Mercury and Venus. After a conjunction (line-up) with the sun, one may watch it each morning before sunrise as it gradually, day by day, works its way westward across the sky, increasing its angle from the sun. When it sets at sunrise, it is in opposition.

BOX 10.4

THE EARTH

(A report from one of Venus' greatest planetologists. Earth time: 1400, well before the invention of radio on earth.)

Ladies, gentlemen, and gorns: as you know, we became aware that there are other planets orbiting our sun when we put telescopes into orbit around our planet, outside our cloudy atmosphere. I am here to report on that oddest of planets which happens to be one of our nearest neighbors. Yes, Earth has similarities to Venus, similar size and a mass 1.2 times that of ours. But I must emphasize that the differences are overwhelming. Earth is really a double planet, one planet has a diameter 4 times that of the other. From orbit around Venus, one could easily see the two objects using only one's naked eyestalk. The larger has a sickly blue color and it rotates fantastically rapidly. I am pleased to report that our Earth lander, *Terra 5*, has landed on the larger planet and has sent back data and pictures. We can at last answer the age-old question, Is there life on Earth?

We chose to land in one of the lowest, least cold regions of the land masses in order to maximize the lifetime of our lander in such a hostile, cold environment. Most of the planet is covered with positively enormous bodies of liquid dihydrogen oxide. The region we landed in has a frigid temperature of 320 K (about 49°C or 120°F) at most! The rest of the planet is even colder. The atmosphere is very thin, only one-hundredth as thick as ours, and consists mostly of poisonous nitrogen and oxygen. There is relatively little life-giving carbon dioxide.

Our pictures show only endless sand hills stretching away in all directions, with seemingly no signs of recognizable life. There is certainly no evidence of intelligent life at the landing site. To be sure, we had expected no civilization, since we receive no radio signals from Earth, radio being presumably the surest sign of nonprimitive life.

The sky is an interesting feature as viewed from the earth. The sun is frequently visible in the clear atmosphere, flooding the landscape with unbearable light. The sky by day is a brilliant blue. The sun seems to race across the sky in only 12 hr, and after it sets one can actually see the stars from ground level. The lander ceased working after the sky filled with clouds one day. Our technicians surmise that the failure was brought on by a combination of the very low temperatures, the low atmospheric pressure, and the possibility that dihydrogen oxide actually fell out of the clouds in liquid form, shorting out our electric circuits and possibly causing a damaging chemical reaction with the lander's metal parts.

In conclusion, we now know that the blue, wispily clouded component of the Earth double-planet has an environment which is strange, cold, and extremely hostile to life. Perhaps some day we will find a way to increase the carbon dioxide in the atmosphere of Earth, thereby increasing the greenhouse effect which makes our planet so comfortable. This process, called Venus-forming, could make it possible to colonize this dead, useless, neighboring world.

Mars Through the Telescope: Areas of Agreement You might think that Mars would be the easiest planet to observe through a telescope. Because it is a superior planet, it always appears in full or nearly full phase (Figure 10.33). Furthermore, it can approach to within 56 million km (35 million mi) of the earth, a separation of only 0.37 AU. Moreover, Mars' atmosphere is thin and usually transparent. Yet, in the past, it has proven to be a subject of controversy among telescope users. We first discuss the observations of Mars about which all astronomers agreed, before the space age.

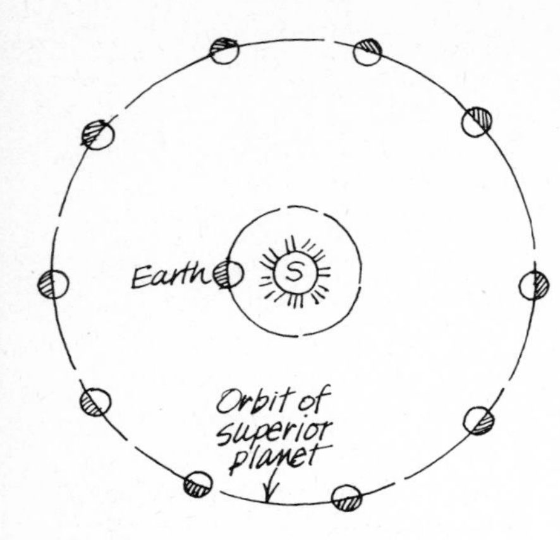

Figure 10.33 Superior planets always appear in full or nearly full phase. The view of a superior planet as seen from earth is always of the sunlit side. The inferior planets can be observed going through all their phases, but the night side of each superior planet is always turned directly or almost directly away from the earth.

The orbit of Mars has an average radius of 1.52 AU; Mars is about half again as far from the sun as we are. Its orbit is considerably more eccentric (flattened) than the earth's. At perihelion, Mars is 1.67 AU from the sun and at aphelion only 1.38 AU, a 19 percent difference. Because of this relatively eccentric orbit, some oppositions of Mars are significantly more favorable for telescope users than others. See Figure 10.34.

Observations of the markings on Mars' surface reveal that its axis of rotation is tipped at an angle similar to Earth's tip. The figures are: Mars, 24.0°; Earth, 23.5°. This implies that Mars should also have seasons analogous to ours. Among the evidence that this is so are changes in the size of its polar caps. See Figure 10.35. When it is winter in the north of Mars, the northern cap grows and the southern cap shrinks because it is summer in the south. One-half Mars year (about one earth year) later, the reverse occurs.

Mars is a small planet; only Mercury and, possibly, Pluto are smaller. It has a diameter of 6800 km (4200 mi), which is about half as large as the earth's diameter. Mars' mass, as determined from the motions of its satellites, is also small, about 11 percent that of Earth's mass. We may compute that Mars has a density of only 3.9 grams per cubic centimeter. For comparison, the earth's density is 5.5 grams per cubic centimeter. Mars' low density leads us to suppose that, if Mars has an iron core, it must be relatively small. Mars also has a small surface gravity, only 38 percent of that of Earth. A 600-newton (135-lb) person on Earth would weigh only 228 newtons (51.3 lb) on Mars.

The surface markings on Mars are not easy to observe, partly because they are so indistinct. An amateur's small telescope may show Mars merely as a small, red disk. At favorable times a polar cap may be glimpsed. Through larger instruments, dusky regions on the planet can be seen. These have been drawn by telescope users with fair amount of agreement. Many of the regions have been named maria (seas), as on the moon, but their actual nature was uncertain until the space age.

The atmosphere, it was agreed by all, must be thin. It is nearly always clear. On rare occasions clouds can be glimpsed. Some are white and are presumed to be caused by ice crystals forming as air rises over mountains.

At times the surface of Mars appears blank, even when viewed with the best telescopes. It is subject to an awesome phenomenon at such times—dust storms that envelop the entire planet. These often occur when Mars is nearest the sun. It is assumed that at such times the extremes of temperature between various places on Mars are most pronounced. This temperature difference gives rise to wind patterns, causing the planet to be engulfed in dust. Apparently, once the dust fills the air, sunlight no longer heats the ground as effectively, the winds die down, and the dust settles once more.

On the next page we mention that Giovanni Schiaparelli (Skyap-ah-rell'-ee, 1835–1910) was an expert observer of the solar system in his day. He demonstrated that some meteor showers are associated with comets, and he discovered an asteroid. Among other things, he asserted that his observations of the faint markings on Mercury demonstrated that Mercury always keeps one face toward the sun.

Mars Through a Telescope: An Area of Disagreement—The Canals It is quite difficult to observe fine details on the surface of a

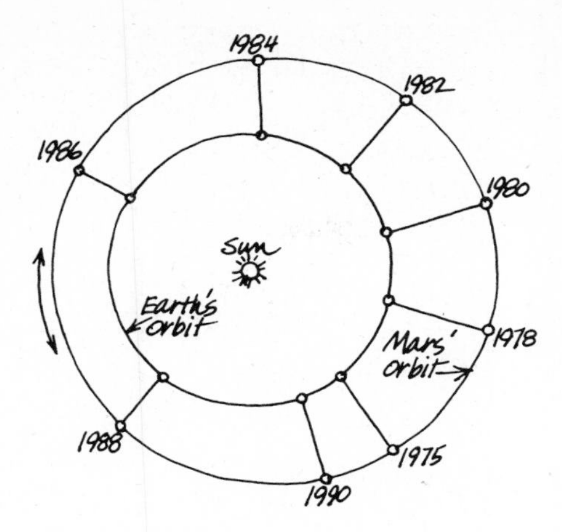

Figure 10.34 (left) Oppositions of Mars. Mars' orbit is the third most eccentric of all the planets, while Earth's is more nearly circular. One effect of this is that Mars is much closer to Earth during some oppositions than during others. The diagram shows all oppositions of Mars from 1975 to 1990. Mars is most heavily studied in telescopes during close oppositions.

planet. The earth's atmosphere is partly to blame. It wavers in a manner similar to the wavering observed above a hot toaster or a hot road. On a bad day, you may as well be observing Mars from the bottom of a swimming pool. There are occasional times though, when "the seeing is good," as astronomers say. On such days there are fleeting instants when the atmosphere seems to be at rest. At such times the markings on Mars seem to stand out in etched detail only to be plunged almost at once into the usual frustrating blurriness. The observer then tries to sketch what he or she remembers. Unfortunately, a camera cannot capture the fine details, because during a time exposure the film is blurred by the intervals of bad seeing. The observer at the telescope must be blessed with excellent eyesight, a good memory for details, drawing skill, and limitless patience. It is not surprising that controversy raged over the observations of the fine details on Mars, even among highly skilled observers. The observations are extremely difficult to make. See Figure 10.36.

Mars came particularly close to the earth in the opposition of 1877. In Italy, an astronomer named Schiaparelli drew a new map of Mars. Many of the names of the martian features Schiaparelli chose are still used to this day. On the map, he drew a number of lines connecting the dusky regions. Schiaparelli called the lines *canali,* which is Italian for "grooves." The term was mistranslated into English as "canals."

When American businessman Percival Lowell (see Box 10.5 and Figure 10.37) heard about the canals on Mars, he determined to study Mars himself. He saw the canals and asserted that each one had a specific width from which it did not deviate throughout its length. While Schiaparelli had mapped 113 canals, Lowell saw over 450 of them. The canals had rather peculiar properties. Both Schiaparelli and Lowell agreed that some were not always visible—they "hibernated."

Lowell went on to passionately defend his own interpretation of the canals. Compared to Earth, Mars was an old planet, he presumed. The dusky regions are ocean beds which are now mostly dried up. As the water supply dwindled, the martians had built at first small and then

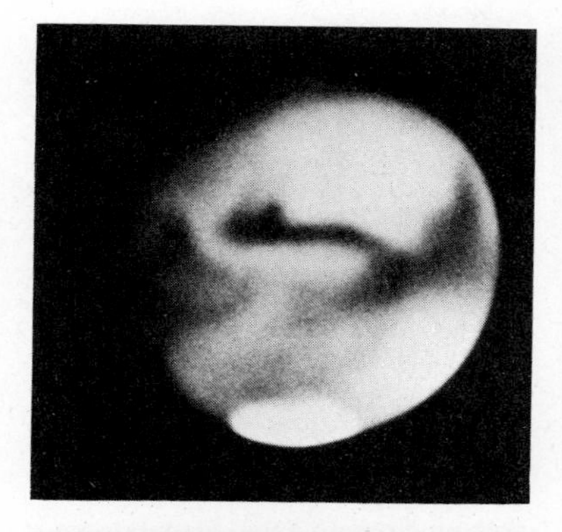

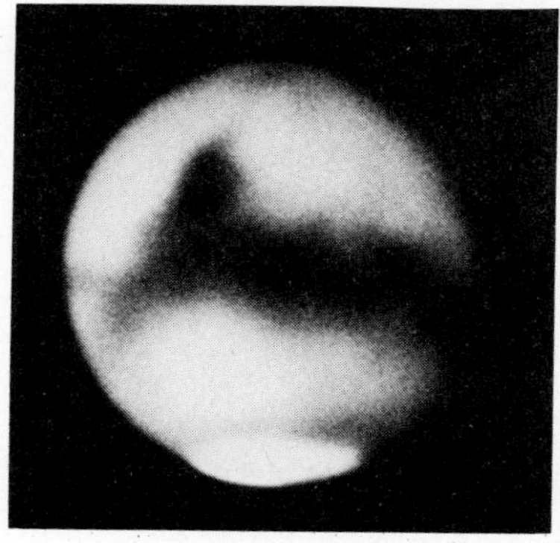

Figure 10.35 Two telescopic views of Mars. These two photographs were made by the 36-inch refractor at Lick Observatory. The southern polar cap of Mars is prominent in both views. The upper photograph shows the dark, elongated feature known as Sabaeus Sinus which resembles a mittened hand on a long, thin arm. The large, pointed, dark feature, Syrtis Major lies to the east of Sabaeus Sinus, is visible in the upper photograph, and is prominent in the lower photograph.

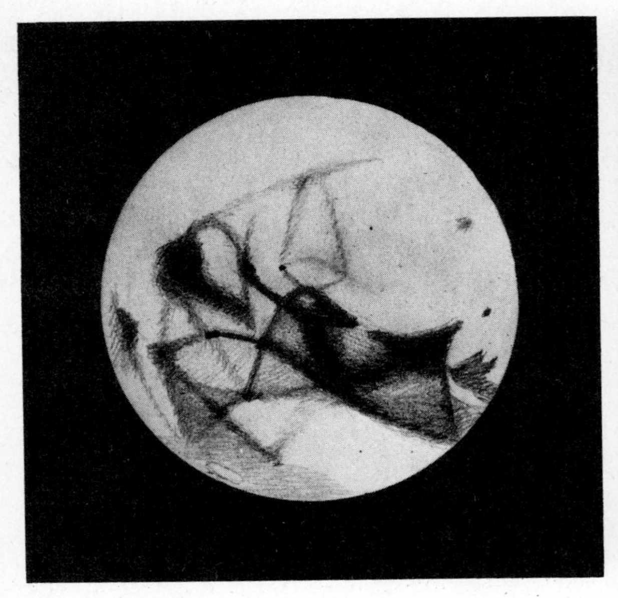

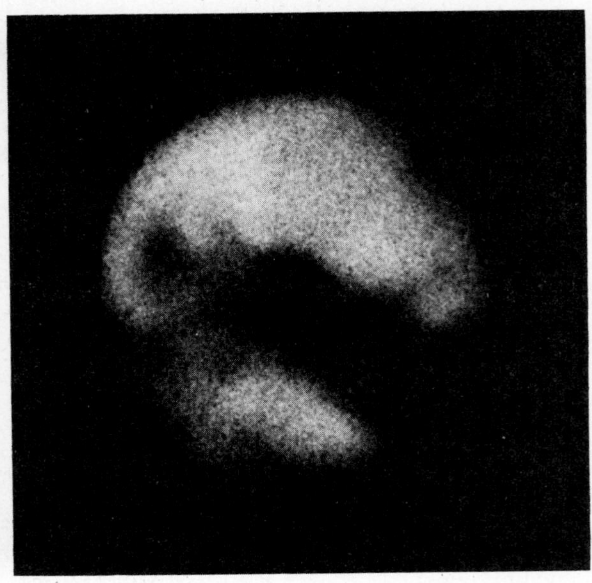

Figure 10.36 Mars: a drawing and a photograph of the same region. Both of these representations of Mars were made during the opposition of 1926. The drawing shows how much more detail visual observers record than photographs. Linear features are seen in the drawing. Are these actually present on the martian surface or only imagined by the observer? As discussed in the text, one had to wait for the space age for a certain answer. The region shown lies to the west of that in the upper photograph in Figure 10.35. The "mitten" of Sabaeus Sinus may be seen at the edge of the drawing at about the 4 o'clock position. To the left (west) of it is the pointed region Margaritifer Sinus, resembling, but smaller than, Syrtis Major (seen in Figure 10.35, lower photograph). The western end of Margaritifer Sinus narrows to a point and points at a feature which, in this drawing, looks something like a bird's head, bill to the right. Coming out of the back of the bird's head is a dark, thick line known as the Coprates region. After the space probes surveyed this region, it was renamed Valles Marineris. It is one of the most remarkable features in the solar system.

more and more lengthy canals to transport water for irrigation. Some of the canals had been doubled when an even larger capacity was required. Canals hibernated when not in use. Lowell was positive that he had clear evidence of extraterrestrial intelligence.

As more and more astronomers observed Mars to look for the canals, a long and at times bitter controversy broke out. Some well-respected astronomers supported Lowell, but many could see no canals. Others saw something, but to them the markings were not so straight or connected as Lowell drew them. The question remained undecided until the space age, although probably the majority opinion was against Lowell well before that time.

Mars As Seen by the Early Space Probes The first successful space probe directed at Mars was *Mariner 4*, which flew by Mars in July 1965. During its brief look at the planet, over 20 photographs of the surface were taken and radioed back to the earth bit by bit. When the resulting pictures were reassembled, they revealed something all but a very few

BOX 10.5

PERCIVAL LOWELL

Although interested in the sky as a youth and highly gifted in mathematics and languages, Lowell pursued a business career in the Far East after graduating from Harvard University. In his spare time he wrote four books on Oriental subjects. He found it hard to concentrate on his job, however, and, after reading about Schiaparelli's observations of Mars, he realized that he was bored with business. Instead, he decided to build an observatory using his wealth, an observatory devoted to observations of Mars and, when Mars was not favorably placed, of the other planets.

Lowell decided to build in Flagstaff, Arizona. Water vapor prevents a clear view of the heavens, and sunny Arizona is noted as a dry region. Lowell originated the idea of placing a telescope at a high altitude; Flagstaff is about a mile above sea level. At such a height there is less air above us and the air is quieter.

Since its opening in 1894, Lowell Observatory has been a source of many important contributions to astronomy. Besides the work on Mars and the search for a planet beyond Neptune, Lowell was interested in the origin of the solar system. This interest led one of his assistants at the observatory to investigate galaxies. As we shall see in Chapter 16, this work produced some of the most important discoveries in the history of science.

---

had expected—craters. This first glance revealed a surface that reminded many of the surface of the moon. The impression the *Mariner 4* photographs left on many people was that Mars must be a dead, barren planet. After all, it seemed to look so much like the moon.

Another interesting piece of data returned by *Mariner 4* was recorded by a detector of magnetic fields carried on the craft. It revealed that Mars' magnetic field is very weak. This ties in well with the density determinations made much earlier. Mars' low density leads us to suspect that, if Mars has an iron core, it must be very small. Thus, if the planets produce magnetic fields by means of convection in an iron core, we would expect Mars to have a tiny magnetic field.

Figure 10.37 Percival Lowell. Aristocratic, dapper, called brilliant by some and misled by others, Lowell stirred up the field of astronomy in more ways than one.

Mars As Viewed from Orbit The goal of orbiting a camera system around Mars was achieved in 1971 when *Mariner 9* was successfully placed in orbit. Soon an avalanche of pictures and other data were pouring back to earth, causing many planet experts to, once again, revise their concepts of Mars. Five years later, in 1976, two more probes, *Viking 1* and *Viking 2*, which were associated with the very successful landing missions described below, were put into orbits around Mars. These enlarged and refined the view we now have of Mars. So much was learned from the orbiters that we have space to mention only a few of the more spectacular discoveries.

In November 1971, *Mariner 9* arrived at Mars and sent back its first photographs. A disappointment: Mars was totally enveloped in one of the most spectacular dust storms it had ever exhibited. Most of the photographs showed no surface features at all. As the dust settled, four craters, which seemed to be at the top of high mountains, appeared. Were these volcanic mountains poking up above the storm and

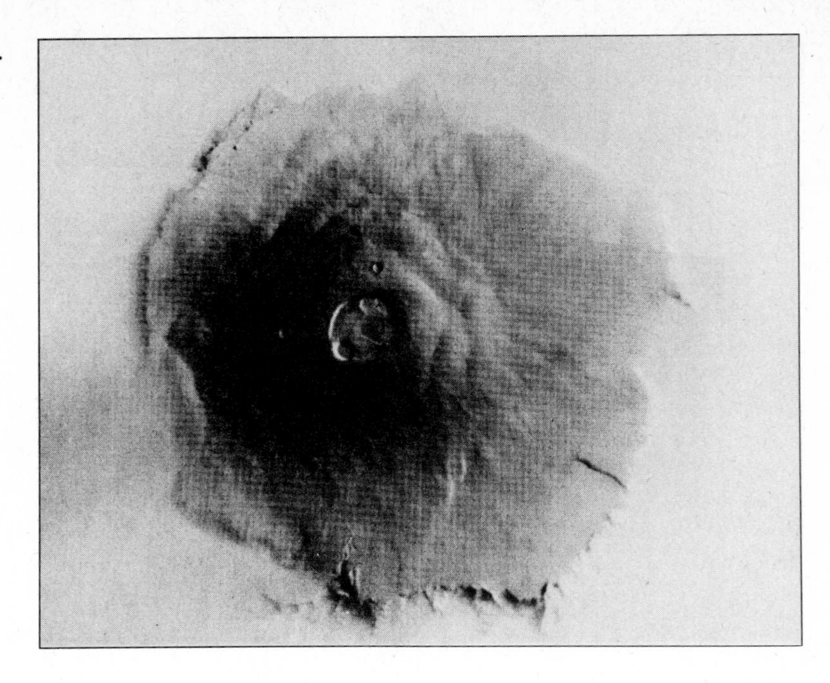

Figure 10.38 Mighty Olympus Mons. The base of this shield volcano would cover the state of Nevada. A *Mariner 9* photograph.

therefore the most tremendous volcanoes known to humans? The mission was designed to last for three months of orbiting (although it ran much longer). Would the dust settle before then?

Shield Volcanoes on Mars By January 1972, the atmosphere of Mars had begun to clear. Before long, photographs began to be transmitted. Astronomers were rocked back on their heels. For one thing, the supposed giant volcanoes had been correctly identified. The most impressive is now called Olympus Mons. It is so large that its location had actually been glimpsed as a dot by earth-based telescope users and named Nix Olympica. See Figure 10.38. It is technically known as a shield volcano, a mountain built up as repeated eruptions of the volcano produced its enormous flanks. Olympus Mons dwarfs any mountain known on earth. It is about 25 km (16 mi) high and about 600 km (370 mi) in diameter at its base. For comparison, the peak of Mount Everest is about 9 km (5.6 mi) above sea level. Mount Everest itself is not this tall—it lies on a high plain. The largest shield volcano on earth is Mauna Kea, one of the Hawaiian islands. It's peak lies about 9 km above the ocean floor. One would need to pile nearly three Mauna Keas on top of one another to equal Olympus Mons in height.

A shield volcano on earth cannot become as large as Olympus Mons. As a terrestrial volcano builds, the plate upon which it rests gradually moves, shifting the volcano away from its source, an underlying hot spot. A new volcano may later begin to form over the hot spot, after the old one has moved on. This concept is used to explain many of the island chains found in the earth's oceans. At least a dozen other shield

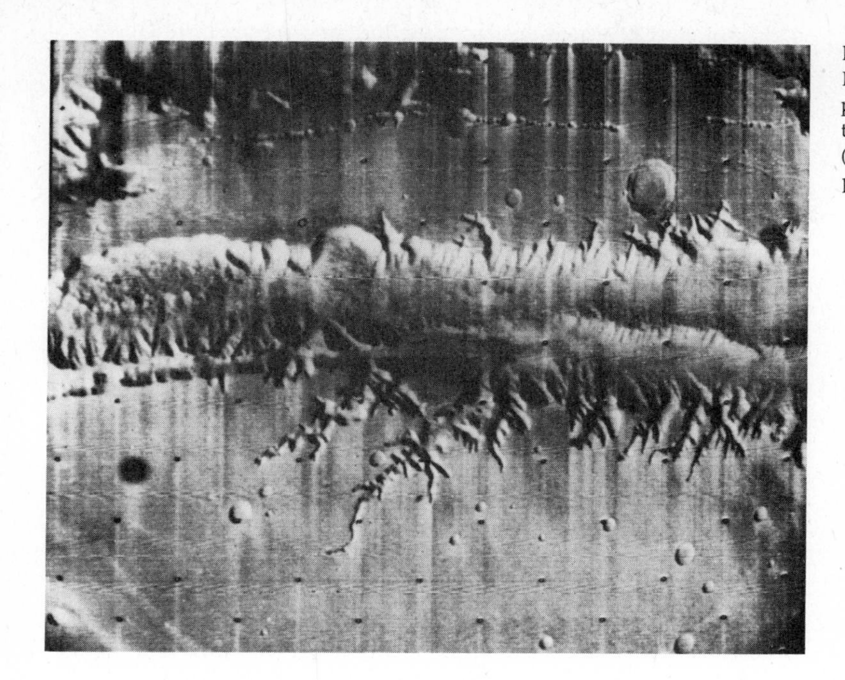

Figure 10.39 A portion of Valles Marineris. Taken by *Mariner 9,* this photograph shows a vast section of the canyon. It extends about 500 km (300 mi) from one margin of the photograph to the other.

volcanoes have been located on Mars, many of them also enormous. This is taken as evidence that plate tectonics (see Section 10.1) has not taken place on Mars to any significant extent. After Olympus Mons, the next three largest shield volcanoes on Mars are named Ascreus Mons, Pavonis Mons, and Arsia Mons, and all three are roughly 20 km (about 12 mi) high with 400-km (250-mi)-diameter bases. They all lie about 1000 km (620 mi) from Olympus Mons and in a straight line on an elevated region known as the Tharsis Rdige. This alignment is unusual on Mars and is not taken to be the result of plate motion.

Possible Evidence of Water in Mars' Past As expected, the nearly complete survey of Mars' surface provided by *Mariner 9* revealed no bodies of water. Well before then, it had been clear that the atmospheric pressure was not sufficient to force water to remain a liquid. However, a related, exciting discovery was made—Mars has numerous channels and chasms.

One of these chasms is another of the most spectacular features in the solar system. Known as Valles Marineris (Valley of the Mariner, named for the spacecraft that discovered it), it is a tremendous system of canyons stretching out roughly parallel to Mars' equator for about 5000 km (about 3000 mi). See Figure 10.39. In places it is about 6 km (3.7 mi) deep and about 500 km (300 mi) wide. One who has stood in awe at the rim of the Grand Canyon in Arizona and stared down into its 1.6-km (1-mi)-deep gorge, will have a hard time imagining Valles Marineris, which would stretch from New York to California and which

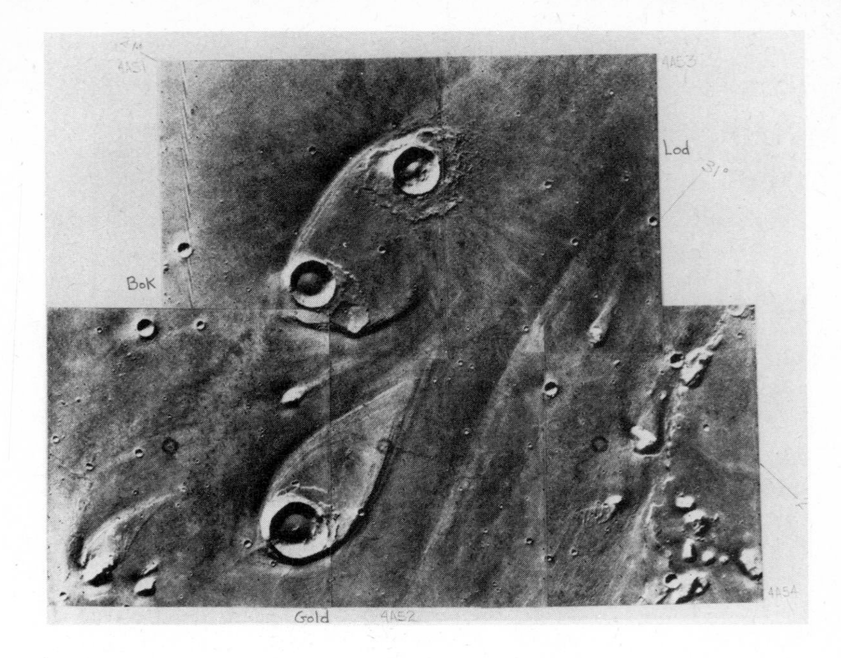

Figure 10.40 Evidence of large amounts of flowing water on Mars? A *Viking 1* orbiter photograph.

is about four times deeper than the Grand Canyon. (The largest tourist attractions on Mars, such as Olympus Mons and Valles Marineris, may be overdone, being too large for a person on the surface to grasp from any one vantage point.) There has been some speculation that such trenches may be evidence of straining at the surface of Mars to produce primitive tectonic plate motions such as are observed on Earth.

Along the rim of such canyons are many channels which remind one of riverbeds. Elsewhere on Mars are many other examples of features which, on earth, are formed by running water: curving channels, tributary systems (branching of channels), teardrop-shaped islands (Figure 10.40), bars, much erosion, and braided patterns. At first reluctantly but with increasing force, many astronomers have come to conclude that, in the past, there were episodes during which large amounts of water flowed on Mars. This conclusion has radically changed the balance of opinion about Mars once more. The earlier *Mariner* flights had left an impression of a dry, dead, moonlike planet, but now there is evidence of huge floods of water some time in the past. This discovery leads immediately to certain questions. Why was Mars a wet planet some time in the past? Where did the water go? Does this mean that life could have existed or even may still exist there? No completely acceptable theories are available yet.

The *Viking* orbiter has revealed that the portion of the northern polar cap remaining in the summer is largely ice. It is also easy to assume that large quantities of water have been incorporated into the soil and permanently frozen there. See Figure 10.41. One expert suggests that the amount of water in the caps and the soil, if melted, would cover the

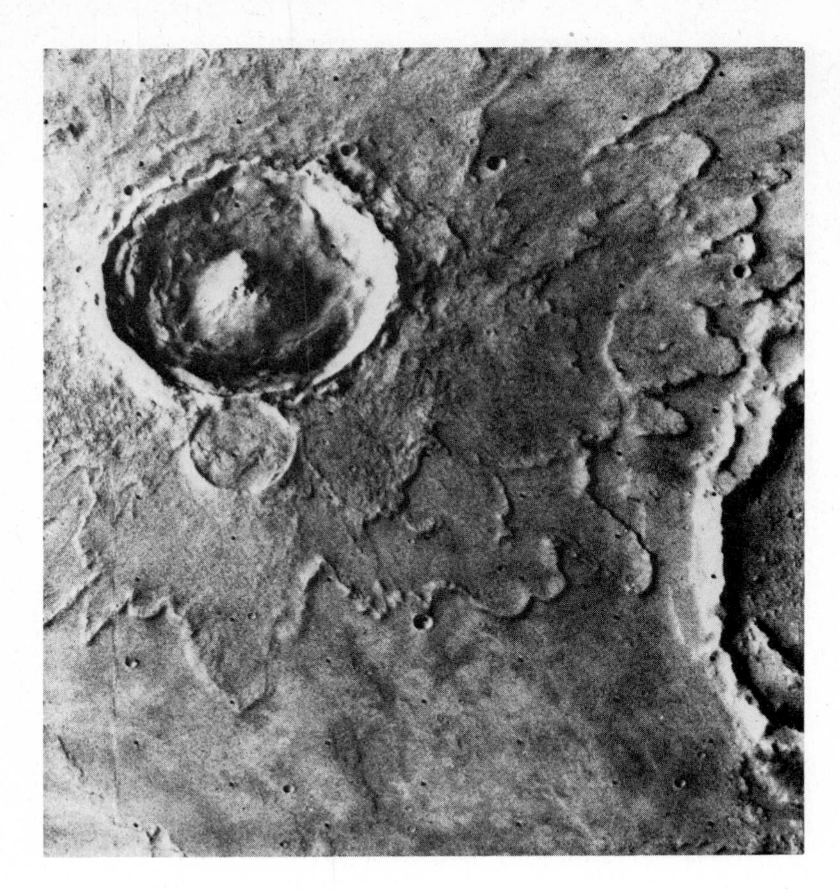

Figure 10.41 Yuty crater and the flows around it. This crater has a diameter of about 18 km (11 mi). Could the impact that formed it have caused underground water to melt and did this help form the lobate (curved) flows? Photograph by *Viking 1* orbiter.

surface of Mars in a layer 20 meters (22 yd) deep. It is only fair to add that some experts interpret the evidence differently and deny that large amounts of water ever flowed on Mars.

Before leaving the orbiters and landing on Mars to search for life, several questions remain. What of the *canali* that Schiaparelli and Lowell were certain they had seen? They do not exist. Many of the *canali* on Lowell's map do not correspond to any features seen by the orbiters. A few *canali* were drawn near large trenches. Other *canali* turned out to be combinations of dark blotches and irregular arrangements of craters. It is now deemed certain that no large, absolutely straight, even lines exist on the surface of Mars. Astronomers have been forced to conclude that Lowell suffered from an advanced case of wishful thinking. Astronomers have learned a lesson. We should be very skeptical when considering any data, but particularly data obtained at the limits of human and instrumental capabilities. It is all too easy for even experienced, skillful observers to delude themselves. Changes in the dark markings on Mars have been observed from earth. They may be due to shifting, wind-blown sand. See Figure 10.42. For an overview of Mars' surface, see Figures 10.43 and 10.44, as well as Plates 6 through 9.

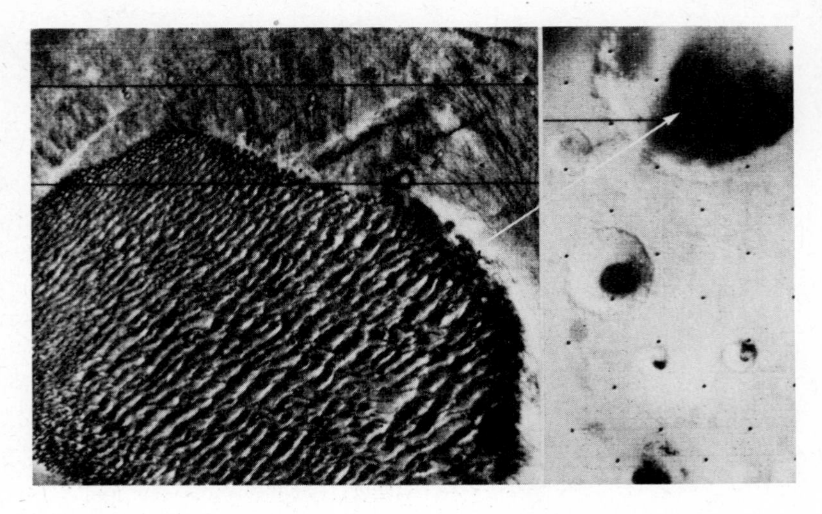

Figure 10.42 Sand dunes in a crater photographed by *Mariner 9.* The right portion of this figure shows a number of craters on Mars that have dark interiors. A close up of the upper crater (arrow) shows that the dark region is a sand dune field.

The Surface of Mars As Viewed from the Surface As late as early 1965, the best astronomers could do when viewing Mars was to strain to glimpse fleeting details which were, under the best conditions, never smaller than about 60 km (about 37 mi). A dozen years later all this was to change dramatically as, in one of the most complex and intricate space missions ever attempted, *Viking 1* and *Viking 2* approached Mars. The years of experience, trial, error, failure, and accumulating success through which NASA had progressed had brought space engineers close to scoring a major breakthrough. Extensions of the human senses in the form of instruments were to be landed on the surface of Mars. But would the billion dollar mission work? After all, all four Soviet landers had failed.

The *Viking 1* space probe went into orbit around Mars on June 19, 1976, ten months after being launched from Earth. Then, on July 20th the lander separated from the orbiter and landed on Mars.

Then there was a wait, nearly an hour long, while the lander's camera scanned the surface, processed the scene, and sent information back. What would the view be? One of the scientists involved, Carl Sagan, had put a guess into print a decade earlier: ". . . rocks, lava flows, sand dunes. An occasional scraggly plant would not be unexpected. But there are other possibilities—fossils, footprints, minarets." The lander was equipped to search for life using a number of sophisticated experiments, but a photograph of a living being would answer the question quickly.

Finally, the first picture was assembled. The initial part of Sagan's guess was correct; the surface showed rocks and sand dunes. The engineers were ecstatic. "The cameras didn't work this well on earth." The geologists (now often called "planetologists") were delighted. "It's just a beautiful collection of boulders. A geologist's delight!" But not even one scraggly weed.

The view from Chryse was fascinating. See Figures 10.45 and 10.46.

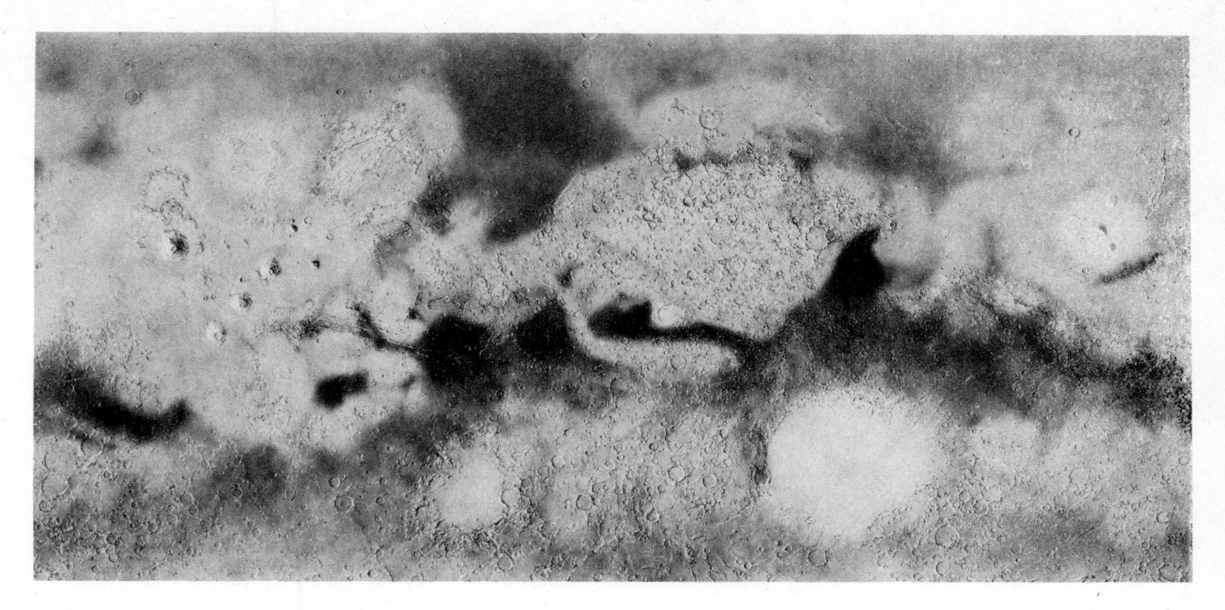

Figure 10.43 A map of Mars: region 60° on either side of the equator. This map was produced by J. L. Inge of Lowell Observatory. He used earth-based telescope observations to draw the dark regions and *Mariner 9* photographs to draw the craters, canyons, volcanoes and other features. We shall point out some of the features in a left to right (west to east) sweep. Olympus Mons is obvious, lying inward a bit from the left-hand margin. Slightly south and east are the three Tharsis shield volcanoes, lined up. Directly east of the most southern volcano is the enormous chasm, Valles Marineris. This merges into the large dark region Aurorae Sinus, called a "bird's head" in Figure 10.36. Next lies the triangular Margaritifer Sinus, pointing a bit east of north. North of these two dark regions is the light region Chryse. *Viking 1* landed in the northwestern region of Chryse. West of Margaritifer Sinus lies first the "mitten" and then the "arm" of Sabaeus Sinus. Farther west lies the large, dark triangular region Syrtis Major. These last two are seen in Figure 10.35. South of Syrtis Major is the large, circular region Hellas, which is remarkably free of craters. Notice how crater-free the most northerly regions on the map are. Roughly half of the surface of Mars has few craters, while the other half is heavily cratered. The reason for this is unknown.

The soil is orange-brown dust. One of the lander's pads was buried in it. The soil apparently contains an iron-bearing mineral which, as had long been guessed, gives Mars its distinctive color. The sky color was a surprise, however. It had long been assumed that the sky would be blue because of the same scattering of light that causes our skies to be blue. Wrong! The martian sky is pink because of the reddish dust remaining in the air long after a dust storm.

Every day the lander watched the sun set. Mars rotates once roughly every $24\frac{1}{2}$ hr, a period of rotation remarkably similar to the earth's. At night, the temperature drops until it reaches a predawn low of about −90°C (−130°F). The midafternoon high is none too comfy, about −30°C (−22°F). During a summer afternoon, the soil, which absorbs more heat than the air, can get up above freezing for a short time.

During its descent and while on the surface, the lander sampled and analyzed the atmosphere. It turned out to be similar to the atmosphere of Venus in that its main ingredient is carbon dioxide, 95 percent by

Figure 10.44 (facing page) A portion of the northern hemisphere of Mars. *Mariner 9* took this photograph in September 1972. It is late spring in this region and the northern polar cap is retreating. Olympus Mons is visible to the right of the lower cutoff region of the photograph. Farther to the right are the three Tharsis volcanoes, the most southerly of which is somewhat indistinct near the edge of the disk. Valles Marineris may be glimpsed near the lower right margin.

Figure 10.45 The first *Viking 1* lander panorama: June 20, 1976. *Viking 1* landed in a dead-looking, rocky desert. The horizon is about 3 km (2 mi) away. Some of the lander's equipment is seen in the foreground. The picture was taken by a rotating camera. The scene faces east on the left, south near the center, west, and then north at the far right. The streaks in the southeastern sky are not real features.

Figure 10.46 Sand dunes viewed by the *Viking 1* lander. Two hours after sunrise on August 3, 1976, the cameras took this striking picture of sand dunes possibly left by the previous wind storms. The boulder on the left was named "Big Joe." It is about 1 meter (3 ft) high and 3 meters (10 ft) long. It lies about 8 meters (25 ft) from the lander. Had the lander struck Big Joe, the mission would most likely have been a failure.

volume. There is about 2.7 percent nitrogen, 1.6 percent argon, and small amounts of carbon monoxide, oxygen, water vapor, and other substances. It is not an atmosphere in which we could live, but it does contain the substances necessary for life as we know it. Do any martians live on it?

Biological tests were designed to answer this question. The lander held an amazingly compact and complex automatic biological laboratory. An arm could be extended to scoop up martian soil for analysis. A number of experiments took place. The soil, and any potential microbes (microscopic organisms) in it, were fed nutrients, given light, water, and heat, and encouraged to grow and give off certain gasses, thereby signifying life.

When the first sample of soil received water, it immediately began to release carbon dioxide and oxygen. The experimenters were shocked. Was life this easy to detect on Mars? "But wait," said others, "Doesn't this rapid response indicate a chemical reaction? Wouldn't life

Figure 10.47 First picture from *Viking 2* lander: September 3, 1976. While landing, the descent engines broke *Viking's* fall. These evidently kicked up an amount of dust. Evidence for this is seen in the dust on the lander's footpad and the brightness variations at the left edge of the photograph. This latter portion of the photograph was evidently taken before the dust had settled from the atmosphere. The rocks are about 20 cm (8 inches) in size. Many show vesicles (holes) perhaps indicating an origin in volcanic activity.

react with more of a delay?'' Naturally, the experiments were continued and repeated many times with many variations. The results are still considered ambiguous to this day. Many experts say that probably only chemical reactions were observed. Others tentatively propose that life may possibly have been observed. The weight of opinion seems to be arrayed on the negative side because of the results of a further experiment. This one analyzed the soil and searched for organic molecules. Organic molecules are carbon-based, very massive molecules out of which all life on earth is formed. None were detected on Mars. Perhaps the detector was not functioning properly? No, it did detect traces of the organic compound used to clean the instrument before the launch.

The landing site for *Viking 2* was chosen to be slightly more risky for a safe landing but, it was hoped, more likely to exhibit a different geology and to harbor life. The chosen region is called Utopia and is on the side of the northern hemisphere opposite Chryse, and farther north, nearer the polar cap. Fortunately, the second lander also made a successful touchdown, although one pad was on a rock, so that *Viking 2* saw Mars on a slant. Unfortunately, the view, Figure 10.47, and the results of the life experiments were not much different from those obtained by *Viking 1.* To this day, no good evidence for the presence of life on Mars has been obtained. See Figure 10.48.

The Moons of Mars As Viewed from Earth In the early 1600s, Kepler guessed that Mars has two moons. Centuries later this lucky guess was confirmed. In 1877, during the same close opposition of Mars during which Schiaparelli was mapping Mars' surface, Asaph Hall was using the 26-inch refractor of the United States National Observatory to look for satellites (moons) in orbit around Mars. After a lengthy search lasting a number of nights, he went home and told his wife that he was

Figure 10.48 Evidence of life on Mars?? The angle of the sun during this photograph was such that one of the rock formations which is about 1.6 km (1 mi) in diameter cast shadows on itself to give the appearance of a human head. Viewed when the sun was at a different angle, the illusion would be destroyed. Alas, this is not a monument built by the martians.

giving up the project. Luckily, she encouraged him to "try it just one more night." And, indeed, the very next night, he saw the first moon of Mars. Some days later he found a second. Hall named them Phobos and Deimos.

Hall's observations indicated that the sizes of these moons and the sizes of their orbits are unusual. The refracting telescope Hall used was, at that time, the largest refractor in the world. Yet, even at maximum usable magnification, the moons appeared merely as points of light. This indicates that the moons are tiny, but how big are they? Until the space age, one could only make a guess based on the moons' known distance from us and their brightness. A typical estimate was that they might each be roughly 10 km (6 mi) in diameter.

The satellites of Mars had been difficult to discover not only because they are small and dim. They are also very close to Mars. Hall had begun by assuming that the radius of a martian moon's orbit divided by the radius of Mars itself might be similar to the radius of our moon's orbit divided by the radius of earth, namely, 60. He found that the numbers for the actual moons are much smaller—2.8 for Phobos and 7.1 for Deimos. Hall had initially been searching much too far from Mars. Because the moons are so close to Mars, they must orbit it rapidly in order to avoid being pulled inward. Phobos orbits Mars in only 7.7 hr, and Deimos in 1.3 days, compared with the one month sidereal (orbital) period of our moon. The rapid motions of the moons of Mars would produce an unusual sight as viewed by a martian.

The Moons of Mars As Viewed from Space Probes *Mariner 9* arrived in orbit around Mars during one of the most severe dust storms

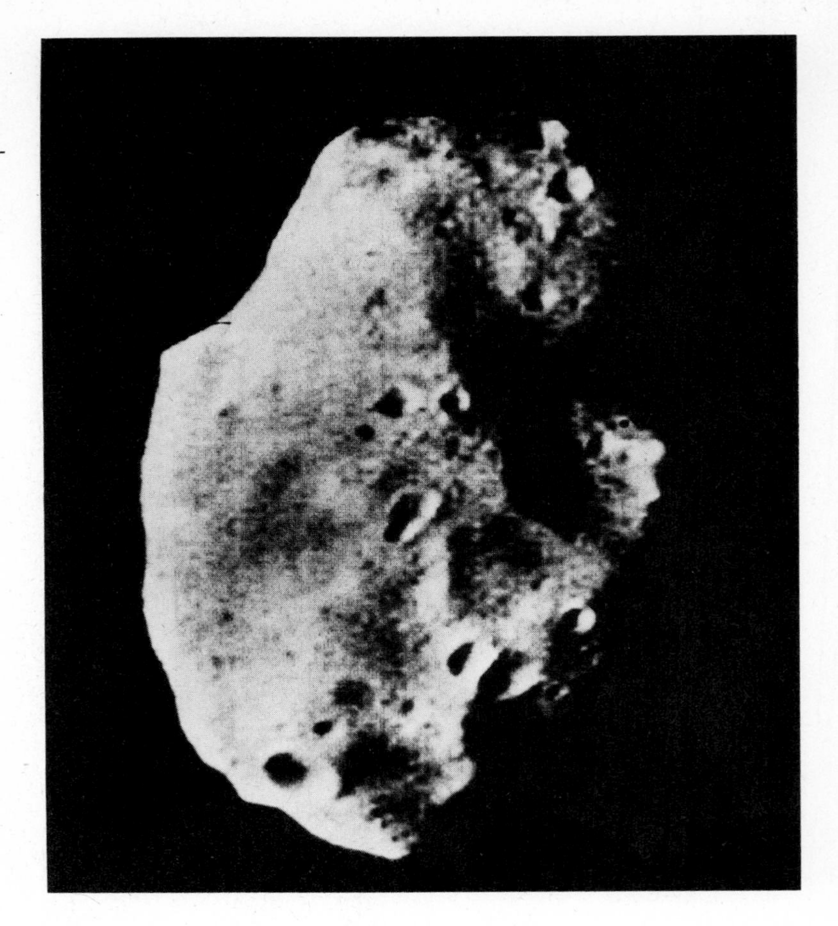

Figure 10.49 Phobos viewed by *Mariner 9:* late 1971. This was the view *Mariner 9* obtained of Mars' largest satellite when it was 5540 km (3440 mi) away. Phobos has been battered by impacts. The piece missing from the upper left may have been knocked off by impact with a small asteroid.

ever seen there. The planet looked nearly blank to the cameras. Rather than let this expensive equipment orbit idly, plans were modified in order to bring the craft closer to the two moons.

The investigators were not disappointed. The moons' reputation for oddness was upheld. Neither moon is even close to being round. Both rather resemble potatoes. The pictures were used to determine the sizes of the moons, and the earlier guesses were not too bad. Phobos is the larger of the two, 28 by 20 km (17 by 12 mi) across, and Deimos, the outer satellite, is 16 by 10 km (10 by 6 mi). Both moons revolve synchronously, keeping one face toward Mars. Both are cratered. The largest crater on Phobos, 8 km (5 mi) in diameter, has been named Stickney, the maiden name of Mrs. Hall. Other features of Phobos are discussed in the captions for Figures 10.49 through 10.51.

It has been pointed out that the gravitational pull at the surface of one of the moons is very small, so small that a person could throw a baseball into orbit around it. The ball would travel so slowly that one would have plenty of time for lunch before the ball came around to be caught, having completed one orbit.

Figure 10.50 Phobos viewed by the *Viking 2* orbiter: September, 1976. This photograph was taken when the orbiter was only 880 km (550 mi) from Phobos. The major discovery was the existence of *striations,* linear features, on the surface of Phobos. They were truly puzzling. The best guess was that they consisted of chains of craters.

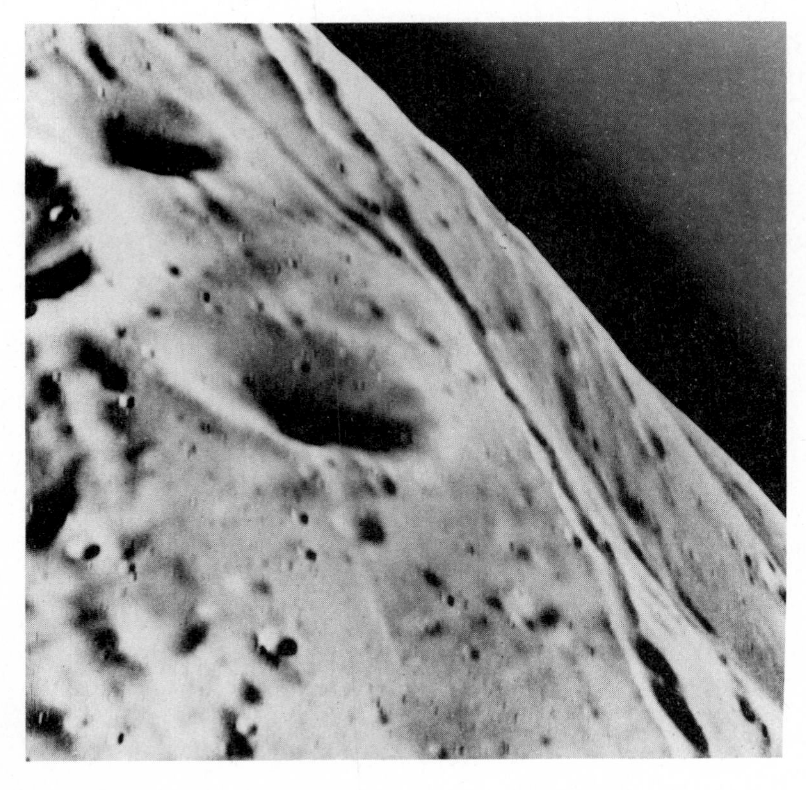

Figure 10.51 *Viking 1* views Phobos: February, 1977. By early 1977, *Viking 1* was maneuvered into a very close encounter with Phobos, only 120 km (75 mi) distant. The striations visible in Figure 10.50 are shown clearly here and one can see that they are grooves and not crater chains. They remain rather mysterious. One hypothesis suggests that they are strain marks caused by the gravitational pull of Mars on Phobos.

SUMMING UP

Four dense, rocky planets and three satellites reside in the inner solar system. We can list the planets in several ways: increasing size (Mercury, Mars, Venus, Earth); increasing surface temperature (Mars, Earth, Mercury and Venus, nearly tied). Mercury has essentially no atmosphere. Mars' is thin. Earth's is ideal (!). And Venus' is remarkably dense. The atmospheres of Mars and Venus are composed chiefly of carbon dioxide, while Earth's is composed chiefly of nitrogen and oxygen. We shall see in the next chapter how very different the outer planets, those beyond Mars, are from the inner planets.

EXERCISES

1. For each of the following planets, name three things that are interesting and unique: Mercury, Venus, Earth, Mars.

2. What is the current theory explaining how mountains are raised on earth?

3. Describe the process that results in a planet's having an iron core and a less dense rocky crust.

4. Sketch a simple map of the moon's near side. Show Mare Crisium, Mare Tranquilitatis, Mare Imbrium, and the craters Copernicus and Tycho.

5. Define the following terms concerning the moon: highlands, mare, earthshine, the moon effect, synchronous rotation, terminator, crater, ray.

6. Briefly describe the environment on the moon. Include the following topics: atmosphere, temperature range, sky color, apparent motions of the sun and earth, length of day.

7. How were the large craters formed on the moon? How were the maria formed? Give some evidence for volcanism on the moon.

8. What is peculiar about the rotation of Mercury? Of Venus? How were these periods discovered?

9. Compare the surface temperatures of Mercury and Venus. What process allows Venus to be so hot even though it is nearly twice as far from the sun as Mercury?

10. What was the result of the *Viking* landers' search for life on Mars?

11. Describe the largest volcanoes and chasms on Mars.

READINGS

Much interesting information on earlier work on the solar system is found in

Watchers of the Skies, by Willy Ley, Viking Press, New York, 1966.

Findings concerning the solar system made since the advent of the space age are well described in many places. Among the best surveys are

Exploration of the Solar System, by William J. Kaufmann, III, Macmillan, New York, 1978.

The Inner Planets, New Light on the Rocky Worlds of Mercury, Venus, Earth, the Moon, Mars, and the Asteroids, by Clark R. Chapman, Scribner's, New York, 1977, by a planetologist.

Closeup: New Worlds, edited by Ben Bova and Trudy E. Bell, St. Martin's Press, New York, 1977, which features accessible, engaging styles of writing. Several of the chapters were written by noted science fiction authors.

The new look of geology due to the theory of plate tectonics is well described on a popular level in

The Restless Earth, by Nigel Calder, Viking Press, New York, 1972.

Among the most interesting of the many books about the moon written since the Apollo program are

The Moon Book, by Bevan M. French, Penguin Books, Baltimore, 1977.

Apollo over the Moon, A View from Orbit, edited by H. Masursky, G. W. Colton, and F. El-Baz, NASA, SP-362, Washington, D. C., 1978, which features splendid photography and a discussion of surface features.

The Moon As Viewed by Lunar Orbiter, by L. J. Kosofsky and F. El-Baz, NASA, SP-200, Washington, D. C., 1970.

The Voyages of Apollo, by Richard S. Lewis, New York Times Book Co., 1974, which chronicles each flight and the scientific findings.

The best book I know that gives the reader a feeling of what it is like to be an astronaut is the delightful

Carrying the Fire, by Michael Collins, Ballantine Books, New York, 1975.

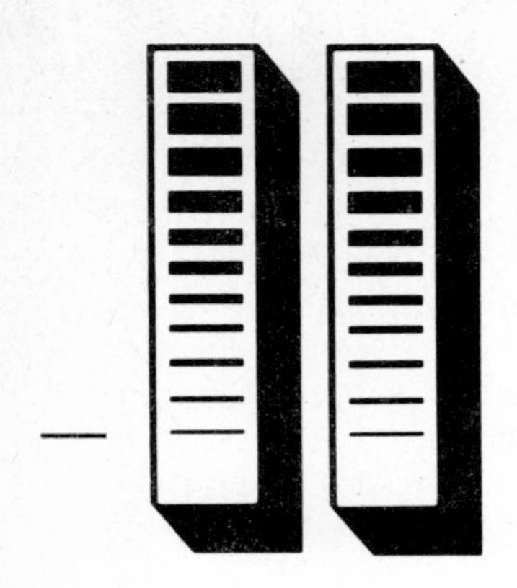

THE OUTER SOLAR SYSTEM: JUPITER AND BEYOND

On the night of March 13, 1781, a self-trained, 43-year-old astronomer named William Herschel was scanning the stars in the constellation Gemini. He was using a homemade 6-inch-diameter reflecting telescope to search for stars that appear to be close together. Herschel was intrigued when he noticed an object that did not look like a star; it looked like a disk. He quickly published a report about the newly discovered object, stating that he "suspected it was a comet." Herschel continued to watch the object, but it failed to develop a tail and did not appear increasingly hazy as most comets do. It looked most definite, rather like a planet. Finally, enough observations of the moving object were made with respect to the fixed stars to allow a Swedish astronomer to calculate its orbit. It moves in a nearly circular orbit with a sidereal period of 84 years. An application of Kepler's third law showed that it is about 19 AU from the sun. It *is* a planet. Herschel named it Georgium Sidus ("George's star") after his king, George III of England, who had an interest in astronomy. The name did not stick, however. Johann Bode suggested the planet be called Uranus, after the mythological father of Saturn, and this name was adopted. But whatever the name, Uranus was an important discovery. It nearly doubled the size of the known solar system and made Herschel internationally famous. To this day, Uranus is the only major planet discovered by accident.

The name of the seventh planet may be pronounced either "Yur'-ah-ness" or "Yur-ay'-ness (rhymes with "stir hay' mess").

In this chapter we describe the planets in the outer solar system. Then we turn our attention to the smaller members of the solar system and conclude with a discussion of the solar system's origin.

11.1 JUPITER: THE GIANT

Jupiter As Viewed with the Naked Eye Copernicus knew that Jupiter is about 5.2 AU from the sun (later found to be equivalent to 778 million km or 483 million mi). When Jupiter is in opposition, on the side of the earth opposite the sun, it is closest to the earth and yet about 630 million km (390 million mi) from us. Mars, when closest, is 11 times closer to Earth. Nevertheless, for all its distance, slightly yellowish, basically white Jupiter usually looks much brighter in our sky than Mars. This is a first hint that Jupiter is an impressive planet.

Mars surpasses Jupiter in brightness as viewed from Earth only when Mars is at one of its rare close oppositions.

Jupiter As Seen Through a Telescope Jupiter is one of the most rewarding planets to study through a telescope. It appears as a disk much larger than Mars even though it is always much farther from us. Indeed, Jupiter's diameter is about 143,000 km (89,000 mi), over 11 times the diameter of Earth. By measuring the radius of the orbit and period of any of Jupiter's moons, astronomers can calculate that Jupiter's mass is 318 times that of Earth. Jupiter has about 2.5 times as much mass as the rest of the planets combined. The solar system is sometimes characterized as "the sun and Jupiter plus debris."

A telescope having a diameter of about 4 inches or larger reveals a series of parallel **belts** on the surface of Jupiter if the seeing is fairly good and Jupiter is not too near the horizon. The belts alternate with lighter regions called **zones** (see Figures 11.1 and 11.2). The belts and zones are arranged parallel to Jupiter's equator. Although the belts do not change much in position, their appearance differs considerably from time to time, sometimes appearing faint and sometimes darker; sometimes appearing uniform and sometimes blotchy. One belt may darken while another gets lighter.

The other major features visible on Jupiter, observed in larger telescopes, are spots. They can be light, dark, or nearly colorless. These ordinary spots are not permanent but appear randomly and then fade. One may occasionally see other round, dark spots on Jupiter's surface, which are shadows cast by one or the other of its moons.

By far the most famous feature on Jupiter is the great red spot. It first became generally recognized in 1878 and appeared brick red at that time. At first it was thought that the spot had just recently appeared, but a check of earlier records showed that G. D. Cassini had drawn it as early as 1665. Now, over 300 years later, it is still visible, even in moderate-sized amateur telescopes. Its color has ranged from very pale pink to brick red. Its width has been nearly constant at about 14,000 km (8700 mi), but its length has varied between 30,000 and 40,000 km (18,600 and 25,000 mi). This latter length is equivalent to more than three earth diameters. From early guesses that the spot is a hot flow of lava about to be spewed into space to form a new moon to the modern suggestion that the spot is a 300-year-old red hurricane, no totally acceptable explanation has yet been given. Even the spot's revolution (motion due to Jupiter's spin) is peculiar. It revolves faster than the rest of the planet, for several years perhaps gaining several laps. Then it slows down and revolves much more slowly.

The belts themselves all move at different speeds, belts near the equator traveling faster, similar to the differential rotation of the sun (see Section 8.1). On the whole, Jupiter rotates with a period of just under 10 hr. The result of such a rapid rate of rotation is that it bulges noticeably at the equator and is flattened at the poles. Such a changeable surface, rotating at different speeds, could not be a solid surface. We see only the tops of clouds.

What is Jupiter composed of? Earth-based studies using the spectroscope have revealed methane and ammonia in its atmosphere. Cold

Figure 11.1 Jupiter as drawn in 1908 by Barnard using the Yerkes Observatory 40-inch retractor. The red spot is prominent in the southern hemisphere. A satellite (far right edge of the planet) casts a dark, round shadow on the planet.

hydrogen is difficult to observe with a spectroscope, yet astronomers before the space age were fairly confident that the bulk of Jupiter was made up of hydrogen. This was believed for several reasons. For one thing, Jupiter's density, its huge mass divided by its enormous volume, was known to be only 1.3 grams per cubic centimeter, just slightly more dense than water. This checks out well with the hypothesis that Jupiter is a ball composed mostly of hydrogen. Furthermore, the theory, explained in Section 11.10, that the whole solar system was formed from a huge cloud composed mostly of hydrogen leads one to expect that a planet as massive as Jupiter could have retained its original hydrogen supply because of its strong gravitational field. These ideas were confirmed when space probes revealed that Jupiter's atmosphere is 82 percent hydrogen and 17 percent helium. Its composition is similar to the Sun's.

The clouds of Jupiter are probably composed of ammonia crystals in the uppermost layer, ammonium hydrosulfide farther down where the temperature is higher, and water-ice crystals even farther down. These ideas are based chiefly on theoretical calculations.

Why does Jupiter have light-colored zones and darker belts? A long-standing conception is that the zones are regions where warm currents of atmosphere are rising. As the gas cools, it spills over into adjacent belts where cooler gasses sink back down. In other words, the

Figure 11.2 Jupiter photographed using the 200-inch reflector at Mount Palomar.

belts and zones result from convection, discussed in Section 10.1. The colors on Jupiter, the reds, blues, and browns, are still unexplained, since the chemicals mentioned above are colorless. Does the sunlight striking the chemicals in the rising zone produce a reaction that causes a color change?

Jupiter As Viewed by Space Probes When the *Pioneer* spacecrafts (Box 11.1 and Figures 11.3 and 11.4) soared past Jupiter, they confirmed earlier observations that Jupiter emits more energy (as infrared radiation) than it gets from sunlight. It may be contracting under its own gravitational field, thereby producing energy, much as envisioned by the old contraction theory of the sun (Section 9.7). Only a millimeter or so of contraction per year would probably suffice to produce the excess infrared radiation. Other experts suppose that Jupiter simply has not yet cooled down from the time when it was formed and is still releasing stored heat.

Radio waves from Jupiter were detected in the mid 1950s. Certain characteristics of the radiation led scientists to conclude that it was due to very rapidly moving electrons spiraling in what must be a strong magnetic field surrounding the planet. This process of producing radiation is known as **synchrotron radiation** and is discussed further in Section 14.8.

BOX 11.1

PIONEER 10 AND *PIONEER 11* TO JUPITER: A PERILOUS JOURNEY

NASA planned a series of flights to Jupiter and Saturn during the 1970s and 1980s. The first two were to be exploratory flights to Jupiter to try out the flight systems and have a good preliminary look at the largest planet. These probes were named *Pioneer 10* and *Pioneer 11* (see Figure 11.3).

Pioneer 10 was launched in March 1972 at a speed of 51,800 km/hr (32,200 mi/hr). It had a long distance to travel, and the sun would be pulling back on it the whole way, so a large initial speed was necessary. *Pioneer 10* reached the orbit of the moon in only 11 hr, rather than the roughly 2.5 days required by the manned *Apollo* flights.

Then, in July 1972, the craft reached the inner edge of the asteroid belt, a region of debris between the orbits of Mars and Jupiter (see Section 11.7). Because it is much too expensive to send a spacecraft out of the plane of the ecliptic, *Pioneer* had to take its chances in the asteroid belt. The craft was aimed so as to clear the known major asteroids, but a fast-moving particle 0.001 to 1.0 mm in size could have dealt *Pioneer* a deadly blow. One preflight estimate was that the chances were 1 in 10 that it would be knocked out. But *Pioneer* made it. Only 75 particles in the range 0.1 to 1.0 mm were seen in the distance, an unexpectedly low number.

After *Pioneer 10* was well on its way, warnings began to come from investigators who predicted that there would be clouds of fast-moving protons and electrons trapped in Jupiter's magnetic fields. It was feared that these particles might be more numerous than previously thought and that the delicate solid-state components in *Pioneer's* electronics could be knocked out. *Pioneer's* brain would be destroyed. There was nothing to do but wait and see.

Finally, in December 1973, 1 year and 10 months after launch, *Pioneer 10* reached Jupiter. The investigators had to wait 46 min after the pass for radio word from Pioneer. Then came the announcement saying, in effect, "*Pioneer* lives!" *Pioneer 10* received a dose of protons and electrons large enough to kill a human 1000 times over. It was later estimated that, had Pioneer 10 passed only one-half of a radius of Jupiter closer to the planet, it would not have survived.

After passing Jupiter, *Pioneer 10* continued to travel away from the sun. It reached the orbit of Uranus in 1979, but was far from that planet at the time. Its speed is so great that it will eventually escape from the solar system with a residual speed of about 12 km/sec. At this slow speed, it will not pass near any stars for millions of years. While thinking about the long time *Pioneer 10* will spend in interstellar space, Carl Sagan had the idea of attaching a plaque to *Pioneer 10* showing when and where the spacecraft originated in case it is ever picked up by intelligent beings (see Figure 11.4).

Even before *Pioneer 10* reached Jupiter, *Pioneer 11* was launched toward the same planet, reaching it in December 1974. It raced past at only one-third the distance from Jupiter that *Pioneer 10* had passed. It survived the proton and electron clouds because it was aimed at a latitude of 45° south, where these particles are less numerous. The strong gravitational field of Jupiter whipped *Pioneer 11* rapidly through the strongest parts of the radiation clouds, and up over the northern hemisphere. This carefully planned maneuver had four benefits: *Pioneer* could get close to Jupiter and still survive the strongest radiation because of the rapid speed; it could photograph both the south and north poles, which are not visible from earth; the slingshot effect flipped *Pioneer 11* up above the plane of the planets to measure conditions there for the first time; and, as planned, *Pioneer* was retargeted by Jupiter's gravitational force so that it would coast to a 1979 rendezvous with Saturn.

Figure 11.3 The *Pioneer 10* spacecraft. The top portion of the craft features the large dish reflector used for radio communications with the very distant earth.

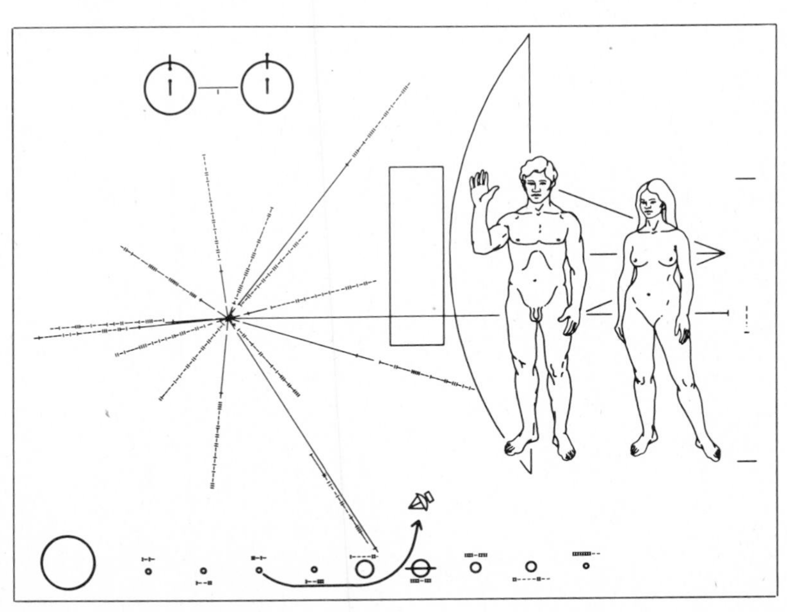

Figure 11.4 The plaque on *Pioneer 10.* The craft is drawn to the same scale as the human figures to show our size. At the bottom is a sketch of our solar system showing that the craft came from the third planet. The radial lines indicate the locations of 14 pulsars (see Section 14.14) with information as to when the craft was launched and which star is our sun. The symbol in the upper left is a hydrogen molecule, used to give a time and distance scale.

Figure 11.5 Jupiter as viewed by *Pioneer 10,* December 2, 1973. Much detail may be seen in Jupiter's belts and zones. The satellite Io, off the picture to the right, casts a shadow on the surface of the giant planet.

Pioneer 10 and *Pioneer 11* confirmed the idea that Jupiter has a magnetic field and is surrounded by a huge volume of trapped charged particles, chiefly protons and electrons.

By making use of earth-based and space probe observations, theories about the internal structure of Jupiter are being evolved. There may be a small, rocky core at its center, which may be roughly the size of the earth. Above this, crushed into a metallic form by the enormous pressure of the overlying mass is likely to be a large layer of metallic hydrogen, a form of hydrogen that has not yet been produced on earth. Metallic hydrogen can conduct electricity. Perhaps electric currents in this layer produce Jupiter's magnetic field. Liquid hydrogen probably forms the next layer, and overlaying this are the layers that produce the cloud tops we see on the surface (see Figures 11.5 and 11.6).

See the color photographs of Jupiter in Plates 10 and 12.

The Moons of Jupiter As related in Section 6.5, Galileo discovered the four largest satellites of Jupiter. The largest, Ganymede, has a diameter 1.4 times larger than our moon's, which makes it larger than the planet Mercury. Any binoculars, steadily supported, will reveal the four largest moons. They are so large that, if the glare due to Jupiter were not present, they would be visible to the naked eye. Occasionally, a very sharp-eyed observer can see one of them without optical aid when Jupiter is near opposition and when the moon is near a greatest elongation.

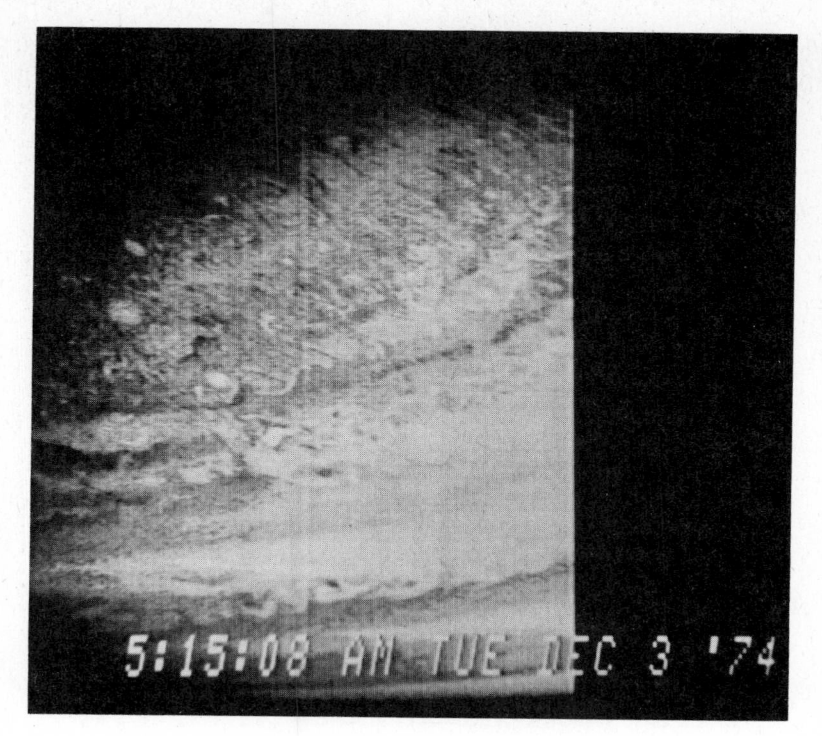

Figure 11.6 Jupiter's north polar region: *Pioneer 11,* December 3, 1974. According to the theory that Jupiter's belts and zones are produced by convection cells drawn into belts and zones by Jupiter's rapid rotation, the polar regions, which do not have such a large rotation rate, should not have belts and zones. *Pioneer 11* arced over the north polar region and confirmed this prediction. The polar regions are mottled rather than belted.

The four large satellites and one small one form an inner group relatively close to Jupiter; all are less than 2 million km (1.2 million mi) from its center. As larger and larger telescopes have been built and trained on Jupiter, the number of known satellites has increased. The thirteenth was added in the 1970s. The four outermost satellites are over 20 million km (12 million mi) from the planet and are weakly held by its gravitational field. Oddly enough, these four revolve around Jupiter in a sense opposite that of all the others. All the satellites beyond the four largest are small. The largest has a diameter of roughly 200 km (125 mi). As is only proper, the king of the planets has a sizable retinue. (See also Plate 11.)

As viewed from its most distant satellite, Jupiter would have an angular diameter smaller than our moon as viewed from earth.

11.2 COMPARING THE TERRESTRIAL AND THE JOVIAN PLANETS

There are significant differences between Jupiter and the planets closer to the sun. We shall see that the planets beyond Jupiter (excluding Pluto) are similar to it in many respects. This has led astronomers to classify the planets into two broad categories—terrestrial planets and jovian planets. The two classes differ in location, size, mass, and density, and in the substances of which they are formed. The four terrestrial planets are Mercury, Venus, Earth, and Mars. They are all relatively close to the sun, the farthest, Mars, being only 1.5 AU away. The jovian

"Jovian" is an adjective meaning "of or pertaining to Jupiter." Words having the same root are found in such expressions as, "He is a jovial chap," and "By Jove, I've got it!" "Terrestrial" means "of or pertaining to the earth."

planets are distant—the nearest, Jupiter, lying at 5.2 AU. The terrestrial planets are small, Earth being the largest. The jovian planets are large; the largest, Jupiter, has a diameter 11.2 times larger than Earth's. The terrestrial planets do not have much mass, the largest mass being that of Earth. The jovian planets are massive; Jupiter's mass is 318 times as great as Earth's. The terrestrial planets are dense; expressed in grams per cubic centimeters, the first three have densities somewhat over 5. The density of Mars is just under 4. The jovian planets are massive but have large volumes, so large that their densities are low, not too different from the density of water.

Strongly related to the differing densities of the planets are the differing substances of which they are composed. One can divide these substances into three classes, rocky, icy, and gaseous. This is a classification based on the ease with which the substances are either melted or turned to gas. Rocky materials are principally iron, calcium, aluminum, and many minerals. The terrestrial planets and our moon are chiefly rocky planets. Gaseous substances, chiefly hydrogen and helium, constitute the bulk of Jupiter and Saturn. Uranus and Neptune, the two outermost jovian planets are thought to be constituted, in large measure, of a rock-ice mixture with an outer layer of gas.

11.3 SATURN: A RINGED PLANET

Saturn As Viewed with the Naked Eye Saturn, which lies roughly twice as far from the sun as Jupiter, nevertheless shines prominently in our night sky. When Saturn is near opposition, only two stars, Sirius and Canopus, are brighter. Saturn appears to move very slowly among the fixed stars, staying in or close to one constellation for more than a year at a time.

Saturn As Viewed Through a Telescope The binoculars user is able to watch Saturn's largest moon, Titan, orbit Saturn about once every 16 days. The nontelescope owner should beg, borrow, or otherwise obtain a look at Saturn through a telescope. Even a small telescope presents a view of Saturn's rings. Words cannot describe nor photographs capture the appearance of this planet floating in the dark sky as viewed through a telescope. One astronomer who has spent many hours viewing Saturn through a large telescope says that the view of Saturn still evokes a "cry of admiration" each time he sees it.

The globe or ball of Saturn is impressively large. Its equatorial diameter is about 120,000 km (75,000 mi), 9.4 times larger than the earth's diameter. It is another giant, similar to Jupiter in size. But it has much less mass, only 95 times that of the earth. As a result, Saturn's density is only 0.7 gram per cubic centimeter—13 percent of the density of Earth and, less than that of water. Astronomers sometimes joke that if there were a filled bathtub large enough, one could float Saturn in it.

Like Jupiter, Saturn exhibits belts parallel to its equator, although

Figure 11.7 Saturn and its rings. Belts and zones are faintly seen. Cassini's division in the rings is evident between rings A and B, the only two visible in this photograph.

these markings are much less obvious and less active than Jupiter's. See Figure 11.7. Only rarely do white spots appear on the planet. These allow a determination of the planet's rotational period. Again, like Jupiter, Saturn spins fast, making one rotation in 10 hr, 14 min. Because its axis is tipped at an angle of 26.7°, we can alternately see its north and south polar regions, each of which is a darker region. Saturn's spectrum indicates the presence of the gasses methane and ammonia. Because of its low density, it is assumed that, like the sun and Jupiter, it is composed chiefly of hydrogen and helium. Saturn is similar to Jupiter in other ways as well. It too emits more infrared radiation than it would if it depended entirely on the sun for energy, and it also emits radio waves.

The Rings of Saturn Huygens was the first to recognize the ring system for what it is, a very flat ring close to Saturn's surface, having a width of about one Saturn radius (Figure 11.8). The ring system is positioned directly above, that is, in the same plane as, Saturn's equator. We see the rings edge-on at times and the north face and south face at other times. This is due to the 26.7° tip of Saturn's axis. At times one pole or the other is tipped toward us, and we see one face of the rings or the other. When the earth is in the plane of the rings, they all but disappear, they are so thin. In 1675, G. D. Cassini saw a gap in the rings between the dimmer, outside A ring (as it is called) and the brighter B ring. This gap is still called Cassini's division. Large telescopes, under perfect seeing conditions, which may occur only once in several years, can reveal many other gaps in the rings, some so narrow they look like dark threads to the observer.

Figure 11.8 Saturn and its three brightest rings, top view. This scale drawing represents the view from high above one of Saturn's poles. The earth is dwarfed by the huge system. Earth would nearly fit between Saturn and ring C. Not shown are ring D, which is inside ring C, and the numerous other fine divisions in the rings.

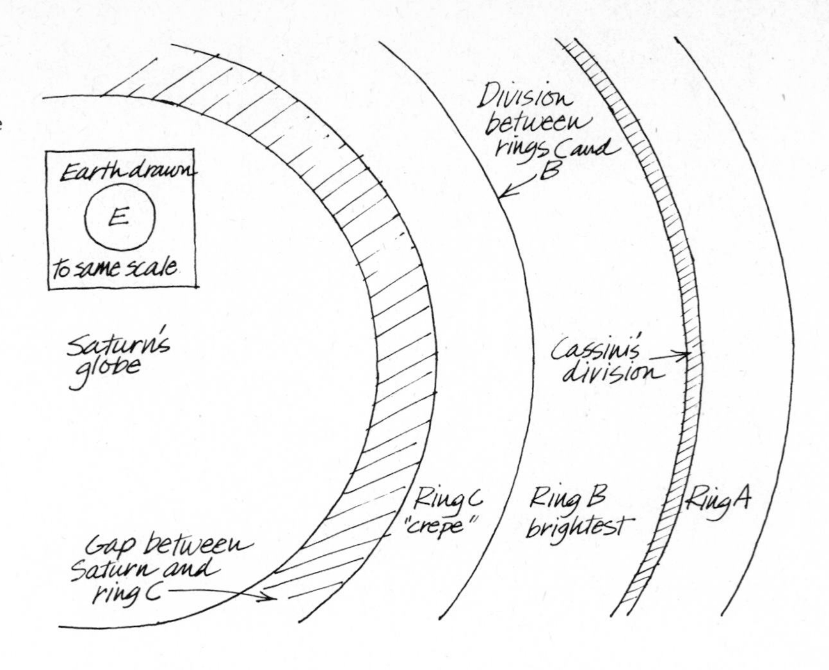

Discoveries about the rings have continued right up to our time. In 1850 Bond discovered a nearly transparent C ring, just inside the B ring. It does not show up on most photographs. Then, as late as 1969, Guerin was observing Saturn from the famous Alpine observatory, Pic du Midi, and found a fourth ring, labeled D, inside the C ring. It is even more transparent than the C ring, and its inner edge lies rather close to Saturn's surface.

Before reading on, please solve Puzzle 11.1.

The rings of Saturn, as you now know, are composed of small particles. The gaps in the rings are places where certain satellites of Saturn have pulled ring particles out of specific orbits by means of the satellite's gravitational pull.

The spectrum of the rings, a spectrum of sunlight reflected from the rings, indicates that the particles are composed of, or coated with, ammonia or ice crystals. In December 1972, a 64-meter- (210-ft)-diameter radio antenna was used to beam a 400-kilowatt radar signal at Saturn. The beam took 2.4 hr to make the 2.4-billion-km (1.5-billion-mi) round trip. The globe of Saturn reflected no measurable signal, but the rings reflected the radar more strongly than expected. The return signal indicated that the rings might be made of pieces as large as 1 meter (3.3 ft) in diameter.

Saturn's Moons Ten satellites of Saturn have been discovered. Titan is perhaps the largest satellite in the solar system and could be considered another planet. It is larger than Mercury, and its surface is covered by reddish-brown clouds. Its atmosphere seems to consist

PUZZLE 11.1

ARE SATURN'S RINGS SOLID?

If the ring system around Saturn were a single solid piece, the outer edge of the ring would have to move faster than the inside edge to keep up, much like a record on a record player. On the other hand, if the ring consists of many individual particles, each in individual orbits around Saturn, Kepler's third law requires that the particles on the inner edge move more rapidly than those on the outer edge. The question has been settled by looking at the ring system in a spectroscope. When the side of the rings moving toward us is so viewed, the inner edge shows a larger blue shift than the outer edge. So—are the rings solid or particles? (You may wish to review Section 7.6.)

largely of hydrogen and methane. Four other satellites are easily visible in moderate-sized telescopes. They lie more or less above Saturn's equator and, as a result of the large tip of Saturn's axis, usually do not appear strung out in a straight line like Jupiter's. (Jupiter's axis is tipped by only 3°.)

Perhaps *Pioneer 11* and the other space probes to follow will help answer many of the questions about Saturn still remaining. The prime question perhaps is, Why does Saturn have its beautiful rings?

11.4 URANUS: THE PLANET LYING ON ITS SIDE

Uranus was discovered in 1781 by William Herschel. At opposition, even though it is then over 18 AU from Earth, Uranus is just barely visible to the naked eye, looking like a very faint star. It could have been discovered before the invention of the telescope, and it is not hard to find with binoculars if one is guided by an astronomical magazine.

When the orbits of Uranus' five satellites were computed, they were found to be peculiar. Each is almost at right angles to the plane of Uranus' orbit, whereas almost all other satellites have orbits more nearly in the plane of the orbit of their parent planet. This brought into question two other recognized patterns in the solar system. All other planets have axes roughly perpendicular to the plane of their orbits, the largest tip being Saturn's, 26.7°. Also, most of the satellites in the solar system orbit in the same plane as their parent planet's equator. (Our moon is something of an exception in that respect.) Given the highly tilted orbits of the satellites of Uranus, one or the other of these patterns had to be broken for Uranus. But which one?

To determine the rotation of a planet, one can try to observe surface markings. A few observers thought they saw an occasional faint spot on the greenish disk of Uranus or the hint of a light band, but photographs show only a blank disk. The question was firmly settled by viewing Uranus in a spectroscope and observing the Doppler shift

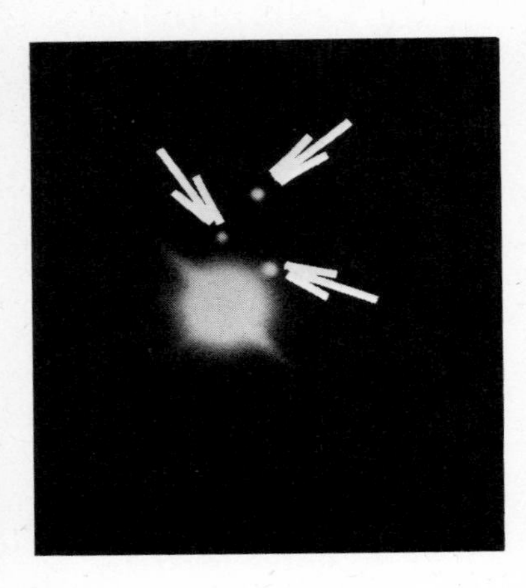

Figure 11.9 The planet Uranus and three of its satellites (arrows). No features are visible on Uranus. The spikes coming from the planet in this photograph are unavoidable defects caused by supports inside the telescope tube. Photograph by the Lick Observatory 120-inch reflector.

from each edge of the planet. Uranus rotates once every 16 hr. In 1902, Deslandres showed, by means of such investigations, that Uranus' lies on its side; its axis is only 8° out of the plane of the planet's orbit. The satellites do orbit over Uranus' equator. This means that at times Uranus' axis is pointed almost directly at us, and at such times the satellite orbits appear nearly circular. This last occurred in 1946 and will next occur in 1985. (How are night and day on Uranus affected when one of its poles is pointed at the sun?)

Uranus is another giant, gaseous planet, having a diameter four times larger than earth's but only 40 percent of Jupiter's. (Figure 11.9.) Its density is about the same as Jupiter's (1.2 grams per cubic centimeter), and its atmosphere is much like those of the other giants, consisting largely of hydrogen, helium, and methane. Any ammonia in the planet has been frozen out of the upper atmosphere by the very low temperature, about 50 K (−220°C or −370°F). Uranus is at a distance of 19.2 AU from the sun. Viewed from Uranus, the sun would have an apparent diameter of only 1.6 minutes of arc. It would not appear to be a disk to the unaided eye, yet it would still look dazzlingly bright.

The Rings of Uranus Both planets and stars have very small angular diameters as viewed from earth. Thus, when a planet passes in front of a star as viewed from earth, a rare event known as an **occultation,** astronomers generally take note. An especially attractive occultation was predicted for March 10, 1977, when Uranus would probably pass in front of a rather faint star.

Because of the uncertainty in the predicted time of occultation, observations began well before the most probable time. Suddenly, well before Uranus reached the star, instruments recording the light from the star indicated a dip in brightness. The amazed observers recorded five dips in brightness before the occultation by Uranus occurred. Then, again, after the occultation, which did take place, five more dips in brightness were observed on the other side of Uranus. Astronomers quickly concluded that Saturn is not the only planet with rings. Uranus has at least five, and possibly nine. They are very thin and narrow and relatively close to the planet and, as a result, had never been directly seen in a telescope. It was one of the most exciting discoveries of the 1970s. Saturn's rings were no longer unique. However, the existence of such rings is still no less difficult to explain than before.

11.5 NEPTUNE: THE SMALLEST GIANT

By 1821, 40 years after its discovery, Uranus was giving astronomers headaches; it did not appear to be obeying Newton's laws. For a while it had rushed ahead of its predicted positions and later had fallen seriously behind. Were Newton's laws inapplicable at such great distances? Was an unseen planet forcing Uranus out of its predicted path by means of its gravitational field?

PUZZLE 11.2

BODE'S LAW

In 1772, Johann Bode, the German who named Uranus and was an author of a catalog of star positions, popularized a law which has become known as Bode's law. The law had been discovered earlier by Johannes Titius, who found an empirical formula which seemed to demonstrate a pattern concerning the distances of the planets from the sun. Here is how it works.

Object number	1	2	3	4	5	6	7	8	9	10	More?
Name	Mercury	Venus	Earth	Mars	?	Jupiter	Saturn	?	?	?	
Start	0	3	6	12							
Add 4	4	7	10	16							
Divide by 10	0.4	0.7	1.0	1.6							
Actual Distance (AU)	0.4	0.7	1.0	1.5							

This table should be completed by the reader. First the row "Start" is completed by doubling each previous number. For the next row, add 4 to the number directly above. Next, divide each number by 10. For the last row, the actual distance of each planet from the sun, rounded to one place after the decimal point, is listed (see Appendix 3). Notice that Bode had to leave a gap at position 5 to make Jupiter fit. Section 11.7 explains how this gap was filled. When position 8 was filled by Uranus in 1781, Bode's law looked all that much better. Small wonder, then, that Adams and LeVerrier assumed that position 9 would be filled following the same law. How well do Neptune and Pluto agree with Bode's law? Do you think it is really a law or just a coincidence? Astronomers still disagree on this last question. (No solution is given in Appendix 2 for this puzzle, so be careful with your arithmetic.) By the way, do you detect a bit of fudging in the law in the "Start" row in going from Mercury to Venus?

In 1843, a 23-year-old student from Newton's alma mater, Cambridge University, decided to search for the unseen planet. His method was unusual, for he used pencil and paper rather than a telescope. He was John Couch Adams, and he relied principally on the ideas of two other men. Adams assumed that Newton's law of gravitation and laws of dynamics were correct. He also relied on what has come to be known as Bode's law. Please work Puzzle 11.2 now.

Adopting Bode's law, Adams assumed that the planet would lie 38.8 AU from the sun. He then spent two years of hard labor trying to calculate the most likely position of a planet that would account for the deviations of Uranus from its calculated orbit. When he finally got an answer, he wrote to the British Astronomer Royal, asking him to search a specific region of the sky to see if a planet could be found there. The astronomer did not look at the sky but rather sent the calculations back, asking for a better explanation. It was a whole year before Greenwich

Observatory began a search for the planet. And even then, the planet was actually seen but not recognized for some months.

In the meantime, a more famous astronomer in France, LeVerrier, had begun an investigation of the wayward Uranus, trying to calculate the position of the new planet. He followed a procedure much like Adams'. A year after Adams obtained his results, LeVerrier published the results which he had arrived at quite independently. Unknown to either of them, the two astronomers had derived quite similar predictions for the position of the new planet. Nine months passed, and no French astronomer had looked for the new planet. Finally, in exasperation, LeVerrier told a German astronomer, Galle, where to look. The same night, Galle checked that part of the sky against a star chart and found the new planet. Galle wrote to LeVerrier, "Dear Sir, the planet whose position you indicated *actually exists.*"

The discovery was a triumph for both Adams and LeVerrier. The first planet to be discovered on paper before it was seen through a telescope was named Neptune, in spite of LeVerrier's desire that it be named after him. The bottom line of this story is as strange as the rest. Neptune was observed and its actual orbit calculated. Its actual distance from the sun is about 30 AU and not the 39 AU Bode's law required. Neptune did not Bode well, so to speak. Oddly enough, the orbits calculated for Neptune by Adams and LeVerrier did not agree with each other, and neither agreed with the actual orbit of Neptune in space. Yet, each man's calculations correctly gave the direction to look for Neptune at the time of discovery.

Neptune Seen Through the Telescope At its enormous distance of 4.5 billion km from the sun, 6 times the distance of Jupiter, Neptune appears as a dim dot of light in all but the largest telescopes. See Figure 11.10. Markings, if any, are very difficult to see. About all that is known about Neptune is that it is large, 3.9 times Earth in diameter, and slightly denser than the other jovian planets. The spectroscope shows that Neptune is spinning about once every 16 hr and has an atmosphere much like that of the other giants.

The two satellites of Neptune that have been found follow very different orbits. Triton is by far the largest of the two, having about the same size as our moon. Its orbit is also about the same size as our moon's orbit, although the stronger gravitational pull of massive Neptune requires Triton to orbit once in only six earth days. The other satellite, Nereid ("Near'-ee-id"), is tiny and so far from Neptune that, on the average, it requires a year and several months for one revolution. The two satellites revolve around Neptune in opposite directions.

11.6 PLUTO: THE FARTHEST PLANET?

The Discovery of Pluto Shortly after the discovery of Neptune, LeVerrier suggested that there might well be another planet beyond Neptune. Perhaps the feat of Adams and LeVerrier could be repeated.

Figure 11.10 The planet Neptune and its major satellite Triton. A Lick Observatory 120-inch reflector photograph.

Then, in about 1905, Percival Lowell (Box 10.5) became interested in the subject. Using his equipment at Flagstaff, Arizona, he made photographs of the sky, which covered the entire ecliptic. He made a pair of photographs of each location, each member of the pair being separated in time by three days. Then Lowell searched each pair of photographic plates with a hand magnifier, looking for an object that had moved relative to the fixed stars during the three-day interval. No planet turned up, and in 1907 the search was stopped.

Meanwhile, Uranus, which had been observed since the early 1900s, was wandering slightly, so it seemed, from the proper orbit, even when the gravitational influence of Neptune was allowed for in the calculations. Many assumed that the apparent error was only due to tiny errors in the observations. Lowell assumed that it was caused by an undiscovered object, planet X.

He proceeded to perform horrendously complex calculations in order to find the orbit of a planet having the proper location to produce the wandering of Uranus. After he obtained a result, Lowell instituted a new photographic search at his observatory in 1914. Lowell dearly wanted to find a new planet. The search failed to turn up a planet, and Lowell died in 1916. Lowell's brother writes, "That X was not found was the sharpest disappointment of my brother's life."

Thirteen years after Lowell's death, another search was begun at Lowell Observatory. An excellent new 13-inch refractor had been prepared to take photographs. Lowell's method of taking pairs of photographs was again used, but with a further improvement. Zeiss, a German optical firm, had built a **blink comparator** which presented both photographs to the eye at once. The illumination switched from one photograph to the other, and the observer looked for an object that seemed to jump from one place to another.

A young astronomer, Clyde Tombaugh, who was too poor to afford a college education, was assigned to carry out the search, beginning in March 1929. It was a monumental task. In less crowded plates, some 50,000 stars were visible, and it took Tombaugh three 7-hr days to compare one pair of plates. He had to be careful, because he did not want to miss the planet if it were there. Asteroids (see the next section) were discovered by the hundreds, and these had to be eliminated. Poor images of stars on one plate or a variable star could also cause images which seemed to jump. Viewing plates for months on end may sound like a dull task, but Tombaugh had the proper outlook. "These plates were a delight to scan," he wrote.

Then, at 4:00 P.M. on February 8, 1930, another jumping object was seen. Tombaugh had a hunch. "That's it!" he said to himself. A careful watch of this object was made for some weeks and, when they were sure, the Lowell Observatory staff announced the discovery of a new planet. The announcement was made on March 13, 1930, 75 years to the day after Lowell's birth and 149 years after the discovery of Uranus.

The discovery made newspaper headlines the world over. The name Pluto was suggested by an 11-year-old British girl. It seemed appropriate, since Jupiter and Neptune were brothers of Pluto in mythology. Furthermore, the first two letters of the name were Percival Lowell's initials.

Nearly every observatory stopped what it was doing to look at the new planet and measure its position. Soon its orbit was computed, and this provided a further surprise. Pluto's orbit is tilted (the technical word is "inclined") with respect to the earth's orbit by 17°, by far the largest inclination of any planet's orbit.

The Mysterious Planet Pluto continued to be a source of interest and frustration. Its orbit is not in particularly good agreement with that predicted by Lowell. Lowell had calculated an average sun-Pluto distance of 43 AU, but the actual value is 39.4 AU. Further, of all the planets, Pluto's orbit turned out to hold the record for eccentricity. Even though, on the average, Pluto is the furthest planet from the sun, its orbit is so flattened that, as viewed from above the plane of the ecliptic, it crosses inside Neptune's orbit. Pluto crossed into the inner portion of its orbit in 1979 and will remain there until the year 1999. During this time, *Neptune* will be the most distant planet from the sun. On the other hand, when near aphelion, Pluto is over 7 billion km (4.6

Figure 11.11 The orbits of the superior planets. As an aid to visualizing the relative sizes of the orbits of the superior planets, the orbits are drawn on a scale model with the sun in Manhattan, New York, and Pluto in Los Angeles, California. (Pluto's orbit is represented as a circular arc with a radius equal to Pluto's average distance from the sun. At present, Pluto's eccentric orbit has carried it inside Neptune's orbit. Pluto is in Albuquerque!) On this model, the sun is just under a kilometer in diameter and so fits into Manhattan easily. Mercury is just across the river, in Newark. Venus is in mid-New Jersey, while Earth is just over the Pennsylvania border.

billion mi) from the sun, that is, 49 AU (see Figure 11.11). The planets Neptune and Pluto cannot collide because of the large tilt of Pluto's orbit. The two orbits do not intersect in space.

Pluto appears very small and faint in a telescope. See Figure 11.12. The Mount Palomar 200-inch telescope was used to estimate roughly its linear diameter, yielding a figure of at most 6000 km (3700 mi), half the earth's diameter. As judged by its gravitational effect on Neptune and Uranus, its mass was guessed to be roughly 10 percent of that of Earth. This small, dense planet is the only outer planet that is unlike the jovian planets. In fact, it is peculiar in almost every way.

No markings can be seen on Pluto, but its rotational period has been measured by changes in its brightness, which have a period of 6.4 days. Since 1953, the variations in Pluto's brightness have become more extreme, suggesting that, earlier, Pluto's pole may possibly have been pointed nearly earthward. Perhaps Pluto also "lies on its side" as Uranus does.

An exciting discovery was announced in 1978. James W. Christy of the U.S. Naval Observatory noticed an odd bump in several photographs of Pluto. Older photographs were examined and the bump was found to be visible on certain of these as well. Astronomers at the observatory have tentatively concluded that the photographs reveal a satellite of Pluto, one so close to the planet that telescopes cannot quite resolve the separate bodies. The satellite's orbital period is 6.4 days, the same as Pluto's period of rotation. As viewed from Pluto's surface, the satellite would not appear to move in the sky.

The reason this discovery is so interesting goes beyond the excitement of finding a new world in space. Knowledge of the satellite's orbital period and its distance from Pluto's center (roughly 20,000 km or 12,500 mi, 5 percent of the distance from the earth to the moon) are the data required to estimate Pluto's mass with more rigor than was previ-

Figure 11.12 Two views of Pluto taken a day apart. A galaxy (see Chapter 16) may be seen in the upper left edge of each view. Compare the position of slow-moving Pluto on each day as compared to the background of distant stars.

ously possible. The result is surprising. Whereas earlier workers had guessed that Pluto's mass might be 10 percent of Earth's mass, the new data indicate that it is nearer to 0.2 percent of the mass of Earth. This mass determination, if correct, indicates that Pluto's mass is only about 4 percent as large as that of the next most massive planet, Mercury. In addition, the new data on Pluto's satellite make possible a new estimate for Pluto's diameter: roughly 3000 km (1900 mi), about the size of our moon.

Two more thoughts about Pluto: Its size and density appear to be typical of many satellites, and it passes inside Neptune's orbit. This has led to speculation that it might have once been a satellite of Neptune which somehow escaped. Finally, Pluto's orbit is quite different from Lowell's prediction. Moreover, its mass is so low that it probably could not have caused the effects on Uranus that Lowell used to make his prediction. This has led to the suspicion that it was pure chance that a planet was found near Lowell's predicted location. It was Lowell's interest and enthusiasm that eventually led to Pluto's discovery, yet it may be that the discovery of Pluto in the direction calculated by Lowell was, at least in part, a wonderful accident.

11.7 ASTEROIDS: FILLING THE GAP

Bode's law (see Puzzle 11.2) had correctly predicted the distance of Uranus from the sun. The law also called for a planet in the gap between Mars and Jupiter at 2.8 AU. That fanatical pattern finder Kepler had also pointed out a gap which he felt lay in the same region.

An organization was founded by an astronomer named Olbers to search for the missing planet, but on January 1, 1801, before the work was underway, Father Giuseppe Piazzi saw a faint object moving along the ecliptic. The resulting orbit was calculated, and it showed that Ceres, Piazzi's name for the new planet, lay at 2.77 AU from the sun, almost exactly where Bode's law had predicted it would be. The gap was filled.

Olbers, while following Ceres for several months, found *another* planet in the gap. Some people were almost annoyed; this second object, Pallas, was not needed. Yet the discoveries continued. Harding found Juno in 1804, and Olbers found his second planet, Vesta, in 1807. By 1850, thirteen of these objects had been found. All these new objects were at about 2.8 AU from the sun. They were so small that the name "planet" did not seem appropriate for them. These objects are now called **minor planets, planetoids,** or **asteroids.**

Although "asteroid" (a term introduced by Herschel) literally means "little star," and thus is not entirely apt, it is the most commonly used term for these bodies.

More and more asteroids were found, but the flood really started when photography began to be used. By 1964 the list of asteroids contained more than 1600 and by now is well over 2000. One authority estimates that the total number of asteroids large enough to be eventually photographed is over 100,000.

Most asteroids remain between the orbits of Mars and Jupiter, but some have very eccentric orbits. Icarus holds the record; it periodically passes within the orbit of Mercury. In 1968 Icarus passed within only 6.5 million km (4 million mi) of the earth (see also Section 8.3). Several asteroids have orbits that could cause them to strike the earth if the earth were in the right place at the right time (if those are the right words). Fortunately, the chances of this occurring are exceedingly small. But such a collision would make a hydrogen bomb explosion seem puny by comparison.

Even the largest asteroids are quite small. Ceres is the largest, with a diameter of 785 km (488 mi), and would just about fit into Mare Imbrium on the moon. Most of the observable asteroids are under 2 km in diameter. Combined, they would make a body having a mass equal to that of the moon, at most. Most experts feel that the total mass would probably be much less. Perhaps Ceres has more mass than the rest of the asteroids combined. When estimating the total mass, one assumes that the asteroids are rocky, having a density similar to that of the rocks on earth or on the moon.

It is easy to suppose that asteroids are debris from a planet that exploded. Perhaps most of the fragments from the explosion have been lost somehow, leaving the relatively small mass of asteroids we have today. This idea is now largely out of favor, because no way is known for a planet to explode. Also, the present theory of planet formation (see Section 11.10) maintains that, as matter was clumping up in the asteroid belt, the effect of Jupiter's strong gravitational field would have been large enough to prevent the smaller pieces from aggregating in a slow, orderly manner. Rather it would have caused them to crash into each other and break up. It is thought that asteroids continue to collide to this day and that many of the meteors arriving at earth (Section 11.9) are fragments from this process.

Binoculars enable one to observe at least three of the four largest asteroids when they are in opposition. Consult an astronomical magazine for help in locating them.

At present, asteroids are studied by examining sunlight reflected from them in a spectroscope and by observing their variations in brightness as they tumble in space. Many asteroids seem to have compositions similar to the compositions of meteors, and many seem to be elongated and perhaps resemble the moons of Mars. Because of the intrinsic interest in asteroids and because they might yield useful data concerning the origin of the solar system, astronomers would be very much interested in sending a space probe to an asteroid.

11.8 COMETS: THE SOLAR SYSTEM'S MOST BEAUTIFUL MEMBERS

My grandmother remembered being taken out on a dark hill in 1911 to view a bright object in the sky which had a long, beautiful tail. "That is Halley's comet," they told her, "and it is going to destroy the earth!" The whole array of behavior that accompanied the past arrival of great

Figure 11.13 Comet Kohoutek, January 15, 1974. This comet turned out to be too dim to be very interesting to the naked eye but was beautiful and scientifically useful when viewed with a telescope. The long, nearly straight gas tail contrasts well with the curved, diffuse dust tail.

comets was reenacted. People trembled, gave away their belongings, bought pills to protect them from "comet gas," and were terrified by headlines such as "End of the World Could Be Today!" Even in 1973, when it was predicted that Comet Kohoutek (Figure 11.13) might become as bright as the full moon, pamphlets were handed out on street corners of my town warning of the impending downfall of America and giving us one month to leave the continent.

It is a pity that some people try to scare us when comets appear, for a great comet is, along with a great aurora display and a total eclipse of the sun, one of the most beautiful sights the heavens can offer. A comet is often found when still far from the sun, between the orbits of Mars and Jupiter. It has been plunging in toward the inner solar system for many years, but only when the **nucleus** (thought to be a "flying iceberg" of ice, pebbles, and dust) develops a coma, is it visible through a telescope, shining by reflected sunlight. The **coma** is a halo of dust and gas that has escaped from the nucleus as the nucleus is warmed by the sun. The coma may be many times larger than the earth. As the comet approaches to within 1 to 2 AU of the sun, a **tail** often forms. It is visible chiefly because it reflects sunlight. Some tails can reach lengths over 1 AU. Two types of tails have been recognized (see Figure 11.14). A **gas tail** is formed when the solar wind (Section 9.6) blows molecules out of the coma. The gas tail is straight and often

Figure 11.14 Comet Bennet, April 4, 1970. The straight, streaked gas tail may be distinguished from the curved, diffuse dust tail. Bennet was easy to observe with the naked eye.

appears blue in color photographs because it emits radiation by fluorescence (discussed in Section 7.3). A **dust tail,** on the other hand, is composed of tiny dust particles pushed away from the comet by the radiation pressure of sunlight. Each particle goes into its own orbit around the sun, resulting in a gracefully curved dust tail. Some comets show both kinds of tails. In all cases, the tail of a comet points away from the sun (Figure 11.15).

Many comets follow orbits carrying them back out of the inner solar system. However, more than 500 comets are **periodic,** that is, they follow elliptical orbits and return at regular intervals of time. The most famous periodic comet is Halley's comet, named for the same Edmund Halley (pronounced to rhyme with "rally") (Figure 11.16) who guided the publication of Newton's *Principia* (see Section 6.7). In 1705, Halley published a paper which noted that the great comets of 1531, 1607, and 1682 had appeared at roughly 76-year intervals. Using Newton's new dynamics, he showed that they all had about the same orbit. He predicted that the comet would return in 1758. It did return, in early 1759, 37 years after Halley's death. Halley's comet (Figure 11.17) is so splendid that historical records of its return have been found dating from as early as 476 B.C. Records of each visit since 240 B.C. have been found. For example, it returned in 1066, just before the Norman invasion of England. The next expected return of Halley's comet is in 1986. Perhaps a great nonperiodic comet will visit the inner solar system before then, and we all will be treated to the spectacular view of a comet hanging in the night sky, slowly changing its position among the stars night by night. Astronomical magazines give advance notice of the approach of comets, and even less than spectacular ones are well worth watching.

Comets do not whiz past the viewer's line of sight as meteors (falling stars) do. Comets are very distant objects and seem to travel slowly with respect to the fixed stars. A bright comet might be visible for several weeks.

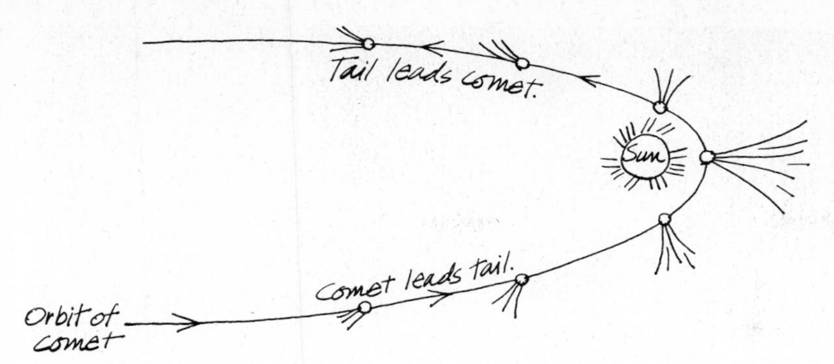

Figure 11.15 A comet's tail points away from the sun. The tail of a comet is formed by pressure from sunlight and from the solar wind. As a result, no matter which direction the comet is traveling, the tail always points away from the sun. As the comet leaves the sun, its tail actually precedes it.

The Origin of Comets Periodic comets present a puzzle to astronomers. Each time one visits the inner solar system, sunlight causes more of it to melt, producing the coma and tail. The material of the coma and tail is lost to the nucleus; thus periodic comets are continually being depleted. Several instances have been observed in which a comet broke up while rounding the sun. Judging from the rate at which most periodic comets are depleted, the average periodic comet will last only a short time, short, that is, as compared to the estimated age of the solar system. The puzzle, then, is this: If periodic comets last only a short time, why do we see them now? The answer must be that the periodic comets we now see are not those that formed in the early stages of the solar system. The stock of periodic comets must somehow be replenished. But how?

Figure 11.16 Edmund Halley, British scientist (1656–1742).

Many hypotheses concerning the origin of comets have been proposed over the years. The most popular was advanced by Jan Oort in 1950. Perhaps there exists a huge swarm, numbering perhaps over a billion, of frozen comet nuclei orbiting the sun at an enormous distance, roughly 1 light year (ly) (over 60,000 AU). These nuclei orbit the sun in all directions and thus constitute a spherical reservoir of comets surrounding the solar system. These nuclei are so far from the sun that the gravitational forces due to the nearest stars have a small but definite influence. Some of the nuclei may eventually be thrown out of the sphere, but others are caused to plunge into the inner solar system. The random orientations of the paths of the nuclei in the sphere explain why most of the comets we see do not follow a path lying in the same plane that the planets follow.

We shall consider another of Oort's many astronomical contributions in Section 15.6.

We shall see in Chapter 12 that the star nearest the sun is about 4.3 ly (270,000 AU) distant.

Most comets that visit us from the sphere of nuclei race around the sun once and return to the far, distant regions of the solar system. If such a comet returns, it will not be for millions of years. A periodic comet results if this visitor chances to pass sufficiently close to a planet and the planet's gravitational force alters its orbit in a particular fashion. If the alteration changes the orbit into a relatively small ellipse, the comet is captured. Jupiter is thought to be the most effective agent in capturing comets, because its gravitational influence is so strong.

Whether Oort's theory will stand the test of time or be replaced by

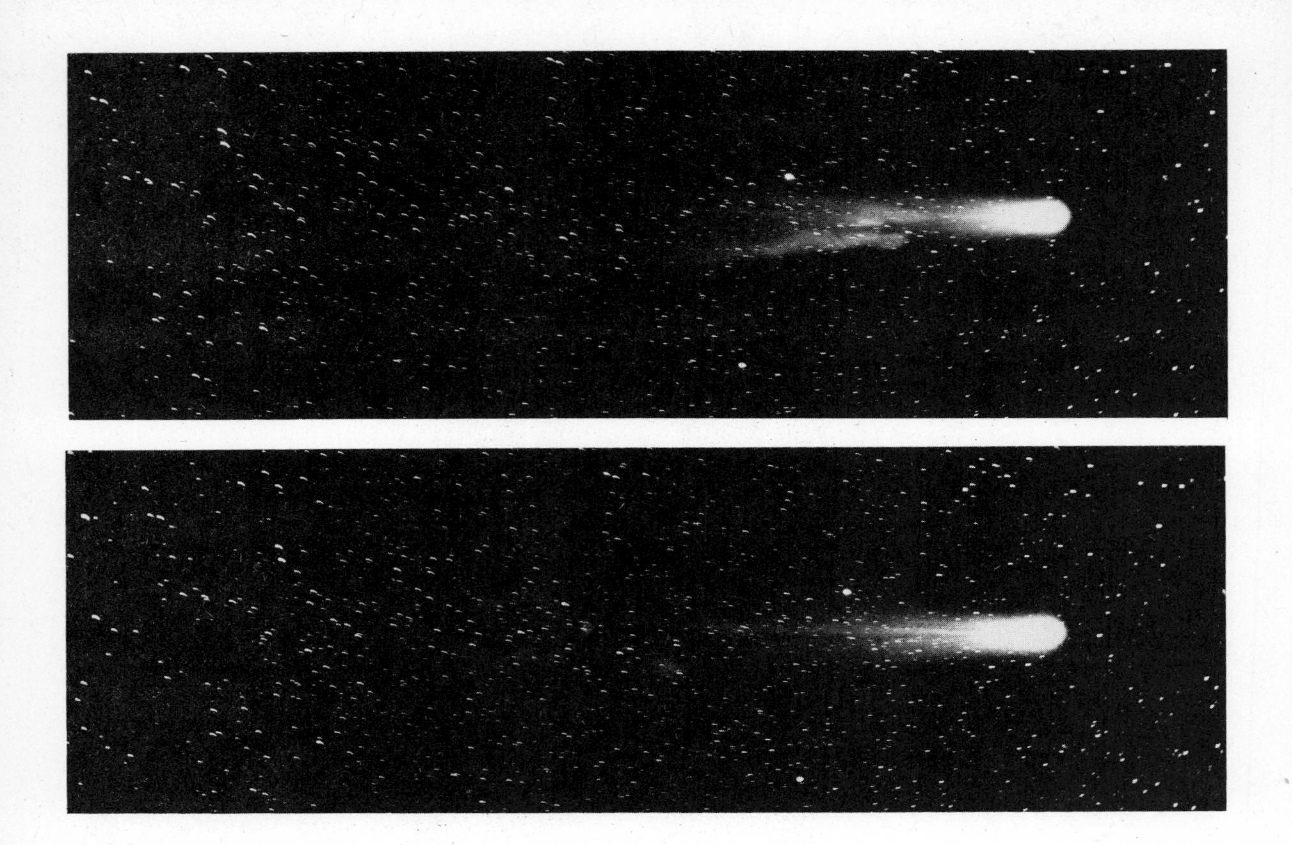

Figure 11.17 Two views of Halley's comet, June 1910. The tail has changed its appearance significantly in only one day.

an even more ingenious theory remains to be seen. The concept of a comet nucleus spending billions of years in the sphere of nuclei and then gradually being reaimed at the very distant sun, gradually falling toward it and then speeding past it, perhaps nearly grazing it, is surely dramatic.

11.9 METEORS: ASTRONOMY BROUGHT DOWN TO EARTH

An observer of the night sky may suddenly be startled to see a brief, bright streak of light in the sky, a meteor. A very lucky observer might even see a more spectacular meteor, one taking several seconds to travel many degrees across the sky and perhaps breaking up into several pieces as it does so, leaving a bright glowing trail that lasts several seconds more.

Many people have never seen a meteor. Yet anyone willing to watch the sky intently for an hour is likely to see five of them or so, fewer, perhaps, in a light-polluted city. One is most likely to see only ordinary meteors, but perhaps, once in a lifetime, a **bolide** or **fireball,** a rare, very brilliant meteor that crosses the whole sky and illuminates the landscape, may appear.

Meteors are commonly called **shooting stars** or **falling stars,** but these are misleading names. Astronomers generally use three terms for such objects, depending on their location: meteoroid, meteor, and meteorite. A **meteoroid** is a rock, grain, or piece of dust (the latter is often called a **micrometeoroid**) in outer space, orbiting the sun.

A meteoroid becomes a **meteor** if its orbit intersects the orbit of the earth and if the meteoroid and the earth arrive at the same place at the same time. As the meteoroid plunges into the atmosphere at speeds up to about 70 km/sec (44 mi/sec), it rubs against the atmosphere. The resulting friction heats up the air until it glows, and it is the glowing air that we see. The meteor itself gets so hot that it crumbles and finally vaporizes. Most meteors become visible at an altitude of about 100 km (60 mi) and are completely vaporized (turned to gas) at about 50 km (30 mi). These altitudes were determined from photographs taken of meteors from two or more locations at once. The distance from the cameras to the meteors can be computed using the method of parallax.

Experts estimate that the average meteor observed is caused by a meteoroid not much larger than a grain of sand. It is the great speed of the meteor that results in such a bright streak. Most meteors are even smaller and cause no effect that is visible to the naked eye. Spacecraft have been orbited to measure the numbers of meteors striking the earth. It has been estimated that roughly 100 tons of meteors enter the earth's atmosphere *daily*.

At rare intervals, a large meteoroid enters the atmosphere, becomes a meteor and, if it is originally basketball-sized or larger, part of it will survive to reach the surface of the earth. It becomes a **meteorite.** Most meteorites strike the oceans and are lost. Those found on land have been classified broadly into three groups. **Stony meteorites** are roughly similar to stones found on earth. It takes an expert to tell the difference. **Stony iron meteorites** are a mixture of stone and iron. **Iron meteorites** are almost entirely iron. Prehistoric iron tools have been found that were shaped from iron meteorites and used by people who had not yet discovered iron ore extraction. The largest iron meteorite ever reported was found in Africa. It measures about 2 by 3 by 3 meters and has a mass of about 60 tons. The date of its fall is unknown. The record for the largest piece of stony meteorite ever recovered from a visual fall was, until recently, held by Kansas. But, in March 1976 a bolide was seen streaking over China. The largest fragment recovered had a mass of 1770 kg (a weight of 3900 lb). Such giant falls, needless to say, are very rare.

We know that there have been large meteorite impacts on the moon, Mars, and Mercury, which caused the many craters found there. Roughly the same numbers of meteorites must have struck the earth, but most of the craters left in the earth's surface have long since been wiped out by erosion. About 40 meteor craters have been identified on earth. For example, craters have been found in Canada that were covered and preserved by the earlier glaciation of the region. Any visitor to Arizona should plan on making a trip to the Barringer Crater, near

Figure 11.18 Barringer meteor crater. Preserved from obliteration by the dry desert climate, this is a major attraction for astronomy-minded tourists in Arizona. It lies just off U.S. 40 between Two Guns and Leupp Corners. The nearest large town is Winslow.

Winslow (Figures 11.18 and 11.19). More than 170 meters (560 ft) deep and over 1.2 km (0.8 mi) in diameter, it is one of the few genuine astronomical sights one can visit without leaving the earth.

Astronomers have studied meteorites closely. Before the moon shots, meteorites were the only matter from space available for close inspection. Some meteorites have been found containing material estimated to be about 4.6 billion years old, presumably dating back to the early days of the solar system. It is assumed that many meteorites are pieces of asteroids that were chipped off when asteroids collided. Many meteorites show signs of having been formed under high pressure such as would exist inside some of the larger asteroids.

Every year the earth passes through swarms of meteoroids, which gives rise to **meteor showers.** Most of these showers may be somewhat disappointing to the beginner who can be misled by the name. Most

Figure 11.19 A meteorite from the Barringer region: A sample from outer space.

showers contribute an additional 5 to 20 meteors per hour to the usual display. Yet, rarely, there is a shower of meteors worthy of the name. Those lucky few who were in the right location and who got up before dawn on November 17, 1966, saw an awesome sight. A friend of mine reported that it was like looking into a bathroom shower head—a continuous display of meteors. One observer estimated that, at its peak, this shower displayed 140 meteors per second. This is the highest rate ever recorded.

The meteors seen in a shower are thought to be quite different in nature and origin from those picked up on the surface of the earth. Shower meteors almost never reach the surface of the earth and seem to be almost entirely fine dust. Also, meteor showers repeat with a period of one year. The most famous and reliable is the Perseid shower, which peaks in mid-August each year. This regular period indicates that the meteors which produce a shower are spread out along a large orbit around the sun, an orbit intersecting the orbit of the earth. Each year, when the earth returns to the same place in its orbit at which the swarm of meteors crosses it, we see the shower again. (The best times to look for meteor showers are given in astronomical magazines.)

For these and other reasons, it is generally accepted that the meteoroids in meteor swarms are produced by disintegrating periodic comets. Strong evidence for this view is that cases have been observed in which periodic comets broke up and failed to reappear. A meteor swarm was seen in its stead, in the same orbit as the comet had followed. Some meteor swarms are associated with periodic comets that still exist.

11.10 THE FORMATION OF THE SOLAR SYSTEM

One of the most challenging problems in the entire field of astronomy is to devise a satisfactory, detailed theory to explain the origin and development of the solar system (Figure 11.20). In spite of the large amount of research, especially since the early 1950s, there is relatively little one can say on the subject that is widely accepted by workers in the field.

Descartes ("Day-cart'," 1596–1650) is usually counted among the major philosophers of all time. He also worked in physiology, mathematics (he united algebra and geometry, a major advance), and astronomy. His most famous statement is "I think, therefore I am."

In 1644, the French Philosopher René Descartes proposed the outline of what may be called the **nebular hypothesis.** He began by considering a huge cloud of dust and gas (a nebula) in space and proposed that it contracted as a result of self-gravitation. As it did so, the sun formed at the center and the planets formed elsewhere in the cloud. When the nebular hypothesis is stated in such broad terms, most modern workers in the field accept it.

Another theory, the **collision hypothesis** first proposed in 1745, envisions a large object, usually a star, passing close to the sun and drawing out from it a huge plume of matter by means of gravitation. The matter in the plume later collected, forming the planets we see today. Since that time, scientific opinion has repeatedly favored first one and then the other of these two hypotheses.

It is intriguing to consider the implications of these hypotheses with regard to the question, Are there other systems of planets similar to ours? If the nebular hypothesis is correct, one might expect that planets would naturally form each time a single star is born. On the other hand, because close encounters between stars must be very improbable, the collision hypothesis implies that planets may be very rare in the universe. This seems to reduce drastically the chances of finding life elsewhere in the universe.

In about 1900, the collision hypothesis was widely held, but since then it has largely fallen out of favor. The largest stumbling block is that no one can see how matter pulled out of the sun could condense into planets. Wouldn't such matter merely disperse?

The nebular hypothesis envisions the original cloud to be spinning slowly. Then, as it contracts, its rate of rotation increases. This effect, that a contracting object spins faster, can be demonstrated by a person standing on a turntable with weights held in her hands. With arms extended horizontally, she is pushed so that she rotates slowly and is then released. She then pulls the weights to her sides producing more rapid rotation. Spinning skaters, divers, and falling cats use this principle to control their rate of spin.

Now, as the nebula contracts and spins faster, it should develop at first an equatorial bulge and eventually flatten out. The center of the nebula is most concentrated and becomes the sun. The outer regions of the nebula condense into the planets, satellites, and other members of the solar system. This flattening of the cloud can account for the fact that the planets, excluding renegade Pluto, all travel in orbits lying in

Figure 11.20 Overconfidence.

nearly the same plane, and also why all the planets orbit the sun in the same direction. This is the same direction in which the original nebula rotated.

The nebular hypothesis has also been successful in explaining some of the differences between the terrestrial and the jovian planets (see Section 11.2). The nebula should have been hottest at its center and coolest at its edges. The terrestrial planets formed in a hot region of the nebula where icy substances could not condense. Only the rocky substances could accrete (lump together). Thus the inner planets are rocky. As they formed, the terrestrial planets had too little mass to produce enough gravitation to retain the abundant hydrogen and helium in the nebula. These gases escaped. On the other hand, the jovian planets were in a relatively cool region of the nebula where both rocky and icy substances could accrete, producing strong enough gravitation to retain large amounts of gasses as well. The theory goes on to explain the formation of most of the satellites of the outer planets in a similar way. (As we have seen, the moon is a unique problem.) Small, flattened nebulas formed around the massive outer planets, producing satellites in orbits above the parent body's equator. There is some evidence that the inner satellites of these systems are denser than the more outlying ones, an analogy to the terrestrial and jovian planets.

So far, so good. The problems with the nebular hypothesis crop up when one asks for more details. We mention one of the troublesome features. Since the sun is the hottest and most massive object in the system, one can easily believe that it would form at the center of the nebula. But why would gas and dust accrete elsewhere in the nebula into neat little planets? Many proposals have been made but few have many followers. One theory suggests that turbulent eddies (like whirlpools in a stream) caused the dust or grains of solid matter to collide and stick together. Perhaps the grains were fluffy and easily formed clumps. Now the theory becomes vague. The details of the process of building from clumps of grains 1 cm (0.39 inch) in diameter to planets are still highly obscure. Perhaps this is a fatal flaw in the theory. However, most astronomers would probably wager that the answers will come to light after enough patience and hard work.

We end this chapter by mentioning a challenge to one portion of the theory of planetary structure. It has been long thought that the earth's iron core and its rotation produce the earth's magnetic field. This is the dynamo theory. The energy to drive the dynamo is derived, in part, from the rotation of the earth. On the basis of this theory, a rapidly rotating planet can have a magnetic field, but a slowly rotating one cannot. All looked well for the dynamo theory during the early space age. *Pioneer 10* verified that rapidly spinning Jupiter has a very strong magnetic field. Other spacecraft revealed that slowly turning Venus has practically no magnetic field. It was confidently predicted that Mercury would not have a magnetic field, but *Mariner 10* revealed that Mercury's field is easily measurable. Although upsetting at first, this observation will probably lead to a better understanding of the planets by requiring new or improved theories.

SUMMING UP

The jovian planets, in contrast to the terrestrial planets, lie far from the sun and are further characterized by their large size and mass, low density, and hydrogen and helium content, with, in the cases of Uranus and Neptune, the addition of a rock-ice mixture. These and other regularities among the planets must be explained by any successful theory of the origin of the solar system.

Uncounted objects orbit the sun in addition to planets. Small, rocky asteroids travel principally between the orbits of Mars and Jupiter. Flying icebergs of pebbles and dust frozen in ice are known as comets. Meteors, small pieces of rock, also orbit the sun.

In the next chapter, we turn our attention to the stars.

EXERCISES

1. For each of the following planets, name three things that are interesting and unique: Jupiter, Saturn, Pluto.

2. What are the principal differences between the terrestrial planets and the jovian planets?

3. What process is thought to account for the belts and zones observed on Jupiter?

4. What evidence is there that the rings of Saturn consist of individual particles in orbit?

5. What distances does Bode's law yield for Uranus, Neptune, and Pluto? What are the actual distances?

6. Briefly describe the circumstances of the discoveries of Uranus, Neptune, and Pluto.

7. What are asteroids? Where are most of them? About how many have been discovered?

8. Define the following terms as they apply to comets: nucleus, coma, gas tail, dust tail, periodic comet.

9. How do meteoroids, meteors, and meteorites differ? What are the two probable sources of meteoroids?

10. Briefly compare the nebular and collision hypotheses of the origin of the solar system.

READINGS

In addition to the general references cited in Chapter 10, see also the following works on particular topics.

Pioneer Odyssey, by R. O. Fimmel, W. Swindell, and Eric Burgess, NASA, Washington, D.C., 1977, SP-396, on the *Pioneer* flights to Jupiter.

Jupiter, by Isaac Asimov, Lothrop, New York, 1976.

Beyond Jupiter, by C. Bonestell and A. C. Clarke, Little Brown, Boston, 1972, features some stunning paintings by Bonestell.

"Asteroids," by David Morrison, *Astronomy,* June 1976, p. 6.

"The Nature of Comets," by Fred L. Whipple, *Scientific American,* February 1974, p. 48.

Theories of the formation of the solar system are discussed in

"The Origin of the Solar System," by H. Reeves, *Mercury,* March/April 1977.

WHAT ARE THE STARS?

On November 12, 1782, John Goodricke stood staring at the constellation Perseus. Somehow it looked changed: but how? Then he realized that the star Algol was at fault. This was the first time he had noticed that it could be much dimmer than normal. Ordinarily, Algol has about the same brightness as Polaris, but that night it had dimmed to about one-third its usual naked-eye brightness. Several earlier astronomers had noticed this unusual behavior, but Goodricke was the first to observe the star regularly enough to recognize the pattern: Algol reaches its minimum brightness regularly once every 2 days, 21 hr. We discuss the importance of this discovery and Goodricke's interesting explanation for Algol's variability in Section 12.8.

Two years later, Goodricke made another discovery involving a different type of variable star. He found that the star named Delta Cephei changes its brightness in a regular way, but that the pattern differs significantly from Algol's behavior. Delta Cephei and others like it which were subsequently discovered are, as we shall see, of enormous importance to astronomy. We devote all of Section 12.9 to them.

Few astronomers have made discoveries as significant as Goodricke's, and no other famous astronomer has labored under his handicaps, for Goodricke was deaf and unable to speak. He was also a prodigy; his paper on Algol appeared in a scientific journal when he was 19 years old. But he had to be a prodigy—he died at the age of 21.

In this chapter we turn our attention from the solar system (the objects in our neighborhood) to the distant, formerly unknowable stars. The reader's attention is drawn to the ingenuity and hard work involved in solving a most difficult problem: observing a distant pinpoint of light and finding out what it is.

12.1 WHERE ARE THE STARS?

Proper Motion In Chapter 2, the stars are called "fixed stars" to distinguish them from the planets, which are "wandering stars." We stated in Chapter 2 that the fixed stars do not change their positions relative to each other as the constellations march across the sky. Like almost every elementary statement in science, this one is not quite correct when closely scrutinized.

The Big Dipper today

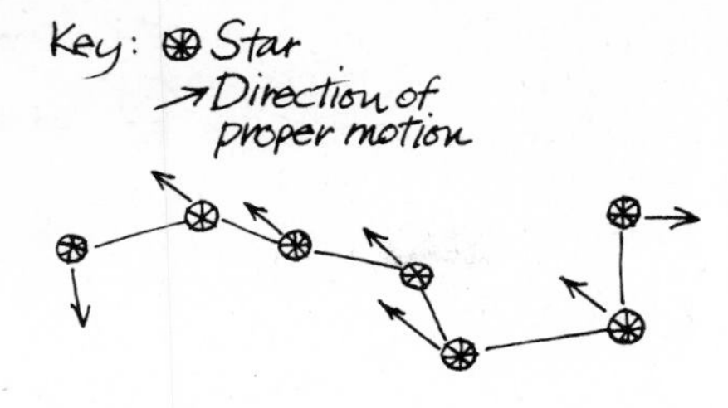

The Big Dipper in 100,000 A.D.

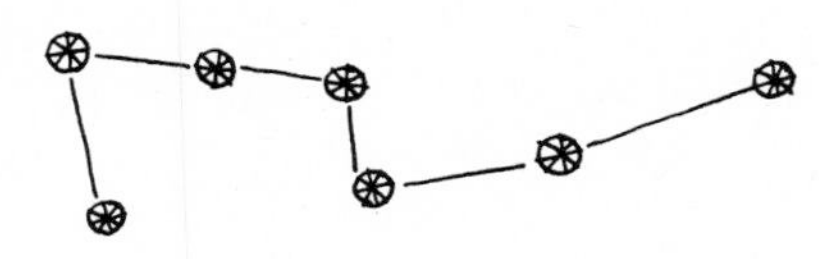

Figure 12.1 The proper motions of the stars in the Big Dipper. Measurements made over a period of many years have revealed that the stars of the Big Dipper have proper motions that are carrying them in the directions shown at the top. These motions will gradually change the Big Dipper's shape until, about 1000 centuries from now, the handle will become the bowl and the bowl will become the handle.

The British astronomer Edmund Halley enters the picture again at this point. In 1718, he observed that several stars, among them Arcturus and Sirius, were not located in the sky in the places given in Ptolemy's 1800-year-old tables. Could the mighty Hipparchus (whose star positions Ptolemy chiefly used) have made such inaccurate measurements? Or had the stars moved? Halley concluded that, over the centuries, at least some of the stars had very gradually changed their positions with respect to the other stars. In the long run, the fixed stars cannot be considered actually fixed.

We have already mentioned Halley's assistance to Newton (Section 6.7) and his prediction concerning "his" comet (Section 11.8). Among many other things, he also prepared a map of the earth's magnetic field, drew up a set of mortality tables, established the first observatory in the southern hemisphere, and was Astronomer Royal of England for 20 years.

This apparent sideways motion of a star with respect to other stars is called **proper motion.** Proper motion is often expressed by giving the number of seconds of arc that a star moves across the celestial sphere in one year. Only about 100 stars have a proper motion greater than 0.1 second of arc per year. Arcturus, one of the fastest bright stars, has a proper motion of 2.3 seconds of arc per year. In the 1800 years between Hipparchus and Halley, Arcturus had shifted its position about 1°, about two angular widths of the full moon, an amount easily noticed. The record for proper motion is held by Barnard's star, which is sometimes called the "runaway star." Too dim to be seen by the naked eye, its proper motion is 10.3 seconds of arc per year, or 2.86° in 1000 years. Nowadays, star positions can be measured to an accuracy of 0.01 seconds of arc or better, so that those having a large proper motion do seem to be running away in that sense. The proper motions of the stars have no noticeable effect on the shapes of the constellations from one year to the next, but a star watcher returning 100,000 years from now would not recognize many of the familiar constellations (see Figure 12.1).

Recall that a second of arc is an extremely small angle, $\frac{1}{3600}$°. A nickel has an angular diameter of 1 second of arc when viewed from a distance of 4.1 km (2.6 mi).

The Distance to the Stars The most difficult question to answer about the position of a star is also the most basic: How far away is it? According to the theory of Copernicus, the motion of the earth around the sun should cause the apparent position of a nearby star to move back and forth slightly with a period of one year as compared to the background of distant stars. As we have seen (Section 5.3), Tycho assumed that the reason he could observe no such motion was that the earth did not actually orbit the sun. But by the 1800s belief in the Copernican theory had become so widespread that many observers were trying to observe the parallax motion they believed the nearest stars must exhibit.

Friedrich Bessel (German astronomer, 1784–1846) made important contributions to both observational astronomy and the mathematical analysis of data. Late in life he began calculations similar to those that eventually led to the discovery of Neptune (Section 11.5). Bessel died before he could commence actual observations.

Credit for publishing the first authentic observation of parallactic motion goes to Fredrich Bessel. Like Tycho and all other great observers, Bessel was a fanatical stickler for detail and accuracy. He used the superb telescopes made by Fraunhofer but had this to say about using telescopes, "Every instrument . . . is made twice, once in the workshop of the artisan, in brass and steel, and then again by the astronomer on paper by means of the list of necessary corrections which he derives by his investigations." Bessel found that he had to make many corrections for each measurement of a star's position. Among these were corrections for the direction the telescope is aimed, the temperature, the weather, the time of day, and atmospheric refraction (see Section 5.2). Without attention to these errors, Bessel would not have been able to measure the distance to a star. He was confident that, by combining Fraunhofer's excellent telescope and his own care, he could measure star positions to an accuracy of about 0.1 seconds of arc. This may be compared to Tycho's naked-eye accuracy of about 2 *minutes* of arc. Bessel made an improvement by a factor of 1200.

The remaining question was which star to measure. An astronomer wishing to measure the distance to a star is in a peculiar position. He or she must choose a relatively nearby star to watch as it appears to change its position with respect to the distant background of stars before he knows how far away it is. Tycho chose bright stars, but Bessel tried a different approach. He chose a star named 61 Cygni which, although rather dim in appearance, has a fairly large proper motion, 3.7 seconds of arc per year. Using the idea that nearby objects appear to move more rapidly, Bessel hoped that 61 Cygni appears to be moving relatively fast because it is relatively close to the earth.

By 1838 Bessel had data clearly showing the parallax motion of 61 Cygni. He published a parallax angle of 0.31 seconds of arc. The formula for converting a star's parallax angle into its distance from the Earth is

$$R = \frac{3.26}{a}$$

where a is the star's parallax angle expressed in seconds of arc and R is the distance to the star in light years.

A light year is a convenient unit of length for expressing distances to stars. A kilometer is a rather small unit for expressing distances between planets. For this reason the astronomical unit was invented. It was soon clear that stars are a very large number of astronomical units away. Thus, in a similar way, the light year came into use. A **light year** (ly) is the distance light travels in one year. If a star's light takes two years to reach us, we say that the star is 2 ly away. Despite its name, a light year is a measure of *distance* and not of time. Since light travels nearly 300,000 km/sec, 1 ly is equivalent to about 1×10^{13} km (about 10 million million km or 6 million million mi), which is also equivalent to about 63,000 AU. A spacecraft traveling at a steady 50 km/sec (110,000 mi/hr, the speed of *Pioneer 11* when nearest Jupiter) would take about 6000 years to travel a distance of 1 ly. Another frequently used measure of stellar distance is the **parsec,** which is equivalent to 3.26 ly.

Using Bessel's result of 0.31 seconds of arc for 61 Cygni and the above formula, the reader should be able to compute the distance to this star: 10.5 ly. As it happened, another astronomer was also successful in measuring the distance to a star at about the same time. Thomas Henderson chose the star Alpha Centauri because it is the third brightest star in the sky, has a large proper motion, and has a companion star which orbits it in an apparently large orbit but a short period. All these facts indicated to him that Alpha Centauri might be close to us. The star is in the southern hemisphere, and Henderson studied it from the Cape of Good Hope. He derived a parallax angle of 0.91 seconds of arc, but the modern value is 0.76 seconds of arc, making the distance to Alpha Centauri 4.3 ly.

Thomas Henderson (Scottish astronomer, 1798–1844) was the director of the Cape of Good Hope Observatory.

Ptolemy had stated that, compared to the distance to the stars, the size of the earth is a mere point. Kepler had supposed that the stars might be 10,000 AU (0.2 ly) from us, while Newton had suggested 100,000 AU (1.5 ly) as more likely. The actual distance outstripped even the imaginations of these great men. Alpha Centauri, the *nearest* star, is 4.3 ly or 270,000 AU from us. We can now see why Tycho was unable to measure any stellar parallax angles. The largest parallax angle ever found, that of Alpha Centauri, is only 0.76 seconds of arc. Tycho, limited to viewing with the naked eye, could measure angles no smaller than roughly 2 *minutes* of arc, over 150 times larger than the largest parallax angle.

If you catch an astronomer off-guard and ask what the nearest star to us is, he or she might say "Alpha Centauri." This is a reasonable answer but technically is not quite right. On the one hand, as we shall see later in this chapter, the sun is a star. On the other, a faint star has been found associated with Alpha Centauri and, since it is slightly closer to us than Alpha Centauri, it has been named Proxima ("near") Centauri. Nevertheless, Alpha Centauri is commonly referred to as "the nearest star."

Where Is That Star and Where Is It Going? An enormous amount of work is involved in establishing the location of a star in space and in determining the direction and speed of its motion. Several years of painstaking observations are required just to find the distance using the method of parallax.

This paragraph uses notions explained in Section 7.6.

In order to determine the direction and speed of the star's motion, one observes its proper motion and Doppler shift. The star's transverse (sideways) velocity may be computed from its proper motion (the rate at which its angular position changes) and its distance. The basic idea is that, for a given proper motion, the more distant a star the faster the transverse velocity. One may determine the star's radial velocity (speed toward or away from us) by observing the Doppler shift of the lines in the star's spectrum. The overall speed and direction may then be computed from the transverse and radial velocities. It is hard work, but necessary for those who wish to understand our universe.

A Limit to the Method of Parallax The more distant a star, the smaller its parallax angle. This places a limit on the usefulness of the method of parallax. If a star is too far away, its parallax angle is too small to measure with much accuracy. Current technology places this limit of a parallax angle at about 0.01 seconds of arc, which implies a distance of roughly 300 ly. Thus, if a star is farther than about 300 ly, its distance cannot be accurately determined using the method of parallax. Beyond that distance, out to several thousand light years, one can merely make a rough estimate. Beyond that, the method of parallax is completely useless.

Other, more indirect methods of measuring the distance of a star have been found. We discuss an important one in Section 12.9.

Consider this method of visualizing the enormous distances to the stars. Suppose we wish to build a scale model of the solar system and want to include the nearest star, Alpha Centauri. We place the sun in New York City, in Manhattan. We place the earth 15 meters (16 yd) from the sun, and Pluto 600 meters (660 yd or a few blocks) from the sun. On this scale, the nearest star would be as far away as Los Angeles. This indicates the enormous amount of empty space between the stars (see Box 12.1).

12.2 HOW BRIGHT ARE THE STARS?

Apparent Brightness Expressed in Magnitudes When Hipparchus was preparing his catalog of stars, he invented a method of describing the **apparent brightness** of a star, that is, its brightness as viewed from earth. He called the brightest stars **first-magnitude** stars, and the dimmest **sixth-magnitude** stars. Every star in his catalog had an associated magnitude; the smaller the number, the brighter the star. Nowadays, the apparent brightness of stars is no longer estimated by eye but rather measured by instruments and photography. One device, a **photometer,** can be attached to a telescope. The photometer changes the light energy from a star into electric energy which in turn deflects the needle on a meter. This is essentially the same device as the light meter used in photography.

The idea of expressing apparent brightness in terms of magnitudes

BOX 12.1

THE TWINS PARADOX

The stars are at an enormous distance. Will our descendants ever visit them? If they do, they will probably need to travel at speeds near the speed of light to cover such vast distances. And,in such a case, they will encounter some of the strange phenomena predicted by Einstein's theory of relativity.

We mentioned in Box 7.1 that, according to Einstein, a spacecraft is limited to speeds less than the speed of light. An additional prediction by Einstein is that anyone making a round trip at such very high speeds will find that they have traveled ahead in time.

There is not enough room here for an examination of these concepts thorough enough that the reader can genuinely understand them. But we do want to mention one of the results of relativity concerning time and round trips to the stars. Known as the "twins paradox," the result asks us to imagine twins; call them Ramona and Bob. When the twins are 20 years old, Ramona takes off for a trip to the stars, leaving Bob behind. Ramona travels at speeds approaching the speed of light and then returns to earth. She has experienced 10 years of her life and is 30 years old when she steps out of the ship. But Einstein predicts that, because of the high speed of her trip, time has been slower for her than for those on earth. The result is that she is in for a shock when she sees her brother, for he is now much older than she is. The faster her trip and the longer the distances, the older he will be in relation to her. Depending on the speed of her trip, the twins could now have ages of 30 and 80. Surprisingly, Ramona noticed nothing odd during her trip. She did not feel as if she were in a slow-motion dream. Her heartbeat seemed normal and clocks all ticked at their usual tempo.

From another point of view, Ramona has traveled forward through time. She may have left in the year 2000 and returns thinking it is 2010. But meanwhile, time has rushed by on earth; it is 2060. A trip at even higher speeds might bring her back in the year 10,000. The closer one approaches the speed of light, the further ahead in time one has traveled by the end of a round trip.

As of yet, there is no known method, even in principle, of traveling backward in time. Doing so could lead to situations that strain the bounds of logic. If you were to go back in time and kill your great-grandfather in his cradle, what then? Then you would not have been born and so you could not have gone back in time and killed your great-grandfather.

---

has been retained, although it has been made more accurate and has been extended and slightly redefined. On the new scale, Alpha Centauri has a magnitude very close to zero. Objects that look brighter than that have negative magnitudes. On the new scale, the dimmest stars the eye can see still have a magnitude of about 6. With the use of large telescopes and photography, objects as faint as magnitude 24 have been glimpsed.

Here are the magnitudes of some familiar objects: sun, −26.8; full moon, −12.6; Venus at its brightest, −4.4; Sirius (the star with the greatest apparent brightness), −1.5; Arcturus, −0.1; Dubhe (the "pointer" in the Big Dipper nearest Polaris), 1.8; Polaris, 2.0; 61 Cygni, 5.2.

Apparent Brightness Expressed in Suns The method of expressing apparent brightness in magnitudes has some awkward features. For example, compare a first-magnitude star with a second-magnitude star. We receive 2.5 times as much light from the first-magnitude star. A difference of five magnitudes corresponds to a change of 100 times in the amount of light received.

We shall utilize another means of expressing apparent brightness. In this method, one compares the amount of light energy (electromagnetic radiation) we receive from an object to the amount of radiation we receive from the sun. Thus, if one sees an object in the sky from which we receive one-tenth as much radiation as the sun sends us (Note: There is no object in the sky that looks nearly so bright), one says it has an apparent brightness of $\frac{1}{10}$ sun. On this scale, the apparent brightness of the sun is 1 sun. The advantage of this scale is its simplicity. If star A has an apparent brightness numerically twice that of star B, then we know that we receive twice as much radiation from star A.

[A technical note: Any object emits thermal radiation over a broad band of wavelengths (see Section 9.3). Thus the total amount of radiation we receive from a star is not confined to the wavelengths of the visible part of the spectrum. Astronomers, when discussing a measurement of apparent brightness, must specify whether they measured only visible light or all the radiation from the star. For simplicity, we ignore this refinement, something a professional astronomer may not do.]

This method of expressing apparent brightness has a disadvantage in that our standard of comparison, the sun, is so bright. As a result, the apparent brightness of any star turns out to be a very small number. For example, the apparent brightness of Sirius is about 8×10^{-11} sun = 0.000,000,000,08 sun. A sixth-magnitude star, the dimmest star visible to the naked eye, has an apparent brightness of 8×10^{-14} sun = 0.000,000,000,000,08 sun. We express apparent brightness in suns henceforth in this book because it is easy to think about.

Luminosity: Actual Brightness The stars look much dimmer than the sun. Are they actually dimmer? The stars appear to differ in brightness. Is this due to actual differences among stars or are all stars the same and just look different because of varying distances? These kinds of questions were answered when astronomers learned how to calculate luminosity. A star's **luminosity** is its actual brightness as compared to the sun. If a star has a luminosity of 2 suns, then it actually is twice as bright as the sun. That is to say, if the star and the sun were the same distance from us, the star would have twice the apparent brightness. The concept of luminosity is similar to the wattage of a light bulb. A 100-watt light bulb emits a certain amount of electromagnetic radiation per second. A 200-watt bulb emits twice as much. Similarly, a star having a luminosity of 2 suns emits twice as much electromagnetic radiation per second as the sun does. It is also important to realize that a star's luminosity is a property of the star itself and does not depend on the distance of the star from us.

How To Determine a Star's Luminosity The apparent brightness, luminosity, and distance of a star are intimately related. What if two stars have the same luminosity but one is twice as far from us as the other? It is tempting to jump to the conclusion that the more distant star

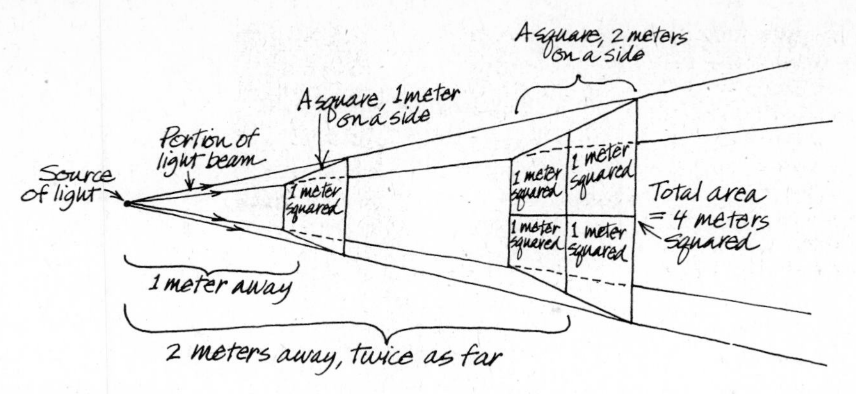

Figure 12.2 The dependence of the apparent brightness on the square of the distance. If a light beam covers an area of 1 square meter at a given distance from a light source, it will cover four times that area at double the distance. This means that a screen of area 1 meter squared placed at a distance of 1 meter from the light source will, in this example, receive all the light in the beam. At twice the distance, it will receive one-quarter as much illumination.

will have half the apparent brightness, but laboratory experiments contradict this. If one measures the apparent brightness of two equal light bulbs, one of which is 1 meter away and the other 2 meters away, the bulb twice as far away looks *four* times dimmer. If the more distant bulb is moved until it is three times farther away than the nearer one, it appears nine times dimmer. One says that the apparent brightness decreases as the square of the distance. Figure 12.2 illustrates the reason for this.

A simple formula indicating the connection between luminosity, apparent brightness, and distance is

$$L = bR^2 \text{ (shorthand for } L = b \times R \times R)$$

where L is the star's luminosity in suns, b is the star's apparent brightness in suns, and R is the distance to the star in astronomical units. (Astronomers use a slightly more complex formula, but this one is sufficient for our purposes.)

$L = bR^2$ can be used to compute the luminosity of a star once its apparent brightness and distance are known. The most difficult step is to determine the distance. After that, the rest is easier, by comparison.

Please work Puzzle 12.1 at this point.

As data concerning star distances accumulated, it became clear that stars have luminosities that vary widely. For example, Krueger 60A is a star with a luminosity of 0.002 sun. It would take 500 stars like Krueger 60A to emit as much electromagnetic energy per second as our sun does.

Most of the familiar bright stars in the sky look bright because they are quite luminous and relatively close to us. Altair and Rigel make an interesting contrast in this respect. Both are among the brightest looking stars in our sky. Altair is fairly luminous, but not exceedingly so. It

The luminosities of some stars, expressed in suns: Eta Cassiopeia B, 0.03; 61 Cygni, 0.08; Alpha Centauri, 1.5; Altair, 11; Sirius, 23; Vega, 53, Arcturus, 100; Spica, 1500; Antares, 9000; Rigel, 53,000.

PUZZLE 12.1

USING $L = bR^2$

The purpose of this puzzle is to help the reader become familiar with the equation $L = bR^2$ as it is used to compute the luminosity of a star once its apparent brightness b and distance R are known. One multiplies the apparent brightness of the star by the square of the distance. We use convenient numbers for b and R so that the arithmetic will not be cumbersome. This means we will use wildly unrealistic values for each imaginary star's apparent brightness and distance. The point is that the procedure illustrates the formula's use.

1. What is the luminosity of a hypothetical star 3 AU from us, which has an apparent brightness of $\frac{1}{3}$ sun?
2. Calculate the luminosity of another star with the same apparent brightness ($\frac{1}{3}$ sun) but at a distance of 9 AU.
3. What is the actual brightness of a star 9 AU away which appears to be only one-ninth as bright as the sun?
4. Suppose we knew by some means that a star with an apparent brightness of $\frac{1}{2}$ sun has a *luminosity* of 8 suns. How far away is such a hypothetical star?

---

has a luminosity of 11 suns. It looks quite bright in our skies principally because it is close to us, as stars go, only 17 ly away. Rigel, on the other hand, looks brighter than Altair, yet its distance is estimated to be roughly 900 ly. Rigel looks so bright to us because of its great luminosity, 53,000 suns. Creatures traveling among the stars may use stars like Rigel as "lighthouses in space."

12.3 THE SPECTRUM OF A STAR

The Spectral Classification of Stars As viewed in a spectroscope, stars have spectra that vary widely in the number and position of dark lines visible and the strengths (width and degree of darkness) of the lines themselves.

Scientists in all fields classify the objects they study into groups as they search for patterns. A modern program of classifying stellar spectra was initiated in 1886 at Harvard Observatory. Every star down to the eighth magnitude (over six times dimmer than the naked eye can see) was studied in the spectroscope and placed in a classification scheme according to the appearance of the spectrum. For example, stars showing the strongest hydrogen lines were called A stars; stars showing the strongest neutral helium lines were called B stars, and so on. The sun fit into the category G, stars with strong calcium lines. The classification scheme became more and more refined until the classes themselves were subdivided into 10 parts (see Figure 12.3). The sun is a G2 star, and Altair is an A7 star, for example.

The world of astronomy owes a great debt to Annie Jump Cannon (1863–1941) who over many years performed much of the painstaking, thankless work of classifying stars. Cannon classified well over 200,000 stars in her lifetime.

The workers at Harvard originated the **spectral classification** scheme used today. Originally the classes were listed in alphabetical order from A to Q, but it was soon recognized that some of them contained too few

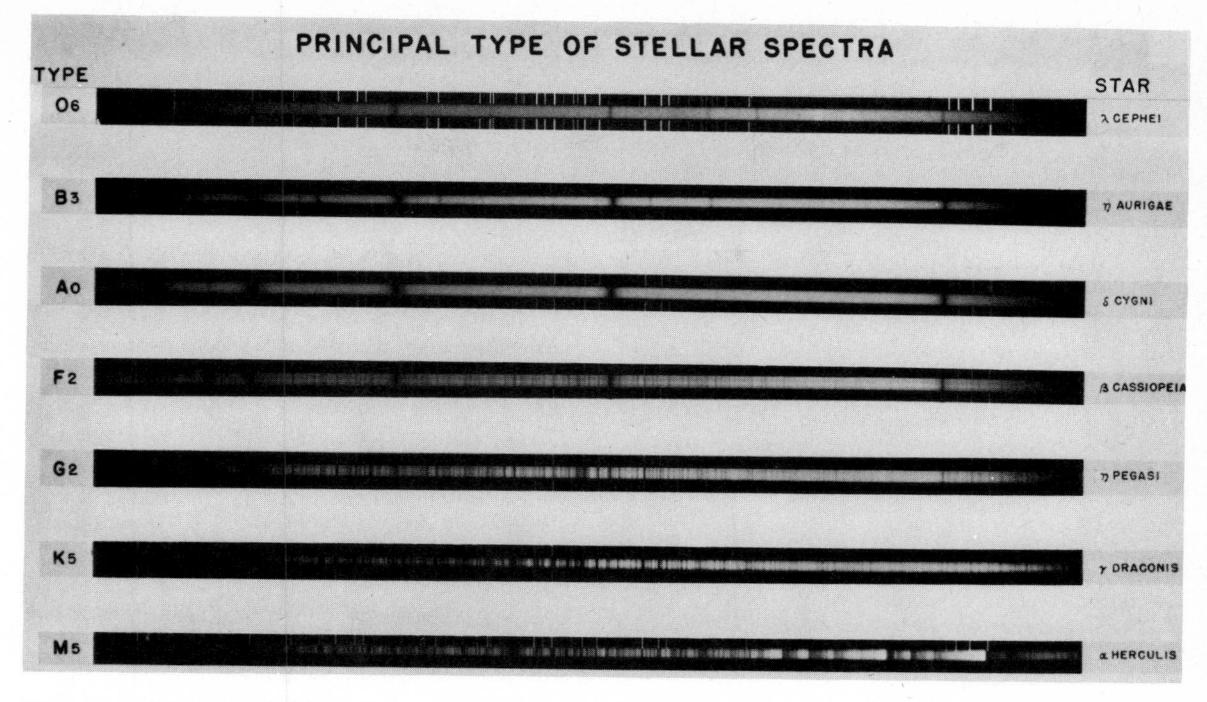

Figure 12.3 The spectra of seven stars. One example drawn from each of the seven spectral classes of stars is shown here. For each star, the class (or type) to which it belongs is shown at left by a letter, followed by a number indicating where in that class it belongs. (Each class has been subdivided into 10 parts.) Next follows the spectrum itself, flanked top and bottom by an emission (bright line) spectrum made in the observatory and used by astronomers for a reference. The sun is a G2 star.

stars, so certain classes were dropped. The remaining classes seemed to make a more logical, continuous progression if they were reordered. The ordering of spectral classes became O, B, A, F, G, K, M. Henry N. Russell (discussed in the next section) devised a memory aid for recalling this order: *Oh Be A Fine Girl (Guy), Kiss Me.*

Spectral Classification and Surface Temperature The classification project was underway. Attention next turned to interpretation of the different classes. At first, it was assumed that the differences could be explained by differences in the chemical compositions of the stars. Thus A stars would have more hydrogen than other stars, and G stars would have more calcium. Before long, however, this idea lost favor, and it became clear that O, B, A, F, G, K, M was an arrangement of the stars in order of *decreasing surface temperature.* For one thing, the colors of the stars differed in the same order, O stars being the bluest, shading down to yellow G stars and red M stars. As the surface temperatures of the stars were measured using Wien's law (Section 9.4) or by other related means, this hypothesis was confirmed. O stars have a surface temperature of about 40,000 K, A stars 10,000 K, G stars 5700 K, down to M stars which are 3000 K on their surface.

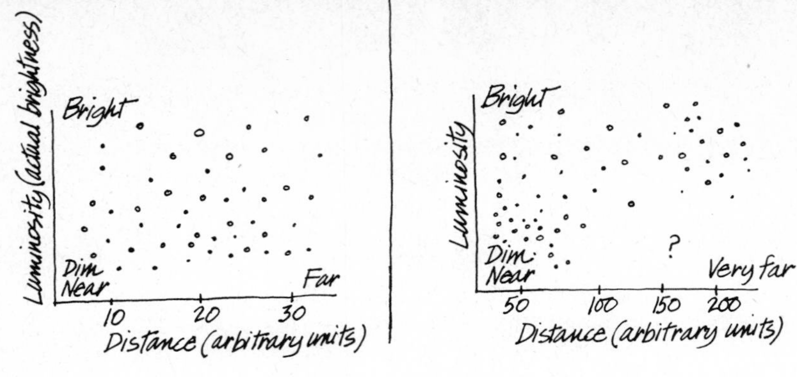

Figure 12.4 Is there a relationship between luminosity and distance? These graphs show how the data for stars might appear if the luminosities of stars were graphed versus distance from the earth. Each dot simultaneously represents a star's luminosity and distance. The near, dim stars lie on the lower left side of the graph, for example. (The graphs do not represent actual data but are only intended to illustrate several ideas.)

The Chemical Composition of the Stars What then about the chemical compositions of the stars? Do the sun and other G stars actually contain more calcium than other stars? Are A stars more abundant in hydrogen than others? The task of answering these questions turned out to be enormously complex. A man who helped to sort out many of these complexities was an astrophysicist from India, Megh Nad Saha. He developed the Saha equation. It is used to interpret the spectra of stars in order to determine their chemical compositions. The result of this analysis is that the temperature of a star strongly influences which lines in a star's spectrum will be strong and which will be weak. For example, if an A star could somehow be cooled down to the temperature of a G star, its hydrogen lines would become weak and its calcium lines strong. Astronomers needed to allow for the effect of temperature on a star's spectrum so that the actual chemical composition of the star could be computed. This is a highly involved research problem which is still being pursued, but it has become clear that most stars have approximately the same chemical composition as the sun. All stars are composed chiefly of hydrogen with some helium. Expressed in terms of the numbers of the various kinds of atoms in a typical star, 91.8 percent of all atoms are hydrogen and 8 percent are helium, leaving only 0.2 percent of other atoms.

12.4 THE HERTZSPRUNG-RUSSELL DIAGRAM

An Example of Observational Selection As the data about stars accumulated, astronomers naturally began to look for patterns. One way of searching for patterns is to make a graph of the data. Before introducing the Hertzsprung-Russell diagram, we first study two schematic graphs. See Figure 12.4. The left-hand graph shows luminosity, the actual brightness of the star, plotted against the distance of each star. Each dot represents the luminosity and distance of one star. For this graph, only stars relatively near the sun are chosen. No particular pattern is observed. All combinations occur: near–bright, far–bright, near–dim, far–dim, and everything in between. In the second graph, data for more distant stars have been added. Now there appears to be a pattern emerging. There is a conspicuous lack of distant stars that are dim. One may be tempted to conclude that dim stars exist only in the vicinity of the sun. But is this a meaningful pattern? Probably not. A more plausible explanation is that dim stars at great distances do exist. They would appear in the lower right portion of the graph, except that they are too dim to be observed from such a great distance. This is an example of **observational selection,** a distortion of data due to a method of observation that overlooks certain objects and emphasizes others. Most likely, there is no connection between the luminosity of stars and their distance. A larger telescope would reveal at least some of the stars belonging in the gap in the second graph. We next turn our attention to another pattern, which, as we shall see, is abundantly meaningful.

Figure 12.5 Henry Norris Russell: dean of American astronomers.

The Hertzsprung-Russell Diagram During the early 1900s, the American astronomer Henry N. Russell (see Box 12.2 and Figure 12.5) was at work measuring the distances of stars using the method of parallax. As he accumulated the data on distances, he used them to compute the luminosity of each star. He prepared a graph of luminosity versus surface temperature using the data he and others had collected, and found that a pattern was evident, one not due to observational selection. This type of graph became very important in astronomy. It was named for Russell and a Danish astronomer, Ejnar Hertzsprung, who had a similar idea at about the same time. Figure 12.6 shows an example of a Hertzsprung-Russell diagram (also called a H-R diagram or a HRD). Rather than being scattered all over the graph, the data tend to cluster in a band stretching from the upper left to the lower right. Well over 90 percent of all stars are on this band, which is known as the **main sequence.** Two other groupings appear on the diagram; the first consists of stars brighter than the main sequence stars, and the second of hot, dim stars with points lying well below the main sequence.

The implications of the H-R diagram are still not completely understood, but much progress has been made. In this chapter and the next we investigate the H-R diagram, and its importance will become increasingly evident.

12.5 HOW BIG ARE THE STARS?

The sun is a huge object, over 100 times larger than the earth in diameter. What about the stars? The sun looks much larger than the stars, but the stars are much farther away.

The most direct way to determine the diameter of a star would be to use the same method used to determine the diameter of a planet: Measure its angular diameter and distance and use these figures to compute

BOX 12.2

HENRY NORRIS RUSSELL, DEAN OF AMERICAN ASTRONOMERS

A transit of Venus occurred in 1882 and was observed by many, including a five-year-old child whose parents used a telescope to show him the black disk of Venus passing across the sun's face. For Henry Norris Russell (1877–1957), this was the beginning of a lifelong fascination with astronomy.

Russell earned his doctorate at Princeton University in 1900 and then spent several years at the British institution with which so many great astronomers of the past and present have been associated, Cambridge University. While at Cambridge, he photographed stars in order to determine their parallax. It was this work that led to his determinations of stellar luminosities and through these to the H-R diagram and his theory of stellar evolution; the latter is discussed in the next chapter.

In 1905, Russell returned to Princeton where he remained until his retirement in 1947. He had begun a study of binary stars (Section 12.6) early in his career. Noticing that the shapes of the light curves of binaries differed widely, he developed a mathematical theory allowing analysis of these curves, thereby obtaining an amazing amount of information concerning these stars. He was joined in this work by a young graduate student named Harlow Shapley who was to become famous in his own right (see Box 15.2).

Russell published his theory of stellar evolution and the first H-R diagram in 1913 and, inspired by the theoretical results of Eddington (see Box 13.2) on the interior structure of stars, later revised this theory. In 1920, the publication of Saha's theory on the spectrum of the sun inspired Russell to extend Saha's work and apply it to a quantitative study of the sun's spectrum. In 1929 Russell announced a detailed analysis of the chemical composition of the sun's atmosphere showing that it consists principally of hydrogen, some helium, and traces of other elements. The high abundance of hydrogen was an unexpected result, but before long it was found that all stars have similar compositions and that, in fact, hydrogen is by far the most common element in the universe.

In addition to this work, Russell studied the uranium found in the earth's crust and used it to obtain an age for the earth of about 4 billion years. At his death in 1957, Russell was honored for his life and work as the "dean of American astronomers."

---

the object's actual diameter. However, even the best telescopes under the highest usable magnification are unable to reveal the actual disks of the stars. Their great distances make their angular diameters too small to measure. Because the direct approach was unavailable, astronomers had to look for an indirect means to reach their goal.

Stefan's Law The key to the puzzle, How big are the stars? was supplied by a law discovered in 1879 by the Austrian physicist Joseph Stefan. Stefan had no particular interest in star sizes. Rather he was concerned with the total amont of thermal radiation emitted by objects.

Not surprisingly, Stefan found that a hotter object emits more radiation than a cool one. One must be careful, however, for this statement is not quite correct as it stands. For example, a white-hot coin emits much less energy in the form of radiation than the earth does. The earth is cooler but has a much larger surface area from which to emit radiation. To make Stefan's law complete, it can be stated:

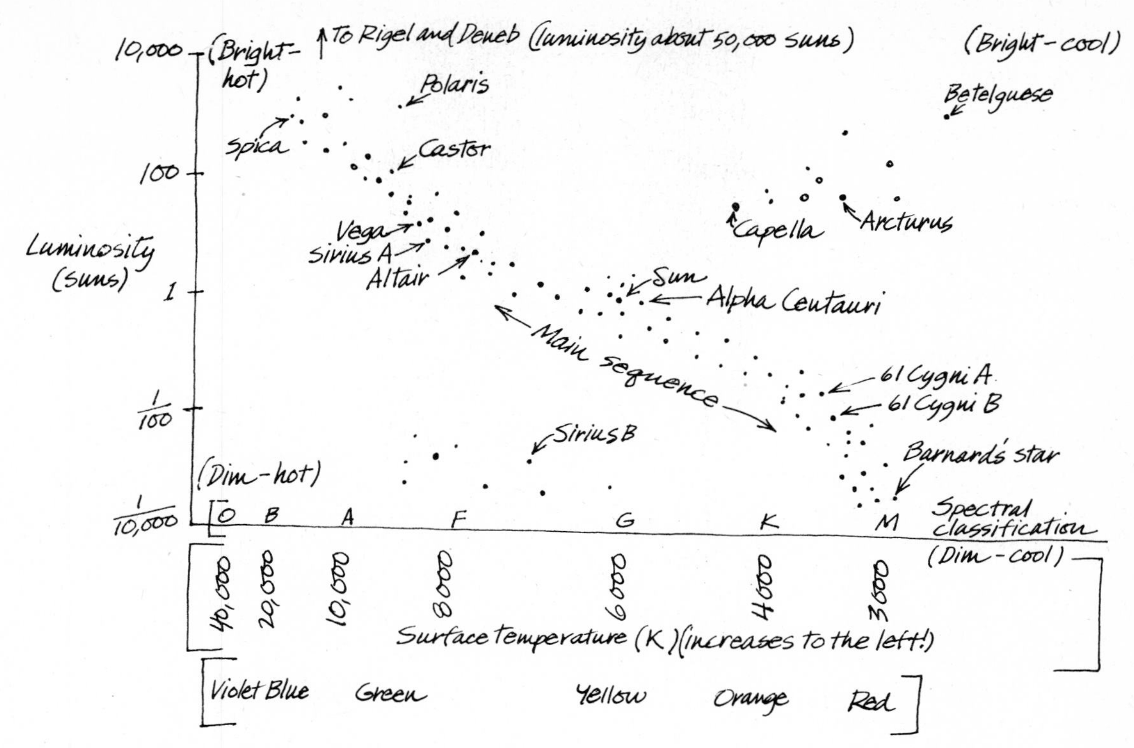

Figure 12.6 The Hertzsprung-Russell diagram. This graph, the most important one in stellar astronomy, has luminosity as the vertical axis. The horizontal axis can have any of three related labels: spectral classification, surface temperature (traditionally drawn with higher temperatures to the left), or star color. This graph does not show how the stars are arranged in space but rather groups them according to luminosity and surface temperature. Both of the stars of the double star 61 Cygni and also Barnard's star are quite close to the sun in space, but they are cooler and dimmer than the sun and so appear lower and further to the right on the diagram.

O stars are very rare. None appear on this diagram.

It is important to understand that this diagram is misleading because of an effect of observational selection. Stars below a certain *apparent brightness* are not visible. This means that we have overemphasized luminous stars and left out many stars having low luminosities. The latter stars will be visible only if they are quite close. Actually, there are *many* more stars in space that are cool and dim than stars that are hot and bright. However, the bright ones are easy to observe over large distances; thus they appear in large numbers on the diagram.

Stefan's Law (Verbal Form)

The hotter an object, the more radiation it emits from each unit area of its surface.

If the hot coin has a surface area of 1 square centimeter (cm^2), we should compare the radiation it emits to that from 1 cm^2 of the earth's surface. In this sense the hotter coin does emit more radiation. Stefan expressed his findings in an empirical formula:

Stefan's Law (Mathematical Form)

$$E = aT^4$$

where E is the total amount of energy emitted as electromagnetic radiation per unit area at all wavelengths, T is the temperature of the surface of the object (in kelvins), and a is a constant. There are reasons for believing that a is nearly the same for most stars. This makes Stefan's law the key to determining the sizes of stars.

The Size of a Star How does this law, the product of experimentation in a laboratory on earth, relate to the sizes of stars? The point is that the law shows there is a relationship between T, the surface temperature of a star, a quantity that is relatively easy to measure using Wien's law (see Section 9.4), and E. After T has been measured for a particular star, Stefan's law permits the computation of E, the amount of radiation emitted by the star from *one unit of area* of its surface. (The numerical value of a, the constant, that applies to stars has been determined in the laboratory.) Next one must find L, the luminosity of the star (see Section 12.2). Recall that L stands for the *total* amount of radiation emitted by the *entire* surface of the star. Because E is the radiation emitted by a unit area of the star, dividing L by E yields the total surface area of the star. Finally, making the reasonable assumption that the star is a sphere, one can use its surface area to compute its diameter. (For those who are interested, the formula is:

$$\text{Diameter} = \sqrt{\frac{\text{area}}{\pi}}$$

where $\pi = 3.142. . . .$) This indirect method makes it possible to determine the diameter of a star by measuring its apparent brightness, distance, and temperature, an amazing accomplishment. Puzzle 12.2 shows how one can determine the relative sizes of stars in simple cases without the use of mathematics. Please work this puzzle before reading on.

The sizes of stars turn out to range from those somewhat smaller than the earth to those much larger. As Puzzle 12.2 points out, stars directly to the left of the sun on the H-R diagram are smaller than the sun. Similarly, stars directly below the sun on the diagram are also smaller. Thus, the stars in the grouping below the main sequence on the H-R diagram must be quite small. As shown in Figure 12.7, the stars in this region are called "white dwarfs." White dwarf stars have diameters of about 20,000 km (12,000 mi), which is somewhat larger than the diameter of the earth. The sun is about 70 times larger than an average white dwarf.

The main sequence stars that are hotter than the sun are above and to the left of the sun. One cannot determine their size relative to the sun without performing a mathematical calculation, because stars directly above the sun on the H-R diagram are larger, while those directly to the

PUZZLE 12.2

THE RELATIVE SIZES OF STARS

It is not very difficult to determine the relative sizes of two stars on the H-R diagram if they both have the same luminosity (both lie on the same horizontal line) or if they both have the same surface temperature (both lie on the same vertical line). Each of the four puzzles below may be solved by answering the following hint questions (A through D) about the pair of stars.

A. Which star has the higher surface temperature? (Keep in mind that, on the H-R diagram, surface temperatures increase to the left.)
B. According to Stefan's law, which star radiates the most energy from, say, 1 square kilometer (km^2) of its surface?
C. Which star radiates more energy when the star is taken as a whole? That is, which star has the greatest luminosity?
D. In view of the answers to questions B and C, which star is the largest? That is, which star has the greatest surface area?

Stellar Size Puzzles When attempting the following puzzles, first sketch an H-R diagram with labeled axes showing the two stars considered. Then answer questions A through D above.

1. Suppose star W lies directly above the sun on the H-R diagram. Which of these objects is larger?
2. Which is larger, the sun or star X which lies directly to its left on the H-R diagram?
3. Compare the relative sizes of the sun and star Y which has the same surface temperature but is less luminous.
4. Star Z is positioned directly to the right of the sun on the H-R diagram. Is it larger than the sun?

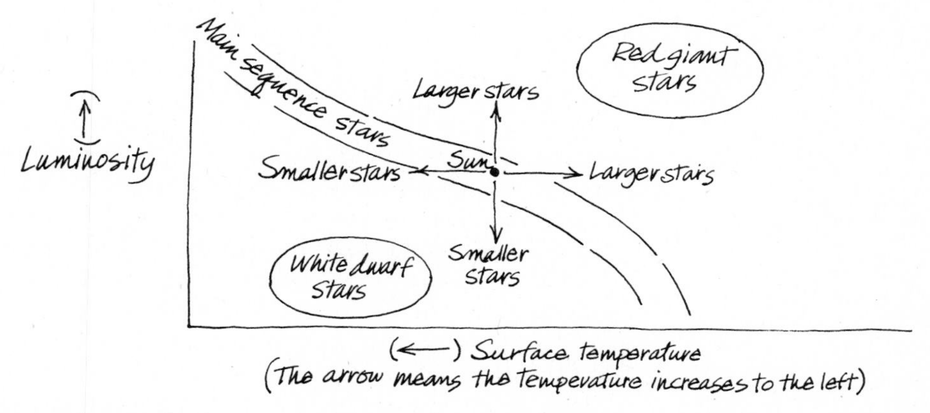

Figure 12.7 Star sizes and the H-R diagram. This graph summarizes the results obtained in Puzzle 10.2. Those results explain why the groupings of stars lying off of the main sequence were named as shown here.

left are smaller. Calculations show that the diameters of stars on the main sequence range from about 20 times the diameter of the sun to one-third the diameter of the sun. Thus most stars are *roughly* the same size as the sun.

But now we come to the relatively few G, K, and M stars in the region *above* the main sequence. Since they lie both above and to the right of the sun on the H-R diagram, they turn out to be very large

In the following list, the spectral class of a main sequence star is followed by the average diameter of stars in that class. The diameter is expressed in terms of the sun's diameter: O, 20; B, 5; A, 2; F, 1.3; G, 1; K, 0.8; M, 0.3.

Figure 12.8 The Big Dipper and an eye test. The names of some of the stars in the Big Dipper are given here. Alcor provides an easy eye test to pass. Mizar was one of the first double stars to be discovered with a telescope.

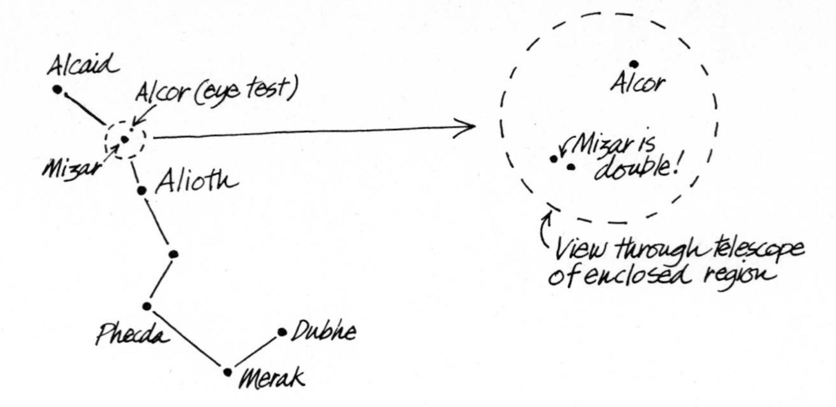

stars. They are labeled "giants" or "red giants" because of their color and size. G-star giants are about 100 times the diameter of the sun, K giants 200 times larger, and M giants about 500 times larger. The red giant star Antares, which is in the constellation Scorpius, has a diameter of about 600 solar diameters. If the sun were to be replaced by Antares, the edge of Antares would lie between the orbit of the asteroids and the orbit of Jupiter.

12.6 BINARIES: STARS WITH COMPANIONS

An applicant to the army in Ptolemy's day had to pass an eye test which anyone can take today. He was instructed to look at Mizar, the second star in from the end of the Big Dipper's handle, and say whether he could see a faint star nearby (see Figure 12.8). It is such an easy test to pass today that some writers speculate that Alcor, the fainter star, may have gotten a bit brighter over the years.

In 1650, Giovanni Riccioli, the man who named the major craters of the moon, turned his telescope on Alcor and Mizar. He must have been surprised to find that, using about 100 power, Mizar, the brighter of the two, also splits into two stars. (See Figure 12.8.) Readers with access to a telescope should not miss this interesting sight. The two stars comprising Mizar are 14 seconds of arc apart.

A pair of stars with a small angular separation in the sky is called a **double star** or a **binary star** if only two stars are involved, and a **multiple star** if there are three or more. William Herschel, whom we have already mentioned as the discoverer of infrared radiation (Section 7.3) and Uranus (see Box 15.1), became interested in binary stars. Herschel had a good reason for being interested in binary stars. He assumed that all stars had about the same luminosity and that dimmer stars were always more distant. (Herschel was working about 60 years before the luminosities of the stars were known.) He therefore concluded that, if a binary consisted of a bright star and a faint star, the faint star was the more distant of the two. He saw such pairs as ideal subjects for measuring stellar parallax. The near star, he hoped, would be seen to move back and forth during one year. The presence of the faint star would make the accurate measurement of the parallax motion easy to perform. During his sweeps he found a large number of binary stars and by 1784 had published the positions of 846 of them.

Herschel checked back on his binaries from time to time during the next 25 years to look for relative motion of the pairs. He found relative motion, but not the yearly back-and-forth parallax motion he had expected. A number of the stars had shifted their positions over the years in such a way that caused Herschel to conclude that they were in orbit around the same point in space. Held together by mutual gravitational attraction, the binary stars performed a slow, stately dance. One of the pairs had gone through 51° of their respective orbits since their discovery. Herschel had to abandon his simplistic assumption that all dim stars were farther away than all bright stars. In fact, it is now realized that most of the stars that appear to be close together in the sky, that is, have a very small angular separation, are actually close together in space. Such stars are called binary stars. The term double star often refers to two stars that are not gravitationally associated with each other but only in the same line of sight, one much farther behind the other.

Binary Stars Classified According to Their Method of Discovery The fascinating work of searching for double stars has continued up to the present day. Stars that can be "resolved" into two or more stars by direct use of the telescope are called **visual binaries.** Amateur astronomers enjoy viewing binary stars. The smaller the angular separation, the greater the challenge to their observing skill.

Astrometric Binaries Some binaries consist of stars so close together or so far from us that they do not appear to be double even under the highest available magnification. In other cases, the companion star may be too dim to be seen. Many of these stars can be recognized as double, however, by using more subtle observational techniques. Some stars, called **astrometric binaries,** appear to be single in a telescope but show a small periodic wobble in their position on the celestial sphere. Such stars must be orbiting a faint companion. In 1844, Bessel (see Section 12.1) reported that Sirius, in executing its proper motion across the sky, did not travel in a straight line but rather wavered slightly back and forth. In 1862, Alvan Clark, one of the greatest lens makers of all time and the one who produced the 40-inch Yerkes lens, was testing a recently finished 18-inch-diameter lens by viewing Sirius. The lens was so superior that the previously unseen companion showed up. The companion, called Sirius B, ceased being classified as an astrometric binary and became a visual binary. In 1915, study of Sirius B's spectrum indicated that, although it is dim, it is very hot. This was one of the first identifications of a white dwarf star.

Spectroscopic Binaries Among the many applications of the versatile spectroscope is in the discovery of binary stars. The spectra of many stars have lines that periodically shift slightly back and forth, first toward the red end of the spectrum and then back toward the violet. This is an example of the Doppler effect. The red shift indicates the star is going away from us, and the violet (or blue) shift indicates it is coming back. See Figure 12.9. The most logical explanation is that the star is orbiting a companion that is so dim that the companion's spectrum is not visible. Binaries discovered by this technique are called **spectroscopic binaries.** In some cases the spectra of both stars are visible. The lines from one star shift toward the violet during the time the lines from the companion shift toward the red, and vice versa. (Why?) One of the components of Mizar turns out to be a spectroscopic binary in its own right. Other famous spectroscopic binaries are Capella, Rigel, and Spica. There are many, many more.

Eclipsing Binaries Another method of discovering that a star is a binary depends on eclipses. The orbits of some binary stars are so situated that one star passes in front of the other as viewed from the earth. When such an eclipse occurs, the seemingly single star appears to get dimmer. Such stars are known by a number of equivalent names, **eclipsing binary, eclipsing variable,** and **photometric binary.** We discuss this category of stars more carefully in Section 12.8.

The Findings Concerning Binaries When all the data on known binary and multiple star systems are compiled, it becomes clear that binary stars are very common. If you point to a star in the sky at random, the chances are about one in three that it is actually a binary or a multiple star. Put another way, about half of all individual stars have at least one star for a companion. This figure does not allow for the high probability that many more stars have companions which have not yet been discovered.

Many interesting binary stars have been found. The faint binary called 36 Ursa Majoris is known to have components about 50,000 AU apart, about 0.8 ly. Their orbital period is too long to be measured. On the other hand, the star system WZ Sagittae has a period of only 81 min. These stars must be touching or almost touching each other.

12.7 HOW MASSIVE ARE THE STARS?

The mass of an object is a measure of the amount of matter in it. Astronomers are always eager to know the mass of any object they study. The mass determines the gravitational interaction of the object with other objects. As we shall see in the next chapter, the mass of a star is the all-important factor in determining its lifetime.

We saw in Chapter 8 how the masses of objects such as the sun and the planets (those with satellites) are determined. You find an object in

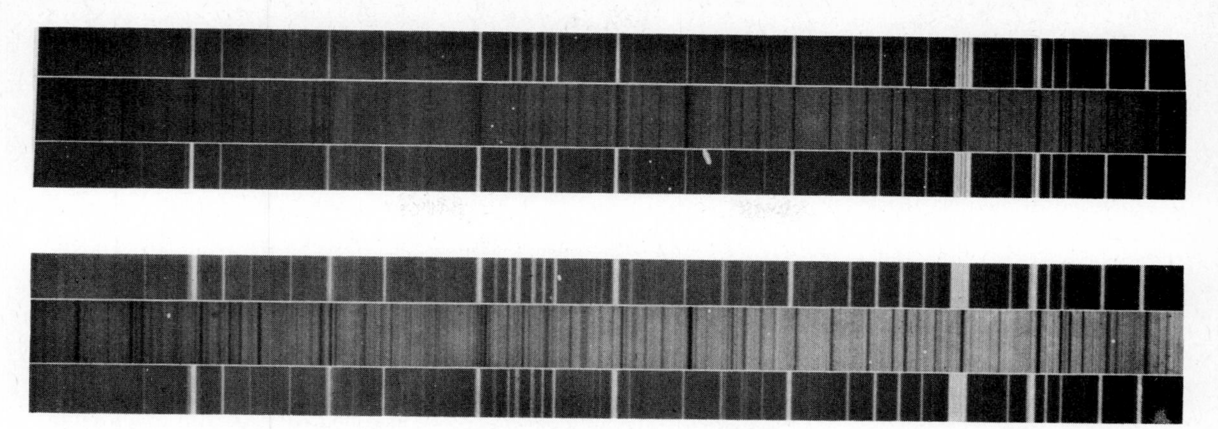

Figure 12.9 The shift of the lines in the spectrum of a spectroscopic binary. The star Mu Orionis is a spectroscopic binary with a period of 4.5 days. Two spectra are shown here, taken roughly two days apart. Each spectrum is flanked by an emission spectrum made in the observatory using a gas at rest. It may be used as a reference to observe that the lines in the lower photograph are shifted considerably to the right.

orbit around an unknown mass, measure the distance between the two objects, and also measure the orbital period. Using these data, you then apply Newton's laws to determine the unknown mass.

Since the stars are much like the sun in other respects, you might suppose that their masses are also in the same general range as the sun's mass. But merely guessing is hardly satisfactory. Astronomers wanted to measure the masses of some stars. An obvious solution seemed to be: Find a planet orbiting a star and then apply the same analysis as was applied to the sun. The catch is in the first phase of this proposal, "find a planet." Before the distances to the stars were known, some astronomers felt certain that the faint visible companions of some stars were planets. But once it was realized that the stars are light years away, this hope was dashed. Remembering how difficult it was for astronomers in our solar system to discover Pluto, you can imagine the difficulty in trying to observe Pluto or any other planet from another star. (See also Box 12.3.)

Is there then no way to determine the mass of any star except our sun? Can the reader think of a method before reading on?

The key to this problem is that we need to observe something orbiting a star. By a surprising coincidence, we discussed just such cases, binary stars, in the previous section. If all stars were solitary, we might never know the mass of any star but our own. Fortunately there are many binary stars in the sky, so the problem is solved, at least in principle. However, a large amount of hard work goes into the determination of the mass of even one star. These are dearly won data.

To begin with, not every binary star is useful. In the case of most visual binaries, the two stars are so far apart that the orbital periods are too long to be measured with sufficient accuracy. For other visual

BOX 12.3

DO OTHER STARS HAVE PLANETS?

As explained in the text, direct observation of a planet orbiting even the nearest star beyond the sun is well beyond the ability of modern telescopes. This is a frustrating situation, since it would be extremely interesting to know whether planets exist elsewhere. If planets are common, one can conclude that planet formation is an ordinary event. The discovery that planets are common would greatly enhance the probability that life as we know it exists elsewhere.

The leader in the search for planets of other stars has been Peter Van de Kamp (born in the Netherlands, 1901), director of Sproul Observatory. An expert on the analysis of the orbits of binary stars, Van de Kamp, over the last 40 years, also has been making photographs of the 40 stars nearest the sun to observe their proper motions. A solitary star traces out a straight path in the sky, but a planet orbiting a star pulls the star first one way and then the other by means of its gravitational attraction. If the planet is massive enough and the star near enough to us, the star will be seen to follow a noticeably wavy path as it pursues its proper motion. We may then estimate the mass of the planet from these observations.

Van de Kamp has announced that his measurements show that at least four of the nearest stars display a wavy path of the type indicating the presence of one or more planets. The stars and the year of the announcement are: 61 Cygni (1943), Lalande 21185 (1960), Barnard's star (1963), and Epsilon Eridani (1973). Perhaps the most interesting of the four is Barnard's star. Van de Kamp's measurements of the star's motion indicated either a single accompanying planet having a mass 1.5 times that of Jupiter or possibly two planets of smaller mass. In 1973 a NASA group published measurements and an analysis of Barnard's star that seemed to confirm the two-planet hypothesis, giving orbital periods of 12 years and 26 years for the two.

Measurements of the waves of these stars are very difficult to make. The angular width of the wave of Barnard's star has been likened to the width of a human hair at a distance of a mile, which is close to the limits of accuracy of modern techniques. Nevertheless, it was surprising that in 1973, the same year as the NASA confirmation, two astronomers in collaboration reported that they had studied 241 photographs of Barnard's star taken from 1916 to 1971 and that they could detect no wavering in the star's path. In the view of some astronomers, no planet of another star has been reliably detected as yet. However, probably most astronomers hope that the evidence in favor of the existence of other planets will win out.

---

binaries, the period may be known, but the system is so far away that its distance is not accurately known; if the distance from us is unknown, the separation of the two stars cannot be computed. If the period and distance *are* measurable, a great deal of work remains. The orbit of each star of the binary must be accurately measured. Very few orbits are seen face-on. Most are tipped at an angle with respect to our line of sight, and this must be taken into account in determining the actual orbits. Although many astronomers have worked for many years, first-rate data on stellar masses are known for only about a dozen visual binaries. Ingenious, complex methods have been worked out to determine the masses of certain other binaries, those discovered by other means (spectroscopic, etc.). In all, the masses of only a few dozen stars have been measured.

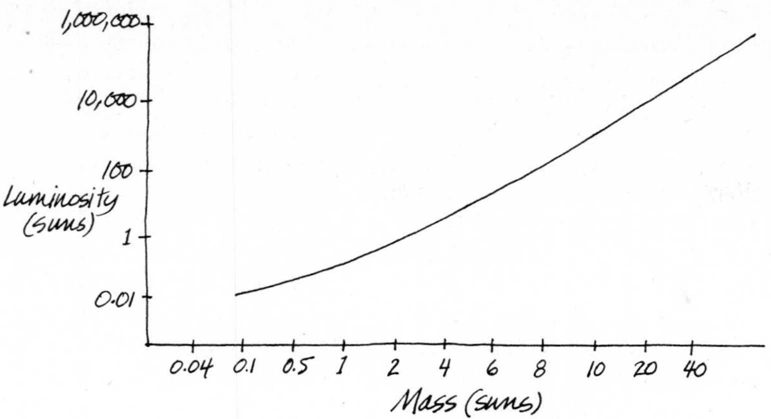

Figure 12.10 The mass-luminosity relation for main sequence stars. This graph displays the general trend discovered for main sequence stars: More massive main sequence stars are more luminous. (A word about the scales on the axes: The data cover such a wide range of values that astronomers find it necessary to compress the upper regions of the scale. Such scales are known as logarithmic scales.)

The data resulting from this labor are interesting. The great majority of stars have a mass in the range from about 0.1 to 4 suns. (A star having a mass of 4 suns has 4 times the mass of the sun.) There are rare examples of stars having masses outside this range. For example, in WZ Sagittae, mentioned in Section 12.6, one of the stars has a mass of only 0.03 sun, and there are only a few rare examples of stars with masses near 60 suns. It is rather striking that stellar luminosities range from more than 50,000 suns to under 0.002 sun, while by comparison the masses have such a small range.

The masses of some familiar stars follow, expressed in suns: 61 Cygni, 0.6 and 0.5, (it is a binary); Alpha Centauri, 1.1; Procyon, 1.8; Sirius, 2.3; Arcturus, 3; Antares, 10.

Earlier it was mentioned that Sirius B, the dim companion of Sirius, was found to be dim yet hot. Stefan's law then implies that Sirius B must be very small as stars go. Numerically, the diameter of Sirius B turns out to be about 30,000 km (19,000 mi), smaller than that of the planet Neptune. After a study of the orbit of Sirius B, its mass was determined to be almost exactly the same as the sun's mass. White dwarf stars must be very strange objects—having the mass of the sun compressed into a sphere the size of a planet. We will discuss this intriguing kind of star further in the next chapter.

The Mass-Luminosity Relation Another interesting fact emerged when a graph was made of luminosity versus mass for the *main sequence stars.* See Figure 12.10. Such a graph expresses the **mass-luminosity relation** which, put generally, says that for stars on the main sequence, the more massive a star, the greater its luminosity. It has been found that main sequence stars range in mass from about 40 suns (O stars) to 0.08 sun (the coolest M stars). Arthur Eddington deduced an explanation for the mass-luminosity relation, which had important consequences for the interpretation of the life of a star. We will delve further into this in the next chapter.

Stars not on the main sequence do not conform well to the mass-luminosity relation. White dwarfs are very dim stars, and red giants are very luminous stars, yet most have masses roughly the same as the sun's. Consider what this means for a red giant. In such a star, the same amount of matter present in the sun is spread over a diameter of about 100 times that of the sun, which means it fills a volume of space about one million times larger than the sun does. One may conclude that the average density of a red giant (its mass divided by its volume) is very low. The outer layers of a red giant, if brought to earth, would constitute a better vacuum than any laboratory could hope to reproduce.

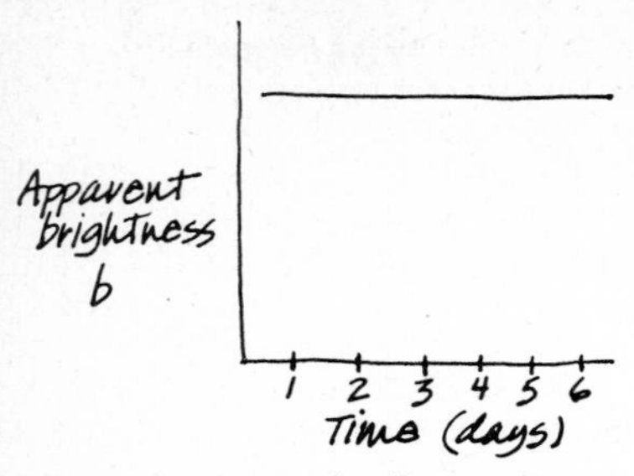

(a) *Light curve of ordinary star*

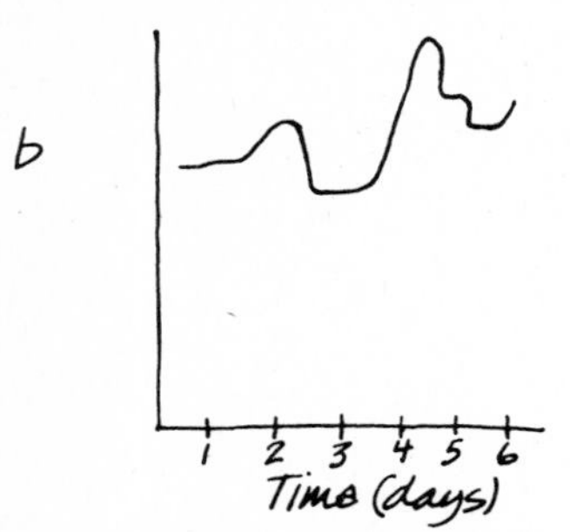

(b) *Light curve of irregular variable, no period*

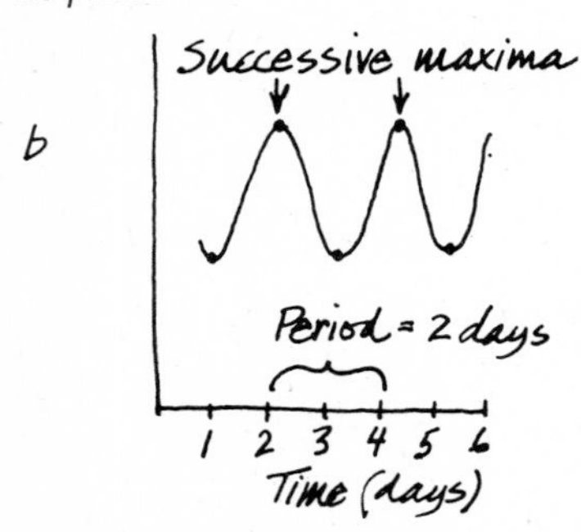

(c) *Light curve of periodic variable*

Figure 12.11 Some examples of light curves. A *light curve* is a graph of a star's apparent brightness *b* versus time. An ordinary nonvariable star does not change its brightness as time passes. Its light curve is a horizontal line (a). Some stars change brightness in an unpredictable, irregular manner as illustrated by the hypothetical light curve in (b). And some stars vary in a predictable, regular manner as exemplified in (c).

12.8 VARIABLE STARS: THE INCONSTANT LIGHTS

We saw in Section 12.1 that, if one takes the long view, the positions of the stars are not really changeless. The constellations that shone over the dinosaurs would be unrecognizable to today's star watchers because of the proper motions of the stars over many millions of years. We will see in the next chapter that some of the brightest stars in our sky were not even shining in the days of the dinosaurs.

On a shorter time scale, we have mentioned that other changes in certain stars have occurred during written history. Several novas have appeared in the sky, among these are the nova of 134 B.C. which inspired Hipparchus and the nova of 1572 which inspired Tycho (Box 5.1). At the beginning of this chapter we came upon Goodricke's work on other types of stars that change. These are all examples of a type of star called a **variable star,** any star that changes its apparent brightness. In this section we describe the three main classes of variable stars: eclipsing binaries, eruptive variables, and pulsating variables.

Following the example of Goodricke, when a variable star is discovered, astronomers work to obtain its light curve. The **light curve** of a star is a graph of its apparent brightness plotted against time. The light curve rises when the star becomes brighter and falls when it becomes dimmer (see Figure 12.11). Some variable stars have light curves in which the same pattern recurs time after time. The length of time required for the pattern to occur once is called the **period** of the variable star. For stars with light curves similar to the third one in Figure 12.11, the period is the length of time from one maximum to the next. In some cases it may be more convenient to measure the period from one minimum to the next. Let us now turn to the three main classes of variable stars.

Eclipsing Binaries In 1672, an astronomer reported that the star Algol in the constellation Perseus undergoes changes in brightness. In 1782 Goodricke, the teenage astronomer mentioned earlier, was the first to study Algol carefully enough and long enough to draw its light curve. The light curve of Algol, shown in Figure 12.12, shows that every 2.85 days the apparent brightness of Algol drops remarkably. Roughly 4 hr after the minimum, Algol regains its former brightness and remains (almost) constant for more than 2½ days. Goodricke offered two possible explanations for this. Either the star is eclipsed "by the interposition of a large body revolving around Algol," or it has a spotty

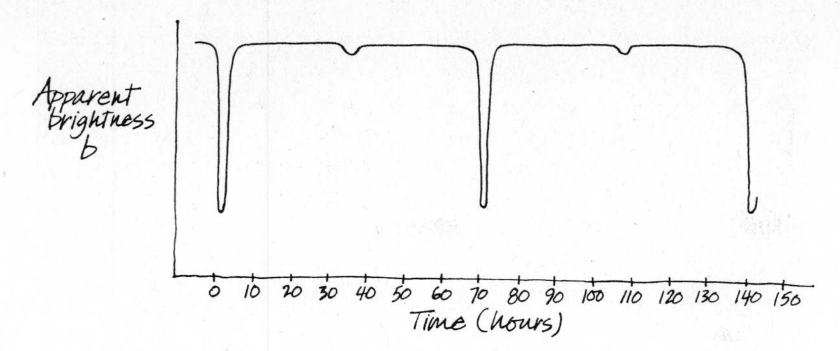

Figure 12.12 Light curve of Algol.

place on its surface which "is periodically turned towards the Earth." Astronomers have accepted the first idea, that a faint star is eclipsing a brighter one, thereby cutting off light from the bright star.

It was later found, in 1888, that Algol is also a spectroscopic binary, thereby confirming Goodricke's first proposal. In 1910, an astronomer using a photometer obtained a very accurate light curve for Algol and discovered the shallow secondary minima shown in Figure 12.12. (The reader is invited to devise an explanation for these secondary minima occurring halfway between the primary minima.)

Over 20,000 examples of star systems similar to Algol have been discovered. A star in this class is called an **eclipsing binary.** Several astronomers, including the pioneering Henry Russell, have devised a method of analyzing the shapes of the light curves of eclipsing binaries to obtain a great deal of information about the stars involved. Most importantly, it is possible to determine the masses of some eclipsing binaries using Russell's method.

Eruptive Variables Among the several classes of variable stars, we shall mention two more. **Eruptive variables** are stars that suddenly and unexpectedly increase their brightness. There are about 10 subgroups of eruptive variables. Some eruptive variables have had only one eruption for as long as they have been watched, while others erupt from time to time on an irregular schedule.

The best known example of a type of eruptive variable is a **nova.** This word, meaning "new star," was coined by Tycho. In all cases a nova is thought to be a star that has suddenly become brighter and perhaps attracted attention for the first time (see Figure 12.13). A typical nova becomes about 60,000 times brighter (an increase of 12 magnitudes) in only a few days. During the next several days it declines swiftly in brightness and then begins a long, slow return to normalcy, which may take a few years or even a century (Figure 12.14). Usually a nova is first noticed after the maximum brightness has occurred, but occasionally an astronomer is able to catch one on the increase. At such times, the star shows absorption lines in its spectrum that are greatly blue-shifted. The speed inferred from the Doppler effect can be as high as 3500 km/sec (about 8 million mi/hr). The standard interpretation is

Figure 12.13 After and during a nova. Nova Herculis changed dramatically in brightness during the two months between these photographs. The first photograph (on the right) was taken during the outburst. The star is so bright that it is overexposed. It did not actually look any larger during the outburst. The spikes on the overexposed image are caused by the telescope and are not real. The second photograph (on the left) shows the star after it had returned more nearly to normal.

that the star has explosively thrown off a shell of matter from its surface. The star increases dramatically in brightness because of the enormously increased surface area from which light is being radiated. Confirming evidence for this theory is that many ex-novas are seen to have expanding clouds of gas surrounding them.

There is no completely accepted theory of why some stars become novas. It is intriguing that many and perhaps all novas are binary stars. Does one star somehow trigger the explosion of the other by losing matter to its neighbor? If this is so, it means that we can breath more easily; the sun will never "go nova."

Within the last 100 years or so, another class of nova has been identified. These stars erupt in furious explosions and destroy themselves in the process. Such stars are called **supernovas** and are so important that we will devote all of Chapter 14 to them.

12.9 PULSATING VARIABLES, CEPHEID VARIABLES, AND LARGE DISTANCES

In 1784, two years before his death at the age of 21, Goodricke discovered another variable star, Delta Cephei. Its brightness changes by one magnitude with a period of about 5.4 days. Its light curve, similar to that in Figure 12.15, is very different from that of an eclipsing variable star. It exhibits a rather smooth, regular, up-and-down pattern which is repeated over and over again like clockwork. Today a large number of similar stars have been found, and they are called **Cepheid** ("Sef'-ee-id") **variables** after the first known example. The period of any given Cepheid variable remains constant. Among the known Cepheids, periods are found that are as short as 1 day and others as long as 55 days. Polaris, the Cepheid with the greatest apparent brightness, has a period of about 4.0 days. The changes in Polaris' brightness are small (a magnitude range of about 0.1), unnoticeable to the naked eye.

Why Do Cepheid Variables Vary? It was discovered in 1894 that the lines in the spectrum of Delta Cephei move back and forth with the

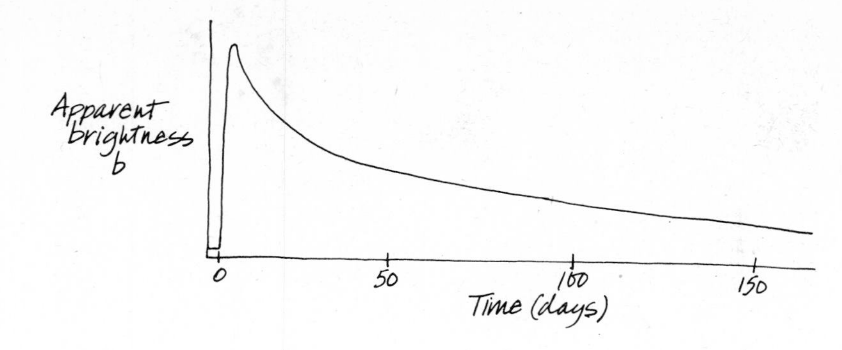

Figure 12.14 Light curve of a nova. A typical nova increases in brightness by about 60,000 times in a short time and then takes anywhere between several years and a century to regain its previous apparent brightness.

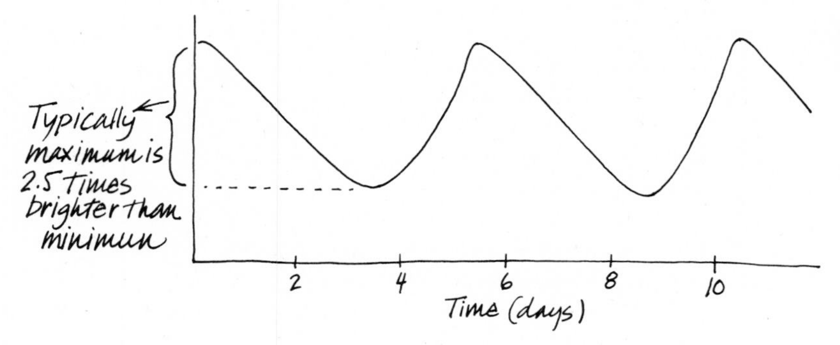

Figure 12.15 Light curve of a cepheid variable.

same period as the visible light variations. At first it was supposed that Delta Cephei was in orbit around an unseen companion. But what caused the light variations? It could not be eclipses; the light curve had the wrong shape.

Gradually, attention turned to the possibility that the Cepheid variables might be pulsating stars, stars that expand and contract, rather as if they were breathing. Arthur Eddington (see Box 13.2), whom we mentioned earlier (Section 9.7), produced a detailed, rather complex theory for pulsating stars. He explained the observations in a satisfactory manner—the changes in brightness are due to a combination of effects, among them, the expansion and contraction, heating and cooling, and ionizing of the star's outer layers. Although a beautiful theory, it is too complicated for us to probe more deeply.

Cepheid variables are giant stars having diameters typically about 25 times larger than that of the sun. Besides Cepheids, about eight other types of **pulsating variables,** stars that change their brightness because of changes in their diameters, have been found.

Using Cepheid Variables to Measure Distance Cepheids have turned out to be of enormous value to astronomy, thanks to a discovery make by Henrietta Leavitt while she was studying Cepheid variables in the Small Magellanic Cloud. This "cloud" is actually a large group of stars lying enormously far from the earth. (For more details on the Magellanic Clouds, see Section 16.8.) In 1912 Leavitt measured the

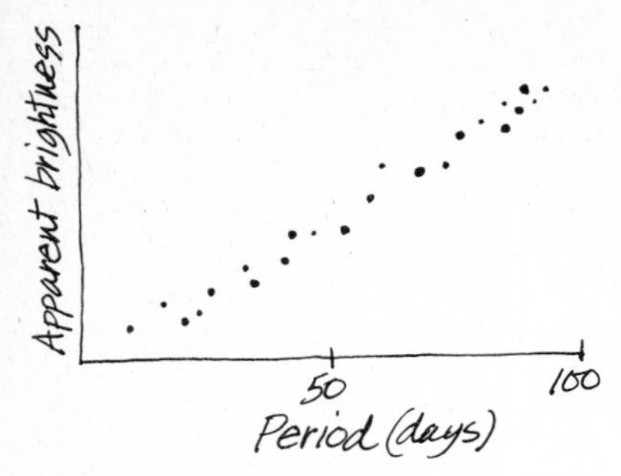

Figure 12.16 Apparent brightness versus period for Cepheid variables in the Small Magellanic Cloud. Studying variable stars in the Small Magellanic Cloud, Leavitt found that certain variable stars show a definite relationship between apparent brightness and period. These stars were later identified as Cepheid variables. This discovery provided one of the astronomers' most valuable tools for measuring the vast distances in space.

average apparent brightness of these variable stars (which she did not identify as Cepheids at the time), as well as their periods. She discovered the remarkable fact, embodied in Figure 12.16, that the longer the period of a Cepheid variable star, the greater its apparent brightness.

Leavitt then pointed out this crucial observation. All these stars are in a group so far from us that they all are approximately the same distance from us. This means that, when she found that Cepheid variable A has a greater *apparent* brightness than Cepheid variable B, she could also infer that A also has a greater *luminosity* than B. Therefore, her discovery could be restated in this way: *The longer the period of a Cepheid variable star, the greater its luminosity.*

Among the first to recognize the enormous potential of this discovery was Hertzsprung, coinventor of the H-R diagram. Hertzsprung identified the variables in the Small Magellanic Cloud as Cepheid variables and conceived a method of measuring the distance to any Cepheid variable we can observe. The key to the method is Leavitt's discovery that there is a relationship between the period of a Cepheid variable and its luminosity. This implies that we should be able to determine the luminosity of a Cepheid without first determining its distance. Then, as explained below, we can calculate its distance.

The catch in this method is that, although Hertzsprung knew there was a relationship between the period and luminosity of Cepheids, he did not know the numerical values involved. He needed to replace the apparent brightness scale on Leavitt's graph (Figure 12.16) by a luminosity scale. If there were even one Cepheid within about 300 ly of us, we could accurately measure its distance and thus obtain its luminosity. Finding the luminosity of even one Cepheid would fix the numerical values on the Cepheid scale but, unfortunately, there are no Cepheids close enough. The closest is Polaris—estimated to be 700 ly away.

This work by Leavitt and Hertzsprung was then taken up by the American astronomer Harlow Shapley (see Box 15.2). Shapley revised and extended Hertzsprung's work by means of a complex, statistical program of measurements. The result of his labors was a graph of *luminosity* versus period known as the period-luminosity relation (Figure 12.17). Since the initial work, Shapley's method has undergone several refinements which we discuss in Chapters 15 and 16.

At last, after Shapley's work, we can find the distance to any Cepheid variable that is visible. Cepheids, because they are very luminous, can be seen over very large distances. Let us next make sure we understand how the method works.

Suppose we desire to know the distance to a given object. If a Cepheid can be found in or near it, we measure the changes in the apparent brightness of the star over a long enough time until the average apparent brightness b and the period of the star are determined. We then use the period-luminosity relation (Figure 12.17) to determine the star's luminosity L. Notice that at this stage we have found the star's luminosity without knowing its distance. Therefore, the equation

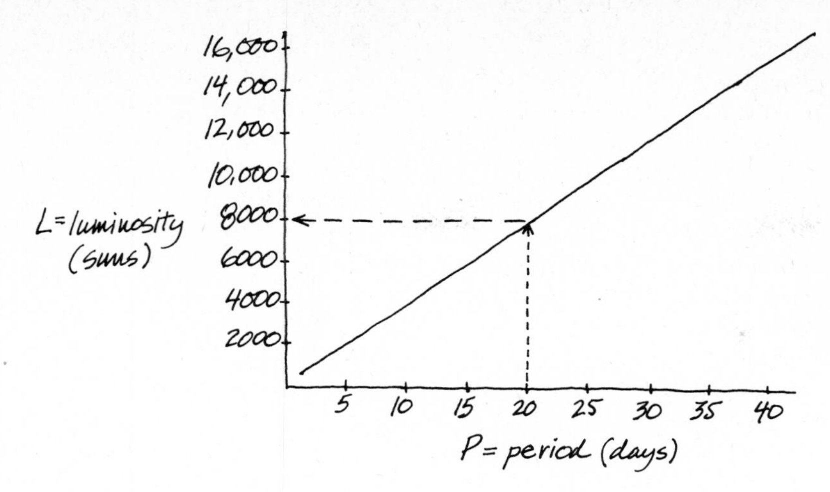

Figure 12.17 The luminosity versus period relation for Cepheid variable stars. Shapley used a statistical analysis of many Cepheid variables to convert Leavitt's result (Figure 12.16) to this one. Leavitt's discovery was that, if the second of two Cepheid variables has a longer period, it also has a greater luminosity. As an example of how to use the graph, the dotted arrows indicate that a Cepheid variable having a period of 20 days has a luminosity of 8000 suns.

$L = bR^2$, which we used earlier to determine a star's luminosity (once b and R, the distance, had been measured), can now be used in reverse. We now know b and L for the star, and so R is the only unknown. A calculation will then yield R, the distance to the star, and therefore the distance to the original object. The former 300-ly limit to our knowledge has been broken.

SUMMING UP

Many basic facts about the stars have been discovered by means of a large number of ingeniously conceived research projects. The results concerning stellar luminosities, spectral classification, surface temperatures, size, and mass (determined by studying certain binary stars) indicate that the sun is an ordinary star in all these respects. Among the many types of variable stars, Cepheid variables are extremely important because they allow the measurement of large distances.

Now that we have pondered some of the known varieties of stars, we next turn our attention to the lives of the stars.

EXERCISES

1. Define the following terms: light year, photometer, period (of a variable star).

2. Define these types of multiple stars: visual binary, astrometric binary, spectroscopic binary, eclipsing binary.

3. Define each of these types of variable stars and sketch a light curve for each: eclipsing binary, nova, Cepheid variable.

4. Which basic property of a star's location and/or motion is determined by measuring its (a) parallax angle, (b) Doppler shift, (c) proper motion?

5. What is the distance from the sun to the nearest bright star, Alpha Centauri?

6. Which looks brighter, a first-magnitude star or a second-magnitude star? What is the

magnitude of the dimmest star visible to the naked eye?

7. (a) What if a star were 12 AU from us and had an apparent brightness of $\frac{1}{6}$ sun? What would its luminosity be? (b) How bright would a star look if it had a luminosity of 8 suns and was 4 AU from us? (Answers: 24 suns; $\frac{1}{2}$ sun.)

8. List the seven spectral classes of stars in order of *decreasing* surface temperature. Which class of stars is actually the brightest? Which class of stars looks reddest? Which class does the sun belong to?

9. Sketch an H-R diagram. Label the vertical axis. Label the horizontal axis in three equivalent ways. Show the locations of main sequence stars, red giants, and white dwarfs.

10. What is Stefan's law? Use it to explain why a star directly to the left of the sun on the H-R diagram is smaller than the sun.

11. What basic type of data concerning stars are obtained by the analysis of binary stars? Sketch the mass-luminosity relation on a graph having labeled axes. To what kinds of stars does this graph apply?

12. What was Leavitt's discovery concerning Cepheid variables? What is the significance of this discovery for astronomy? What must one measure concerning a Cepheid variable in order to determine its distance?

13. Suppose that two stars, A and B, are both Cepheid variables and that both have the same apparent brightness. Suppose further that star A has a period of 5 days and star B has a period of 10 days. Which star has the greater luminosity? Which star is farther from us?

READINGS

The following books are valuable references for this and many of the next chapters.

Astronomy of the Twentieth Century, by Otto Struve and Velta Zebergs, Macmillan, New York, 1962, covers developments up through the mid-1950s very well even if a bit too technically in places for some general readers.

The Amazing Universe, by Herbert Friedman, National Geographic Society, Washington, D.C., 1975, is a lavishly illustrated survey of twentieth-century astronomy.

See also the works by Wyatt and Pasachoff and Kutner cited in Chapter 7, which are very valuable concerning stars.

THE LIVES OF THE STARS

Although we humans vary widely in our looks, lives, and attitudes, we all have two things in common: Each is born and each must die. Much of our life, literature, and philosophy is conditioned by these two facts. Birth and death are perhaps the two most poignant facts about us, yet we often forget, or even try to forget, about them in our daily lives.

The sun rises each morning looking much as it did the day before, and we tend to take it for granted. Yet astronomers believe that the sun and the other stars have lives. They are born, they live, and they die. The story of the life of a star is fascinating, as is the story of how astronomers arrived at their present conception of the life of a star. We explore these themes in this chapter.

13.1 DO STARS HAVE LIVES?

Why do astronomers believe that stars have lives? Couldn't it be that they always were and will remain the same as we see them now?

In one sense, this entire chapter is an answer to this question. Yet it is possible to give a shorter answer at once. Stars must be born, live, and die if the principle of the conservation of energy is correct. We discussed this firmly entrenched law of physics in Section 9.7. The important ideas here are that the stars radiate electromagnetic energy into space and that energy cannot be created or destroyed. If the stars are losing energy, they are analogous to a ticking clock. They had to be wound up (born) somehow and eventually have to run down (die). The energy they radiate must come from some source within them, and that supply cannot last forever.

Why Studying the Lives of Stars Is Difficult William Herschel suggested an analogy between an astronomer studying the stars and a botanist studying a forest. Suppose the botanist is not allowed to see any trees except during one day of his life. On that day, he is allowed to ramble through a forest all he likes. Then he must construct a theory on the life of a tree. The astronomer interested in stars is in a similar position. He cannot watch stars being born, living, and dying; they change much too slowly compared to the time he can spend studying them.

It would be interesting to know how the reader reacted to his or her first look at the H-R diagram. Did you ask yourself what the pattern in that diagram might mean? Did you perhaps begin to wonder if hot stars became cool ones or if big ones became small ones? If so, you had the same reaction Henry Russell did when he prepared his famous diagram. He was inspired to develop a theory concerning the life of a star. We now take up this first important theory of stellar evolution.

13.2 THE SLIDE THEORY OF STELLAR EVOLUTION

We mentioned in the previous chapter that Henry Russell spent a number of years measuring the distances of stars, from which he derived their luminosities. He noticed that, among other things, some K and M stars have low luminosities and others have high luminosities. This showed up clearly on the H-R diagram. During this work he began to formulate a theory to explain his diagram in terms of lives of the stars. This "slide theory of stellar evolution" was formally introduced by Russell in 1913 in a paper which included the first H-R diagram and made the terms "giant" and "dwarf" popular. He adapted the ideas of some earlier astronomers and modified them in light of the H-R diagram and his own original conceptions.

Astronomers use the phrase "stellar evolution" to refer to a theory of the life of a star. This is using the word "evolution" in a sense somewhat different from that familiar to biologists.

The term "slide theory" is not found in most accounts of Russell's theory, but it is an appropriately descriptive nickname.

Many books refer to the stars *on* the lower main sequence as "dwarf stars" in contrast to the giants. To avoid confusion we reserve the term "dwarf" for the white dwarfs, those stars lying well *below* the main sequence.

At the time Russell was working on this theory, information about a large number of stars was still being assembled. No detailed theory concerning the interior of a star existed. The Helmholtz theory stating that the stars produce energy by contraction (Section 9.7) was still current. Some astronomers were beginning to wonder if the recently discovered phenomenon of radioactivity might contribute to the energy supply of a star, but Bethe's detailed theory of nuclear fusion in the sun was still a quarter of a century in the future.

As the foundation of his theory, Russell stated that two things about a star seem obvious beyond debate: (1) Over the lifetime of a star, gravity causes it to contract. (2) In the long run, a star, a hot object immersed in the emptiness of outer space, must eventually cool down. Thus broadly stated, these two ideas are still accepted today.

The Slide Theory: A Star's Early Life Russell's theory begins with red giant stars. Such stars are large, cool, and the least dense of any stars on the H-R diagram. A red giant has been formed from a much larger, cooler cloud of gas which contracted. According to the slide theory, it is reasonable to assume that red giant stars are young. They are cool because gravitational contraction has not yet heated them up.

We shall not repeat the words, "according to the slide theory" in each sentence. It would be appropriate, but tiresome.

As time goes on, the self-gravitation of the star causes it to continue to contract, and this contraction further increases its temperature. Thus, if we could come back and observe the star after an appropriate number of years, it would have changed its position on the H-R diagram and would now lie further to the left. It will eventually become an O-type star (see Box 13.1).

It is important to keep in mind that, on the H-R diagram, surface temperature increases to the left.

BOX 13.1

MOVING ON THE H-R DIAGRAM

We will be using a bit of astronomical jargon which could lead to confusion without some warning. Astronomers find it convenient to think of an evolving star as *moving* on the H-R diagram. For example, one might say, "After having been stationary for billions of years, the star then moves up and to the right." This should not be taken to mean that the star began to move through space after standing still for a long time. Rather, "after having been stationary" means that the star's position as graphed on an H-R diagram did not change, which is equivalent to saying that the luminosity and surface temperature of the star did not change. Also, "the star then moves up and to the right" means the star's luminosity then increases while its surface temperature decreases. This is a handy shorthand way of speaking. The complete history of a star's luminosity and temperature can be summarized by a line on the H-R diagram. This line is known as the **track** of a star's life.

On an H-R diagram there are very few stars in the region between the red giants and the O stars. This region is known as the Hertzsprung gap. See Figure 13.1. Yet Russell believed that every star must cross this gap. Why are so few stars found there? The explanation was that a red giant star contracts and heats up slowly at first. Then, later on, the contraction speeds up and the star quickly crosses the gap. Few stars are found in this gap because each one crosses it so quickly.

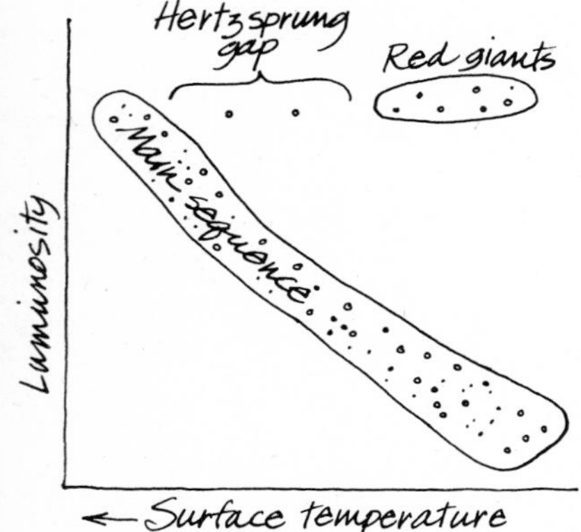

Figure 13.1 The Hertzsprung gap. A small number of stars are found on the H-R diagram between the red giant region and the O stars of the main sequence. The slide theory assumed that red giants evolve into O stars and that each star makes the transition very rapidly. This accounts for the scarcity of stars in the Hertzsprung gap according to this theory.

The Slide Theory: Main Sequence Stars Since there are no stars hotter than O stars on the H-R diagram, something drastic must happen to them to slow down further contraction and heating. During the relatively few years of contraction after leaving the red giant stage, the star's diameter decreased from about 100 suns to about 20 suns. According to the slide theory, the atoms near the central region of the star are about as compacted together as possible. The central region is no longer an elastic (springy) gas and stops contracting. This robs the star of its source of energy—contraction—and, having reached the peak of its luminosity and temperature, it begins the slow decline toward death.

As the star cools, its luminosity decreases. Thus an O star moves down (dimmer) and to the right (cooler) on the H-R diagram. In other words, it becomes a B star. The reason for the main sequence on the H-R diagram, according to the slide theory, is simply that the hotter, brighter stars are cooling off, becoming dimmer in the process. They are "sliding" down the main sequence, so to speak.

As the star continues to cool, its outer layers continue to contract until they too become as compacted as possible. This accounts for the decreasing size of stars as one goes down the main sequence. After passing through each spectral class, O, B, A, F, G, K, and M on the main sequence, the star becomes too dim for us to observe and may be considered dead.

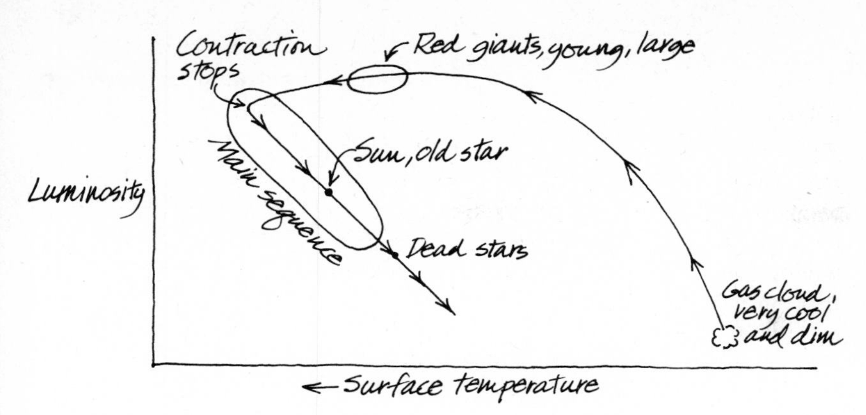

Figure 13.2 Complete track of a star according to the slide theory. The slide theory envisioned two main stages in the life of a star. First it contracts, heats up, passes through the red giant stage, and finally reaches the top of the main sequence. Then, since the star is almost completely contracted at this point, it cools off and gets dimmer, thereby executing a slide down the main sequence. Eventually, the star becomes too dim to be visible and can be considered dead.

We conclude this description of the life of a star according to the slide theory by referring to Figure 13.2. There we see that, according to the slide theory, the sun is a relatively old star. Humans have come on the scene in time to witness the sun's old age. In about only another 10 million years, the theory goes, the sun will have passed off the end of the main sequence and be dead.

The slide theory is attractive for its simplicity and its ability to account for many of the known properties of the stars. It was current for about 11 years until, as is the probable fate of all theories, it had to be set aside. The reasons for the demise of the slide theory are as interesting as the theory itself.

13.3 THE DOWNFALL OF THE SLIDE THEORY

How Do White Dwarfs Affect the Slide Theory? The reader may have noticed that white dwarf stars are not mentioned in the slide theory's description of the life of a star. There was one star on Russell's original H-R diagram that lay to the left of and far below the main sequence. Russell ignored it, saying that its ". . . spectrum is very doubtful." Gradually, more white dwarfs were discovered and placed on the diagram. The fact that the theory did not account for them did not *disprove* it, but certainly made it seem less elegant than before.

The Slide Theory and the Mass-Luminosity Relation During the decade following Russell's announcement of the slide theory, the data being assembled made the mass-luminosity relation more and more evident and further strained the theory. The mass-luminosity relation (see Section 12.7) states that, for main sequence stars, the more luminous stars are considerably more massive. If the slide theory were correct, this would imply that, as a star moves down the main sequence, it loses mass rather rapidly. It was found that O stars have masses of about 40 suns. If the sun was once an O star, how could it have lost 98

The later theory of fusion (Section 9.7) does state that the sun converts mass into energy. However, the rate of mass loss in this theory is very slow. It could never account for the large mass losses required by the slide theory in any reasonable length of time.

Figure 13.3 Arthur Stanley Eddington.

percent of its mass in the meantime? Attempts conceivably could have been made to save the slide theory, but it was up against even more serious difficulties.

The Slide Theory Confronts the New Theory of Stellar Structure In 1924, Arthur Eddington (see Box 13.2 and Figure 13.3) published a new theory concerning the interiors of stars—a theory of stellar structure. One result of this theory was to cast doubt upon the slide theory.

Eddington's theory resulted from a sophisticated application of the physics of his time to the structure of stars. Although the theory is highly mathematical, some of the important results can be expressed in words. The sun, a normal star, is not presently contracting. But gravity is trying to crush it. Why doesn't the sun collapse? Eddington's answer is that the inward force of gravity is counterbalanced by the outward force of gas pressure.

Gas pressure may be felt by holding a finger over the end of the hose of a tire pump and then pushing down on the handle. You can feel the gas pressure pushing back on the handle. According to the atomic theory, a gas consists of individual atoms rushing about, colliding, and rebounding from each other and the walls of the container. As you push down the pump handle, the compression heats the atoms, that is, makes them move faster. Thus the atoms hit the piston in the pump harder and more often. You feel the result as increased gas pressure.

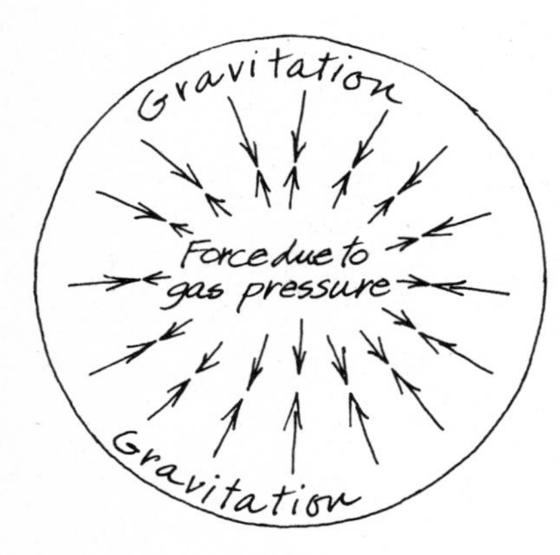

Figure 13.4 The forces on a star. Every star feels an inward squeeze due to gravitation. This inward pressure results from the gravitational attraction of each particle of the star for every other particle. The more massive the star and the more compacted it becomes, the greater the gravitational squeeze. This inward force is opposed by the force due to the gas pressure of the star, which results from the particles of the star colliding with each other. Gas pressure depends on the compactness of the star and its temperature.

In any given star, the force of gravity and the force due to gas pressure are in conflict. If the gravitational force is stronger, it will cause the star to contract. If the force due to gas pressure is stronger, the star will expand. In a stable star, the two forces balance each other (see Figure 13.4).

Using the concept of a balance in a main sequence star, Eddington derived a theoretical explanation of the mass-luminosity relation. We can relate, in a general way, how Eddington reasoned. Of course, *his* reasoning was expressed mathematically.

Compare two main sequence stars and concentrate on the one with the greater mass. Because of its higher mass, it experiences a stronger inward force of gravitation. In order to balance this, it needs a greater outward force due to gas pressure. This greater pressure is supplied by more rapidly moving atoms. Faster-moving atoms imply a higher temperature, and a higher temperature implies greater luminosity for stars having similar sizes. In other words, as the mass-luminosity relation states, a more massive star must be hotter and brighter.

Eddington gathered the available data on the mass-luminosity relation and compared them to the mathematical description he had derived. To everyone's surprise, the fit between his theory and the data was very good, right down the main sequence to the coolest, most dense M stars. The unexpected implication is that *all* the stars on the main sequence consist of gas, even down to their centers. This contradicts the slide theory which requires that the interiors of the stars on

BOX 13.2

ARTHUR STANLEY EDDINGTON—A TWENTIETH-CENTURY KEPLER

The Cambridge astronomer, Arthur Eddington, had some characteristics in common with Johannes Kepler. Both were deeply religious men. Eddington was born a Quaker and remained so throughout his life. Both were tireless investigators of the universe. Eddington said, "It is the search that matters. . . . You will understand the true spirit neither of science nor of religion unless seeking is placed in the forefront." Like Kepler he combined outstanding mathematical skill with deep insight and creativity.

Born in England in 1882, Eddington graduated from Cambridge University in 1905. By 1916 he had evolved a theory of stellar structure based on the opposing forces in a star. His theory implied that the temperatures in the centers of stars must be many millions of degrees. By 1924 he had shown that any star on the main sequence is gaseous down to its center, a result that led to the downfall of the slide theory. Earlier, in 1919, he had worked out a theory explaining the variations in the apparent brightness of Cepheid variable stars (see Section 12.9).

In a book published in 1927, Eddington suggested that the stars might produce energy by converting hydrogen into helium (see Section 9.7). One of Eddington's most famous flashes of wit is contained in this book. Critics of his idea maintained that the centers of stars were not hot enough for such a process to occur. Eddington pointed out that helium had to be formed *somewhere* and if stars were not hot enough his critics should "go and find *a hotter place*." Today it is generally agreed that the center region of a star is sufficiently hot for fusion to occur.

Late in life another parallel with Kepler became apparent. Eddington worked on what he called his fundamental theory, an elaborate attempt to harmonize the major ideas of physics and astronomy in one all-encompassing theory. The parallel to Kepler's *Harmony of the World* is striking (see Section 5.8). The fundamental theory requires some steps that can best be described as leaps of faith. It has semimystical qualities about it that some have called numerology. It arrives at a number of astounding conclusions, for example, that the number of electrons in the universe can be calculated using only measurements made in a laboratory on earth. Eddington never finished his fundamental theory, which has led to its being called an "unfinished symphony," and a "challenge to musicians among natural philosophers." Like Kepler's *Harmony of the World*, Eddington's fundamental theory has been bypassed by the mainstream of succeeding generations of scientists.

---

the main sequence be composed of matter so compressed that it can no longer be considered a gas. Eddington's theory implies that even the M stars on the main sequence are far from reaching such a fully compacted state.

At this point, the slide theory had to be abandoned. Although various new hypotheses were proposed, real progress in the understanding of stellar evolution had to await a series of developments on several fronts. It was not very long ago that all the tools necessary to construct our current theory became available. Yet, the slide theory had not been proposed in vain. A great deal of thought went into building up and tearing down the theory—a procedure that always leads to a deepened understanding.

13.4 INTRODUCTION TO THE CONTEMPORARY THEORY OF STELLAR EVOLUTION

The Difficulty in Obtaining Data Concerning Stars Astronomers in the mid-1920s were greatly hampered in their attempts to work out a new theory of stellar evolution. One problem concerned collecting data about the stars. We can see only one star, the sun, in any detail. Worse yet, we receive information only from the *surface* of the stars. The gas in a star is opaque below its photosphere, so that we have no direct data about the internal structure or behavior of the star. Everything about the inside of the star must be deduced using data derived from the star's surface. One enlists the laws of physics to draw inferences about the star's interior.

A Breakthrough in Physics and Computing We now realize that the known laws of physics were inadequate in the mid-1920s. The theory of the behavior of atoms, known as quantum mechanics, was just maturing then. It would be some time before it could be applied to stars. As one astronomer said, "How can we expect to understand the light from the stars until we understand the light from a candle."

The main hurdle in the mid-1940s was the complexity of the physical theories. The appropriate laws of physics were now available, but they were difficult to apply. The problem was that the laws were expressed in complex mathematical expressions which were very difficult to utilize. It was a frustrating situation. The physics and mathematics were at hand, but the number of calculations necessary to compute the life of a star were so enormous that 100 Keplers could not have scratched the surface of the problem in 100 years.

Then came the last necessary breakthrough, the high-speed electronic computer. The computer is changing our lives, and it surely has increased the ease with which the theory of stellar evolution has advanced.

In 1964 a computer was programmed to repeat the calculations Kepler performed during his 4-year search for the shape of Mars' orbit. The computer finished the job in 8 sec.

The rest of this chapter constitutes a description of the contemporary theory. Multivolume works have been written on the subject, and this account is highly simplified because parts of the theory are difficult to put into words and can be completely described only in the language of higher mathematics. Parts of the theory seem to be firmly established, especially the theory of main sequence stars. Other parts, concerning the birth and death of a star, for example, are still somewhat vague and are the subject of on-going research. Of course, even the most firmly entrenched parts of the theory are subject to revision. It would be interesting to be able to compare the following sections with those in a similar book written, say, 30 years from now.

For convenience, we divide the contemporary theory of the life of a star into five sections: birth, prime of life, aging, old age, and death.

Figure 13.5 The Great nebula in Orion, M42. This magnificent cloud of gas is located in the sword of Orion (see Appendix 1.). It looks slightly fuzzy to the naked eye, very fuzzy in binoculars, and is one of the most fascinating objects to study in a telescope. It is so easy to see because it is one of the nearest large nebulas. It is roughly 1500 ly from us and is about 13 ly across. It is an example of an emission nebula, one that exhibits an emission spectrum in a spectroscope (see Section 9.5). To the eye aided by a telescope, many emission nebulas look green because of light emitted by the oxygen atoms in the nebula. (See also Plate 18.)

13.5 STAGE I, THE BIRTH OF A STAR: THE PROTOSTAR

Nebulas: Birthplaces of the Stars Stars are spheres composed chiefly of hydrogen gas. Where are they most likely to form? Surely in the huge clouds of gas and dust seen in large numbers in space. Such clouds are composed chiefly of hydrogen. Some of these nebulas (from the Latin word for "clouds") are shown in Figures 13.5 through 13.7. Nebulas illuminated by O or B stars imbedded within them make spectacular photographic subjects. Some are very interesting when observed through binoculars or a telescope.

Figure 13.6 The Lagoon nebula, M8. This is an emission nebula found in the constellation Sagittarius. The M number refers to a list of over 100 objects in a catalog prepared by Messier (see also Section 16.1.). M8 is roughly 4500 ly distant and about 50 ly across. Very likely, stars are forming in this cloud of gas and dust at the present time. (See also Plate 17.)

The theory of the birth of a star is not fully established in all its details, but the broad outline is thought to be as follows. The cloud may have been in existence for millions of years. (The question of how the matter in the nebulas originated in the first place is an intriguing one which we will address in Chapter 17.) Occasionally pockets of more concentrated gas and dust form within the cloud. If the concentration is strong enough, the pocket's self-gravitation will cause it to contract. Probably, one huge pocket of gas will break up into smaller globules, and it is these globules that eventually contract to form stars. This is why almost all the stars thought to be young are found in star clusters. It is an indication that they originated from the same huge cloud which broke up as it contracted. More on these clusters of stars will be found in Section 13.7.

The Contraction of a Protostar We now concentrate on one of the globules which will eventually become a star. Its inner region contracts rapidly until it has roughly the same diameter as Jupiter's orbit, 10 AU. At this point the gas in the inner regions has reached a rather high temperature because of the contraction. As a result, it is producing a large amount of radiation. The object may now be called a **protostar,** a new star producing energy by contraction.

The Helmholtz contraction theory has been dropped as an explanation of the present source of the sun's energy. We see here, however, that this work was not wasted. The idea is used to explain the source of a star's energy at other stages in its life.

Computer calculations reveal a significant change in the star at this stage. Formerly, when it was less dense, the radiation it produced in its interior escaped directly into space. Now, however, this is no longer true. The inner, hot regions of the protostar are so dense that a light beam is very likely to strike an atom of the star and be absorbed before it can escape. We say that the gas is opaque. Soon after being

Figure 13.7 The Trifid nebula, M20. This emission nebula is roughly 3500 ly distant and about 15 ly across. Its mass is estimated to be roughly 1000 suns. Its central star is an O star.

absorbed, this light is reemitted by the atom, only to be reabsorbed by another atom. In this way, the light gradually works its way out to the edge of the new star.

An important effect of the gas' becoming opaque is to drastically slow down the contraction of the star. It is important to understand why.

All stars are being squeezed by self-gravitation. They resist this squeeze by means of gas pressure. When the gas atoms of a star emit radiation, they consume their energy of motion to produce the energy in the radiation. This causes the atoms to slow down. That is, they become cooler. If the protostar were to radiate but not contract, it would gradually cool off, reducing the pressure. On the other hand, if the protostar were to contract without radiating, it would heat up, increasing the pressure. What really happens is that, as the radiation streams into space, the protostar contracts because of reduced gas pressure. (Calculations indicate that the contraction more than makes up for the radiation loss, in the sense that the temperature and pressure rise as the star contracts.) The point here is that the rate at which the star contracts is governed by its rate of energy loss due to radiation. Before the gas became opaque, the protostar could lose radiation freely throughout its volume, and contraction was rapid. After the gas becomes opaque, radiation leaks away only at the surface and the rate of contraction is drastically slowed.

The protostar is not yet visible in the sky because there is still an enormous, cooler envelope of gas and dust surrounding it. Light leaving its hot inner regions is absorbed by the envelope. This heats the envelope, but to a much lower temperature than the inner regions. The

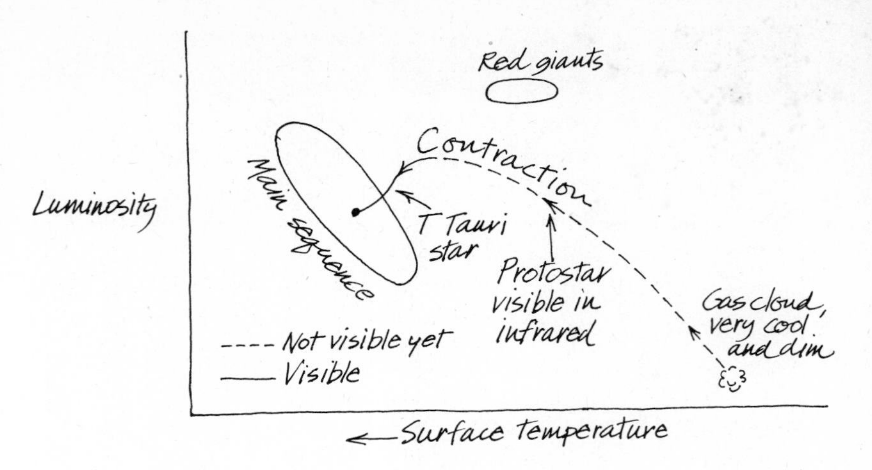

Figure 13.8 The birth of a star: contemporary theory. The path of the star shown here is schematically representative of the paths calculated by astronomers using computers. Depending on the assumptions the astronomer builds into the computer program, the path can have various curves and kinks. The details of the path are not yet fully agreed upon.

envelope becomes merely hot enough to emit infrared radiation. Within the last few decades the new field of infrared astronomy has opened up. Sensitive detectors of infrared radiation can now be attached to telescopes. One of the most interesting discoveries made in this way is that there are small objects in the sky that emit infrared radiation. Many astronomers believe that some of these infrared objects, usually found in or near nebulas, are the envelopes of protostars.

The Protostar Becomes Visible and Approaches the Main Sequence As its radiation leaks into space, the protostar continues to contract and get hotter. The next important change is that it somehow clears away the enormous envelope of cooler dust and gas that has concealed it until then. Just how the star dissipates this envelope is not certain. It may be that the envelope merely falls into the star. Or, possibly, the star develops a strong wind analogous to the solar wind discussed in Section 9.6 and blows the envelope away. At any rate, a new star then appears in the sky.

The newly visible star continues to contract, raising its temperature. As a result, it moves to the left on the H-R diagram. Computer calculations indicate that at this stage the star also grows *less* luminous because of its decreasing size. This causes it to move down on the H-R diagram. See Figure 13.8. The overall effect is that the star approaches the main sequence on the H-R diagram.

There is a kind of star that is subject to sudden changes in luminosity, as if it were unsettled. Called T Tauri stars, such stars lie just above the main sequence on the H-R diagram. It is hypothesized that they are stars that have almost reached the main sequence.

The contraction of the protostar has generally caused it to drift to the left on the H-R diagram, toward the main sequence. Computations show that, when the star's temperature and luminosity become those typical of a main sequence star, these changes cease. The star then indeed becomes a main sequence star.

13.6 STAGE II, PRIME OF LIFE: THE MAIN SEQUENCE STAR

How a Protostar Becomes a Main Sequence Star This changeover from protostar to main sequence star is a major event in the life of a star. It represents the star's arrival at the prime of life, so to speak. The major change that occurs is that fusion begins in the central regions of the star. The inner region where fusion occurs is called the **core** of the star. The layer between the core and the photosphere is known as the **envelope,** not to be confused with the protostar's cool envelope of dust and gas. See Figure 13.9. The key event that triggers fusion in the core is the heating of the core to a temperature above about 10 million K. At or above this temperature the hydrogen nuclei are moving rapidly enough so that their energy is sufficient to allow them to begin to fuse, producing helium nuclei. (Fusion is discussed in Section 9.7.)

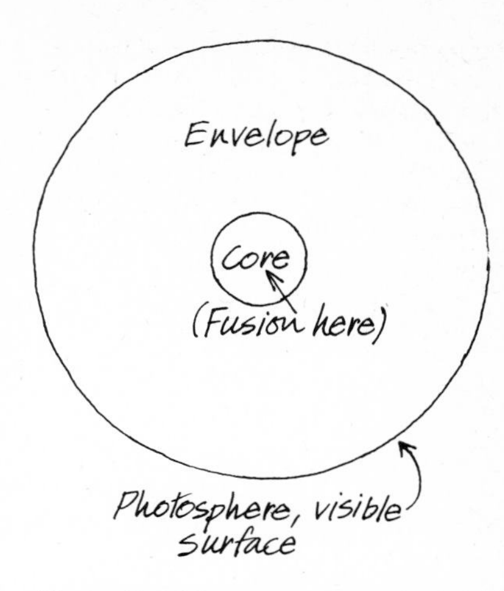

Figure 13.9 The core and envelope of a star. The central, densest, hottest region of a main sequence star is known as the core. It is here that fusion occurs. The photosphere is the surface of the star that is visible from the outside. Between the core and the photosphere is the envelope, the region through which radiation produced in the core passes.

Once the fusion reaction is fully underway the contraction of the star ceases. The star becomes stable; the end of contraction means the end of rising temperature and changing luminosity for the time being. But how does the fusion process manage to frustrate gravity's squeeze? The loss of radiation from the surface of the *protostar* allows gravitation to contract the star so that a drop in pressure will not occur. In a *main sequence star,* the loss of radiation from the surface is exactly compensated for by the radiation produced by fusion in the blazing hot core. The star no longer contracts. Rather, fusion takes over the task of maintaining the star's radiation supply, thereby maintaining the temperature and pressure.

Why Is There a Main Sequence on the H-R Diagram? In short, the distinguishing feature of a main sequence star is that the fusion of hydrogen into helium is occurring in its core. One may now ask why the main sequence stars appear strung out in a line on the H-R diagram rather than all being at about the same luminosity and surface temperature. Calculations show that almost the entire course of a star's life is determined by its mass.

Readers who look at other works are warned that it is a part of astronomical jargon to use the words "cooking" and "burning" as slang terms for "fusion." Of course fusion has little in common with these other processes.

Let us consider a protostar with a mass greater than 1 sun. Because of its larger mass it experiences a greater gravitational squeeze which it must oppose with greater gas pressure. Therefore, the star must maintain a higher temperature. The result is that, when the star settles down into the hydrogen fusion stage, it has a higher surface temperature; that is, it lies to the left of the sun on the H-R diagram. Furthermore, calculations show that a more massive star is also somewhat larger than the sun. The higher temperature (and somewhat larger surface area) of the star cause it to have a higher luminosity than the sun; that is, it lies higher than the sun on the H-R diagram. The overall effect on both surface temperature and luminosity places the star higher on the main sequence. Similarly, lower-mass stars need not maintain as high a temperature to restrain the force of gravity; they are lower on the main sequence of the H-R diagram. The main sequence may be seen as a way

Figure 13.10 Mass and the main sequence star. The mass of a star determines its eventual position on the main sequence. (The paths shown here are merely schematic. Detailed theories show more kinks and curves than shown here.)

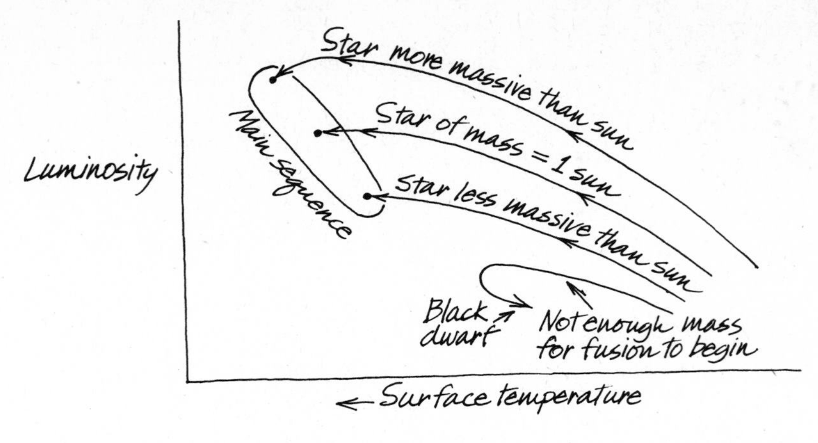

of displaying all hydrogen fusion stars in order of decreasing mass. See Figure 13.10.

Main sequence stars range in mass from roughly 60 suns to 0.05 sun. Why doesn't the main sequence extend to even higher- or lower-mass stars? Eddington was the first to explain that electromagnetic radiation emitted near the center of a star exerts **radiation pressure** on the outer layers of the star as it works its way to the surface. The higher the temperature, the greater the outward radiation pressure. Radiation pressure is not an important factor in the sun, but in very massive, very hot stars it is comparable to gas pressure. According to calculations, if a cloud of gas having a mass of about 100 suns began to contract, the radiation pressure would become so intense that the star would be blown apart. No stars with such large masses can form. To quote Eddington, "The force of gravitation collects together nebulous and chaotic material; the force of radiation pressure chops it off into suitably sized lumps." Even the few stars known to have masses of about 60 suns appear to be very unstable—expanding, contracting, and throwing mass off into space.

At the other end of the main sequence, the theory indicates that no protostar can reach the main sequence if its mass is less than about 0.05 sun. Such a low mass cannot achieve the core temperature necessary for fusion to begin. The protostar heats up a bit and then cools down, becoming a cold, dark ball of gas, a **black dwarf.** See Figure 13.10. The planet Jupiter is a body composed chiefly of hydrogen gas. It has a mass of 0.001 sun, well below the limit. Perhaps Jupiter is a minor example of a black dwarf. It is not dead yet. As mentioned in Section 11.1, it emits more radiation than it receives from the sun. Possibly the energy producing this radiation is derived from a slow contraction of the planet.

How Long Does a Star Remain a Protostar? How long does it take for a star to contract from the globule stage until it reaches the main sequence? The answer, perhaps not surprisingly, is that this too depends on the initial mass of the star. A lower-mass star experiences a

PUZZLE 13.1

ESTIMATING A STAR'S LIFETIME ON THE MAIN SEQUENCE

Suppose two stars, a B star and a K star arrive on the main sequence at the same time. Estimate which one will leave the main sequence first by answering these questions. (Caution, the following unsophisticated analysis merely leads to an educated guess.)

1. The first star to leave the main sequence will be the first one to run out of fuel. The star with the greater mass has more fuel. Consult Appendix 5. Which star has more mass? Approximately how many times more mass?
2. One cannot say that merely because a star has more fuel that it will last longer. The rate of consumption of fuel also matters. For example, a motorcycle might go farther on a full tank of gasoline than a luxury car can go on a full tank, even though the car carries more fuel. The car uses its fuel more rapidly. As discussed in Section 13.6, a star loses energy by radiation and, for a star on the main sequence, this loss is exactly balanced by the energy produced by fusion in its core. Thus we may take a star's luminosity as a good indication of its rate of fuel consumption. Consult Appendix 5 again. Which star, a B or a K star, has a larger luminosity? How many times greater is the larger luminosity?
3. Which star will use up its fuel first in your estimation? That is, which star will leave the main sequence first?

---

more gentle gravitational squeeze than more massive stars and as a result takes a longer time to develop. A potential O star rushes through the protostar stage virtually in an instant, as star lifetimes go. Table 13.1 gives the numerical estimates obtained from calculations.

Table 13.1 The Lifetime of Protostars

Typical mass of star (suns)	*Spectral class upon reaching main sequence*	*Estimated time spent by a star as a protostar (years)*
30	O	30,000
10	B	100,000
3	A	1 million
1.5	F	10 million
1	G	90 million
0.7	K	120 million
0.2	M	150 million

Life As a Main Sequence Star Once the core temperature exceeds the temperature needed for the fusion of hydrogen to helium, a star is called a main sequence star. When hydrogen fusion is fully underway in a star, it produces a high degree of stability. The star no longer contracts; gravitation is temporarily thwarted. Thus the surface temperature and luminosity of the star no longer change. This means, in turn, that the position of the star on the H-R diagram no longer varies, in striking contrast to the central hypothesis of the slide theory. A main sequence F star, for example, never becomes a main sequence G star.

The main sequence stage of a star's life must be its longest stage. The evidence for this is that well over 90 percent of all stars are main sequence stars. For a star with planets, the long, stable, main sequence stage is the best time for life to appear.

How Long Does a Star Remain a Main Sequence Star? Nothing lasts forever. A star must eventually cease its existence as a main sequence star. A main sequence star maintains itself by fusion of the hydrogen fuel in its core, and this fuel eventually runs out. The section following this one deals with the fate of a star that has run out of fuel. We close this section by asking a vital question: How long does a star remain a main sequence star? Before we answer this question, the reader is invited to solve Puzzle 13.1.

Table 13.2 Length of Time A Star Remains on the Main Sequence (Residence Time)

Typical mass of star (suns)	*Spectral class*	*Estimated residence time (years)*
30	O	2 million
10	B	20 million
3	A	500 million
1.5	F	4 billion
1	G	10 billion
0.7	K	100 billion[a]
0.2	M	200 billion[a]

[a] Uncertain.

The puzzle's result indicates that the more massive a star, the shorter its life on the main sequence. The results of the puzzle may be obtained in a more rigorous fashion by a detailed computation of the rate at which each star uses fuel. The theory states that the rate of fuel consumption is greatly affected by the temperature in the star's core. The core temperature of, say, a B star is roughly 30 million K, and so it consumes its fuel extremely rapidly. The result is that the hottest stars emit a great deal of radiation (they are very luminous) but can keep up this torrid activity for a relatively short time. In contrast to these prodigal sons, the least massive stars are analogous to misers before their hearth, waiting for the fire to go out and, just before it does, adding a tiny piece of coal.

Table 13.2 gives the results of calculations concerning the length of time each type of star remains a main sequence star. Comparing these data with Table 13.1, we may conclude that, indeed, a star spends relatively little of its life as a protostar.

According to Table 13.2, O and B stars consume the hydrogen in their cores in only a few million years. Any O or B star on the main sequence can be considered young, as stars go. The O and B stars in the sky were formed just a short while ago, an indication that star formation still occurs. At the other end of the scale are the M stars. They remain main sequence stars for hundreds of billions of years. If recent theories are correct, the universe is roughly 20 billion years old. (See Section 17.2.) Even the first M stars ever formed are still on the main sequence.

Perhaps of most interest to us in Table 13.2 is the result for our sun. The table shows that our sun has an expected main sequence lifetime of about 10 billion years. But how old is it now? Among the oldest known objects discovered in the solar system are meteorites and some of the minerals in earth rocks and moon rocks which, when tested by the method of radioactive dating, are found to have an age of about 5 billion years. If the sun was formed at the same time as these objects were, it is about halfway through its life. We have 5 billion more years of sunshine left. This is considerably more elbow room than the 10 million years the Helmholtz contraction theory (Section 9.7) allowed.

13.7 STAGE III, AGING: THE STAR LEAVES THE MAIN SEQUENCE

It is becoming common usage among scientists to call 1 billion years an *eon* ("ee'-on"). The sun has only 5 eons left as a main sequence star.

A Sense of Time If the calculations are correct, the sun has roughly 5 billion years left as a main sequence star. Five billion years seems like forever only until you study a subject such as stellar astronomy. Then you must stretch your sense of time. Five billion years is merely half of the life of the sun as a main sequence star. An M star remains a main sequence star for approximately 30 times that long.

According to calculations, the sun has so far converted about one-third of the hydrogen fuel in its core to helium. As it continues to shine, it will use this fuel at a slowly increasing rate until, 5 billion years from now, most of the hydrogen in the core will have been converted to helium. The fusion of hydrogen to helium in the sun's core will be essentially finished forever. What fate awaits the sun and other stars at this stage?

A star with a helium core, the "ashes" from the previous life-sustaining fusion of hydrogen, cannot remain a main sequence star. Only stars undergoing fusion of hydrogen in their cores can remain main sequence stars. It is believed that a narrow shell of hydrogen fusion may surround the helium core and that this shell supplies much of the energy to keep the star shining. Nevertheless, the *core* of the star now lacks the support it received from fusion. Recall that, at the onset of fusion the self-gravitation of the star was balanced. Now that fusion has run out, the inexorable grip of gravitation dominates once more in the core.

Gravitation needs no fuel to cause its effects. It can be restrained only temporarily. Even on the earth, the destructive work of gravity is apparent. In defiance of gravity, humans erect buildings and other structures on the earth's surface. Unless these structures are constantly maintained, however, they come tumbling down in surprisingly short order. Even the mountains, structures lifted by the enormous forces working under the earth's surface, will be worn down and leveled in a comparatively short time, leveled by the work of gravity aided by the wind and rain.

In a star, when hydrogen fusion in the core ceases, contraction of the core resumes. Once again, the core's temperature, which has been constant for so long, rises. As a result it begins to emit more radiation than previously.

According to computer calculations, the effect of this heating and brightening of its core has a surprising effect on the star's envelope. The lowest layers of the envelope experience an increased flow of heat and increased light pressure from the core. This leads to an increased outward pressure on the envelope as a whole, and the envelope expands. The seemingly paradoxical effect of the core's contraction is that the star gets larger.

The reasons given here for the expansion of the envelope are not really satisfying if examined in depth. The reasons for the expansion of the envelope are complex, differ in different mass stars, and cannot easily be put into words. As an astronomer friend of mine put it, "The envelope expands because the computer says it does."

Before reading on, please work through Puzzle 13.2, which gives practice in applying the principles we have been using and arrives at an important result.

The point of the preceding puzzle is that, when a main sequence star runs out of hydrogen fuel, it becomes a red giant star. That is, as the star expands, the surface temperature drops, the color reddens, and the luminosity increases. The star's position on the H-R diagram changes, moving toward the upper right. The change is slow at first and then quite rapid. This rapid change accounts for the Hertzsprung gap mentioned earlier. Russell felt that red giants were youths. Today we think of them as aging stars, past their prime, as it were.

Computer calculations indicate that, as a star leaves the main sequence, its motion on the H-R diagram again depends on its mass. Massive stars do not become much more luminous, but they do become

PUZZLE 13.2

LEAVING THE MAIN SEQUENCE

The purpose of this puzzle is to help you think about what happens to a star after it leaves the main sequence. In other words, we inquire as to how a star's surface temperature and luminosity change when the core's hydrogen is consumed. Our inquiry consists of seven related problems.

1. The computer says that when a star's core has run out of hydrogen, it will cause the envelope to expand. Based on this, choose the most reasonable alternative in this statement: As a star leaves the main sequence its *surface* temperature changes to approximately (double, half) the temperature it had when it was a main sequence star. Hint: What happens when a gas expands? (Note: You need only select the most reasonable answer. You need not know why the *numerical* value chosen is roughly correct. This comes from the computer.)
2. This temperature change at the surface causes the star's color to change. It will become (bluer, redder). Hint: Use Wien's law.
3. Now consider the star's changing luminosity (total light output). This too changes. First think about the temperature change at the surface. *Taken alone,* this change will *tend* to make the star (more, less) luminous.
4. The computer indicates that the star will become, say, 100 times larger. This means that its surface area will increase 10,000 times. *Taken alone,* this increase in size will *tend* to make the star (more, less) luminous.
5. You should have found in problems 3 and 4 that there are two opposing tendencies. The temperature change tends to cause a change in luminosity in one direction, and the change in surface area tends to cause a change in the other direction. Reread problems 3 and 4 and then make an educated guess as to which tendency will dominate. In other words, will the star *actually* become more or less luminous as it leaves the main sequence? (A rigorous solution requires a numerical calculation. All we are asking for here is a reasonable guess based on the information given in the above problems.)
6. Sketch an H-R diagram with labeled axes. In which direction will a star move on it as it leaves the main sequence? Use your answers to problems 1 and 5.
7. After a star leaves the main sequence, it becomes a __________ __________ star.

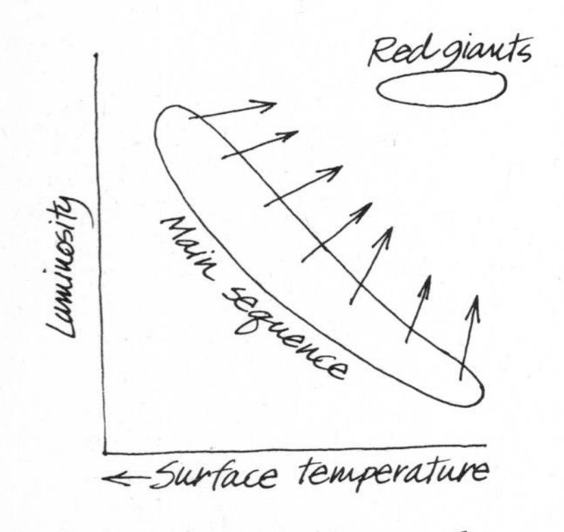

Figure 13.11 Tracks of stars as they leave the main sequence. Computer simulations indicate that any star leaving the main sequence as it runs low on fuel tends toward the red giant region on the H-R diagram.

considerably cooler on their surfaces. Also, less massive stars brighten considerably but do not become much cooler. The overall effect is shown in Figure 13.11. All stars change in such a way that their positions on the H-R diagram tend toward the same region. The most recent calculations show that the tracks taken by individual stars are much more complex than those shown in Figure 13.11. We leave these details to more advanced books.

Evidence Confirming the Theory: Open Clusters The preceding two sections are based on complex computer calculations. Is there any evidence that the results obtained bear any relation to reality? The answer is yes. In Section 13.5, we mentioned that, as huge nebulas contract, they are thought to give rise to a number of stars at about the same time. This is thought to be the explanation for the more than 1000 open clusters that have been discovered. The average **open cluster** is roughly 10 ly across and consists of a loose, irregular concentration of stars. There may be as many as several hundred stars in a major open cluster. Others consist of fewer than 10 stars. The two nearest open

Figure 13.12 The Pleiades, M45, also known as the Seven Sisters. This object is easy to locate in the fall and winter sky in the constellation Taurus, the Bull. Viewed with the naked eye, most people see a tiny group of six or seven stars arranged in a dipper shape. The Pleiades ("Plee'-ah-dees") is an open cluster and is a favorite object for binoculars users. Roughly 30 stars are visible with binoculars. About 120 stars are visible in a large telescope. The wisps of nebula seen in this photograph are very faint and not visible in most telescopes. They are an example of a reflection nebula (see Section 16.4) and are thought to represent gas and dust left over from the original nebula from which the stars formed. The cluster is about 400 ly from us and is the second closest open cluster. It is estimated to be roughly 40 million years old and is about 14 ly across. (See also Plate 21.)

clusters, the Pleiades and the Hyades, both in the constellation Taurus, the Bull, are easily visible to the naked eye and well worth looking for. Several open clusters are displayed in Figures 13.12 and 13.13, but they look much more impressive when viewed directly through a telescope.

The important aspect of open clusters for our present purpose is that the stars in a cluster all originated from the same nebula and at about the same time. Since they are all about the same age, the stars in a young open cluster should all still be on the main sequence. (See Figure

If the sun was once a member of an open cluster, it must have moved out of it long ago. None of the stars in the sky seem likely to have been ancient nest mates of the sun, judging by their present motions.

Figure 13.13 The Double Cluster in Perseus. These two open clusters are two favorite objects for binoculars users. They are both about 8000 ly from us. Each cluster contains roughly 250 stars. They are estimated to be about 10 million years old.

13.14.) If the contemporary theory is correct, during the next few millions of years, the first stars to leave the main sequence will be the O stars, followed by the B stars. Also, these stars will move toward the red giant region on the H-R diagram, as shown in Figure 13.11. Much later, when the cluster is old, even more stars will have "peeled off" the main sequence, leaving only the coolest ones still in place. Since they cannot wait around to see what will happen to a single cluster, astronomers have plotted the H-R diagrams of many open clusters to see if the various types of H-R diagrams in Figure 13.14 can be found. The results are encouraging. For example, the double cluster in Perseus is almost as young as the one in Figure 13.14(a). The Pleiades more nearly approximate the cluster in (b), while the Hyades are more like the one in (c). This agreement between theory and observation is very encouraging.

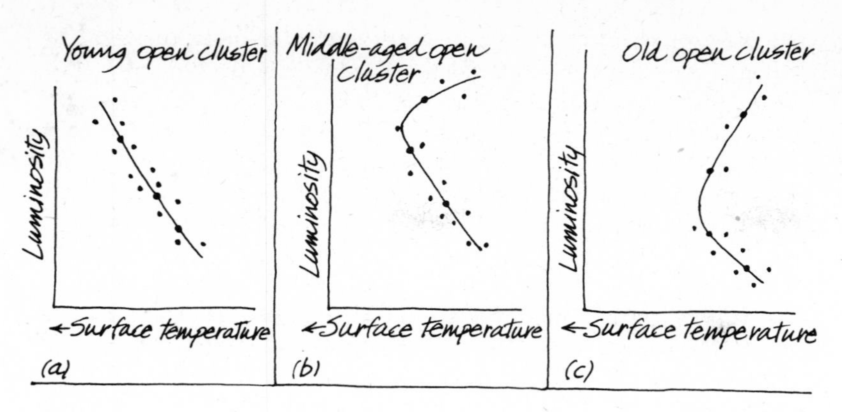

Figure 13.14 Open clusters as evidence for the contemporary theory of stellar evolution.

If the calculations are correct, the sun will one day begin to grow larger, perhaps becoming 50 times larger, making a disk with an angular diameter of 25° in the sky as viewed from earth. It's luminosity will soar to several hundred times its present value, scorching the planets and boiling away our oceans. If any of our descendants survive, they will have had to move out of the solar system by then. Perhaps they will even have forgotten the former beautiful blue cradle where their kind began their journey billions of years earlier.

13.8 STAGE IV, OLD AGE: THE RED GIANT STAR

A Star Becomes a Red Giant: Helium Fusion Begins As the aging star expands its envelope and gravity contracts and heats its core, the end may seem near. But, according to the theory, about one billion years after a star like the sun leaves the main sequence, a new energy source develops which grants it a reprieve. This new source seems to develop in all stars having a mass of 1 sun or larger. Less than the average amount of research has been done on stars having a mass much lower than that, largely because not enough time has elapsed since the start of the universe for such a star to leave the main sequence.

(We pass over many of the details of the next developments in the star, even at the risk of oversimplification. More advanced books are available for those who wish to pursue the subject further.)

A star like the sun experiences a continued contraction of its core, raising the core temperature from the original 15 million K to roughly 100 million K. At this elevated temperature, the helium nuclei are rushing about so rapidly that occasionally three of them collide nearly simultaneously and a fusion reaction occurs, producing energy and a carbon nucleus. This new source of energy apparently causes the core to expand. Gravity is once again thwarted for the time being.

A Red Giant at the End of Helium Fusion The helium fusion stage does not last very long. In a relatively short time (perhaps several million years for stars like the sun), the helium in the core is converted to carbon and contraction sets in for the third time in the life of the star.

The next events to occur depend on the mass of the star. A star with a mass of about 1 sun has at last reached the end of the line. Once it achieves a carbon core, gravitation is unimpeded by any further fusion. The star can no longer remain a red giant. The death of such stars is described in Section 13.9. We first discuss very massive stars.

How Very Massive Stars Prolong the Red Giant Stage Stars more massive than the sun that have exhausted their helium supply do not die yet. Because of their high mass, the force of gravitation is strong enough to contract and heat the carbon core to about 800 million K. At such a temperature, the carbon in turn undergoes fusion, the principal products being oxygen, neon, and magnesium, as well as other less abundant products. In stars of moderate mass, this is the last of the fusion, but even more massive stars again contract violently and further expand their envelopes until their core temperatures reach about 2 billion K. At such extreme heat the heavy, complex nuclei of oxygen fuse to form even heavier, more complex nuclei such as silicon and sulfur. Ultramassive stars can even go beyond this point and reach a core temperature of 4 billion K. By this time their envelopes have expanded to larger than 1 AU in radius. The nuclear reaction in the core at this point is principally the fusion of silicon into iron. At each stage along the way, the earlier stages of fusion proceed in shells surrounding the core. We can begin to see why a computer is required to follow the evolution of objects as complex as very massive red giants and why the results are still considered tentative.

An Ultramassive Star with an Iron Core We have seen how very massive stars can postpone death by a series of alternating contractions and more advanced types of fusion. But even a supergiant star must die when its core becomes iron. The reason is that, no matter how high the temperature, iron cannot produce energy by undergoing fusion.

In order to understand this, you must understand the difference between fission and fusion. The 92 elements found in nature have been numbered from 1 to 92. The number indicates the number of protons in the nucleus of the atom. As the numbers rise, the atoms increase in mass and complexity. Hydrogen is element number 1; it has a nucleus consisting of only 1 proton. Helium, number 2, has 2 protons and 2 neutrons. Iron, with 26 protons and 30 neutrons, is number 26. Uranium, number 92, has 92 protons and 146 neutrons. Nuclei having numbers under 26 can undergo fusion, and when doing so yield energy because some of the mass disappears (see Section 9.7). Uranium, on the other hand, is a bulky nucleus which is very easy to break apart. When a massive nucleus is split, the process is known as **fission.** See Figure 13.15. The products of fission consist of two less massive nuclei and some remaining mass which is converted into energy. Either uranium fission or plutonium fission may be used as the energy source for atom

For each atom in this list, the number of neutrons given is that found in most atoms of that type. In most cases, the number of neutrons may vary. Two atoms having the same number of protons but differing numbers of neutrons are said to be *isotopes*.

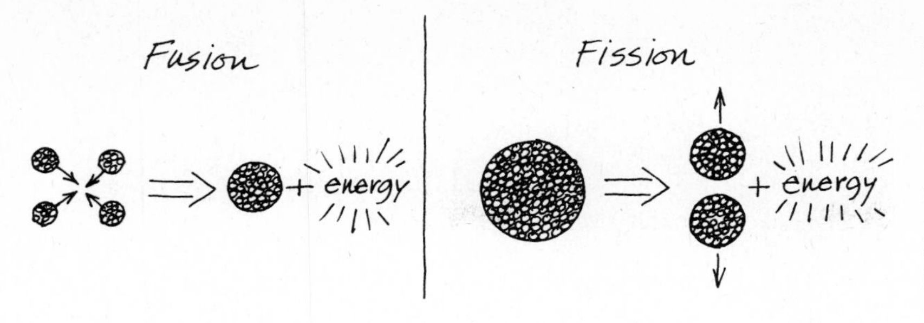

Figure 13.15 Fusion and fission. If the pressure and temperature are high enough, nuclei less massive than iron nuclei will combine and produce energy in a process known as fusion. Nuclei more massive than iron nuclei can be induced to fall apart, releasing energy in a process known as fission. Iron nuclei cannot produce energy while undergoing either process.

bombs and nuclear fission power plants. The point here is that nuclei less massive than iron undergo fusion into more massive nuclei, while nuclei more massive than iron undergo fission into less massive nuclei. Iron is the dividing point; it does not produce energy by means of fission or fusion.

Thus every star sooner or later reaches the end of its red giant stage, at which point it can no longer draw energy from fusion. Its death warrant is now signed. In the next section we describe some of the current hypotheses on the painful subject of death.

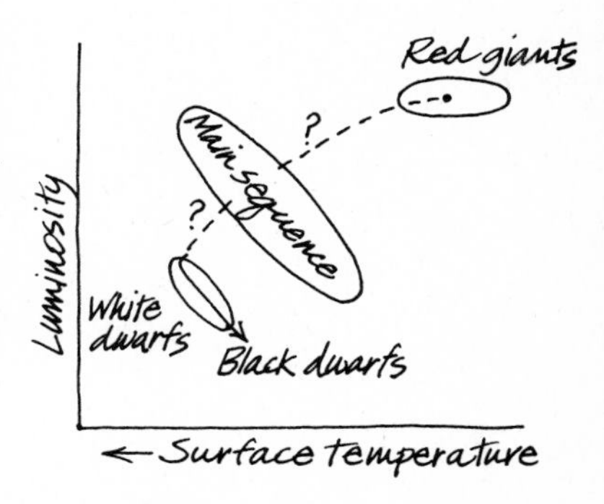

Figure 13.16 The death of a star having a mass of about 1 sun. Stars having a mass roughly the same as that of the sun are thought to pass into the white dwarf stage. The theoretical details of this transition are not clear at this time. White dwarf stars, having no source of energy, cool and grow dimmer until they are no longer visible.

13.9 STAGE V, DEATH: WHITE DWARF STAR OR . . . ?

The Death of a Star Having a Mass of 1 Sun Just as a star's mass determines the course of its life, so too does its mass determine its manner of death. We first turn our attention to dying stars having a mass of about 1 sun. More massive stars are discussed later in this section. Roughly six or seven billion years from now the sun will be a red giant with a carbon core. Because of its relatively low mass, its relatively weak gravitation will be unable to heat the core to a high enough temperature to ignite carbon fusion.

Astronomers feel quite sure that such a star next becomes a white dwarf star, but the details of the transition from red giant to white dwarf are somewhat uncertain. The envelope may collapse down around the core. This contraction would cause the surface temperature to increase and the luminosity to decrease drastically. The result would be a change in the star's position toward the lower left on the H-R diagram. Thus, the star's track crosses the main sequence and enters the white dwarf region (Figure 13.16).

Another conceivable way for a star to transform into a white dwarf besides having its envelope collapse is by forming a planetary nebula. A **planetary nebula** consists of a huge shell of gas, having a typical diameter of 0.6 ly (about 40,000 AU), surrounding a central star which is dim but very hot. About 1000 of these objects have been found. At first glance in a telescope, the planetary nebula exhibits a planetlike disk, hence the name which was coined by William Herschel. Figures

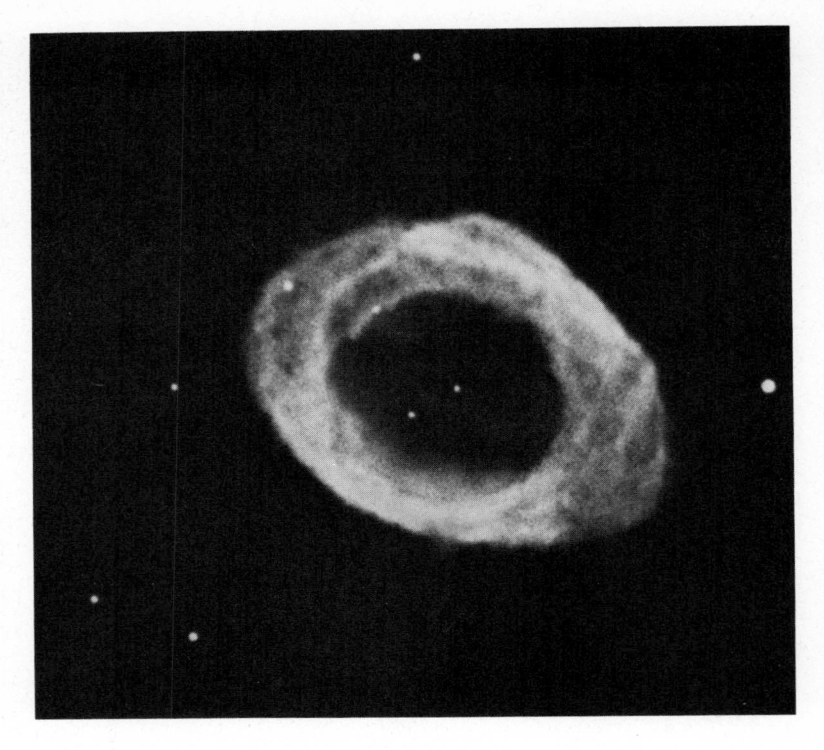

Figure 13.17 The Ring Nebula in Lyra, M57. In a moderate-sized telescope, this planetary nebula looks like a tiny smoke ring. It is an eerie sight. It may consist of the former envelope of a star which was cast off. The "naked core" is visible in this photograph at the center. This nebula is about 5000 ly distant and about 1.8 ly across.

13.17 through 13.19 reveal objects of wondrous beauty. One common interpretation of these objects is that they represent a star in the transition between the red giant stage and the white dwarf stage. If this is correct, the star has somehow expelled its former envelope into space, and the naked core is revealed at the center of the nebula.

Once the star has become a white dwarf, the theory becomes less controversial once more. A white dwarf is believed to have reached a state of maximum compression. Since neither source of energy—neither fusion nor contraction—is now available to the star, it must cool down at last. This cooling process is a very slow one, lasting billions upon billions of years. A white dwarf cools very slowly because it is very small, about the size of the earth, and so it has only a tiny surface area from which to radiate its energy. As the white dwarf cools, its temperature and luminosity drop correspondingly, and it slides down and to the right in the H-R diagram. The analogy of this concept with the slide theory is almost exact: A fully compacted star cools and therefore slides downward on the H-R diagram. The major difference is that it now is thought that *white dwarf stars* slide, *not main sequence stars.* Eventually a white dwarf cools off and becomes a cold sphere of dark gas. Such stars are called black dwarfs (see Figure 13.16).

The term "black dwarf" is used here in the sense of a cold former white dwarf. The same term is also used to refer to a sphere of gas with too little mass to begin fusion (Section 13.6). Do not confuse these objects with black holes (Section 17.5).

As explained in Section 13.2, the slide theory was held from about 1913 to 1924, at which time Eddington's theory convinced astronomers that main sequence stars were far from fully compacted. During the

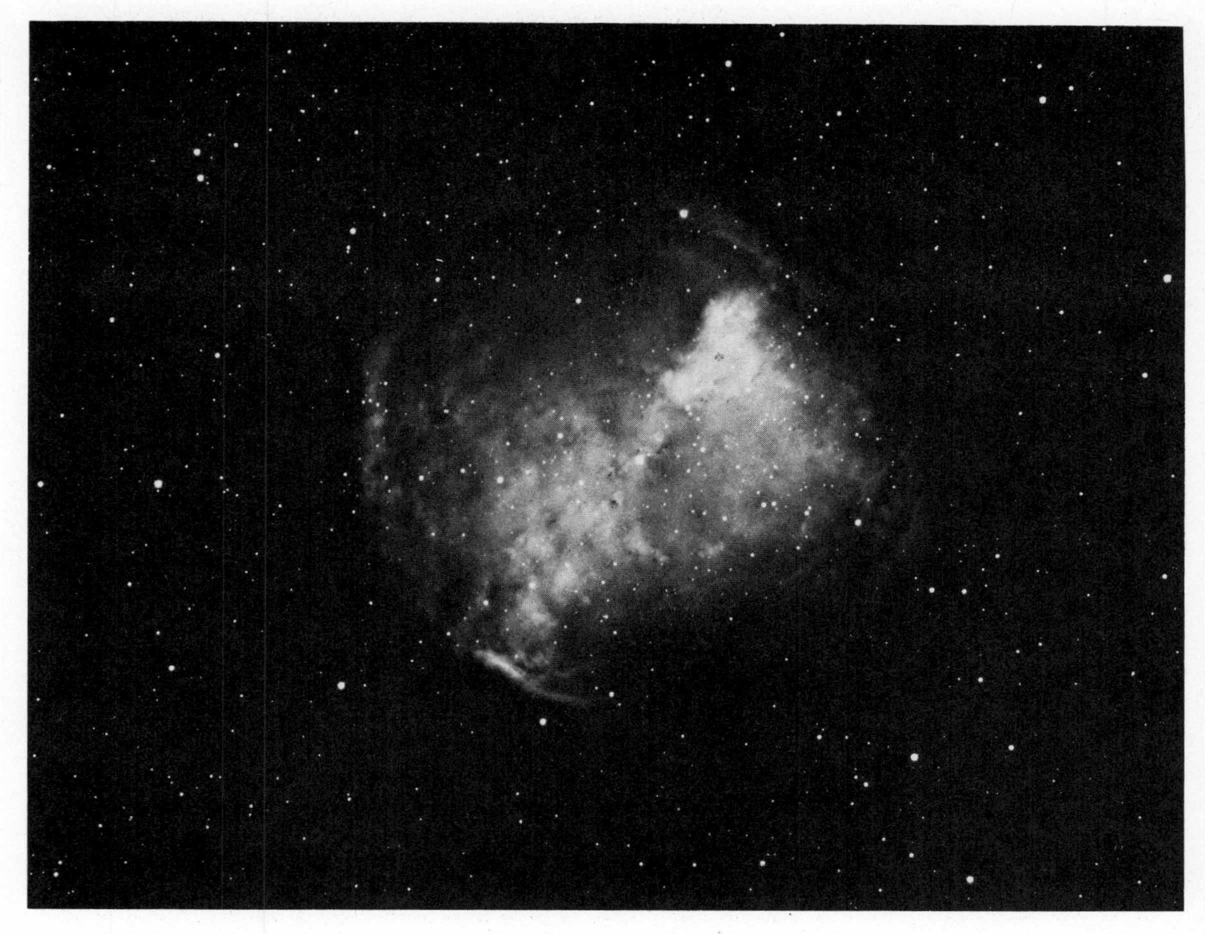

Figure 13.18 The Dumbell nebula, M27. This planetary nebula is faint but visible in larger amateur telescopes. It makes a splendid photographic subject, however. It is roughly 3500 ly distant and about 7 ly across, making it one of the larger known planetary nebulas. Its central star is visible.

1920s, astronomers came to realize that it is actually the white dwarf stars that are fully compacted. The astronomers of the 1920s used the then current ideas about the atom to explain this situation in a neat pictorial manner.

An Older Theory of the Atom Explains Why White Dwarfs Are Fully Compacted The model of the atom used in the 1920s consisted of electrons, negatively charged particles, in orbit around a tiny, dense nucleus with a positive charge. The electrons were pictured as tiny, hard spheres and the nucleus as an even smaller hard sphere. The diameters of the objects involved were thought to be approximately: atom, 1×10^{-10} meter = 1 Å; electron, 10^{-15} meter = 0.00001 Å; nucleus, 1×10^{-16} meters = 0.000001 Å or one-tenth the diameter of the electron. These figures indicate an atom composed primarily of empty

Figure 13.19 The Helix nebula. This planetary nebula is roughly 450 ly away and about 2 ly across. Apparently the central star expelled the shell of gas rather more violently than is typical for a planetary nebula, as witnessed by the more jagged appearance of the surrounding gas.

space. To envision this, imagine a scale model of an atom consisting of a nucleus represented by a 200-meter sphere in Kansas City (200 meters = 660 ft, about the height of a six-story building). The electrons orbiting at the edge of the atom would pass close to New York and Los Angeles.

Now, here is how this model of the atom was applied to the stars in the 1920s. In the previous decade, the decade of the slide theory, stars were thought to be composed of complete atoms. It was computed that the inner layers of a main sequence star consisted of complete atoms in contact with each other, making the star incompressible, more like a liquid than a gas. When Eddington announced that the stars of the main sequence are gaseous even in their central regions and therefore not fully contracted, he explained that the interiors of stars are so hot that the atoms in the star are ionized. Any atom that is missing one or more of its full number of electrons is said to be an **ion** and to be **ionized.** The electrons are knocked off their atoms by violent collisions between the atoms of the hot gas in the star's interior. Thus the stars on the main sequence are said to be composed of **plasma,** a mixture of free electrons and nuclei. Main sequence stars are not composed of large atoms in contact but rather individual electrons and nuclei which are 100,000 and 1 million times smaller, respectively, than atoms. Small wonder, then, that main sequence stars are compressible; there is abundant space between their individual particles.

Now the white dwarf stars are a very different story. According to the physics of the 1920s, they consist of matter that has become so compressed that the electrons themselves are in contact, although the 10 times smaller nuclei are still relatively free. The hard electrons cannot be further compressed and therefore neither can the white dwarf.

Quantum Mechanics and the White Dwarf—Degenerate Matter All through this era, the theory of the behavior of atoms, called quantum mechanics, was being perfected. Based on the idea that electrons and nuclei exhibit the properties of both particles and waves, the model of electrons as hard spheres of a definite size had to be dropped.

Even though the simple picture of the white dwarfs is no longer in use, the newer theory still maintains that the electrons in a white dwarf are sufficiently crowded together to supply a powerful outward pressure which balances the force of gravitation. This type of matter, of which white dwarfs are composed, is known as (electron) **degenerate matter,** matter in which the electrons will not allow further compaction. In a white dwarf, the force due to electron pressure and the force of gravitation are in a final, stable balance. Degenerate matter is incredibly dense. If a fragment of a white dwarf about the size of a matchbox could somehow be brought unaltered to the surface of the earth, it would weigh more than a ton! Of course we could not remove degenerate matter from a white dwarf and have it remain degenerate. Released from the intense pressure inside the star, the sample would return to its ordinary state.

We owe a great deal of our understanding of white dwarfs to the astrophysicist Subrahmanyan Chandrasekhar and his theory of white dwarfs. He showed, for example, that the more massive a white dwarf, the smaller its radius. The idea is that a more massive white dwarf must contract further until the electron pressure rises high enough to balance the larger force of gravitation. In the course of this study, Chandrasekhar also showed that, if the mass of a star were more than about 1.4 suns, the electron pressure would never be sufficient to balance gravitation, no matter how small the star became. This mass limit of 1.4 suns is known as **Chandrasekhar's limit.** No white dwarf having a mass greater than 1.4 suns has ever been found.

Subrahmanyan Chandrasekhar ("Chahn-drah-seek'-arr," born in 1910 in Lahore, India, now Pakistan) specializes in clarifying and consolidating various areas of physics.

The Death of Stars More Massive than the Sun This brings us to our next question, how do stars more massive than the sun die? Chan-

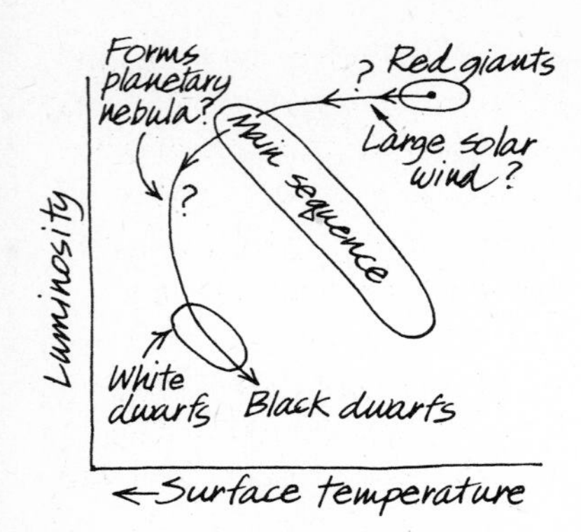

Figure 13.20 Possible track showing the death of a very massive star. A white dwarf cannot have a mass of more than about 1.4 suns. Can a massive star lose enough mass to become a white dwarf?

drasekhar's limit poses a serious problem here; it implies that stars with a mass of more than 1.4 suns cannot become white dwarfs directly. But what other possibility is there? No other major regions of the H-R diagram remain to be explained. This question has not yet been settled to anyone's total satisfaction.

A number of hypotheses for solving this problem will be mentioned in succeeding chapters (see Chapter 14 and Section 17.5), but let us end this section by discussing a proposal due to Chandrasekhar. He suggested that, if a massive dying star is not too much over the 1.4 suns mass limit, it might eject sufficient mass into space to bring itself under the limit. After this crash weight reduction program, the remains of the star could settle down into the white dwarf stage (see Figure 13.20). Among the proposed mechanisms for losing the excess mass is this: The fusion of oxygen in more massive stars may occur explosively, supplying the necessary energy to eject a large amount of mass. Other possible mechanisms exist for mass reduction. Some massive red giants show evidence in their spectra that they are blowing matter into space at a significant rate. This may be a large-scale version of our sun's solar wind. Also, some experts on stellar evolution suggest that planetary nebulas are actually massive stars in the process of shedding enough mass to drop under the Chandrasekhar limit.

SUMMING UP

The principle of the conservation of energy implies that a star must be born, live, and die. The contemporary theory of stellar evolution, which has replaced the slide theory, envisions a star being born by contraction from a gas cloud, becoming a main sequence star when hydrogen fusion begins, and becoming a red giant star after hydrogen fusion ceases in the core.

The fate of a red giant star depends upon its mass. Stars having a mass of roughly 1 sun become white dwarfs which evolve into black dwarfs. In the mass range from about 1.4 to 5 suns, stars may die by shedding enough mass to become white dwarfs. The death of a star having more mass than about 5 suns can be very spectacular; such a star may become a supernova, the subject of the next chapter.

EXERCISES

1. Define the following terms: protostar, core (of a star), envelope (of a star), black dwarf (two meanings, see Sections 13.6 and 13.9), open cluster, fission, planetary nebula, (electron) degenerate matter, Chandrasekhar's limit.

2. In what way does the principle of the conservation of energy imply that stars must be born, live, and die rather than shine eternally?

3. Summarize the life of a star according to the slide theory. Draw an H-R diagram showing the entire track of a star during its life according to this theory.

4. What three advances in astronomical knowledge or theory

led to the slide theory's demise?

Answer the following questions on the basis of the contemporary theory of stellar evolution:

5. Describe the life of a protostar. What controls the rate at which a protostar can contract? What signifies that the star has become a main sequence star?

6. Why don't main sequence stars contract? How do they produce energy? What basic property determines where they lie on the main sequence? Why do O stars have short lives on the main sequence compared to M stars?

7. How old is the sun? How many years remain before it leaves the main sequence? What event will cause it to leave the main sequence?

8. What changes occur in a star as it leaves the main sequence stage and evolves toward the red giant stage? Discuss the changes both in the core and in the envelope.

9. Suppose a grade school child announces that he thinks it will be "neat when the sun leaves the other main sequence stars and drags the earth along with it over to the region of the universe where all of the red giant stars live." Help him to understand his misconception which may have been brought on by loose phrases such as "leaves the main sequence."

10. How do the data from open clusters help verify the theory of stellar evolution when these data are plotted on an H-R diagram?

11. Describe the life of a red giant star (a) if it has a mass of 1 sun and (b) if it is extremely massive.

12. How, according to current ideas, do stars die if they have a mass of about (a) 1 sun, (b) 1.4 to 5 suns, (c) more than 5 suns?

13. Describe a white dwarf star. How does it resist the squeeze of gravity? What will eventually become of it?

14. On an H-R diagram, draw a sketch of the track of the entire life of a star having a mass of about 1 sun.

READINGS

The following two references are valuable for many of the remaining topics in this book.

The Universe, by Isaac Asimov, Walker, New York, 1971.

Black Holes in Space, by Iain Nicolson and Patrick Moore, Norton, New York, 1974, chap. 3.

Both of these books cover stellar evolution on an accessible level. See also the works by Wyatt and Pasachoff and Kutner cited in Chapter 7.

THE SUPERNOVA PUZZLE

It was 1054, during the Northern Sung dynasty in China. Astronomers at the Sung National Observatory at K'aifeng were startled by an unexpected and wonderful event in the sky on July 4 of that year. A brilliant new star, a "guest star" as they called it, appeared in the sky. It was much brighter than Venus, so bright, in fact, that even after the sun rose it was still visible in the sky. Astronomers recorded the position of the star as being just to the southeast of another star, Zeta Tauri, which is in the constellation Taurus, the Bull. They also recorded the guest star's gradual decrease in brightness. For 23 days it could be seen during the day and remained visible in the night sky for about 2 years after it appeared. Then it faded from view. One might have said that the guest star had gone home.

14.1 THE PIECES OF A PUZZLE

By the end of this chapter, we shall be able to summarize the contemporary astronomer's picture of a supernova. But that is the end of the tale, the substance of which is somewhat reminiscent of the task of assembling a jigsaw puzzle. In order to explain to the reader how current ideas came about, we shall lay out the 10 pieces of the puzzle one by one and only gradually assemble them into a whole. In this way you will not know any more than the astronomers did at any given time. This should help convey the excitement the puzzle solvers felt.

The first piece of the puzzle was described in the introduction to this chapter.

★ *Piece 1: The guest star of 1054*

The second piece was mentioned in Section 13.9: It is widely held that stars having a mass of roughly 1 sun die by becoming white dwarfs. Stars having a mass ranging from 1.4 suns to roughly 4 or 5 suns may die by shedding enough mass to get in under the 1.4-sun mass limit set by the theory of Chandrasekhar. Such stars also become white dwarfs. But this leads to what we shall now call

★ *Piece 2: Very massive stars cannot become white dwarfs.*

By "very massive," we mean stars having a mass in the approximate mass range from 5 to 30 suns. Such stars are thought to be too massive to become white dwarfs. No method has been conceived for them to get below Chandrasekhar's limit.

14.2 THE CRAB NEBULA

By 1845, the Third Earl of Rosse (see Box 16.1) had built the world's largest telescope featuring a 72-inch diameter mirror. His instrument, which gathered light so effectively, was especially suited for studying nebulas. One nebula particularly concerns us in this chapter. Located in the constellation Taurus, the Bull, this nebula lies just northwest of the star Zeta Tauri. Earlier, smaller telescopes had revealed only a small, fuzzy patch of light; however, Rosse could see a tangle of stringy filaments in the nebula (see Figure 14.1). To his eye the nebula resembled a crab, and so it became known as the Crab nebula. It figures prominently in our puzzle.

Figure 14.1 Rosse's drawing of the Crab nebula, 1844. Rosse drew on white paper with black ink so the colors are reversed, much as a photographic negative would be. Although the nebula is not easy to find using most amateur telescopes, Rosse's giant mirror made it an "easy object." He reports, "I have shown it to many and all have been at once struck with its remarkable aspect." Rosse believed it to be a cluster of stars. Today's opinion is quite different.

★ *Piece 3: The Crab nebula*

Are the Guest Star of 1054 and the Crab Nebula Related? The alert reader will have already started trying to fit the pieces together. The guest star of 1054 and the Crab nebula are located in the same region of the sky. Could it be that a star exploded there, giving rise to the light seen in 1054? Is the Crab nebula the remains of this catastrophe, seen about 900 years later? The first astronomer to make this suggestion did so in 1921. He was Knut Lundmark of the University of Lund, Sweden.

The very alert reader will have noticed a slight discrepancy, however. The guest star's recorded position with respect to the fixed stars was southeast of Zeta Tauri, while the Crab nebula is to the northwest of Zeta Tauri. Did the ancient astronomers err slightly? Are the two objects related? More evidence was needed.

Gathering Data About the Crab Nebula During the decades following Lundmark's suggestion, enough data were accumulated by a number of workers so that a judgment could be made. Among others, the following measurements were reported. One can compare a recent photograph (Figure 14.2) with those of earlier times and observe that the Crab nebula is expanding at a rate of 22 seconds of arc per year and that its present angular diameter is about 6 minutes of arc. The spectroscope reveals that the gas is expanding at a rate of about 1400 km/sec (870 mi/sec).

These three pieces of data are all that is needed to compute the Crab nebula's distance, diameter, and age. Its distance is roughly 6000 ly. This means that we see light emitted 6000 years ago. Furthermore, if the Crab nebula is indeed the wreckage of the explosion of the guest star, the Chinese astronomers saw the explosion roughly 6000 years after it actually occurred in about the year 5000 B.C.

Figure 14.2 The Crab nebula, the most famous supernova remnant. (See also Plate 20.)

The diameter of the Crab nebula as presently seen can be computed from the known data; it is about 6 ly. Knowing this and the rate of expansion of the nebula, we can calculate how long ago the object appeared to be as small as a star. The approximate answer is 900 years ago. This is very strong evidence that Lundmark was right; the guest star of 1054 was an explosion that resulted in what we today call the Crab nebula. Most people assume that the ancient astronomers accidentally turned things around when writing down the guest star's location and wrote "southeast" when they meant "northwest."

14.3 THE ANDROMEDA "NOVA" OF 1885, S ANDROMEDAE

Our attention shifts to the year 1885 and another nebula. This fuzzy patch of light is much brighter than the Crab nebula and visible to the naked eye on clear, dark nights. It has long been called the Andromeda nebula, after the constellation in which it appears.

In 1885, a new star was observed in the Andromeda nebula, but it did not cause much excitement at the time. Even at its brightest, it would have required a very sharp eye to see it under the very best of conditions. It was named S Andromedae. In less than a year it had faded until it was too dim to be seen in even the largest telescopes. Everyone assumed that it was an ordinary nova (see Section 12.8). Although no one knew it at the time, it is our

★ *Piece 4: S Andromedae*

It was not until the mid-1920s that the incredible nature of this event was recognized. By then, the Andromeda nebula had been renamed the Andromeda galaxy. It had been shown to be an enormous disk of stars lying at a very great distance. (More details are given in Section 16.6.) We now know that this galaxy is about 2 million ly from us. With this knowledge, it became clear that S Andromedae was very different from an ordinary nova. After all, it was a single star, yet nearly visible to the naked eye from a distance of 2 million ly. Since it was in the Andromeda galaxy, its distance R was now known. Using the earlier measurement of b, its peak apparent brightness, we could use the formula $L = bR^2$ (Section 12.2) to calculate its peak luminosity (maximum actual brightness). The answer was astounding: The exploding star in the Andromeda galaxy had a peak luminosity of $L = 10$ billion suns. It would take 10 billion stars like our sun to emit as much light. At its peak, this single star emitted as much light in one second as our sun does in about 300 years. A luminous nova might approach a luminosity of 100,000 suns; the star in Andromeda clearly was not a nova. The astronomers Fritz Zwicky and Walter Baade (see Section 16.7) coined the term "supernova" for such an object.

Fritz Zwicky ("Tsvee'-kee," Swiss-American astronomer born in 1898) is a specialist in supernovas. His name is mentioned several times in this chapter.

14.4 THE SEARCH FOR OTHER SUPERNOVAS

Supernovas in Other Galaxies Astronomers were eager to obtain more data on supernovas. Had there ever been any others or was S Andromedae unique? By checking back through earlier photographs taken of the many galaxies in space, other, previously unrecognized supernovas were found. By 1939, 21 supernovas had been located in the old photographs.

Eager for fresh data, Zwicky began a pioneering effort to catch supernovas in the act. Starting in 1934, he repeatedly photographed the same galaxies, attempting to determine if any new stars had appeared. After five years he had discovered 12 supernovas.

Inspired by Zwicky's efforts, a number of observatories now contribute to a supernova patrol of the galaxies. Some of these observations have automated telescopes run by a computer. Hundreds of galaxies can be checked in one night by this method. The effort has

Figure 14.3 A supernova (arrow) seen in a spiral galaxy in 1961. The galaxy (galaxies are discussed in Chapter 16) and its supernova are about 60 million ly from us. The other bright dots in the photograph are stars in our own galaxy and thus are *much* closer to us than the galaxy and its supernova.

paid off (see Figure 14.3). Over 400 supernovas have been recorded in galaxies. On the average, one is discovered every few months by using modern techniques.

Among other things, these efforts have shown that, as a rough average, a supernova occurs in a typical galaxy once every 50 years. The luminosities of supernovas typically range from 30 million to several billion suns, but S Andromedae still holds the record.

Supernovas in Our Own Galaxy As we shall discuss in Chapter 15, astronomers believe that our sun is a star in a galaxy. We call our galaxy the Milky Way galaxy. Have any supernovas been seen in it?

The answer is yes. A careful search of records from many countries has turned up seven historical supernovas in our own galaxy, which were seen by the naked eye. The criterion used to distinguish a supernova from the much more frequent novas was that the new star must have been visible for six months or more. The most famous of the seven is the guest star of 1054. The earliest of the seven was seen in 185.

We have already mentioned in Chapter 5 one other "new star" now thought to have been a supernova. It was seen in 1572. Careful records of it were kept by Tycho. We recall that it helped steer him back to astronomy from his studies in chemistry. Another was seen in 1604 and was carefully observed by Kepler, who had also witnessed Tycho's supernova. Galileo first turned his telescope to the sky in 1609 (see Section 6.4), five years after Kepler's supernova. Strangely, no more supernovas have been observed in our galaxy since then. If it is true that a typical galaxy experiences a supernova once every 50 years, we seem to be long overdue. (It may be that some supernovas have exploded in distant regions of our galaxy and have been obscured by the matter separating them from us.)

A very close supernova, exploding within, say, 30 ly of the earth, would be a sight to see. It would be prominent in the daytime and would light up the landscape at night. It would have an apparent brightness one million times brighter than the brightest star, Sirius. We would receive a blast of ultraviolet radiation and x-rays that might damage our ozone layer and endanger life on earth. Many thousands of years later the expanding envelope of gases would reach the earth. The effect of this encounter is difficult to predict but, no doubt, it would not be pleasant. We can hope for a supernova at a safe distance, say, 1000 ly. It would still be very spectacular. Some estimate statistically that, during the life of the earth, six supernovas may have occurred within 30 ly of it. Some of these may have caused mutations in the earth's living beings, mutations that permanently altered the course of life on this planet.

Supernova Remnants The Crab nebula is now thought to be a **supernova remnant,** an expanding cloud of gas left over from a supernova explosion. Are there other supernova remnants in our galaxy visible from the earth?

Tycho and Kepler pinned down the locations of their respective supernovas with such precision that the remnants have been found. In 1943, Walter Baade, seeking Kepler's supernova remnant, located a small nebula in the proper place. It looked like a young supernova remnant. Both Tycho's and Kepler's supernova remnants have been mapped using the radio waves they emit. The radio telescope shows that each one consists of a shell of gas.

About 100 supernova remnants have been located using radio telescopes, and about 25 of them have been photographed. Some of these objects make spectacular subjects for the astronomical photographer (see Figures 14.4 through 14.7).

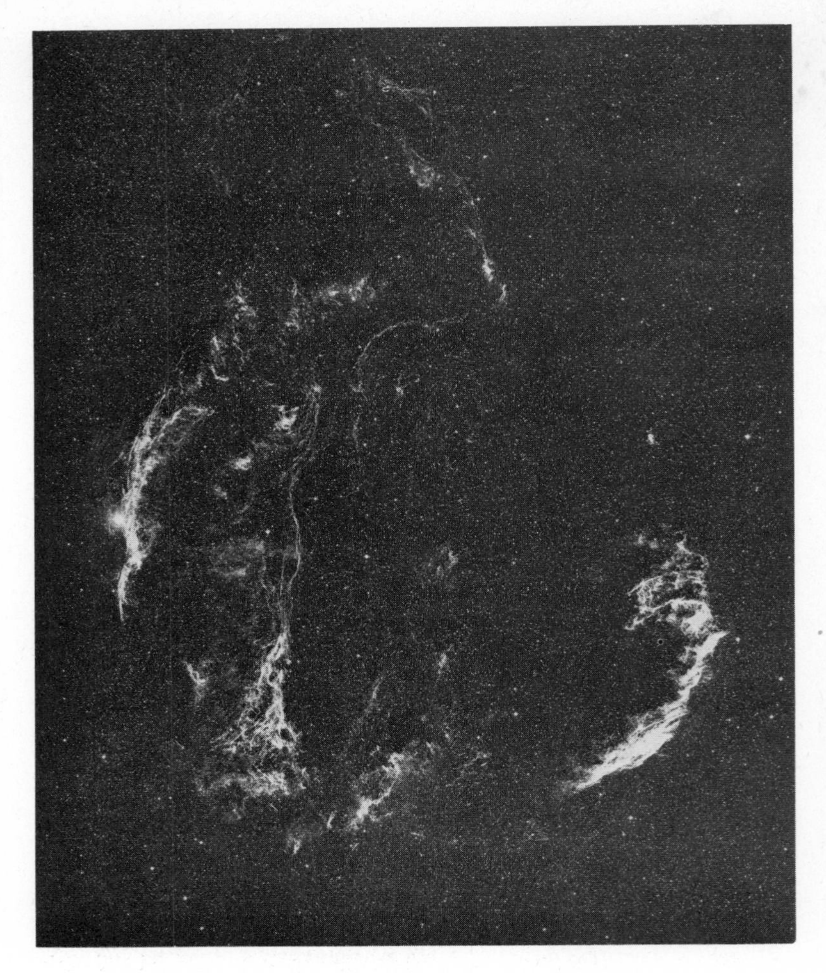

Figure 14.4 The Cygnus loop nebula: a supernova remnant. Also called the Loop nebula and the Veil nebula, this is one of the oldest supernova remnants known. Its estimated age is 50,000 years. Only 2500 ly from us, its delicate structure shows clearly. It has a diameter of about 70 ly which is about 16 times the distance between the sun and Alpha Centauri. Its *angular* diameter is 3°, equivalent to six full moons.

14.5 THE SPECTRUM OF THE CRAB NEBULA

As discussed in Section 9.3, the sun and other stars emit radiation from their surfaces by the process known as thermal radiation. The sun emits radiation most strongly in the visible part of the spectrum, but it also emits radiation in most other regions of the electromagnetic spectrum. A graph representing the intensity spectrum of the sun is shown in Figure 14.8 (see also Figure 9.2). Most stars reveal similar patterns, differing only in the height of the graph (luminosity) and location of the peak (surface temperature).

When the age of radio astronomy was well underway, the Crab nebula began to figure even more prominently in astronomy. The sun emits radio waves only weakly and is detectable only because it is so close. Most other stars are undetectable in a radio telescope. But the

Figure 14.5 The mermaid in the Cygnus loop. This is a close-up of a portion of the left edge of the Cygnus loop as seen in Figure 14.4. The light from the nebula is produced as the expanding gas from the explosion meets gas atoms in space. A shock effect occurs, which heats the gas so that its atoms lose electrons. As the electrons rejoin the atoms after the shock wave has passed, radiation is emitted. Some people can imagine a mermaid in this photograph. She has flowing hair and arms raised over her head.

Crab nebula is the fifth brightest object in the sky when observed by means of microwaves (short-wavelength radio waves). Now, this leads to a problem. If the Crab nebula emits thermal radiation, one would expect that, since it is so bright at microwave wavelengths, it would be even brighter at visible wavelengths. But the Crab nebula is actually difficult to locate in many amateur telescopes.

The puzzle deepens when one learns that rather than emitting *less* radiation at longer radio wavelengths, as one would expect a star to do, the Crab nebula emits *more* (see Figure 14.8). The peak in the spectrum occurs at long-wavelength radio waves.

This strange behavior of the Crab nebula constitutes the fifth piece of our puzzle.

★ *Piece 5: The radiation from the Crab nebula is not thermal radiation.*

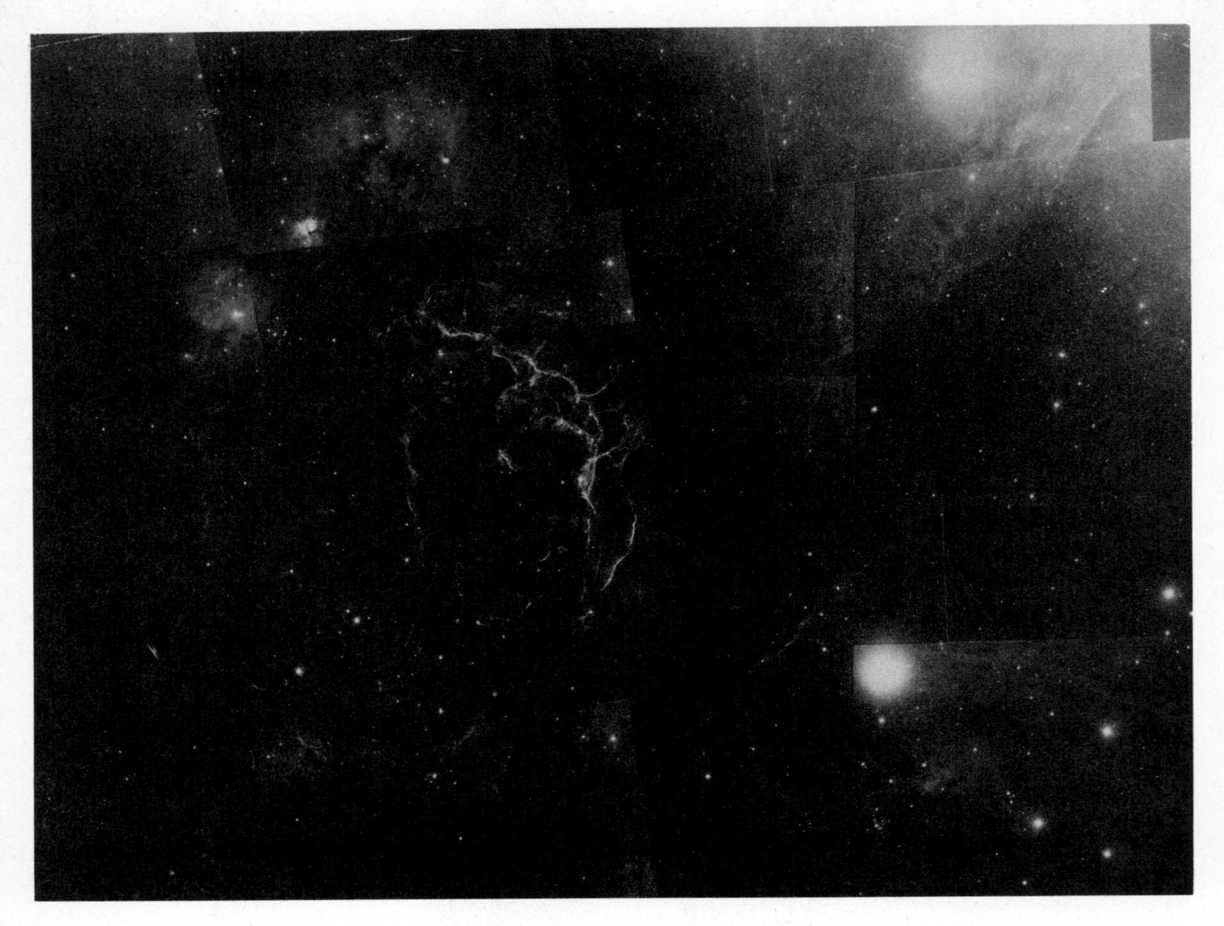

Figure 14.6 A portion of the Gum nebula. It is difficult to be certain of the boundaries of the Gum nebula, but it may have an angular diameter as large as 45° (and perhaps larger in some portions). Thus, if it were visible to the naked eye, we might see one edge touching the horizon and the other edge reaching halfway to the zenith. The Gum nebula is the largest nebula known in our galaxy. It has a diameter estimated to be about 2300 ly, implying a radius of 1150 ly. The estimated distance to the center of the nebula is about 1500 ly, which indicates that its nearest edge is not very far away. See Figure 14.7. The wreathlike nebulosity to the left of center was long thought to be a supernova remnant. This was confirmed when a pulsar (see section 14.14), the Vela pulsar, was found in this region.

14.6 COSMIC RAYS: NUCLEI FROM SPACE

As early as 1912, physicists began to suspect that light is not the only kind of radiation we receive from space. When "ray detectors" were flown aloft in balloons, it was found that the weak effect noticed at sea level became stronger the higher the detector was raised. It was concluded that the rays come from outer space and that the earth's atmosphere acts as a shield, absorbing most of them before they reach its surface.

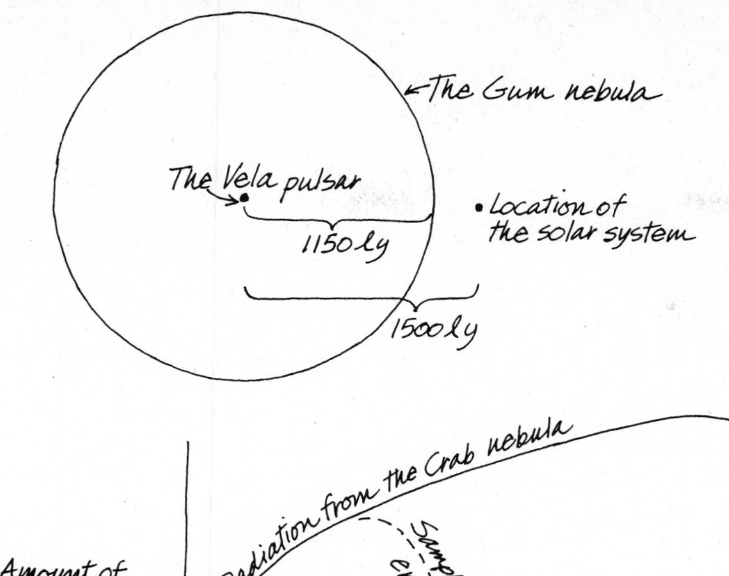

Figure 14.7 A scale drawing of the Gum nebula and its distance from the solar system.

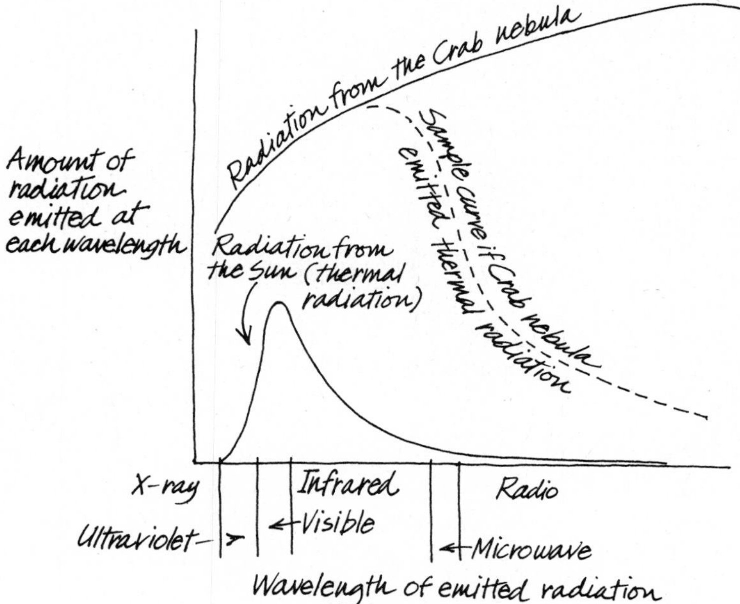

Figure 14.8 The radiation from the sun and the Crab nebula. The sun, which emits radiation by the process known as thermal radiation, is compared to the Crab nebula, which does not. Both curves are intended only to be schematic. The actual curves are more complex than shown here but preserve the general features we wish to indicate.

Since those early days, a great deal of research has been performed on **cosmic rays,** as they are now called. It has revealed that they consist chiefly of isolated protons. It has also been found that some cosmic rays are helium nuclei and that most of the other elements are also represented, although in relatively small numbers. Among the heavier nuclei found among cosmic rays, a significant number are iron nuclei, a fact worth remembering.

Recall that the proton constitutes the nucleus of the hydrogen atom.

It has also been found that these nuclei are totally ionized; that is, they have lost their normal clouds of electrons.

A further surprising fact about cosmic rays is their speed. It is not unusual to detect cosmic rays moving faster than 99.99 percent of the speed of light.

Although it was little suspected in the early days of cosmic research, cosmic rays are also a piece of the puzzle.

★ *Piece 6: Cosmic rays*

What process could be propelling these atoms with such force that they lose their electrons and rush through the galaxy at nearly the speed of light?

For the time being we shall leave this and several other pieces of the puzzle unassembled and turn to yet another piece.

14.7 THE ORIGIN OF THE ELEMENTS, PART ONE

According to the atomic theory, all ordinary matter is composed of atoms and there are only 92 essentially different atoms found in nature. As the theory evolved in the latter eighteenth century, chemists came to recognize that there are certain substances, each of which is composed entirely of only 1 of the 92 kinds of atoms. Such a substance is called an **element.** An element is characterized by the fact that no chemical reaction can break it down into simpler substances. Thus, since water can be broken down into hydrogen and oxygen, it is not an element. However, hydrogen, oxygen, gold, lead, silver, iron, sulfur, and calcium are all familiar examples of elements.

Further study of atoms showed that one can distinguish the atoms of one element from those of another by means of the number of protons in each atom's nucleus. One of the most basic, profound questions asked by scientists is, How did the elements come to be? This is another piece of our puzzle.

★ *Piece 7: The elements exist.*

The Ylem Theory of the Origin of the Elements What is the explanation for the origin of the elements? We next describe an important early attempt to answer this question.

This theory begins by assuming that, billions of years ago, all the matter in the universe was jammed together in an object sometimes known as the "cosmic egg." The egg exploded, and its contents expanded outward. This is the basis of the "big bang theory." The task of the scientist interested in such things is to fill in the details occurring between the original explosion and the universe as we see it today. In particular, one of the tasks is to show how the elements as we see them today came about. Taken as a whole, it appears that about 75 percent of the universe is composed of hydrogen, another 24 percent is helium, and the remaining, heavier elements constitute 1 percent.

As we shall see in Chapter 17, there are good reasons for believing in a cosmic egg and its subsequent explosion.

One of the most interesting early theories was proposed by George Gamow and his co-workers in 1946. He proposed that, just a few minutes after the explosion of the cosmic egg began, the universe consisted of what he called ylem ("ī-lem," rhymes with "buy them"), a very hot "soup" consisting of such ingredients as electromagnetic radiation, electrons, protons, and neutrons. Gamow worked out detailed nuclear reactions to show how, during the first hour of the explosion, the particles of the ylem combined in various ways to produce the 92 known elements.

George Gamow ("Gam'-off," Russian-American physicist, (1904–1968) made important contributions in a number of fields. He was an important pioneer in much of the early work in nuclear physics. He was the first to show that, when a star runs out of hydrogen fuel, its core will become hotter rather than cooler. He was one of the most prominent supporters of the big bang theory and a pioneer in suggesting that nucleic acids act as a genetic code in living beings. Gamow also wrote well for the layperson, and his many witty and informative books have been widely enjoyed.

This theory, that the elements were produced in only 1 hr many billions of years ago, was worked out in great detail by Gamow and his associates. It was almost completely successful in explaining the known amounts of the elements, but unfortunately, there was a problem. It showed convincingly how protons (hydrogen nuclei) had combined with other particles of the ylem to produce helium, element number 2. Once element number 4, beryllium, was produced, the theory showed how the rest of the heavier elements could have been formed. But neither Gamow nor other nuclear physicists could find a suitable nuclear reaction that could have taken place in the ylem to keep the process of element building going beyond helium. Hope was held out for awhile that some discovery would supply the missing link, but none has been found.

The ylem theory was not a complete success, but the work was not wasted. It is still believed that the original helium in the universe was produced from hydrogen during the explosion of the cosmic egg, as Gamow had proposed. But how did the heavier elements arise?

14.8 SYNCHROTRON RADIATION

We next go from our consideration of the possible beginnings of the universe to Schenectady, New York, in the year 1945. An event occurred there which is part of our tale.

Scientists of the General Electric Company had built a *synchrotron* ("sin'-crow-tron"). The purpose of the device was to cause electrons to travel in a circle at high speed. The technique used is based on the physical principle that one can deflect a moving charged particle from a straight path by means of a magnetic field.

A magnet placed near the screen of a *black-and-white* television set will deflect the electrons that produce the picture and have a noticeable and sometimes comical effect on the picture. *Do not* try this on a color set. It is too sensitive to magnetic fields.

The machine consisted essentially of a large, hollow, metal doughnut-shaped tube. The electrons traveled inside this evacuated chamber (see Figure 14.9). Surrounding the tube were coils of wire used to produce the necessary magnetic fields to deflect the electrons into a circle and also to cause them to speed up. When it was turned on, the synchrotron worked perfectly.

The observers watched the interior of the chamber through a window. No one expected to see the electrons; they are much too small. However, it was expected that they would emit visible radiation. At slower speeds, a dull red light was seen and, as the electrons were forced to travel faster, the radiation became brighter and shifted to shorter wavelengths. Eventually, a beautiful blue light poured into the laboratory. The light was also viewed through a polarizing filter, much like those found in Polaroid sunglasses, and it was verified that it was polarized: The filter blocked the light or allowed it to pass, depending on how the filter was positioned.

Figure 14.9 A synchrotron.

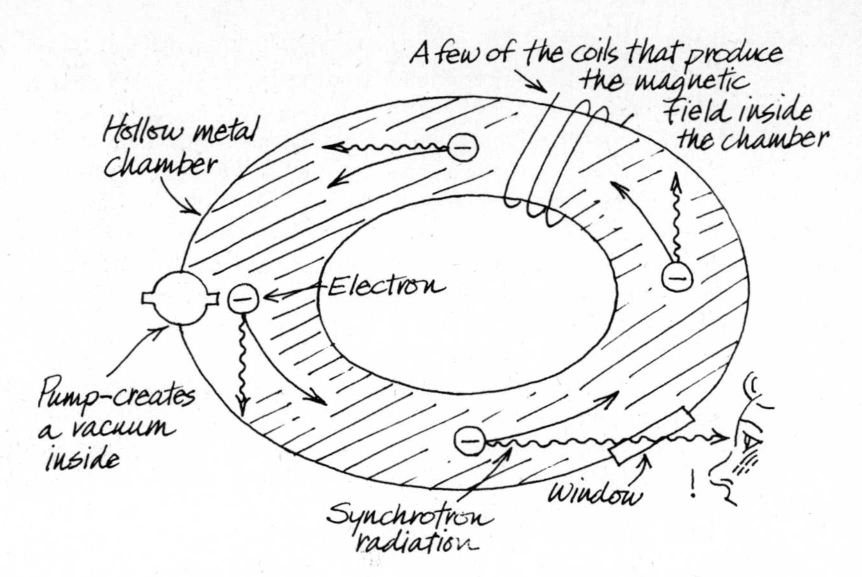

Table 14.1 The First Eight Pieces of the Puzzle

1. The guest star of 1054
2. Very massive stars cannot become white dwarfs
3. The Crab nebula
4. S Andromedae
5. The radiation from the Crab nebula is not thermal radiation
6. Cosmic rays
7. The elements exist
8. Synchrotron radiation

The workers had verified three ideas that are relevant to our puzzle.

1. Electrons spiraling very rapidly in a magnetic field emit electromagnetic radiation. The process is now called **synchrotron radiation.**
2. The faster the electrons travel, the shorter the wavelength of the radiation they emit.
3. Synchrotron radiation is polarized.

The scientists and engineers at General Electric may not have realized it, but they were dealing with

★ *Piece 8: Synchrotron radiation*

We recall that hot objects, like stars and incandescent lights, emit radiation because of their temperature. The rapid vibrations of the electrons in these objects produce radiation by the process called thermal radiation. In synchrotron radiation, the electrons are not vibrating randomly but spiraling through a magnetic field.

For the reader's convenience, the first eight pieces of the puzzle are listed in Table 14.1.

Iosif Samuilovich Shklovsky (Russian astronomer born in 1916) has dealt with a wide-ranging body of topics. He is a leading exponent of the theory of supernovas. His book, *Intelligent Life in the Universe,* written in collaboration with Carl Sagan, is a highly recommended work on a subject about which much of dubious value has been written elsewhere.

14.9 EXPLAINING THE RADIATION FROM THE CRAB NEBULA

We now come to 1953 in our tale, a point at which some of the pieces are fitted into the puzzle. One of the important figures in this work was the Soviet physicist Shklovsky. We can attempt to describe Shklovsky's reasoning in the following (simplified) way.

If the Crab nebula resulted from a star which exploded, two things are likely. First, it is believed that most stars have magnetic fields. As the guest star exploded, its magnetic field would, according to the laws of physics, have expanded with it. Second, since a star consists, in part, of countless electrons, the tremendous explosion that produced the Crab nebula must have produced many rapidly moving electrons. The electrons could not have easily escaped from the resulting nebula but rather would have encountered the twisted magnetic field and looped and spiraled around inside the nebula. Therefore, the conditions for synchrotron radiation are present in the Crab nebula. Aha! Perhaps the strange radiation from the Crab nebula is synchrotron radiation. Based on this insight, astronomers observed the Crab nebula through a polarizing filter and, as Shklovsky had predicted, found that the light was indeed polarized.

Shklovsky and others followed up this insight with involved theoretical calculations of the amount of radiation to be expected from the Crab nebula at various wavelengths. It was predicted that there would be larger numbers of slowly moving electrons than more rapid ones. Since slower electrons emit longer-wavelength radiation, this would explain why the Crab nebula's radiation was most intense at long radio wavelengths. Quantitatively, the resulting calculations matched the observations quite well. Piece 5 of the puzzle had found its place along with numbers 1, 3, 4, and 8.

X-rays from the Crab Nebula—A Problem In 1962 it was discovered that the Crab nebula is a strong source of x-rays, that is, very short electromagnetic radiation. This highlighted a problem that had been bothering astronomers for some time. As we now see it, the Crab nebula is 900 years old. However, electrons lose energy and slow down as they emit synchrotron radiation. The initial explosion pumped a great deal of energy into the electrons, but how can so many rapidly moving electrons be left 900 years later? The worst case was for the extremely fast electrons emitting x-rays. According to calculations, such an electron would slow down and stop emitting x-rays in only one year. The x-ray emissions from the Crab nebula should have died out only one year after explosion, but they didn't. Astronomers suspected that there must be a continuing source of energy still in the Crab nebula. But what could it be?

14.10 A POSSIBLE SOURCE OF COSMIC RAYS

In 1933, not long after S Andromedae was reinterpreted to be a supernova, Fritz Zwicky proposed that cosmic rays originate in supernovas. Over the years this idea has been pursued and elaborated by Zwicky, Shklovsky, and many others. Today, the theory is held by a majority of astronomers.

The proposal is attractive at first sight. The cosmic rays are nuclei—mostly hydrogen, some helium, and a scattering of other elements. This could just as well be a description of the chemical composition of a star. The nuclei are totally ionized and travel extremely swiftly, just as if a tremendous amount of energy had propelled them on their way. A supernova explosion is after all a source of enormous amounts of energy. If supernovas result from very massive stars with iron cores, this even explains the fact that the abundance of iron nuclei is emphasized among the heavier nuclei of cosmic rays. Whereas the electrons in a supernova find it difficult to escape because they curve sharply in the magnetic field, protons are nearly 1900 times more massive than electrons; even though the magnetic field causes the protons to follow curved paths, the curves are gentle. In other words, protons and other heavier nuclei, having more inertia than electrons, are able to plow through the magnetic field and escape.

All this stacks up very favorably for the supernova theory. The last and most crucial task is to describe how cosmic rays travel between the supernova and the earth. Cosmic rays reach the earth from all directions. Furthermore, they arrive at a steady rate, even though supernovas occur intermittently.

The theory states that a supernova explosion sends a burst of cosmic rays out into the galaxy. These particles do not travel in straight lines but rather are gently curved this way and that as they interact with the tangled magnetic field existing between the stars. Thus, they follow huge loops and spirals inside the galaxy. The results are twofold. The cosmic rays are contained within the galaxy, only slowly leaking out through the boundary. Also, the original directions of travel of cosmic rays are gradually scrambled until cosmic rays are found throughout the galaxy traveling in all directions. Analogously, if a group of flies is released from a box at one point in a room, they will buzz around in random loops and spirals and soon be found throughout the room traveling in random directions. The fly that lands on you may not approach you from the direction of the box at all. Flies will approach you from all sides. (Shudder.)

The researchers next went beyond qualitative words and produced a quantitative theory to show the balance between production and depletion of the number of cosmic rays. The calculations are difficult to perform and depend on a number of assumptions that are not easy to verify. Yet the numerical results work out well enough so that large numbers of astronomers today endorse the idea that cosmic rays originate in supernova explosions. Piece number 6 has been fitted into place with numbers 1, 3 through 5, and 8.

14.11 THE ORIGIN OF THE ELEMENTS, PART TWO

All that is Earth has once been sky
Down from the sun of old she came. — *C.S. Lewis*

In the mid-1940s, Gamow and other workers were researching the details of the big bang theory. As mentioned in Section 14.7, it appeared that the theory might be able to explain the origin of the elements, if only a seemingly small but important problem, getting beyond helium, was solved.

At about the same time, Fred Hoyle (see Box 17.1), a very well-known British astronomer, decided that he did not accept the big bang theory. He became the advocate of another scheme, called the "steady state theory," which we discuss in Section 17.3. This theory did not involve an exploding cosmic egg in which to produce the elements. Hoyle needed an alternative factory that could produce oxygen, carbon, iron, and the other heavy elements. He decided that the stars are the required factories.

Working with E. Margaret Burbidge, her husband, Geoffrey Burbidge, and William A. Fowler, Hoyle began what was to be a long, complicated research project. The resulting theory of element building—often called **nucleosynthesis**—begins with a universe consisting of hydrogen and some helium spread throughout space. The big bang theory, which eventually appropriated Hoyle's nucleosynthesis scheme, can supply the hydrogen and helium by means of explosion of the primeval egg. Hoyle's steady state theory supplies it in a different way (see Section 17.3).

The hydrogen and helium condense into stars which, once their cores get hot enough, produce helium from the hydrogen by means of fusion. In the next phase of its life, the star produces carbon by means of helium fusion. If the star is very massive, it will later produce oxygen by causing carbon fusion, and then neon and later magnesium, and so on until the core is composed of iron. In this way, the elements up to iron make their first appearance in the universe in the cores of stars. Notice that, according to this theory, the cores of stars are sufficiently hot and dense to bridge the gap between helium and the heavier elements.

So far, so good. But there seem to be two big holes in this theory. First, how do these newly produced elements wind up as part of planets and people if they are formed in the interior of a star? Second, how are the elements heavier than iron formed? Fusion in the core cannot proceed further than the element iron.

Research showed that, when a star is nearing the end of its life, a process they called the **s process** produces some elements heavier than iron. Atoms of the star slowly (thus the "s") capture neutrons. But this was only a partial solution; not all known elements are produced by the s process and, besides, the atoms are still inside the star.

To complete the solution, the researchers brought supernovas into the picture. What if a massive star with an iron core were to become a supernova? Such a star would contain iron and the remains of the other elements it has produced during its life. These elements would be spewed out into space during the explosion, later to be incorporated into other objects such as stars and planets.

Furthermore, the intense heat, pressure, and violence of the explosion create an environment in which more element building can take place. In the chaos of the explosion, the rest of the elements found in nature can be produced by the rapid capture of neutrons, a process the researchers called the **r process** ("r" for "rapid"). This theory has been elaborated upon by later workers and shown to be capable of explaining not only the existence of all known elements but also the *relative abundance* of most elements. Hoyle and his co-workers initiated one of the most impressive theories of the twentieth century.

This theory leads to an amazing realization. The theory of nucleosynthesis implies that the solar system was formed out of matter "polluted" by earlier supernova explosions. A reason to believe that the solar system was formed out of material mixed with the debris from earlier supernovas is as plain as the nose on your face. This conclusion is confirmed by the very existence of such objects as the moon, the earth, copper wire, gold rings and, indeed, the nose on your face. Had the solar system formed out of pure, primeval hydrogen and helium, the sun might have had jovian planets in orbit around it (they are mostly hydrogen and helium) but no earth or other terrestrial planets and satellites. The earth and the other planets are composed of complex atoms which must have been formed deep in the cores of earlier stars (if the theory holds true). It is sobering to consider the history of the atoms heavier than helium in, say, one's finger. Now these atoms are quietly turning the pages of this book, but in the past they were built up and modified in the crushing pressure and infernal heat of a star's core. And they were blasted back into space again when their star became a supernova. In fact, some of these atoms may have been reprocessed many times in successive stars.

This theory provides another example of our growing awareness that "everything is connected to everything else." We owe the very atoms in our bodies to stars that destroyed themselves billions of years ago. Pieces 1 and 3 through 8 have been assembled into a fascinating arrangement.

14.12 THE INVENTION OF THE NEUTRON STAR

The existence of neutrons was discovered in 1932. When a neutron, a particle having about the same mass as a proton but no charge, exists inside the nucleus of an atom, it can last indefinitely. However, when isolated from other particles, it lasts only about 10 min. It then decays by changing into a proton and an electron.

In 1933, only one year after the discovery of the neutron, Zwicky, already mentioned several times in this chapter, invented a new kind of star based on this discovery. He suggested that it might be possible that stars exist that are composed almost entirely of neutrons. Such a star would be called a neutron star and would consist of a new kind of matter, neutron degenerate matter.

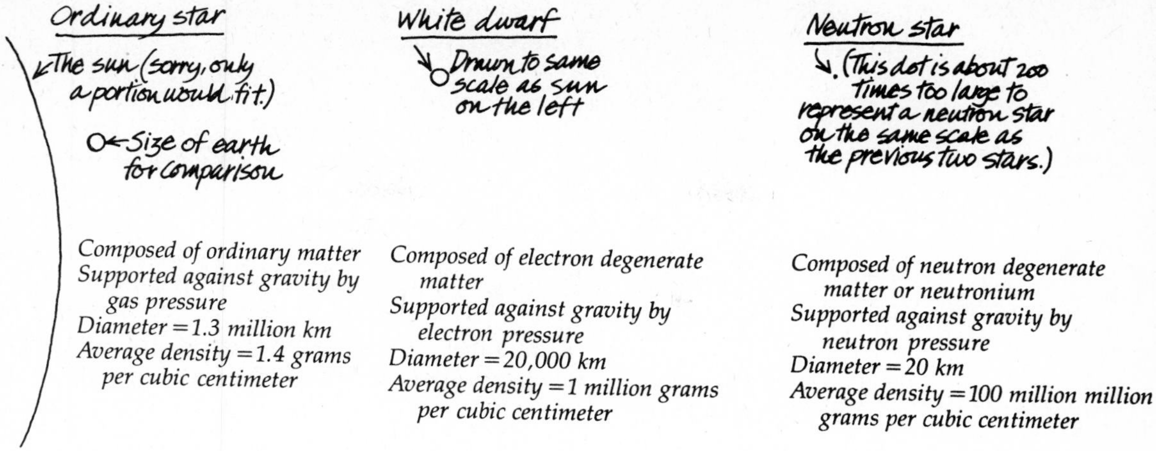

Ordinary star

The sun (sorry, only a portion would fit.)

Size of earth for comparison

Composed of ordinary matter
Supported against gravity by gas pressure
Diameter = 1.3 million km
Average density = 1.4 grams per cubic centimeter

White dwarf

Drawn to same scale as sun on the left

Composed of electron degenerate matter
Supported against gravity by electron pressure
Diameter = 20,000 km
Average density = 1 million grams per cubic centimeter

Neutron star

(This dot is about 200 times too large to represent a neutron star on the same scale as the previous two stars.)

Composed of neutron degenerate matter or neutronium
Supported against gravity by neutron pressure
Diameter = 20 km
Average density = 100 million million grams per cubic centimeter

Figure 14.10 Three types of stars. Keep in mind that all three of these stars have approximately the same mass.

By 1939, the theory of neutron stars had been worked out by Oppenheimer using the laws of nuclear physics and Einstein's general theory of relativity. The basic idea is that, if a star could somehow be subjected to a sufficient squeezing force, the electron pressure in a white dwarf (see Section 13.9) could be overcome. As a result the electrons and protons would combine, leaving a neutron as the product. This is the exact reverse of the decay scheme of an isolated neutron.

J. Robert Oppenheimer (American physicist, 1904–1967) is known for his important work in nuclear physics and even better known as the former head of the Los Alamos Laboratories where the theory, design, and testing of the first atomic bomb were successfully completed.

Now, if such a force were available, the resulting neutron star would have some strange properties. It would collapse dramatically because the electron pressure would be relieved, the electrons having been consumed. The **neutron degenerate matter** would be compressed until the neutrons begin to supply enough pressure to resist gravity. According to calculations, this pressure would be large enough to resist gravity only when the star had reached the incredibly small diameter of about 20 km (12 mi), small enough to fit inside a city. (Recall that a white dwarf is about the size of the earth.) See Figure 14.10.

A neutron star, composed of neutron degenerate matter (to distinguish it from the *electron* degenerate matter in a white dwarf), would be incredibly dense. A matchbox full of such matter would weigh roughly 100 million tons!

This was all fine physical speculation. But do neutron stars exist? What could possibly cause the enormous force necessary to produce one? Zwicky suggested that it might occur in a supernova explosion—that a supernova might produce a neutron star. But how could we check if this is correct? Observers looked for neutron stars at the time and found none. Yet hindsight tells us that they were seeking

★ *Piece 9: Neutron stars*

14.13 WHY DO VERY MASSIVE STARS BECOME SUPERNOVAS?

Piece number 2 of our puzzle, "Very massive stars cannot become white dwarfs," has been fitted into the puzzle by today's astronomers with a fair degree of confidence. A star having a mass in the range from roughly 5 to 30 suns cannot lose enough mass to die slowly by becoming a white dwarf. Rather, it is subject to the sudden and dramatic end known as a supernova. In a sudden burst of overwhelming violence, the massive star destroys itself, spewing most of its former contents into space.

A more difficult question to answer is, What occurs in a massive star to *produce* the supernova explosion? As in the case of the problems involved in trying to explain the earth's recurring ice ages or the disappearance of the dinosaurs, there are too many theories available. Or, perhaps it would be more accurate to say that there have been many proposals, none of which is sufficiently compelling to persuade a large majority of experts. For example, it might be that when a massive star begins the fusion of carbon, the reaction sometimes becomes uncontrolled, causing the supernova explosion.

Among the many proposals for the mechanism of a supernova explosion, one of the most important is:

The massive star reaches the ultimate stage of having an iron core before any major explosion occurs. By this time its core is extremely hot. Huge numbers of gamma rays are produced. Some of them are absorbed by the iron nuclei which break down into lighter nuclei in the process. This results in a *loss* of energy from the core. Thus the core contracts, gets hotter, and produces more gamma rays which break down more iron, which causes an accelerated loss of energy and therefore an accelerated collapse. The whole process of the collapse of the core may be over in *seconds.* The resulting release of energy blasts the outer layers of the star, blowing them away and producing all the other effects we have been discussing in this chapter.

This proposal has an interesting feature. In the final stages of the process, the core is rapidly collapsing. What will its fate be? There are two possibilities. The collapse may continue until the electrons and protons merge to produce neutrons, and the resulting intense neutron pressure might halt the collapse. Here we have a mechanism for producing a neutron star and indeed doing so in the course of a supernova explosion, as Zwicky had conjectured.

A further possibility is that, in some cases, even the neutron pressure will be insufficient to halt the collapse and the core will go on collapsing forever. We say that the core will become a black hole. We shall take up this remarkable idea again in Chapter 17.

Again, theory has led to the possibility that neutron stars might be found associated with supernova remnants. Yet, the problem of how to recognize a neutron star so as to confirm this possibility remained unsolved until the late 1960s.

14.14 THE DISCOVERY OF PULSARS

As children we all learn to recite, "Twinkle, twinkle, little star. How I wonder what you are." A group of radio astronomers at Cambridge University, England, may have found themselves repeating this rhyme to themselves in 1967 while they investigated one of the most exciting discoveries in astronomy in many years.

"Serendipity" is the faculty for finding something accidentally while searching for something else. The Cambridge group, headed by Anthony Hewish, who in 1974 won a Nobel Prize for his work, had designed and built a radio telescope to search for twinkling galaxies. They found them, but the project will always be remembered for the accidental discovery of stars that twinkle in a most unusual manner. Serendipity.

Ordinary stars **twinkle**—dance irregularly about their average positions, flash brighter and dimmer, and flash various colors—because of random refractions caused by the turbulence of our atmosphere. Seen from above our atmosphere, stars shine steadily. Twinkling is similar to the effect one gets when viewing the outside world from the bottom of a swimming pool.

The Cambridge project was designed to discover very distant galaxies that scintillate (twinkle) when observed through a radio telescope. Radio waves are not affected by the earth's atmosphere but are refracted this way and that by random turbulence in the solar wind (see Section 9.6). Only very distant galaxies, those appearing to be points in the sky, scintillate; nearer ones, which appear to have an appreciable angular size, do not. The hope was that the Cambridge instrument would be able to discover quasars (discussed in Section 16.15) by this technique.

The radio telescope designed to view galaxy scintillation differed in two important ways from the other radio telescopes of the day. Most previous telescopes had been tuned in on radio waves having a wavelength between a centimeter and a meter. Hewish expected the scintillation would be most apparent at a longer wavelength, and so the Cambridge radio telescope was built to handle radiation at a wavelength of 3.7 meters. The other innovation involved the electronic equipment that processed the radio signals. Radio signals from outer space are rather weak. They can easily be swamped by random static from the atmosphere and from random signals generated by the electronic equipment itself. To overcome the problem due to this "noise," the electronics of most telescopes are designed to accumulate the radio signal for a predetermined time period, average the results, and then report out this averaged signal. This process tends to cancel out unwanted random signals, leaving only the steady signal from space. Hewish could not use such an electronic setup because it would average out the very scintillation effect he was looking for. These two changes are where the serendipity came in, as you'll see.

After the instrument was built and in operation, it began to produce data. One of Hewish's graduate students, Jocelyn Bell, had the task of analyzing the data the telescope produced. She stayed alert for any radio signals that changed rapidly, indicating a scintillating galaxy, and many were found. Bell looked for patterns in the data and made the discovery that galaxies near the position of the sun in the sky scintillate most strongly. This, she explained, should be expected because the solar wind is most concentrated and turbulent near the sun.

In the midst of her research, Bell noticed that one of the objects she had been observing scintillated even when it was in a part of the sky far from the sun. She studied this peculiar object more carefully and called Hewish's attention to it. This object was not scintillating, at least not in the irregular way the distant galaxies did. This radio source was emitting short pulses of radio waves in a regular pattern, one pulse every 1.3 sec. This was truly unprecedented—an object sending out radio signals that go "beep," instead of the normal, irregular static of an ordinary radio source.

The research team concentrated on the object to find out everything they could. One priority was to check that the pulses were really coming from outer space. Hewish says that his initial reaction was, "It was simply unbelievable that such signals could be from a genuine astronomical object." It would be very embarrassing to publish such sensational observations only to find out that they had been picking up a rotating radar beacon. The object's apparent motion was timed, and they found that it crossed the meridian once every 23 hr, 56 min, just as a star does (see Section 2.1). This made them more confident that they were observing a celestial body.

The pulses from the object are very short. Each pulse lasts only about 1/10,000 second. The researchers next measured the time from one pulse to the next and found that the object behaves like an amazingly good clock. The pulses are spaced 1.33730113 sec apart. It was determined that the time interval between the pulses repeats so regularly that the object constitutes a clock accurate to one part in a billion.

Several other important discoveries were made. The Cambridge scientists studied the comparative timing of the pulses at differing wavelengths and inferred that the object is about 400 ly away, putting it well inside our galaxy. As astronomical distances go, it is a rather close neighbor. Then they were pleased when they found three more objects pulsing in much the same manner. Such objects are now called **pulsars.**

Radio telescopes had already been in use for many years when pulsars were discovered by Hewish's group. Why had they gone unnoticed for so long? The reason lay in the unique design of the Cambridge telescope. Pulsars radiate more strongly at the longer wavelengths Hewish was observing than at the more often observed shorter wavelengths. Furthermore, the electronic apparatus of Hewish's telescope was designed to detect rapid variations in the radio signal rather than smoothing out any variations as former equipment had done.

Finally, satisfied that their discovery was well worth publishing, they told the world about it. They were unaware that they had uncovered what was to become

★ *Piece 10: Pulsars*

14.15 WHAT ARE PULSARS?

Soon there was a flood of articles in astronomical journals as other radio telescopes found more pulsars and explanations of these strange objects were published.

Some researchers immediately jumped to the conclusion that, at last, evidence for the existence of an extraterrestrial civilization had been found. Were pulsars directional beacons used by a civilization that travels between the stars? But skeptics objected. Pulsars broadcast over a very wide range of wavelengths. This would be a poor design for a directional beacon, because it is so wasteful of energy. The believers responded that the civilization might have found a very cheap and powerful source of energy. Or, perhaps these were actually signals meant to catch the attention of others; thus it would be wise to broadcast over a large range in wavelength. The skeptics could come back with this: No Doppler effect has been found from the pulsar signals as would be expected from a beacon situated on an orbiting planet. A believer might say that perhaps these are stations, even colonies, in deep space which are not orbiting a star. And so on and so on. . . . Such discussions are interesting but rarely reach any conclusions satisfying all the parties concerned.

If one were to take a poll of astronomers asking, (1) Is it likely that extraterrestrial life exists? and, (2) Do you know of any compelling evidence of extraterrestrial intelligent life? the probable response would be (1) yes, and (2) no. This points up an important attitude taken by most trained scientists. It is sometimes known as Occam's razor or the principle of parsimony and can be stated: Always seek the simplest answer. As applied to all research, including pulsars and any other supposed evidence of extraterrestrial beings, the scientist first tries to think of some ordinary, simple explanation of the evidence. In any particular case, only if every less sensational possibility can be excluded, will a scientist come to believe that life has been found. A scientist does not reject the obvious no matter how much he or she wants to believe in the marvelous.

And so the idea of extraterrestrial life was put aside, and a search for a possible natural phenomenon to explain pulsars was undertaken. The first task was to explain the timing of the pulses. Each pulsar's pulses are timed very accurately. Pulse rates have been found in the range from about 30 each second to 1 every few seconds. What sort of mechanism could exhibit such rates? What sort of physical system could produce such accurate timing? Three possibilities come to mind at once: orbiting, expansion and contraction, and spinning.

Could the pulses be timed by the orbiting of one object around another? An object orbiting even a white dwarf star once each second would have to be inside the star. Two neutron stars, if such exist, could orbit each other with a period of 1 sec, but not for long. Einstein's general theory of relativity predicts that such a system would lose energy very rapidly by emitting gravity waves (see Section 17.4). The neutron stars would fall into each other. No, orbital motion was out.

Could the timing mechanism be the expansion and contraction of a star? Such vibrations, of the type Cepheid variables perform (see Section 12.9), are very regular. Could any star vibrate at a rate of about once every second? Ordinary stars pulsate much too slowly. The fastest known pulsating star has a period of a few hours. A white dwarf might possibly pulsate in the same fashion, but its period can be calculated to be about 1 min. Could a neutron star vibrate? If so, its period would be too *fast*, amounting to perhaps 100 pulses per second. Expansion and contraction did not work either.

Rotation is also a phenomenon that is steady and repeats with great regularity. Perhaps the pulses are regulated by a spinning object. If so, the star cannot be an ordinary star. The sun, or even so compact a star as a white dwarf, would be ripped apart if forced to spin at a rate of one revolution per second. But what about a neutron star? Such a star would be so tightly compressed that the force of gravitation at its surface would be enormous. A human on the surface of the star would weigh about 400 billion lb. (The poor human would be squashed into a film one atom thick.) A theoretical calculation by Thomas Gold, which takes this powerful force into account, suggests that a neutron star could spin even as rapidly as 30 times a second and still not be disrupted. This is pleasing because a neutron star would be expected to be rotating rapidly. As it collapses down to a tiny sphere, a neutron star should experience a very large increase in its spin rate, a phenomenon already discussed in Section 11.10.

Thomas Gold (Austrian-American astronomer born in 1920) is best known as one of the originators of the steady state theory, discussed in Section 17.3, and for his theory that pulsars are neutron stars.

Aha! Two more pieces of the puzzle, numbers 9 and 10, fit together. Zwicky's vision has been vindicated. It is now quite generally accepted that the discovery of pulsars was, at the same time, the first evidence for the existence of neutron stars. We have yet to place these two pieces firmly into the supernova puzzle, however.

14.16 WHY DO PULSARS PULSE?

Once it had been decided that pulsars are spinning neutron stars, theorists turned to a much more difficult question. How does a rapidly rotating neutron star produce pulses of radiation? Could the star have a hot spot on its surface emitting intense radiation in a beam (see Figure 14.11)? Does the neutron star have a tilted magnetic field so that the poles of the field do not coincide with the star's axis? This could cause a beam to be emitted along the magnetic field's axis (Figure 14.12). Does the star emit a solar wind which rotates rapidly with it? When the

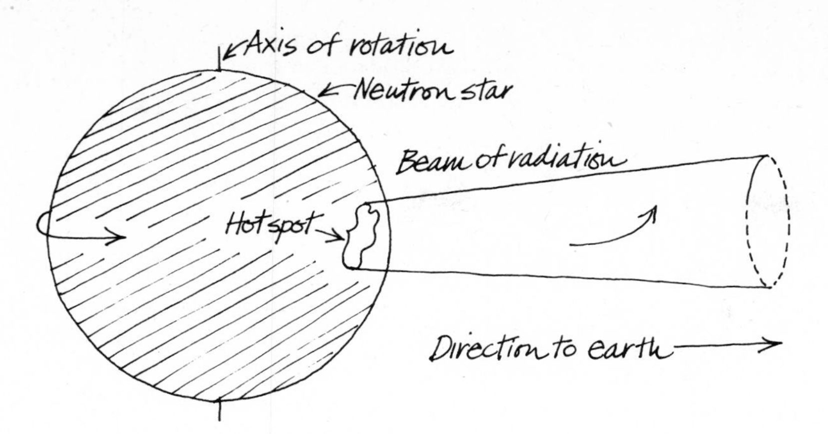

Figure 14.11 The hot spot hypothesis of pulsar radiation.

material in the wind speeds up to nearly the speed of light, the electrons could emit a beam of radiation (see Figure 14.13). All three of these propositions would result in a narrow rotating beam of radiation being sent out like the beam from a lighthouse. Each time the beam sweeps past the earth, we receive a burst of radiation from the pulsar. We see that, if any of these ideas is correct, the earth has to be in the correct direction to receive the radiation. There could be many pulsars that we do not detect because they are aimed the wrong way. In spite of the large amount of thought put into the subject, no theory to explain pulsars has yet won out.

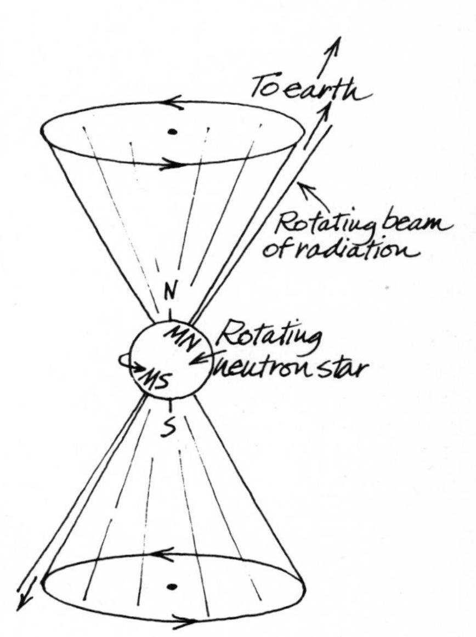

Figure 14.12 The magnetic axis beam hypothesis of pulsar radiation. The magnetic poles (MN and MS) of this neutron star do not coincide with the axis of spin (poles marked N and S). Thus the magnetic axis sweeps out a cone in space. Radiation may escape along the axis. If the earth happens to lie on the surface of the cone, we will observe pulses of radiation.

14.17 THE ASSOCIATION OF PULSARS WITH SUPERNOVAS

Radio telescopes had found pulsars, and now astronomers were eager to view them using an ordinary optical telescope. But there was a problem. Radio telescopes could not pin down a pulsar's location in the sky with respect to the fixed stars with much accuracy. When optical telescopes were pointed in the pulsar's general direction as determined by a radio telescope, they could see many stars. Which, if any, were pulsars?

In 1969, a clever technique was worked out at Lick Observatory, which solved the problem for at least one pulsar. It was known that one of the new pulsars had been detected in the constellation Taurus, in the general direction of the Crab nebula. A photograph of the region showed many stars. To find out which was the pulsar, the designer of the technique had to make several assumptions. The first was that the pulsar near the Crab nebula emits not only radio pulses but also visible pulses instead of glowing steadily as normal stars do. Second, it was assumed that the visible pulses come at the same rate the observed radio pulses do (about 30 pulses per second, which makes this still the fastest pulsar known). With this in mind, an apparatus was constructed for use with the telescope, which in effect was a set of fan

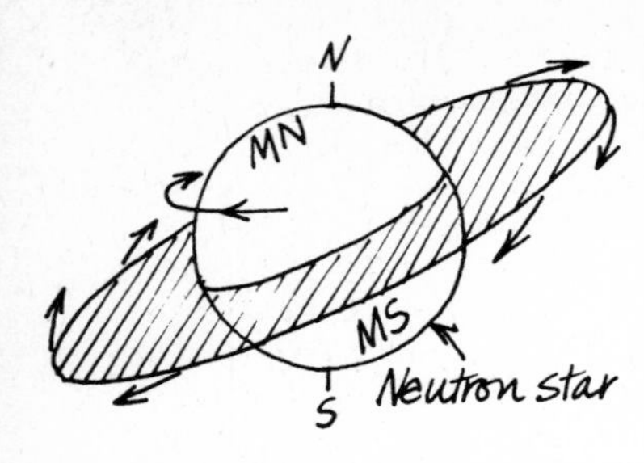

Figure 14.13 The tilted disk hypothesis for a pulsar. Again the star's magnetic field is tilted. Electrons are emitted from the magnetic equator and rotate with the star. The further they go from the star, the faster they must move, until, at very high speeds, they emit radiation in the same plane as the magnetic equator. An observer properly placed will "see" the disk edge-on twice during each rotation of the star and will receive a pulse of radiation each time.

blades that whirled in front of the telescope's eyepiece. The fan was made to turn at such a speed that the blades interrupted the view 30 times per second. It worked. When viewed with the blades in place, and with their speed adjusted properly, one of the stars in the Crab nebula nearly disappeared. Whenever the star was "on," the fan blocked the incoming light. The rest of the stars could be seen with or without the fan apparatus (see Figure 14.14).

One could describe the discovery this way. A variable star had been found with a period of 0.03 sec; that is, it flashes 30 times each second. When viewed with the eye, the star seems steady, since the eye cannot perceive such rapid variations. This was a very significant discovery because of the *location* of the optical pulsar.

This was a very pleasing discovery, to say the least; theory and observation meshed so beautifully. A neutron star might be produced in a supernova explosion. Pulsars are believed to be neutron stars. A pulsar has been found in the Crab nebula. The Crab nebula is believed to be a supernova remnant. Check and double check. The picture puzzle of a supernova is now complete. All 10 pieces fit together in a very persuasive way.

The Source of the Energy of the Crab Nebula: A Problem Solved In Section 14.9 we discussed a problem associated with the Crab nebula: How can it continue to emit synchrotron radiation 900 years after the supernova explosion? It is now believed that the whirling neutron star in the Crab nebula supplies the required energy to drive the electrons in the nebula so that they can continue radiating. The electrons somehow gain energy from the neutron star and lose energy again by radiating it as synchrotron radiation.

This theory can be checked. If the neutron star is losing rotational energy to the electrons, it should be slowing down. The rate of slowdown can be calculated; it should be very gradual but measurable. Indeed, it has been found that all pulsars are slowing down; that is, the time between pulses is gradually lengthening and at the calculated rate. Again, the theory that pulsars are neutron stars is quite persuasive.

Something To Worry About The case we have described for the association of pulsars and supernovas is very strong. Other supernova remnants have been searched with radio telescopes and optical telescopes in the hope of finding pulsars associated with them. The Gum nebula (Figure 14.6), has a pulsar in it known as the Vela pulsar. It was discovered using radio telescopes and was recently found to be an optical pulsar. The Cygnus loop has a source of x-rays in it which may possibly be a pulsar.

However, that is it. Of the over 300 pulsars discovered by radio telescopes and the over 100 supernova remnants found so far, in only two certain cases have a pulsar and a remnant been found together. This is admittedly a bit bothersome. Among the reasons advanced to account for this are:

Figure 14.14 The Crab nebula pulsar—an optical pulsar. The view through the telescope was displayed on a television screen. On the left, the fan blades have been removed and the pulsar is visible. On the right, the blades are rotating at the proper speed so that, when the pulsar is bright, a blade blocks the view, only to allow light through when the pulsar is at a minimum. (The tiny speckles on the screen are not stars but "snow" in the television system.) The pulsar is the dot of light missing from the right-hand photograph.

1. Perhaps the pulsar mechanism is highly directional. Most remnants produce pulsars, but the radio pulses are not aimed at the earth.
2. The terrific explosion of a supernova might expel the newborn pulsar from the supernova remnant. It then would become a runaway star.
3. Pulsars may somehow be formed without a supernova explosion.
4. Pulsars may last much longer than supernova remnants. That is, the remnant eventually disperses, but the pulsar still broadcasts.

No one is sure what the best answer is. Perhaps some combination of these or other proposals will be found satisfactory. No doubt there will be more to learn about the supernova puzzle and more pieces to add in the future. One cannot even exclude the possibility that some of the pieces we have inserted will have to be removed at some future date. date.

SUMMING UP

When very massive stars, those ranging in mass from roughly 5 to 30 suns, reach the end of their lives because they have used up all their fuel, they die the destructive death known as a supernova. The exploding star may temporarily outshine the combined output of all the other stars in its galaxy. The mechanism causing the outburst is not understood, but it may be triggered by a rapid collapse of the star's core, at least in some cases. The guest star of 1054 and S Andromedae are examples of supernova explosions which have been observed. The energy from the explosion blasts the outer layers of the star into space, producing a supernova remnant such as the Crab nebula and also producing such effects as cosmic rays and very rapid electrons which emit synchrotron radiation. In the process of exploding, the elements produced during the star's life are expelled into space and new elements are created as well. Thus supernovas are the source of all the elements heavier than hydrogen and helium found in the universe. If neutron pressure can halt the collapse of the core, a neutron star will result, a star that is enormously dense, tiny, and spinning rapidly. A neutron star can produce a rotating beam of electromagnetic radiation which, when observed on earth, is called a pulsar, a source of very regularly timed pulses of radiation.

In the next chapter we shall see the wonderful discoveries astronomers made when they investigated the sun's place among the other stars.

EXERCISES

1. Define the following terms: guest star, supernova, supernova remnant, cosmic ray, element, ylem, synchrotron radiation, nucleosynthesis.

2. What is the evidence that the guest star of 1054 was the explosion that produced the Crab nebula as we now see it?

3. How often do supernovas occur in a typical galaxy? When was the last one seen in our galaxy? How can this discrepancy be accounted for?

4. What is the evidence that the radiation we receive from the Crab nebula is not thermal radiation? According to current theory, how is the radiation from the Crab nebula produced?

5. How, in the opinion of most astronomers, are cosmic rays produced? What is the significance in this theory of the fact that the element iron is emphasized in cosmic rays?

6. Write a paragraph or two contrasting the theories of Gamow and Hoyle as to the origin of the elements.

7. Describe a neutron star. What supports it against the squeeze of gravity? What is the best evidence found that neutron stars do indeed exist? How could a neutron star be formed?

8. A very massive star cannot die by becoming a white dwarf. By what process does it die? Describe the process of death of a very massive star, which might produce a neutron star.

9. Describe the data received from a pulsar. What are pulsars thought to be? Describe one of the methods by which a pulsar might produce pulses.

10. What is the evidence linking pulsars and supernovas?

11. How can the Crab nebula continue to produce x-rays 900 years after the explosion if the explosion could have produced x-rays for only about 1 year?

READINGS

The Supernova, A Stellar Spectacle, NASA, Washington D.C., 1976, NAS 1.19:126.

"After the Supernova, What?," by J. Wheeler, *American Scientist*, January/February 1973, p. 42. In spite of the title, this article covers supernovas, their mechanisms, and much more in some detail.

See also the books by Asimov and Nicolson and Moore cited in Chapter 13.

Larry Niven, a science fiction writer, uses astronomy discoveries in his writing. For starters, see the title story in the collection.

Neutron Star, by Larry Niven, Ballantine Books, New York, 1968.

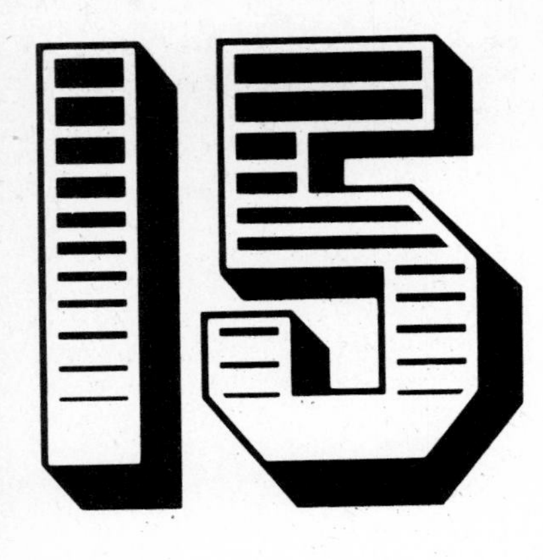

THE MILKY WAY

On clear, dark nights, the hazy band of light known as the Milky Way still glows clearly over Mount Wilson. That mile-high mountain, 13 km (8 mi) from Pasadena, California, was an ideal place for an observatory. The telescope established there in 1907 features a 60-inch mirror and at that time was the world's largest active telescope. In 1914, a young astronomer named Harlow Shapley joined the Mount Wilson staff. In a few years he showed himself to be a gifted hand with a telescope. He wrote, "I realized that I could do things other people could not or would not do." He gradually had more time to pursue his own research and began to use the 60 inch to photograph Cepheid variables, those distance markers in the sky. As Shapley worked at the telescope night after night, he probed the nature of the Milky Way, which from earliest times had been seen gleaming mysteriously in the night sky.

15.1 THE ANCIENT MILKY WAY

If modern urban children were asked about the Milky Way, it would not be surprising if a large percentage said it is a snack. The gift of electric lighting has robbed many of us of one of the sky's most beautiful sights, the irregular band of light reaching completely around the celestial sphere. We call it the Milky Way (see Figure 15.1). Far from the cities on a summer night, it catches the eye easily, stretching up from Sagittarius in the south, through the Summer Triangle, and over to Cassiopeia in the north. On winter evenings another large portion of the Milky Way is visible, extending beyond Cassiopeia, through Perseus and Auriga, passing between Gemini and Orion, and extending to Canis Major in the south.

What Is the Milky Way? Tycho Brahe (see Chapter 5) suggested that the Milky Way is a region of glowing gas. This gas, he suggested, can congeal and detach itself, resulting in a star. Tycho explained his supernova of 1572 as a star that did not survive because it was imperfectly formed.

Tycho and all earlier astronomers could only guess at the nature of the Milky Way because they had few data to work with. The first advance in thousands of years concerning the Milky Way was made in

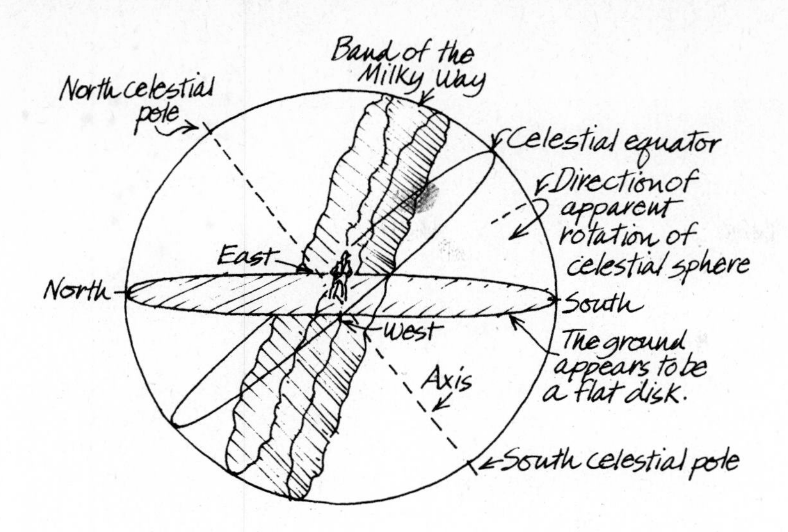

Figure 15.1 The band of the Milky Way on the celestial sphere. This is a figure similar in spirit to those like it in Section 2.1. Here we show how the band of the Milky Way appears to lie on the celestial sphere. As the celestial sphere rotates on its axis from east to west, the Milky Way rotates with it, as if it were attached. The observer here is viewing the Milky Way on a summer evening when it is high in the sky.

about 1610 when Galileo saw it through his new telescope (see Section 6.5). He saw that the Milky Way consists of countless stars.

Not much was made of this new information until Immanuel Kant (1724–1804) was influenced by a newspaper article he read in about 1750. Kant, a German, is best known for his philosophical writings such as the *Critique of Pure Reason,* but his early interest and education were in physics and astronomy. The newspaper article concerned a confusing hypothesis on the construction of the universe, a hypothesis originated by an English astronomer and one which the newspaper managed to confuse further. However, the article brought an image to Kant's mind of the stars in the Milky Way arranged in a flat disk. Kant drew an analogy with the rings of Saturn, which had been seen clearly for the first time in 1655 by Huygens. Kant assumed that the rings of Saturn consist of particles in orbit around Saturn. Analogously, in Kant's view, the stars in the disk we call the Milky Way galaxy orbit around the center of the disk, held in orbit by mutual gravitational attraction. The stars near us appear as individuals. However, as we look along the length of the disk, the distant stars appear to be packed close together. These stars are not actually close to each other, but rather only seem so because of the perspective. When looking at right angles to the plane of the disk, we see relatively few stars in the line of sight and can easily look between them into the depths of empty space (see Figure 15.2).

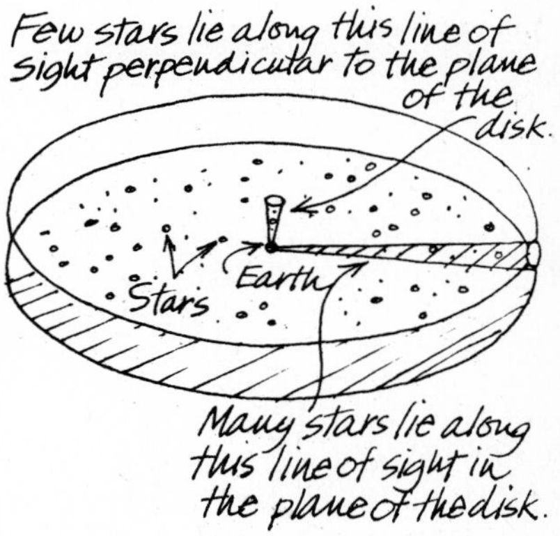

Few stars lie along this line of sight perpendicular to the plane of the disk.

Stars

Earth

Many stars lie along this line of sight in the plane of the disk.

Figure 15.2 Kant's disk of stars, explaining the appearance of the Milky Way.

15.2 MEASURING THE MILKY WAY USING STAR GAUGES

Further advances in the understanding of the Milky Way required that someone make a huge number of observations. The Herschel family (Figure 15.3) supplied the necessary painstaking, hardworking ob-

Figure 15.3 William, Caroline, and John Herschel.

servers (see Box 15.1 and Figures 15.4 and 15.5). William Herschel came to astronomy relatively late in his life; he began his first review of the sky at the age of 35. Yet it has been estimated that he spent more time at the telescope than any other astronomer, before or since. Herschel started his second review of the sky in 1779, a task that was to require the next two years. During that time he concentrated on discovering double stars, for reasons discussed in Section 12.6. It was also during this review that he happened to discover the planet Uranus. Another objective of the review was to perform what Herschel called star gauges.

Counting Stars William Herschel had independently come to the conclusion that the appearance of the Milky Way is caused by the fact that the stars are arranged in space in the shape of a grindstone or disk. It was his objective to explore the boundaries of this disk in order to discover its size and also to discover the location of the sun within it. Herschel convinced himself he could do this if the following two assumptions were correct. He assumed that all stars are of roughly the same luminosity, which would mean that, the dimmer a star looks, the greater its distance from us. He also assumed that his telescope, an 18-inch (45.8-cm) reflector, collected sufficient light to show him the most distant, dimmest-appearing stars at the edge of the disk.

When Herschel looked at any location in the sky, he saw both bright and dim stars. The more distant the edge of the disk lay in that direction, the more stars and dimmer stars he expected to see. Attacking the problem, he chose a square region of the sky 15 by 15 minutes of arc on a side and then counted the number of stars visible in that region. This constituted what he called a **star gauge.** Then he turned the telescope slightly, chose a new square, and counted again. Before long, it became clear that he could never gauge the whole sky even in several lifetimes, so he gauged principally in a complete circle on the celestial

BOX 15.1

THE HERSCHEL FAMILY

Then the Lord led him outside and said, "Look up at the sky and, if you can, count the stars."

Genesis 15:5

Fifteen-year-old William Herschel (1738–1822) joined his father's army band in 1753. He could not have guessed the two major changes in his life that lay ahead. Four years later the Seven Years' War came to his native Hannover, Germany, as the French invaded. In order to avoid military service, William escaped to England and settled down.

The other major change occurred when he was 26. He read a book about the telescope and the amazing views it gave of the heavens. From then on he felt an ever-increasing urge to see these wonders for himself.

Because he did not have enough money to buy a telescope, he set about making his own. After 200 attempts he finally made his first satisfactory example, satisfactory, that is, to his own exacting standards.

In 1772 he sent for his sister Caroline Herschel (1750–1848) to help him in his work. She was a talented singer but dropped her career to help William with his astronomical work, reading to him and even feeding him as he spent long hours grinding telescope mirrors.

After the Herschels had made an excellent 4-inch reflecting telescope in 1774 (all their telescopes were reflectors), William began his first review of the sky, looking at everything he could see with the instrument.

Two of his telescopes are particularly famous. His favorite was the 18-inch-diameter reflector with which he did most of his great work (Figure 15.4). In 1789 he completed a huge 48-inch reflector, a project that had been financed by the king. It was considered one of the wonders of the world. Yet, the 48 inch was so large and difficult, even hazardous, to use, that Herschel returned to the 18 inch for most of his work (Figure 15.5).

The second review of the sky was finished in 1781, and the third begun in the same year. Some of his many discoveries and projects are described throughout this book. Although not especially noted for his skill at measuring positions in the sky, he vastly widened the horizons of astronomers with his excellent telescopes and immense appetite for hard work.

Today Caroline Herschel is primarily remembered as the person who wrote down William's observations, kept his house, did his paper work, and entertained visitors. Caroline was also honored as an excellent astronomer during her 98-year life. She too ground mirrors and searched the skies. She found eight comets in the course of this work.

In 1788, the 50-year old William Herschel married Mary Pitt. Their only son, John Herschel (1792–1871), was born in the twelfth year of their marriage. John inherited the talent of his forebears and led an incredibly active, productive life. His most important work was to extend the work of his father (who died when John was 30) in many areas, continuing the search of the sky and the compiling of catalogs of heavenly objects. He took one of his father's telescopes to the southern hemisphere and began one of the first surveys of the southern skies.

---

sphere and used that as a representative sample. The patience and will power required for such a self-imposed task are not available to most of us.

William Herschel's Version of the Milky Way Galaxy The results of his star gauges were as he had expected. The disk of the stars appeared to extend farthest in the direction of the band of light in the sky called the Milky Way. When viewing regions of the sky progressively

Figure 15.4 William Herschel's best telescope, the 18 inch. This was Herschel's workhorse. Since the mirror had a 20-ft (6.1-meter) focal length, the tube was very long. One rode an elevator to reach the observing position. The tube was aimed by raising it with a block and tackle and then rotating the entire framework which rested on wheels.

Figure 15.5 William Herschel's largest telescope. This telescope had a 4-ft-diameter (1.22 meter) mirror and a tube 40 ft (12.2 meters) long. King George III of England financed the project of building it, to the tune of 4000 pounds. The drawing reveals that the observer's cage is at the tube's lip. Herschel had tilted the mirror so that the focal point was at the lower lip of the tube. This eliminated the need for a flat diagonal mirror, as required in Newton's design. This telescope was difficult to keep pointed, and the high-riding observer was at some risk of falling. Herschel never did, but quite a few astronomers have fallen from their telescopes.

The portion of the Milky Way visible in the south on summer evenings looks brighter to the eye than other portions. Yet, Herschel's star gauges led him to conclude that the number of stars visible in a telescope was roughly the same in any direction along the Milky Way. He concluded that the sun is only slightly offset from the center of the galaxy.

further from the band, he counted successively fewer stars, indicating that the edge of the disk is less distant in that direction. This disk of stars, the Milky Way galaxy, had, in Herschel's estimation, a diameter of about 10,000 ly and a thickness of about 2000 ly. Herschel also estimated the position of the sun in the galaxy. He found that the stars seem to be distributed around the Milky Way in more-or-less equal numbers, indicating that the sun lies rather near the center of the galaxy (see Figure 15.6).

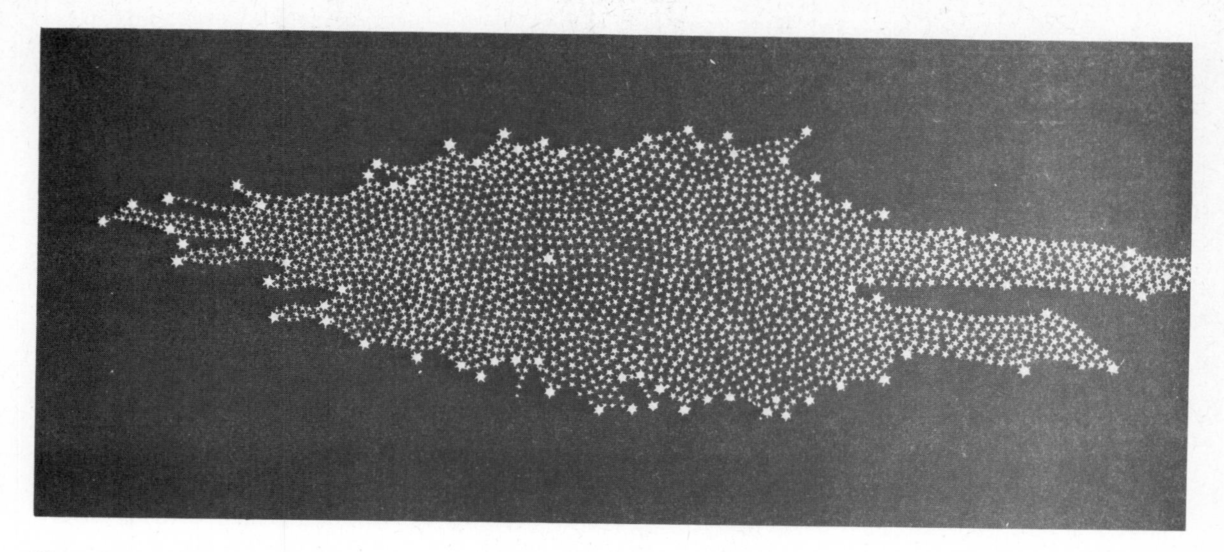

Figure 15.6 Herschel's galactic system. The bright star near the center represents the approximate position of the sun. Toward the right is the region of the Milky Way we see in the Summer Triangle. Herschel indicated the dark lane we see there by a notch in the galaxy. He also indicated that the Milky Way looks brightest in the summer sky by placing the sun slightly off-center, to the left.

In the later stages of William Herschel's career, he made several observations which shocked him because they contradicted both of the assumptions he had formerly relied upon. We already mentioned in Section 12.6 that Herschel found that at least some double stars orbit each other. If a dim star and a bright star are orbiting each other, it means they are both the same distance from us. This undermines Herschel's assumption that all dim-looking stars are farther away than all bright-looking ones. The other shock came when he began to use his new, 48-inch-diameter (122 cm) telescope. The vastly improved light-collecting power of this instrument made it possible for him to see many fainter stars than before. He concluded that, at least in some directions in the sky, he had not actually seen to the edge of the galaxy.

Figure 15.7 Jacobus Kapteyn, who extended William Herschel's star-counting methods into the twentieth century.

Kapteyn Measures the Milky Way Galaxy Herschel's star gauge methods were taken up once more in the early 1900s. Kapteyn (Figure 15.7) used the results of a current, sophisticated star count to estimate the size of the Milky Way galaxy anew. He found a diameter of about 23,000 ly. Not satisfied with this, Kapteyn organized a number of astronomers in 1906 to make star counts in more than 200 regions of the sky, taking into account such data as the apparent brightness, spectral class, radial velocity, and proper motion of the stars. By the early 1920s, when the mass of data had been compiled and analyzed, Kapteyn announced what is called the **Kapteyn model** of the Milky Way galaxy, a disk of stars which is most populated with stars nearest the center, with a gradually thinning population of stars as one approaches the edge. The overall diameter of the disk was thought to be about

Jacobus Kapteyn ("Cap-tine'," Dutch astronomer, 1851–1922), who came from a family of 15 children, spent most of his adult life amassing data on the stars and using them to investigate the structure of the heavens. He was a pioneer in attempting to organize international cooperation among astronomers. An expert on the proper motions of stars (Section 12.1), he discovered the star with the second largest proper motion. It is called Kapteyn's star.

50,000 ly—more than double Kapteyn's previous estimate—and the thickness of the disk to be 6000 ly. The Kapteyn model agreed with Herschel's finding that the sun resides relatively close to the center of the galaxy. Kapteyn died in 1922 at the age of 71. Before he died he was to see his galaxy model challenged by Shapley, a young radical who was in his early thirties.

15.3 SHAPLEY AND THE NEW MILKY WAY GALAXY

The reader will recall that in Section 12.9 we discussed a breakthrough in measuring large distances, a breakthrough that had begun in 1912 when Leavitt demonstrated that there is a relationship between the period and the luminosity of any Cepheid variable. Hertzsprung pointed out how valuable this tool could be: Cepheid variables are very luminous and therefore can be seen across enormous distances. Furthermore, they could be used to measure distances well beyond the 300-ly limitation imposed by the method of parallax. The credit for making Cepheid variables a practical astronomical tool goes to Harlow Shapley (see Figure 15.8 and Box 15.2).

Globular Clusters Shapley's work at the Mount Wilson Observatory made him a leading expert on globular clusters. A typical **globular cluster** consists of 100,000 stars existing in a spherical region having a diameter of 100 ly. Near the center of a more compacted globular cluster the stars may be separated by an average of only 2000 AU (or 0.03 ly) compared to an average separation of roughly 10 ly in our vicinity. Each star follows its own orbit around the center of the cluster, yet it is thought that collisions between stars in the cluster are rare events. As compared with a typical open cluster (see Section 13.7), globular clusters are about 10 times larger in diameter and are composed of roughly 1000 times as many stars. Furthermore, globular clusters are quite spherical, while open clusters are irregular in shape.

About 130 globular clusters are known at present. One need only glance at Figures 15.9 through 15.11 to appreciate one reason why Shapley was drawn to study these objects. And yet, a photograph cannot do them justice. Words are not adequate, but imagine a ball of wet sand thrown against a wall. Now imagine each grain of sand glowing brightly against a dark background. My first view of the Hercules cluster through a 12-inch reflecting telescope is something I shall never forget.

The Hercules cluster was the first globular cluster to be resolved into stars, by none other than William Herschel. "A most beautiful cluster of stars," he wrote, "exceedingly compressed in the middle and very rich." In the next generation, John Herschel pointed out the very significant fact that the globular clusters seem crowded into one region of the sky. More than one-quarter of the known globular clusters are in the constellation Sagittarius.

Figure 15.8 Harlow Shapley, who remeasured the Milky Way galaxy.

BOX 15.2

HARLOW SHAPLEY, THE MAN WHO MEASURED THE MILKY WAY

Harlow Shapley (1885–1972) was drawn to journalism at an early age. By the age of 16, he was a reporter for the *Daily Sun* in Chanute, Kansas. Later he decided that, if he wanted to continue his career, he would need a college education. After some problems, he was accepted at the University of Missouri in 1907. He arrived there, intending to study journalism, of course, but was surprised to learn that the school of journalism would not open for a year. Deciding to take some courses to fill in the year, Shapley reports that he began at the top of the alphabetical listing of those available. He rejected archeology because he was not sure he could pronounce it. So, he settled for the second course on the list—astronomy. Shapley became engrossed in the subject and never left it.

In 1910, Shapley went to Princeton University to study for his Ph.D. There he became a student of Henry Russell (see Box 12.2) and began to collaborate with him in a study of eclipsing binary stars. Shapley married Martha Betz who also became a recognized expert on eclipsing binaries.

Shapley completed his Ph.D. in 1913. He had made such a good impression that he obtained a position as astronomer at the Mount Wilson Observatory in California. An older astronomer then suggested to Shapley that he investigate variable stars in globular clusters, a fateful suggestion. Shapley then began his long, arduous project of first coverting the discovery of Henrietta Leavitt concerning Cepheid variables into a practical technique and then using this tool to estimate the size of the Milky Way galaxy.

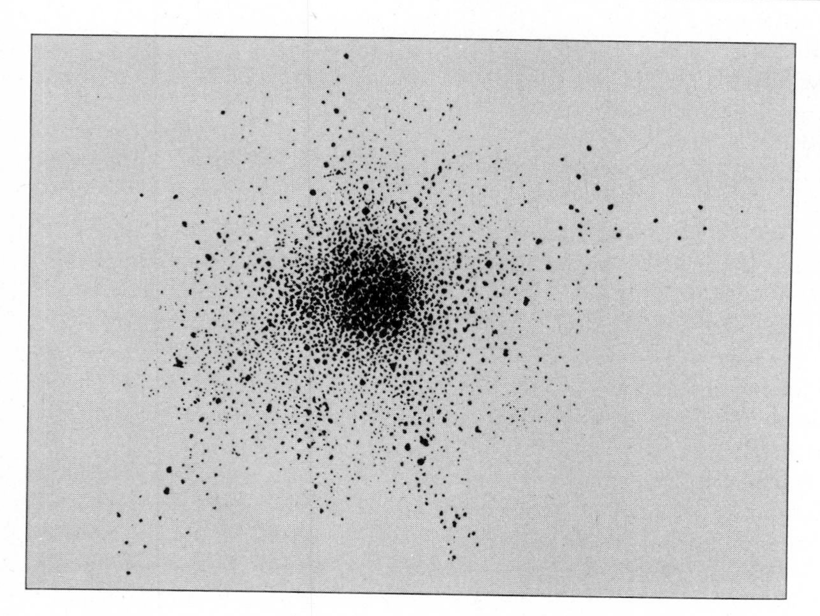

Figure 15.9 Rosse's drawing of M2, a globular cluster. Using his 72-inch-diameter telescope (see Box 16.1), the third Earl of Rosse drew this excellent representation of the globular cluster known as M2. It is a very beautiful object even in an 8-inch telescope. Located in the sky in the constellation Aquarius, M2 is about 40,000 ly from us and about 80 ly in diameter.

Shapley Measures the Distances of Globular Clusters In the twentieth century Shapley decided to measure the location in space of each known globular cluster. To do so, he had to measure their positions in the sky with respect to the fixed stars; this gives the line of sight to each cluster. He also had to determine the distance of each cluster so as to

Figure 15.10 The globular cluster known as M4. Photographed by the 3.86-meter (152-inch) Anglo-Australian telescope in Australia, this globular cluster in the constellation Scorpius is easy to find through a telescope because it lies so close in the sky to the bright red star Antares. M4 is about 15,000 ly from us, one of the nearest globular clusters known. It has a diameter of roughly 90 ly.

know its position along the line of sight. He had hoped that this latter task would be possible if he could turn Cepheid variables into useful tools.

Shapley was well aware of the mountain of labor that lay ahead of him. First he had to gather data on the Cepheid variables so that he could improve on Hertzsprung's work regarding the relationship between the period of a Cepheid and its luminosity. Shapley devised an ingenious, indirect scheme for doing this and then performed the photography and calculations necessary to carry it out. The work was hard and tedious, but the reward was great. For the first time, astronomers could measure distances greater than 300 ly.

This phase of Shapley's work was also discussed in Section 12.9.

Now Shapley could turn his attention to measuring the distances to globular clusters. The method was to photograph each globular cluster

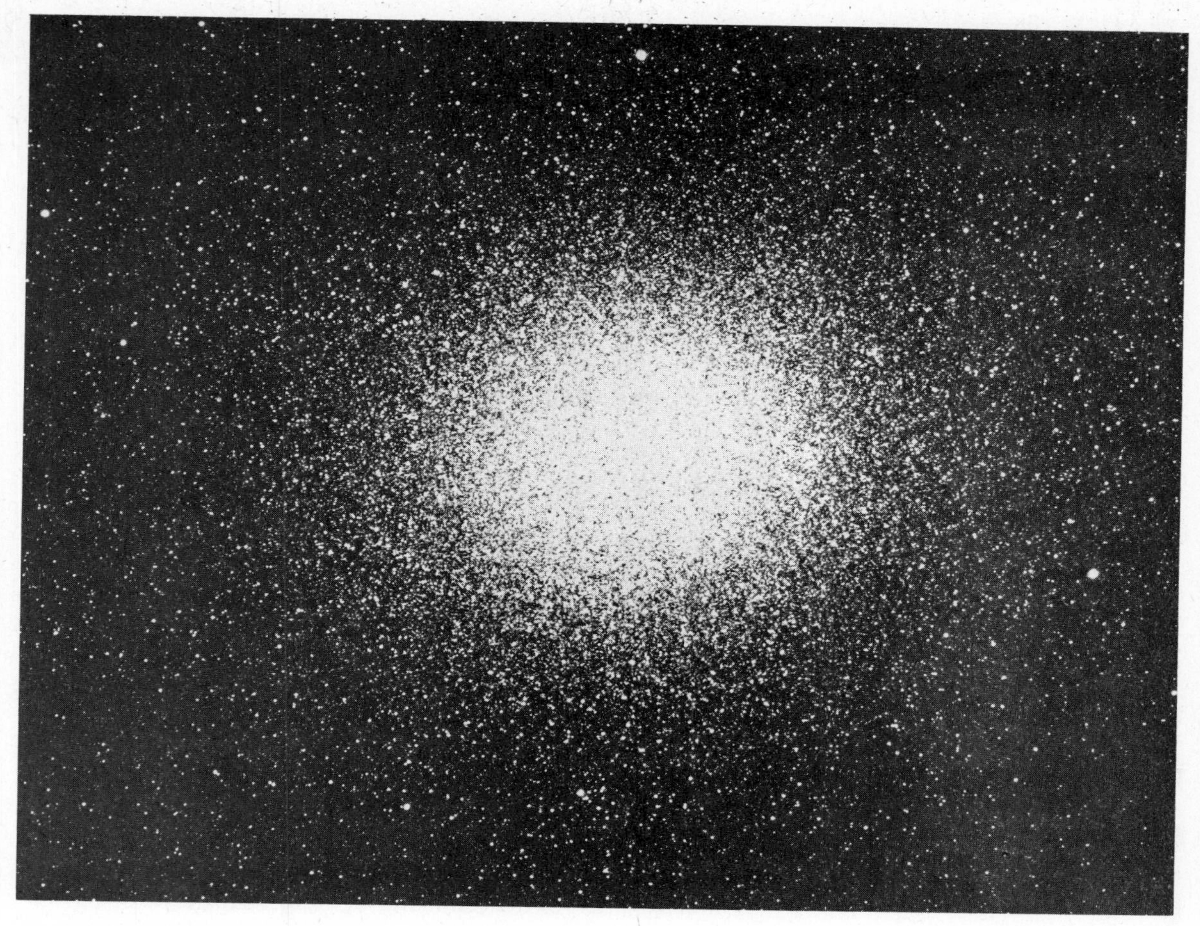

Figure 15.11 The splendid globular cluster known as Omega Centauri. Omega Centauri has an angular diameter in the sky of more than 1°. It lies about 17,000 ly from us and has a diameter of about 300 ly, much larger than that of most globular clusters. Unfortunately it lies too far south to be seen by most observers in the continental United States.

a sufficient number of times so that the Cepheids in the cluster could be discovered and their periods measured. Then the distance to the Cepheid could be calculated, yielding the distance to the cluster itself.

Shapley spent an extraordinary amount of time on this work. He wrote to a friend that it required "a painful amount of stupid labor." But the results he obtained justified it all. He now knew the arrangement in space of the globular clusters. He found that they are arranged in a huge sphere having a diameter of roughly 300,000 ly. Even more extraordinary, the center of the sphere of globular clusters is about 60,000 ly from the sun, according to his results, in the direction of the constellation Sagittarius. When Shapley placed the Kapteyn model of the Milky Way galaxy together with his globular clusters (see Figure 15.12a) he obtained a result he found difficult to believe. He boldly

Figure 15.12 Globular clusters and the Milky Way galaxy. Shapley believed that the globular clusters were arranged in a large spherical region centered far from the sun (X). From this he concluded that the Milky Way Galaxy could not be as small as Kapteyn had thought (a). Shapley believed that the center of the Milky Way galaxy and the center of the sphere of globular clusters were both at the same location (b). Observe that Shapley found no globular cluster in the central horizontal slab to the right of the sun. This "zone of avoidance" is explained in Section 15.4.

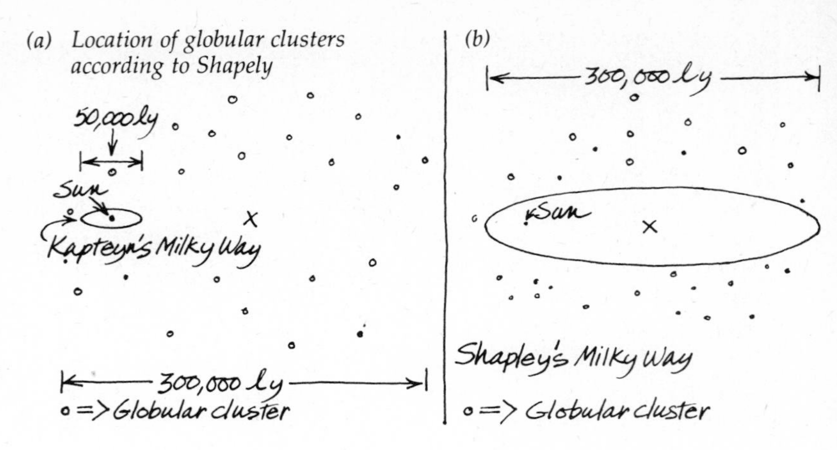

asserted that his results could best be understood by assuming that the center of the galaxy coincides with the center of the sphere of globular clusters (see Figure 15.12b). This implies that the galaxy itself has a diameter of roughly 300,000 ly and that the sun, far from being near the center as Kapteyn's star counts implied, was about 60,000 ly from the center. Now, if Shapley were right, it would be necessary to explain how such an established name as Jacobus Kapteyn could be wrong about the size of the galaxy and the sun's location and how he, Shapley, a young upstart, could be right.

15.4 GIANT CLOUDS IN THE GALAXY: INTERSTELLAR DUST

An explanation of the conflicting results obtained by Shapley and Kapteyn involves the dark lanes and blobs seen in the Milky Way. There are many regions in the sky where relatively few stars are visible, even along the band of the Milky Way. Prior to the twentieth century, most astronomers assumed that these dark regions were places that lacked stars—holes in the galaxy through which one could see the empty space beyond the Milky Way galaxy. As William Herschel declared of one such region, "Surely, this is a hole in the heavens."

Edward Barnard (American astronomer, 1857–1923) was born into extreme poverty and rose to become one of astronomy's finest observers. He was famous for his keen eyesight and was one of those who did not see canals on Mars. He discovered the fifth satellite of Jupiter and felt he could see craters on Mars, an observation he never published. He discovered the star having the largest proper motion, Barnard's star (see Section 12.1). His study of dark clouds in our galaxy was his most important contribution.

Edward Barnard (1857–1923) was not so sure. At Yerkes Observatory, he made a search of the sky for dark regions and by 1919 had found over 180 of them. He expressed doubt that there could be so many dark holes among the randomly distributed stars and decided instead that "there are obscuring masses of matter in space" which block our view of the stars lying behind the clouds. Astronomers had known of other clouds of matter before this, for many of them are illuminated, at least in part, by the stars in or near them. These illuminated clouds, or **diffuse nebulas,** are discussed in Box 15.3 (see also Figures 15.13 through 15.16).

BOX 15.3

DIFFUSE NEBULAS

Many clouds of interstellar matter are dark and observable only because they block our view of the stars behind them. Others, however, are illuminated by stars that are in or near them, and some of these clouds are interesting in binoculars or a telescope and make spectacular photographic subjects. If the illuminating stars are cooler than about 20,000 K, the dust in the cloud reflects starlight, giving rise to a **reflection nebula.** The spectrum of a reflection nebula is the same as that of its illuminating stars.

Stars hotter than 20,000 K emit sufficient ultraviolet radiation to cause the gas in the cloud to fluoresce, producing an **emission nebula** also called an **H II region** (see Figures 13.5 through 13.7, and 15.13). The spectra of emission nebulas consist of colored lines separated by black spaces (see Section 9.5). In some regions of the sky, dark nebulas and emission nebulas are seen together, often producing striking results (see Figures 15.14 through 15.16).

Figure 15.13 M17, The Omega nebula, an H II region. This emission nebula is in the constellation Sagittarius and is easily visible as a line of light in binoculars on clear summer nights in the southern sky. Through a telescope it resembles a horseshoe or the Greek letter "omega." It is a mere 3000 ly from us and has a diameter of about 18 ly.

Figure 15.14 The North America nebula, an H II region. This nebula's nickname comes from its resemblance to the East Coast and Gulf regions of the United States and Central America. To the lower right of this emission nebula is a good example of a dark cloud of the type studied by Barnard. The North America nebula is about 3500 ly away and about 100 ly across.

Figure 15.15 M16, an H II region in the constellation Serpens. Located just north of M17 (Figure 15.13) in the sky, this nebula has an open cluster embedded in it. A nearer dark cloud covers parts of the emission nebula, giving this interesting view. M16 is about 6000 ly from us and about 26 ly across. It is easily visible in almost any telescope.

Henry Russell was among the first to seize upon clouds of matter in the galaxy as the reason that Kapteyn had been misled. These clouds seemed to be largely confined to the central plane of the disk of the Milky Way galaxy. This would explain why Shapley had seen few globular clusters in the disk of the galaxy itself, as shown in Figure 15.12. This "zone of avoidance" was a region beyond which no objects could be seen because of the intervening clouds. Russell also was among the first to suggest that the clouds contain dust—tiny, solid particles of matter. He wrote, in a letter to Shapley: "Dust . . . is a far more powerful absorber . . . than gas." It is interesting that Shapley did not accept the idea of obscuring clouds for many years.

To summarize, then, the idea was that these clouds of dust lying in the disk of the Milky Way galaxy block our view of the greater part of

Figure 15.16 The Horsehead nebula, an H II region in Orion. This dark cloud covers part of an H II region, giving the appearance of a horse's head. It is a tiny, difficult object to view through small telescopes. This photograph illustrates nicely how a dark cloud blocks out the objects behind it. (See also Plate 16.)

the galaxy. When looking toward the center of the disk, Herschel and Kapteyn saw few stars beyond a certain distance and interpreted this to mean that they had observed the edge of the disk. There are stars beyond that distance, but their light is absorbed by the intervening clouds of dust. Shapley was able to see even very distant globular clusters as long as they were not in the disk, because they were in a spherical region around the disk, out of the line of sight of the obscuring clouds (see Figure 15.17).

15.5 THE SHAPLEY-CURTIS DEBATE, PART ONE

As Shapley's results became known in 1918, there was a reluctance on the part of many of his fellow astronomers to accept them. In 1920 the weight of opinion was still on the side of Kapteyn's smaller galaxy.

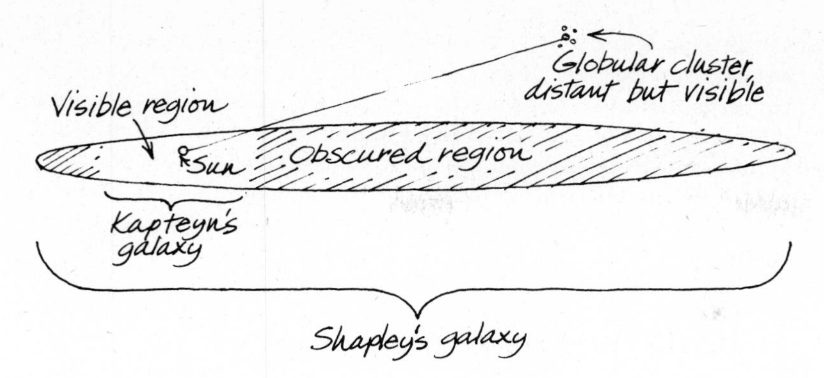

Figure 15.17 Obscuring clouds in the disk of the galaxy. Clouds of dust lie in the central plane of the galaxy and block our view of distant (shaded) regions of the galaxy. Shapley could see globular clusters much more distant than the most distant stars Kapteyn could see because he observed globular clusters lying out of the plane of the galaxy.

In April of the same year, a debate was staged which had as one of its purposes to thrash out this interesting question. The 34-year-old Shapley was chosen to present his radical views. The side of the establishment, so to speak, was presented by Heber D. Curtis, director of the Allegheny Observatory, who was then nearing the age of 50.

Another phase of the Shapley-Curtis debate is described in Section 16.5.

Are Shapley's Distance Determinations Much Too Large? Curtis expressed doubts about the method of determining distances to the globular clusters by means of Cepheid variables. He agreed that Shapley had found the *relative* distances to the globular clusters correctly. If Shapley said cluster A was twice as distant as cluster B, Curtis was willing to accept the result. Curtis' doubt centered on the method Shapley had used to determine the actual or *absolute* distances to the globular clusters. It seemed to Curtis that Shapley had overestimated the distances by a factor of 10 and that the galaxy had a diameter nearer 30,000 ly, as Kapteyn's method estimated, rather than Shapley's 300,000 ly. Curtis pointed out that many variable stars in the Milky Way do not conform to Shapley's luminosity-period relation. Shapley responded by explaining and defending his methods and by pointing out that his relation applied only to Cepheid variables and not to other types of variable stars.

A good deal of discussion centered around M13, the globular cluster also known as the Hercules cluster. Shapley's method gave its distance as 36,000 ly, while a different method used by Curtis yielded a distance of 3600 ly. If Curtis' method were correct, it would reduce Shapley's distance to M13 by one-tenth and would thus reduce all other distances, including the size of the galaxy, by the same factor.

The Proper Motion of Globular Clusters Besides defending his methods, Shapley had another line of argument. The *radial* velocities of a number of globular clusters had been determined by means of the Doppler effect. They yielded an average value of about 150 km/sec. Assume that this is a good indication of the normal speed of a globular cluster. Then the average globular cluster should also have a *tangential* velocity of the same amount. If the globular clusters were as close as

The terms "radial velocity" and "tangential velocity" were described in Section 7.6.

Curtis expected, they should show an average proper motion of 0.04 second of arc per year. If they were as distant as Shapley deduced, the proper motions would be much less. Shapley reported a large number of observations of globular clusters indicating, at most, a tiny proper motion, much smaller than the figure of 0.04 second of arc per year.

The debate was considered a standoff. Few minds were changed. All agreed more data were needed, and the new data were not long in coming.

15.6 THE ROTATION OF THE GALAXY

Jan Oort (rhymes with "fort," Dutch astronomer born in 1900) studied with Kapteyn and followed interests very similar to his: the motions of stars and the size and structure of our galaxy. We discussed Oort's theory of the origin of comets in Section 11.8 and will explain some of his work in radio astronomy in Section 15.8.

In his book in 1755, Kant had supposed that the entire galaxy of stars rotates about the center. This would prevent the stars from collapsing into the center of the disk. In turn, they would be prevented from flying off into space by mutual gravitational attraction. The best evidence that the Milky Way galaxy actually does rotate was obtained by Jan Oort.

Oort assumed that the sun and most of the stars in the vicinity of the sun rotate in roughly circular orbits about the center of the galaxy. According to Newton's laws of motion, stars nearer the center travel faster in their orbits than stars farther out, much as planets nearer the sun travel faster in their orbits. If this were so, stars farther from the center of the galaxy than the sun would fall behind the sun, while those closer would pass the sun.

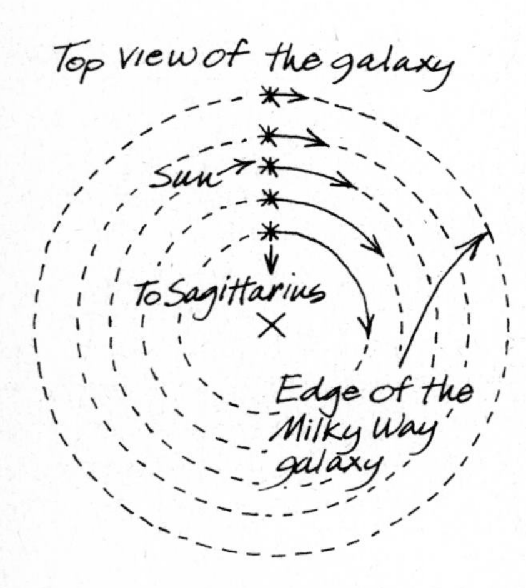

Figure 15.18 The direction of the center of the galaxy as determined by Oort. By studying the motions of the stars in the sun's neighborhood, Oort determined that the stars move as shown here: The closer a star is to the center of the galaxy, the faster its orbital motion. From his data, he concluded that the center of the galaxy lies in the direction of the constellation Sagittarius.

Oort undertook a detailed study of the stars in the neighborhood of the sun. He found that each one has its own individual motion but that, taken in large numbers, the randomness of the stars' motions can be averaged out. Thus you can seek patterns in the small but definite overall motions of the stars with respect to the sun. Oort studied the proper motions and radial velocities of the stars in the sun's neighborhood and found that, on the average, the stars in the direction of the constellation Sagittarius are passing the sun, while those in the opposite direction are falling behind (see Figure 15.18).

Recall that Shapley had interpreted the crowding of globular clusters into the constellation Sagittarius to indicate that the center of our galaxy lies in this direction. Oort's results, published in 1926, are independent of globular clusters, and yet they confirm Shapley's idea. This agreement between the two independent methods had the general effect of winning approval for Shapley's work.

But, there still was a discrepancy. Oort used his data on the motions of stars not only to determine the *direction* to the center of the galaxy but also the *distance* to the center. In other words, he determined the radius of the sun's orbit as it travels around the galaxy's center. He obtained a radius of about 30,000 ly, one-half as large as the 60,000-ly distance Shapley had found. The two methods agreed on the direction to the center of the galaxy, but not on the distance. If Oort were right, the diameter of the galaxy would be only about 100,000 ly and not

300,000 ly as Shapley claimed. Curtis and Kapteyn had been wrong, but perhaps Shapley had not been completely right. By the mid-1920s, at any rate, it seemed clear that Oort had turned astronomical opinion away from that of his recently deceased teacher, Kapteyn.

15.7 INTERSTELLAR ABSORPTION: THE FINAL CLUE

Harlow Shapley had specialized in globular clusters. Robert Trumpler is best remembered for his detailed, painstaking observations of the stars in open clusters.

Robert Trumpler (Swiss-American astronomer, 1886–1956) devoted most of his professional life to studying the roughly 1000 open clusters (see Section 13.7) known in our galaxy. His work provided major contributions to the study of stellar evolution and the knowledge of the structure of our galaxy.

Trumpler devised a method of determining the distance to open clusters. The method of Cepheid variables was not available to him, since Cepheid variables are not found in most open clusters. Instead, he tried to estimate the luminosity of each star in the cluster from the appearance of its spectrum. He assumed, for example, that a star having the spectrum of a G star would have a luminosity of about 1 sun. Then, he measured the star's apparent brightness and used the equation $L = bR^2$ (see Section 12.2) to evaluate R, the distance to the star and thus to the cluster to which the star belongs.

When Trumpler had determined the distances to open clusters by this means, he used them to compute the actual sizes of the clusters based on their angular diameters. For those who are interested, the formula is:

$$\text{Actual size} = 0.01745 \times \text{distance} \times \text{angular diameter (in degrees)}$$

The results he obtained by this method were puzzling. He found that, the more distant a cluster, the larger it seemed to be. According to the computation, the most distant clusters were twice as large as the nearest ones. What was wrong? "None of the observational errors . . . offers a possible explanation," Trumpler wrote in 1920. He went on, ". . . there are only two alternatives left; either to admit an actual change in the sizes of open clusters with increasing distance or to assume the existence of an absorption of light within our stellar system."

Trumpler, and everyone else, rejected the first alternative and accepted the second. Up to that time, most workers in the field had assumed that the dust between the stars existed only in huge clouds known as dark nebulas and diffuse nebulas. Trumpler proposed that one could best explain his results by assuming that the space between the stars contains thinly dispersed amounts of dust. The reasoning ran like this: If the distant open clusters are actually about the same size as the nearby ones, then the distant clusters must not be as far away as Trumpler's original method had estimated. Rather, interstellar dust blocks some of the light from the clusters, making them look dimmer than they would otherwise.

Trumpler assumed that all clusters have the same approximate size,

Figure 15.19 Absorption of light by dust in the galaxy's disk.

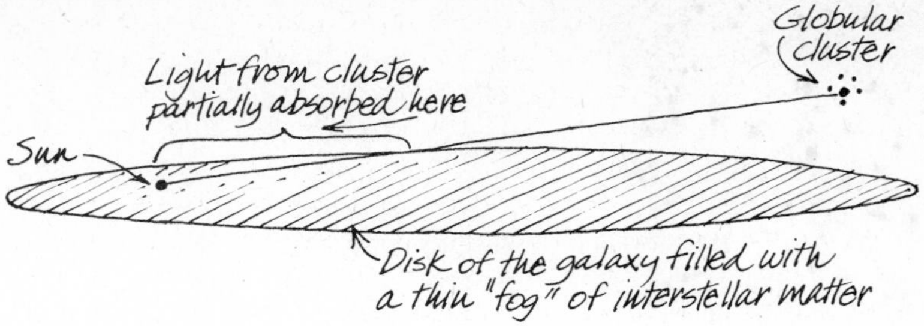

and from his data calculated how drastically the interstellar dust reduces the apparent brightness of a star. He found that a star's brightness is dimmed by about 0.2 magnitude for each 1000 ly of distance. In all future work, one could not simply use $L = bR^2$ but would have to allow for the presence of interstellar absorption as well.

This finding resolved the discrepancy between the results obtained by Oort and by Shapley. Oort had studied only the radial velocities and proper motions of stars, not their brightness; interstellar dust would not have affected his data. Shapley, however, had been looking at globular clusters at enormous distances, tens of thousands of light years away, in some cases. These distant globular clusters, seen through the interstellar version of smog, were actually brighter than had been thought. This caused Shapley to overestimate their distances (Figure 15.19). When the interstellar absorption of light was taken into account, the new figures resulting from the study of globular clusters were: diameter of the galaxy, 100,000 ly; distance of the sun from the center of the galaxy, 30,000 ly. These results agree nicely with the results of Oort's work and have not been changed significantly by any research since then.

15.8 THE CONTEMPORARY VIEW OF THE MILKY WAY GALAXY

The task of discovering the nature of the Milky Way galaxy has been compared to preparing a street map of Los Angeles by observing on a smoggy day from a rooftop in the suburbs. Nevertheless, progress has been made. The current conception of the Milky Way galaxy is a huge, flat disk of stars. It has a diameter of about 100,000 ly and a thickness of about 3000 ly. Within the disk of stars is a thinner disk of dust and gas having a thickness of roughly 700 ly.

The Position and Motion of the Sun in the Milky Way Galaxy The sun seems to lie about 30,000 ly from the center of the galaxy, both according to Oort's method of measuring the motions of nearby stars and according to Shapley's method of observing globular clusters, once Trumpler's correction for interstellar dust is made. Oort's method is very valuable because it yields other interesting facts. The speed of the sun in its orbit around the galactic center comes out to about 250 km/sec (about 150 mi/sec or about 560,000 mi/hr). This may be compared to the orbital speed of the earth around the sun, about 30 km/sec. The sun's orbit in the galaxy is assumed to be nearly circular. The sun travels about 190,000 ly as it completes one orbit. At this rate, it takes about 250 million years or $\frac{1}{4}$ billion years to orbit the galactic center once. If this is correct, the last time the sun was in its present position in the galaxy the earth was in the Permian geological period, dominated by insects and early reptiles, and dinosaurs and mammals were still far in the future. (This assumes that the geologist's time scale is also correct.) One "galactic year" or 250 million years before that, the earth was entering the Ordovician period in which animal life existed only in the ocean; no creatures having a backbone, such as fish, had yet appeared. If the sun was formed 5 billion years ago, one might say that it is now 20 galactic years old.

The Mass of the Milky Way Galaxy Oort's analysis of the orbital motion of the sun in the galaxy yields another important bonus. The orbital period and radius of the orbit of the sun are known. This is exactly the type of data that we have seen are necessary for calculating the masses of the sun (Section 8.4), the stars (Section 12.7), and planets having satellites (Box 8.4).

The sun moves in a nearly circular orbit around the center of our galaxy under the gravitational influence of every other star in the galaxy. These gravitational forces average out to produce a force on the sun which points toward the center of the galaxy. By analyzing the radius of the sun's orbit and its orbital period, astronomers conclude that the Milky Way galaxy has a mass of about 1×10^{11} suns (100 billion suns or 2×10^{41} kg). They also estimate that most of this mass is in stars, leaving only small amounts in dust and gas. It is also thought that the average star has a mass of about 0.5 sun (there are *many* K and M stars on the main sequence). Thus we may infer that there are about 200 billion stars in our galaxy.

The Halo of the Milky Way Galaxy Although the greatest concentration of matter in the galaxy is in the flattened disk, there also seems to be a more spherical, less populated region surrounding the disk, a region called the **galactic halo.** Although the stars in the halo are much more dispersed than those in the disk, the halo has a large volume and may contain a large fraction of the mass of the galaxy. Whether the halo is significant in mass or not is still under investigation.

The halo is centered on the galactic center and has a diameter of roughly 150,000 ly. It is the realm of the globular clusters and an unknown number of individual stars. Each globular cluster and star in the halo follows its own orbit around the galactic center. Many of the orbits seem to be highly eccentric, which means that the clusters travel far out to the edge of the halo and then plunge into the crowded regions near the center of the galaxy, a motion analogous to the motion of many comets in the solar system (see Figure 15.20).

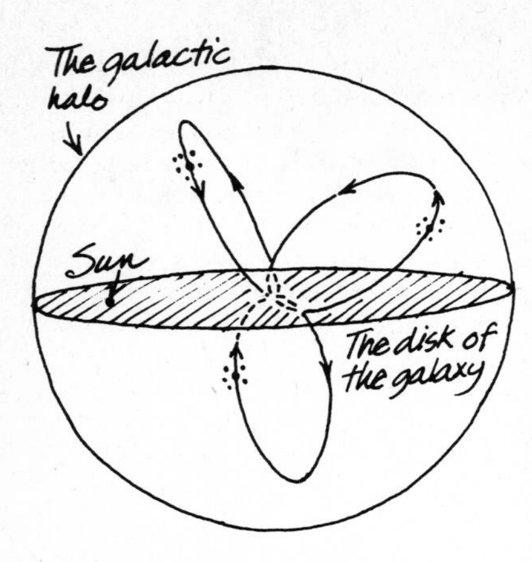

Figure 15.20 The galactic halo and typical orbits of globular clusters.

Radio Telescopes Probe the Milky Way Galaxy We have seen in this chapter that interstellar dust blots out our view of most of the Milky Way galaxy, making it impossible to view the potentially most interesting portion of the galaxy, its center. However, this barrier has also been breached. It started in 1931 when Karl Jansky began researching a problem for Bell Telephone Laboratories. He was to investigate the source of radio interference (static) which occasionally interrupted radio transmissions. Jansky built a large radio antenna (Figure 15.21) and soon located two types of source. The first was thunderstorms, but the second seemed to be located in outer space. Since the second source of radio waves seemed to pass around the earth once a day, Jansky at first assumed it was the sun. A year later, however, he had determined that the period of revolution was 23 hr, 56 min. This is the important number we learned in Section 2.1, the length of time required for the celestial sphere to appear to rotate once. This told Jansky that the source of the radio waves lies outside the solar system. His antenna was not very directional, so he could not be sure where the source was, but he proposed that it might be the galactic center.

No astronomers took up this discovery. It was up to a young ham radio operator to pursue Jansky's momentous discovery. He was Grote Reber, a gifted amateur in his midtwenties. When Reber heard of Jansky's work, he designed and built an impressive disk-shaped radio antenna. It worked like a reflecting telescope but was 31 ft (9.5 meters) in diameter. After initial difficulties, Reber confirmed that Jansky's source was indeed the galactic center. Radio waves, because of their long wavelength, can penetrate interstellar dust and gas with little absorption. Here was a way to investigate the *entire* galaxy. The strong radio source detected by Jansky and Reber lies at the same position in the constellation Sagittarius that Shapley and Oort had claimed to be the galactic center.

Reber went on to determine that the entire band of the Milky Way emits radio waves, although not as strongly as the center. He also found that the sun is a weak radio source most of the time. Later it was found that, during a solar flare, the sun temporarily becomes a strong radio source.

The 21-Cm Line of Hydrogen Reber was the only radio astronomer throughout those years. When he published his results in 1942, no immediate follow-up was made, principally because so many astronomers were absorbed in the war effort. However, Oort read Reber's report and at once saw the importance of this new means of observing the galaxy. Oort commented to one of his fellow workers, Hendrik van de Hulst, that it would be of great importance if a line could be found in the radio spectrum. By a "line" Oort meant a single definite wavelength of radiation comparable to the lines found in visi-

Figure 15.21 Karl Jansky and the first radio telescope. This radio antenna, so different from the giant dish reflectors used today, was designed to receive radio waves only from directions near the horizon. It could be rotated on the wheels shown to view a radio source only as the source was rising or setting. Had the center of our galaxy been so near the celestial pole as to be a circumpolar object, Jansky might not have discovered its radio emissions with this apparatus.

ble spectra. This set van de Hulst thinking about the hydrogen atoms in space. Before long he evolved a theory that cold, neutral (un-ionized) hydrogen atoms should emit radio waves at a single wavelength of about 21.1 cm. This is usually termed "the 21-cm line."

When a scientist says that atoms are "un-ionized," it means that an atom has its full number of electrons in its electron cloud. An atom lacking one or more electrons is said to be "ionized."

After World War II, the new radio technology was put to use by astronomers who constructed ever larger, more refined radio telescopes. (The largest to date is shown in Figure 15.22.) In the forefront of these developments was the Dutch group under Oort, which was one of the first actually to detect the 21-cm line, the "song of hydrogen," that had been predicted.

The 21-cm line became a powerful tool, as Oort had foreseen. With it, astronomers have been able to observe neutral hydrogen in all parts of the galaxy. The principal method is to point the radio telescope in a particular direction in the Milky Way and observe the radio waves at and near a wavelength of 21 cm. Some of the hydrogen in any particular direction may have a radial velocity toward us because of the rotation of the galaxy, while other clouds of hydrogen farther off may be moving away. In such cases, the astronomers observe two signals, perhaps one Doppler-shifted slightly above 21 cm and the other Doppler-shifted slightly below that. By making a number of assumptions (including circular motion of the hydrogen about the center of the galaxy), the astronomers can estimate the location of each hydrogen cloud in the galaxy. By repeating this analysis all along the Milky Way, an interesting if tentative picture of the location of the hydrogen in the galaxy emerges. This is shown schematically in Figure 15.23. The hydrogen seems to be concentrated in irregular tubular regions, called **arms,** which wind around the galactic center. In the regions of the sky

Figure 15.22 The largest radio telescope: Arecibo in Puerto Rico. A naturally occurring valley was modified to hold this reflector which is about 300 meters (1000 ft) in diameter. The radio waves from space are focused by the dish onto the suspended antenna. The dish points permanently at the zenith but, by moving the antenna, any point on a strip 40° wide in the sky can be scanned as the earth's rotation brings it into view.

Figure 15.23 Schematic representation of the structure of our galaxy from radio telescope data: The 21-cm line of hydrogen. Three arms in our vicinity have been named. We appear to be situated on the inner edge of the Orion arm. Typically, arms are about 6000 ly apart. The arms have been named for the constellation of stars past which we look to see these arms.

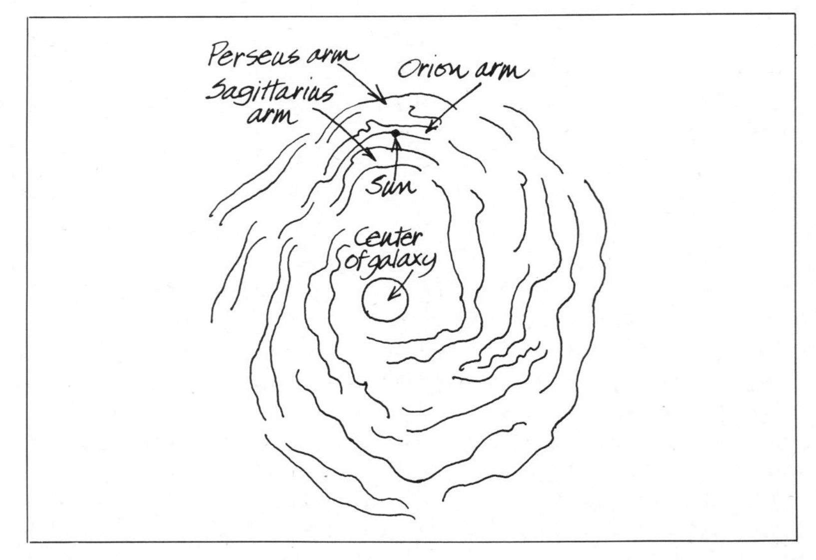

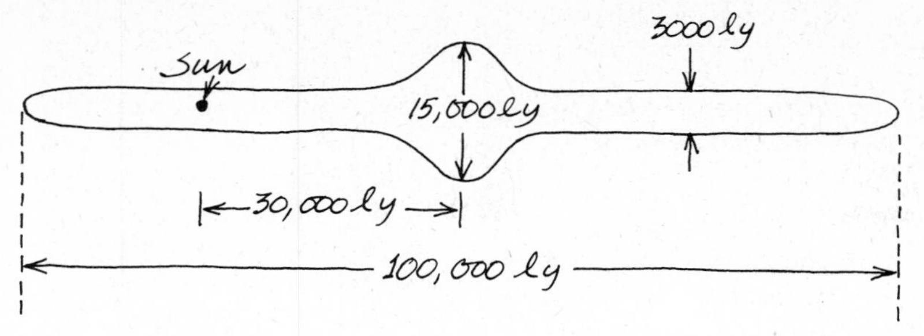

Figure 15.24 The contemporary conception of the Milky Way galaxy, edge view. It is not easy to grasp the size of our galaxy. Suppose this drawing were enlarged so that it stretched across the United States, one edge of the galaxy in Los Angeles and the other in New York City. The sun and Alpha Centauri, which are actually separated by a distance of 4.3 ly, would be about 200 meters (about 600 ft) apart. The orbit of Pluto (actual diameter, 80 AU) would be represented by a circle of diameter 5 cm (2 inches). The drawing on this page is approximately to scale, except for the dot labeled "sun." On this drawing, Pluto's orbit would be about as small as a molecule of the paper, about 10 Å (or about 0.00000005 inch).

visible through the ordinary telescope, stars, gas, and dust seem to be concentrated in these arms. In addition to the radio discovery of galactic arms, it is comforting to have radio confirmation that the galaxy really is out there beyond the gas and dust just as Shapley and Oort had predicted.

Technology has advanced rapidly since the war. Since the time radio waves were discovered coming from the galactic center, it has been found to emit infrared radiation as well as ultraviolet radiation and x-rays. Clearly, the galactic center is an active, mysterious region, and it is presently under investigation.

Radio maps of the galactic disk have been made at many radio wavelengths. They show a feature that had been suspected from visual observations, the **central bulge** of the galaxy. Whereas the galactic disk is about 3000 ly thick in the outer regions, there appears to be a bulge 15,000 ly thick around its center. This gives the galaxy the shape shown in Figure 15.24 when viewed edge-on.

SUMMING UP

Just before 1920, Shapley's measurements of the locations of the globular clusters brought the sun-centered model (which was based on star counts) of the Milky Way galaxy into question. After being corrected for interstellar absorption, Shapley's result has been accepted. It is now believed that the Milky Way galaxy is a disk of roughly 200 billion stars. It has a diameter of roughly 100,000 ly, and the sun lies about 30,000 ly from its center. A central lane of dust in the disk obscures our direct view of most of the galaxy, but radio telescopes reveal it and radio observations imply that it has spiral arms and a central bulge. The center of the galaxy is a strong radio source. The galaxy rotates in such a way as to prevent collapse. The sun requires roughly 250 million years to orbit the center of the galaxy once. The galaxy is surrounded by a halo in which some stars and globular clusters reside. It is not clear how much mass is present in the halo.

In the next chapter we investigate the notion that other galaxies besides our own may exist.

EXERCISES

1. Define the following terms: globular cluster, zone of avoidance, reflection nebula, emission nebula or H II region.

2. Explain how Kant's idea that the stars are arranged in a disk can explain the appearance of the Milky Way in the sky.

3. Why did William Herschel conclude that the sun is near the center of the Milky Way galaxy?

4. (a) Draw a diagram explaining why so many globular clusters are seen in the sky in one particular region. (b) Why did Shapley see so few globular clusters along the Milky Way galaxy in the sky?

5. Why was Kapteyn misled by star counts into thinking that the sun is rather close to the center of our galaxy?

6. Oort studied our galaxy by studying the motions of the stars in our vicinity. In what respect did his results agree with those of Shapley? In what respect did his results disagree with Shapley's original results?

7. What observations led to the discovery of interstellar absorption? What causes this phenomenon? How did this discovery alter the results obtained by Shapley concerning the size of our galaxy?

8. (a) Draw a sketch of our galaxy and its halo seen edge-on. Indicate the numerical values of the diameter, the thickness of the disk, the diameter of the central bulge, and the sun's position. (b) Draw a sketch of our galaxy seen face-on, that is, from a position far above the galaxy along the axis of rotation. Sketch in the sun's orbit.

9. How was the mass of our galaxy determined? How many stars are estimated to be in our galaxy?

10. We cannot see very far in toward the inner regions of our galaxy because of the dust clouds blocking our view. What discovery made it possible to sketch out the distant regions of the galaxy?

READINGS

Two excellent books describing how our present views of our galaxy came about are

The Discovery of Our Galaxy, by Charles A. Whitney, Knopf, New York, 1971, highly recommended.

Man Discovers the Galaxies, by Richard Berendzen, Richard Hart, and Daniel Seeley, Science History Publications, New York, 1976, sec. 2, pp. 51–100.

The Shapley-Curtis debate is described in

"A Historic Debate about the Universe," by Otto Struve, *Sky and Telescope,* May 1960, p. 398.

See also

"Updating Galactic Spiral Structure," by B. J. Bok, *American Scientist,* November/December 1972, p. 208.

"Interstellar Smog," by George H. Herbig, *American Scientist,* March/April 1972, p. 200, on interstellar absorption.

Radio astronomy is very well discussed in the following books, all of which cover many of the topics of this and the next two chapters.

Radio Universe, by J. Hey, Pergamon, New York, 1971.

The Invisible Universe, by Gerrit L. Verschuur, Springer-Verlag, New York, 1974.

The Evolution of Radio Astronomy, by J. Hey, Neale Watson, New York, 1973.

The interesting memoirs of a person who figures prominently in this chapter are presented in

Through Rugged Ways to the Stars, by Harlow Shapley, Scribners, New York, 1969.

THE GALAXIES: ISLANDS OF STARS

George Hale (Figure 16.1) was to be the first to look through the eyepiece of the new 100-inch-diameter (254 cm) Hooker reflecting telescope on Mount Wilson. It had been decided that Jupiter was to be the telescope's first subject on that long-awaited November evening in 1917. Hale took one look and turned away in silent horror. He had seen six blurred, overlapping disks of Jupiter. Hoping that the mirror simply had not yet cooled down from the high daytime temperature, Hale and his companions decided to sleep awhile and then return for another try in three hours. Hale got into bed but could not sleep. Was he reviewing the history of his efforts to build this, the first telescope to be larger than the 72 inch built by Lord Rosse?

Much of Hale's professional life had been concerned with constructing large telescopes. He had organized an observatory in Chicago in his early years. In 1892, the 24-year-old Hale, persuasive even then, had convinced a wealthy man named Charles Yerkes, who had made his money from the Chicago trolley system, to finance construction of the 40-inch (102-cm) refractor at Williams Bay, Wisconsin, still the largest refractor ever built. The telescope was completed in 1897. Hale then decided to attempt a still larger telescope.

Realizing the limitations on even larger refractors (see Section 6.6), he decided to use a 60-inch (152-cm) disk of glass and to fashion a mirror for a reflecting telescope. In 1903, Hale obtained a grant from the Carnegie Foundation to build the telescope, and by late 1908 the 60 inch was at work. This telescope was used to obtain the spectrum of Sirius B, demonstrating that it is a white dwarf (Section 12.6). It was also used by Shapley to locate the center of the Milky Way galaxy (Section 15.3). Even before the 60 inch was ready, Hale had found a Los Angeles businessman, J. D. Hooker, who was willing to finance a 100-inch reflector. But when the huge glass dish arrived from France, it was found to contain countless tiny bubbles and other defects. Hale ordered a new disk to be poured. In 1910 the second disk developed defects during the cooling process. Hooker was dead by then, and his grant had been spent. The stress caused Hale to have a nervous breakdown. Upon recovery, he reinspected the 100-inch disk and decided to gamble that it might be usable. He ordered the polishing of the disk to begin. Finally the 100 inch was readied. Then the Carnegie Foundation donated the money for construction. And now, as Hale lay in bed, sleepless, the distorted image of Jupiter must have come to his mind again and again.

At last, at 2:30 in the morning, the small group reassembled in the dark dome of the 100 inch. The telescope was pointed at Vega, and Hale looked into the eyepiece for a second time. This time he uttered a cry of delight. The image was perfect. "The agony had not been wasted," he later said.

Hale then went on to set the process in motion that led to the construction of the mighty 200-inch (308-cm) reflector at Mount Palomar. It was finished in 1948, 10 years after his death, and was named the Hale telescope. But the 100-inch Hooker had proved its worth well before then.

In this chapter we will see how the Hooker telescope helped answer the question, Are we alone in the universe? This question can be asked in many senses. Here, it refers to the earlier contention of some astronomers that nothing lies beyond the Milky Way galaxy, that beyond the borders of our galaxy lie only the infinite, dark depths of empty space. The 100 inch collected sufficient light to provide an astounding answer, one that carried implications beyond anything Hale could have imagined.

Figure 16.1 George E. Hale.

16.1 HOW COMET HUNTING LED TO THE STUDY OF NEBULAS

We shall see in this section how the search for new comets led to an investigation of the nature of nebulas. Later sections show how this eventually led up to our present understanding of galaxies.

Comet Hunting Comet hunting (see also Section 11.8) is a hobby shared by many amateur astronomers today. In modern times, few professional astronomers spend much time in the search for new comets, although they may snare one from time to time in a photograph. In previous times, however, when data concerning comets were badly needed, comet hunting was a pursuit of many professional astronomers. We shall be particularly interested in the comet hunter Messier.

A difficulty experienced when one is comet hunting is that comets are usually quite faint when first seen. To be sure that one actually has seen a comet, one must wait for it to move with respect to the fixed stars. It is easy to confuse comets with other faint, fuzzy-looking objects. We have seen how William Herschel initially assumed he had found a comet when he had actually found the planet Uranus.

Charles Messier ("Mess-yay'," French astronomer, 1730–1817), led to astronomy by an eclipse of the sun which he witnessed at the age of 18, became, in the words of Louis XV, "a ferret of comets." It was a difficult profession. Messier had been eager to be the first to see Halley's comet after Halley's prediction that it would return. Messier was bitterly disappointed when a farmer saw it first, a month before he did. Worse yet, the farmer had merely used his naked eye.

Messier's List of Nebulas Another source of confusion for comet hunters is the presence in the sky of nebulas. The Latin word **nebula** means "fog, mist, or cloud," and this is a good description of the appearance of a nebula through binoculars or a small telescope. But it is also a good description of a distant comet, especially before it develops a tail. There is always the possibility of temporarily misidentifying a nebula as a distant comet.

Several times in his early comet-hunting days, Messier had been temporarily distracted by one or another nebula he had watched for several days before he was sure it was not moving relative to the stars. This gave him the idea of compiling a list of all these nuisances (as Messier thought of them) so that he would not be misled again. Eventually his list of nebulas came to 102 objects which have become known as Messier objects or M objects. We have already seen some M objects in earlier chapters and will see more in what follows. For example, the Crab nebula (Figure 14.2) is M1, the Great nebula in Orion (Figure 13.5) is M45, and the Ring nebula (Figure 13.17) is M57.

William Herschel Studies Nebulas Messier's list of nebulas was published in 1781, and a copy was sent to William Herschel (Box 15.1). Herschel decided to observe all the M objects, a task that was easy for him because of his far superior telescopes. Then, during his third review of the sky, he decided to make an expanded list of nebulas. His method of "sweeping the sky" consisted of setting the telescope at a stationary position, pointed at some location on the meridian. He then let the spinning of the earth cause the celestial objects to pass in review and recorded each one on a sky chart. Begun in 1783, this program lasted 20 years and raised the total number of known nebulas to 2500. Herschel's method of numbering nebulas has been replaced by the more recent New General Catalogue (NGC) which has expanded and updated his work. Now, each nebula has an NGC number: The Crab nebula is M1 or NGC 1952, for example. More than 7800 items are now designated by NGC numbers.

Herschel did more than merely locate nebulas. He also was able to see that a large majority of them were not actually "clouds of light" at all but rather clusters of stars so crowded together that they had appeared to be a blur to Messier. Technically, one says Herschel was able to **resolve** some of the nebulas into stars. In such cases, the object's classification was changed from "nebula" to "star cluster." In addition to star clusters, Herschel recognized some genuine clouds of gas, round in appearance, for which he coined the term **planetary nebulas** (see Section 13.9). This left a number of nebulas that still appeared to be unresolved into stars by Herschel's best efforts. He supposed that telescopes larger than his would show that most of them were also star clusters, and was partly correct. A much larger telescope was constructed in 1845, 23 years after his death. It resolved some of the nebulas that had eluded Herschel, and it did much more.

16.2 LEVIATHAN SEES THE SPIRAL NEBULAS

We have already seen, in Section 14.2, how Rosse used his mighty 72-inch telescope, Leviathan, to investigate the Crab nebula (see also Figure 16.2 and Box 16.1). Among his other projects using the tele-

BOX 16.1

LORD ROSSE AND LEVIATHAN

Can you pull out Leviathan with a hook? . . . Will it enter into an agreement with you to become your slave for life?

Job 41

William Parsons (1800–1867) was born into the aristocracy and a life of wealth and ease. When he was 21, he entered Parliament. When his father died in 1841, he became the Third Earl of Rosse, inheriting his father's wealth and estate in central Ireland, including Birr Castle and its magnificent grounds.

Rosse (the name by which he is best known) had a knack for machinery and construction. He also had the money to pursue his interests. He began to experiment with telescope construction in 1827 and gradually developed the necessary skill and equipment.

By the early 1840s, Rosse felt ready to break Herschel's record by producing a mirror twice the diameter of his own previous success, a new mirror 72 inches (1.8 meters) in diameter. Merely pouring the molten metal (glass mirrors were not yet used) for the mirror into the mold proved to be a considerable challenge, and four attempts failed. By the end of 1842, the fifth was cooling satisfactorily.

Rosse then turned to the construction of the mounting for the telescope. He built a 56-ft-long (17 meter) wooden tube for the telescope and two walls of masonry 56 ft (17 meters) high. The tube was raised between these two walls by means of a chain passing under the tube (see Figure 16.2). Moving the telescope up and down required two strong assistants.

By early 1845 the 4-ton (3600 kg) mirror was in place, and the telescope could begin its work. It was so huge that it received a nickname taken from the Book of Job—Leviathan ("Lev-eye'-a-thin"), a huge sea monster or whale.

Rosse may have asked himself whether or not Leviathan would indeed become his "slave for life," because it was not easy to use. It was limited by the masonry walls and so it could be pointed only along the meridian and roughly 10° on either side. It was heavy and ponderous to move and was so large that it required the finest seeing (stillness of the atmosphere) to do useful work. Here, indeed, was the greatest problem, for it was located on Rosse's estate in Ireland. This is as far north as Edmonton, Canada, or Moscow but much milder in climate, partly because of the heat delivered by the Gulf Stream. However, the Gulf Stream also delivers the famous cloudy, drizzly weather of the British Isles, and Ireland is one of the worst possible places for a large telescope. But Rosse seized his scattered chances to observe and used Leviathan skillfully, producing a surprising amount of fascinating work.

---

scope, he observed and sketched as many of the nebulas from Herschel's catalog as he could, resolving many more of them into stars.

One of the objects Rosse studied most closely in 1848 was M51, which is found not far from the end of the Big Dipper's handle. Messier had reported it to have a double structure. Rosse confirmed this and found that he could see a wisp of the larger nebula reaching out to the smaller one. As he tried higher magnifications, the detail he could see became more complex. The overall pattern was one of a "spiral or whirlpool" in his words (see Figure 16.3). Rosse found himself totally at a loss to explain this spiral in the sky. In a large, modern telescope M51 is one of the wonders of the sky (see Figure 16.4).

Rosse found other nebulas that exhibited a spiral structure. He reported about 15 **spiral nebulas,** as they came to be called, in the first

Figure 16.2 Rosse's 72-inch telescope.

Figure 16.3 The Whirlpool galaxy, M51: drawing by Rosse. This was one of the first of the spiral nebulas discovered by Rosse. He wrote of the "difficulty of forming any conceivable hypothesis" concerning the nature of this object. He did say that it was not probable that "such a system should exist without internal movement," presumably meaning rotation.

five years of observation. Perhaps by analogy with a terrestrial whirlpool, he assumed that the spiral nebulas must be rotating, but, try as he might, he never detected any motion.

Thus astronomers learned of the mysterious spiral nebulas. How they investigate these nebulas is the subject of this chapter.

Figure 16.4 The Whirlpool galaxy, M51: A modern photograph. This galaxy is roughly 14 million ly from us and about 30,000 ly in diameter. It is famed for the wealth of detail it exhibits and its apparent connection to another, smaller galaxy which appears to have distorted the whirlpool's shape by means of gravitation.

16.3 EARLY IDEAS ABOUT SPIRAL NEBULAS

The Island Universe Hypothesis of Immanuel Kant Even before Rosse had found that some nebulas are spiral in form, they had been the subject of much speculation. Kant's book on the Milky Way (Section 15.1) concluded that the Milky Way is a galaxy of stars. As evidence that this was not a frivolous suggestion, Kant pointed to the nebulas, especially those with a round or elliptical outline (see Figure 16.5). He proposed that these are other enormous disks of stars seen face-on or more or less tilted to our line of sight. Although they were too distant to be resolved into separate stars, Kant boldly asserted that they are huge galaxies, "island universes" like our own, seen at enormous distances.

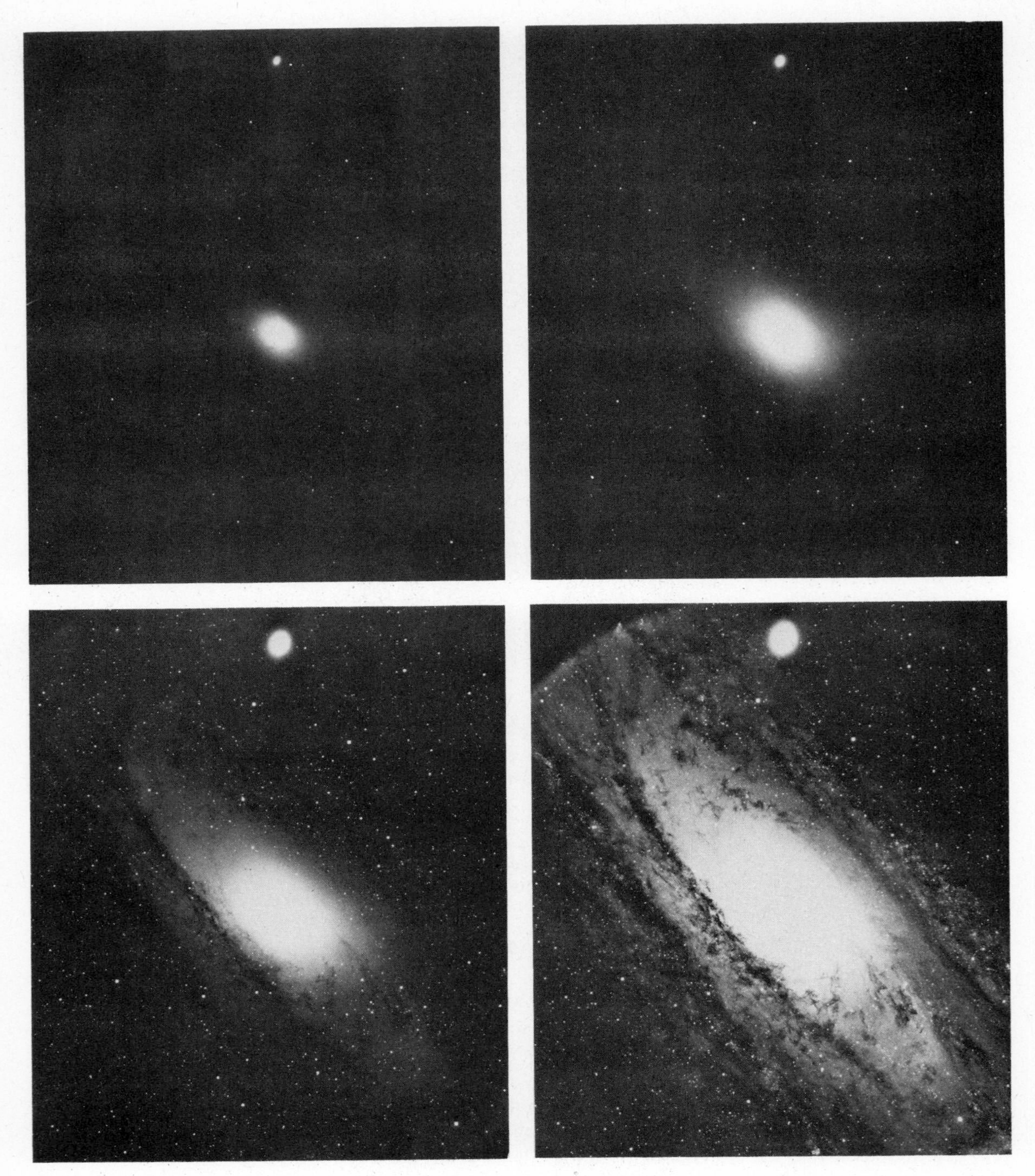

Figure 16.5 Four exposures of the Andromeda "nebula," M31. The Andromeda nebula was photographed four times by the same telescope. The exposure times were 1, 5, 30, and 45 min, respectively. The point was to show how the nebula appears through binoculars, through a small telescope, through a larger telescope when viewed with the eye on a good night, and as photographed using a very large telescope.

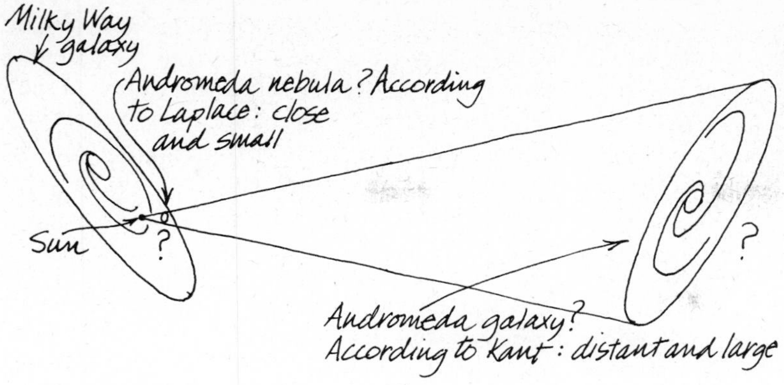

Figure 16.6 Is the Andromeda object a small nebula or a huge galaxy? The same lines of sight could encompass a small nearby object or a large distant one. How does one decide which is correct? This long-standing puzzle was eventually resolved in 1923, as related in Section 16.6

The Nebular Hypothesis of Pierre-Simon Laplace Often in science, the same evidence is used by several scientists to bolster opposing points of view. Laplace took a quite different view of nebulas. He proposed a theory to explain the origin of the solar system, a theory that became known as the **nebular hypothesis** (see Section 11.10). The theory suggests that, as the sun was forming, the planets condensed out of its huge, flattened, spinning atmosphere. As evidence that such a proposal was a logical one, he pointed to the round and elliptically shaped nebulas Herschel had been unable to resolve into stars. Most of these nebulas have a bright, starlike nucleus (central region), and Laplace was led to conclude that they are young stars surrounded by a flattened, rotating atmosphere. According to Laplace, these nebulas are examples of future solar systems whose planets have not yet condensed.

Pierre-Simon Laplace ("La-plahss'," French mathematician and physicist, 1749–1827), was highly gifted in mathematics. He built upon Newton's mathematics and dynamics and produced a monumental five-volume work on the motions of the heavens.

The interpretations of nebulas by Kant and Laplace could not be further apart. Laplace's theory makes them relatively small objects, no larger than our solar system and therefore relatively nearby. Kant saw them as huge galaxies and therefore almost unthinkably distant. No one could judge who was right merely by observing the apparent sizes of the nebulas (see Figure 16.6).

The discovery by Rosse that some nebulas are spiral in nature did not settle the question. Both Kant and Laplace required their nebulas to be rotating, and the evidence of rotation exhibited by the spiral structure did little to solve the dilemma.

16.4 FURTHER STUDIES OF NEBULAS, 1850–1920

Astronomical opinion continued to swing on the question of the spiral nebulas. Are they giant galaxies of stars scattered throughout the depths of space or are they inside our galaxy, implying that we are alone in an almost empty universe? Agnes Clerke, astronomer and historian of nineteenth-century astronomy, had a firm opinion which followed the general consensus. She wrote in 1890, "The question whether nebulas are external galaxies hardly any longer needs discussion. It has been answered by the progress of discovery. No competent thinker, with the whole of the available evidence before him, can now, it is safe to say, maintain any single nebula to be a star system of coordinate rank with the Milky Way. A practical certainty has been attained that the entire contents, stellar and nebular, of the [universe] belong to one mighty aggregation. . . ."

William Huggins (English astronomer, 1824–1910) and his wife built an observatory near London and were pioneers in applying the spectroscope to astronomy. Among their list of firsts were measuring the radial velocity of a star, observing the emission spectra of nebulas, showing that a nova has an envelope of hydrogen, identifying carbon in a comet, photographing spectra, and starting to classify stars by means of their spectra.

Distinguishing Between Types of Nebulas Using the Spectroscope But then what was one to say of the discoveries of William Huggins concerning nebulas? In the early 1860s he and his wife had studied stars with their new spectroscope. Then Huggins turned it to nebulas. The first nebula he inspected was a planetary nebula. He was used to observing the continuous, colored spectra of the stars, spectra crossed by numerous thin, dark lines, and the nebula provided a surprise. "No such spectrum as I expected! A single bright line!" When Huggins got over his amazement and studied the nebula's spectrum again, he found other, fainter glowing lines. He knew how to interpret this, for he was familiar with Kirchoff's laws of spectroscopy (Section 9.5). Law number 2 applied here: A mass of glowing gas at low pressure will emit an emission spectrum (bright lines on a dark background). Huggins quickly checked other nebulas and found that the spectroscope revealed that they too are masses of gas at low pressure. Famous examples of emission nebulas appear in Figures 13.5 through 13.7 and 15.13 through 15.16.

But Huggins also reports, "A few days later I turned the telescope [with an attached spectroscope] to the Great Nebula in Andromeda. Its light was continuous throughout." That is to say, the spectrum of the Andromeda nebula is a continuous band of color like that of a star. Huggins took this as evidence that this and other spiral nebulas like it are actually composed of numerous individual stars too far away to be resolved with a telescope. True, Huggins could see no dark lines in the spectra, but this was thought to be because the light from the nebula is a mixture of the light from its many stars. The various spectra overlap and make the dark lines indistinct.

Once again the spectroscope had shown its worth. Spiral nebulas have continuous spectra, and so are made up of stars. They could not be resolved into stars through a telescope and so are very distant. But are they outside our galaxy? Some said yes, but most said no.

Evidence That Spiral Nebulas May be Close and Small Vesto Melvin Slipher (Box 16.2) made an observation that tipped the scales of opinion again. He studied the nebula seen in the Pleiades (see Figure 13.2), one of the nearest open clusters. Slipher used a spectroscope and saw a continuous spectrum. Surely the nebulosity in the Pleiades cannot be composed of stars; we can see stars in the nebulosity itself. Slipher concluded that the gas and dust in the cluster was simply reflecting the light of the nearby stars. Other examples of such **reflec-**

BOX 16.2

SLIPHER, MASTER OF THE SPECTROSCOPE

In 1900, Percival Lowell (see Box 10.5) ordered a new, superbly made spectroscope for the 24-inch refracting telescope at his observatory. The following year he hired a 26-year-old astronomer named Vesto Melvin Slipher ("Slai'-fer," 1875–1969) to install the new spectroscope and use it to observe the spectra of heavenly objects. By 1902 Slipher had installed the instrument and had began observations, which were to continue until his retirement in 1952. Because of the excellent instruments at the observatory, the excellent atmospheric conditions at Flagstaff, and his outstanding skill as an observer of celestial spectra, Slipher's career consisted of a string of discoveries of fundamental importance.

One of the first projects Lowell assigned to Slipher was to observe the atmospheres of the planets with the spectroscope. Using the Doppler effect, he determined the periods of rotation of the planets and found, among other things, that Venus rotates so slowly that its period cannot be measured accurately by that method. Slipher also obtained the first spectra showing absorption bands in the atmospheres of Jupiter, Saturn, Uranus, and Neptune. The cause of these bands was unknown for almost 30 years until it was recognized that they indicate the presence of methane and ammonia in these planets' atmospheres.

As we saw in Section 16.13, it was Slipher's work on the spectra of galaxies that produced his most sensational results.

---

tion nebulas, as they are called (see also Box 15.3), were found and this, in turn, threw the interpretation of the spiral nebulas into doubt. Slipher suggested that the Andromeda nebula, for example, might only be a reflection nebula so distant that we are unable to see the stars in it but do see the light reflected from its clouds of dust. The implication is that spiral nebulas are clouds of dust and gas with imbedded stars and therefore that they are inside our galaxy.

16.5 THE SHAPLEY–CURTIS DEBATE, PART TWO

There was a second phase of the Shapley-Curtis debate of 1920. In Section 15.5, we saw how Shapley had upheld the then radical opinion that our galaxy is huge and extends well beyond the portions we can see. Another question debated was the location of the spiral nebulas: Are they in our galaxy? This time, Shapley took the part of the astronomical establishment and answered yes. Curtis was one of the few arguing that spiral nebulas are actually huge galaxies outside the Milky Way and, indeed, are analogous to the Milky Way galaxy.

Curtis' main line of attack was a series of observations he had made of ordinary novas (see Section 12.8). He had accumulated data concerning novas appearing to be inside the Andromeda nebula. By comparing their average apparent brightness with the novas known to be in the Milky Way galaxy and using $L = bR^2$, he had derived a distance of 500,000 ly for the Andromeda nebula (or more properly, the "Andromeda galaxy," assuming he was right).

Shapley discussed a nova that had been seen in the Andromeda nebula in 1885. It was estimated that this nova, named S Andromedae, had reached an apparent brightness nearly great enough to be visible to the naked eye (see also Section 14.3). How could a nova appear that bright if it were in a galaxy 500,000 ly away? Curtis replied that S Andromedae may be an exception. The other novas in Andromeda were *much* dimmer. Perhaps it was an unusual nova or even a close nova lying in the same line of sight as the Andromeda nebula.

At the time of the debate, the concept of a supernova had not yet been clarified.

Shapley summarized the evidence concerning spiral nebulas we have discussed in earlier sections and then referred to a report by an astronomer named van Maanen, a report that seemingly clinched the debate for him. Van Maanen had studied the spiral nebula M101. He had carefully measured the positions of various features of the nebula in an attempt to detect its rotation. In 1916 he had reported that each feature had a proper motion of 0.02 second of arc per year and that the pattern was one of circular motion within the nebula. He obtained similar results for several other spiral nebulas. To Shapley this was clear evidence that the spirals must be close. If the spirals were as distant as Curtis required, the speed necessary to cause van Maanen's proper motions would be improbably great, in some cases greater than the speed of light.

As in the first part of the debate, these and many other lines of argument failed to win many converts. Most people left the debate with their original opinion unaltered. One thing was agreed upon, more research was necessary. Eddington (see Box 13.2) had decided that spiral nebulas must be galaxies. He quipped that he had no positive evidence for his opinion, but he felt it must be so because it was so much more interesting that way.

16.6 EDWIN HUBBLE AND THE ANDROMEDA GALAXY

In early 1917, the 100-inch Hooker telescope was nearing completion, and Hale began a search for astronomers to staff the observatory. He made an offer to a 27-year-old man who was on the staff at Yerkes and about to earn his Ph.D. The astronomer was Edwin Hubble (see Box 16.3). By 1919 Hubble and the Hooker telescope, the first telescope larger than Rosse's Leviathan, had become a team. Hubble began to photograph nebulas and to improve photographic techniques.

In 1923, the question of the location and nature of the spiral nebulas was finally answered. After many trials, Hubble had found a way to obtain ultrasharp photographs of the Andromeda nebula using the 100-inch telescope. His best results showed the outer fringes of the nebula resolved into countless individual white dots—stars. Aha! At long last the Andromeda nebula had been resolved—from now on it was to be called the *Andromeda galaxy* (see Figure 16.7).

Hubble searched among the individual stars he now could see and

BOX 16.3

EDWIN HUBBLE AND THE GALAXIES

Equipped with his five senses, man explores the universe around him and calls the adventure science.

Edwin Hubble

Edwin Hubble (born in Marshfield, Missouri, in 1889) could have had a number of very different careers. He was an excellent law student and used a Rhodes scholarship to obtain a law degree at Oxford University. After returning to the United States and practicing law for less than a year, the lure of the sky became so strong that he obtained a position at Yerkes Observatory in Wisconsin; he received his Ph.D. in astronomy in 1917.

By that time he had earned such a high reputation that Hale offered him a position at the Mount Wilson Observatory where the 100-inch telescope was being constructed. Hubble saw it as his duty to enlist in the infantry first, however, and he served during World War I in France. Having survived the war, he joined the staff at Mount Wilson where he remained until his death in 1953 at the age of 64.

The 100-inch telescope became operational on Mount Wilson at the time Hubble began his work there. The text recounts the magnificant work he did with that instrument, bringing the galaxies into the realm of astronomy. Hubble is best known both as the man who solved the age-old question of the nature of the Andromeda and other galaxies and as the discoverer of the invaluable red shift, or Hubble's, law.

---

located first one and eventually 35 Cepheid variables. Now he could estimate the distance to the Andromeda galaxy with some confidence. After obtaining the light curves of the Cepheids and measuring their apparent brightness, Hubble announced the distance to the Andromeda galaxy: His method yielded a distance of about 1 million ly. Curtis must have been pleased, for this was roughly the distance he had estimated for the same object.

But what of all the earlier evidence Shapley and others had used to conclude that the spirals were close and small? Huggins saw a continuous spectrum, because the spirals are indeed composed of stars. Slipher's reflection nebulas were simply very different objects with similar spectra. The "nova" S Andromedae was later recognized to be a phenomenon vastly different from an ordinary nova; it was a supernova. It was entirely possible that a supernova could become nearly visible to the naked eye from a distance of 1 million ly. It was not until 1935 that the puzzling proper motions seen by van Maanen were explained away. Hubble repeated measurements similar to van Maanen's and could find no evidence of proper motion within galaxies. To this day it is not clear what had led van Maanen so far astray. One must always be skeptical of even seemingly valid data, especially if it is difficult to obtain. (See Section 10.3 on Mercury's rotation, which provides another striking example.)

At last it seemed safe to say that we are not alone in the universe, in the sense that space is populated with many millions of spiral galaxies which are, at least superficially, similar to our own Milky Way galaxy.

Figure 16.7 The Andromeda galaxy, M31. A subject of much controversy over the ages, its nature was definitively described by Hubble. (See also Plate 22.)

The Size of the Andromeda Galaxy: A Problem? It was a simple matter for Hubble to compute the size of the Andromeda galaxy. Judging from its 3° angular diameter and 1-million-ly distance, it should have an actual diameter of about 50,000 ly. Hubble also measured the distance to M33 (Figure 16.12), another spiral galaxy lying not too far from the Andromeda galaxy in the sky. Hubble found that it appears to be about the same distance from us as the Andromeda galaxy but only about one-half its size. As judged by their large angular diameters, these two galaxies are clearly the two nearest large galaxies.

These results concerning the sizes of these two galaxies made many astronomers uneasy, since the diameter of the Milky Way galaxy originally measured by Shapley was 300,000 ly. Many astronomers preferred to believe that all spiral galaxies were roughly the same size, and yet measurements implied that the Milky Way galaxy is 6 times larger than the Andromeda galaxy and 12 times larger than M33. Could something still be wrong with the methods used?

By 1930 Trumpler had discovered interstellar absorption, and this reduced the estimated size of our galaxy to the presently held value, 100,000 ly (see Section 15.7). But still our galaxy seemed to be twice as large as the Andromeda galaxy. Was there still a problem? The answer was yes, as we shall see in the next section.

But in the meantime, our view of the universe had changed dramatically. We now could picture a universe populated with huge disks of stars separated by huge distances. We live rather near the edge of one of these disks which is one of millions upon millions of similar objects. One writer has pointed out that three users of the telescope were the principal pioneers in three, ever expanding, realms. As Galileo was to the solar system, so Shapley was to the Milky Way, and so Hubble was to the galaxies. Will a fourth ever be added to this exalted company?

16.7 WALTER BAADE AND STELLAR POPULATIONS

Baade Resolves the Inner Regions of the Andromeda Galaxy The early 1920s were exciting times for astronomers. For example, the research projects of Shapley and Oort were revealing the size of our galaxy while Hubble was unveiling the nature of the other galaxies. During this time, a young German astronomer destined to further this progress was earning his spurs in Germany. He was Walter Baade.

In 1929 Baade moved to the United States and spent many years working with the telescopes at Mount Wilson and Mount Palomar. By the later 1930s, he had become interested in the Andromeda galaxy. Not only did it appear too small compared to our galaxy, but the globular clusters seen surrounding it had been found to be about four times less luminous than those surrounding our galaxy. This result was based on the clusters' apparent brightness and on Hubble's determination of the distance to the Andromeda galaxy. Of course, it might simply be

Walter Baade ("Baa'-deh," German-American astronomer, 1893–1960) discovered the asteroid Hidalgo which long held the record for the asteroid that travels the greatest distance from the sun, about 10 AU. Besides his discovery of stellar populations, he also did observational work on supernovas, much of it with Fritz Zwicky who figured prominently in Chapter 14.

true that the Andromeda galaxy is a smaller galaxy with faint globular clusters, but Baade's intuition told him otherwise. He decided to inspect the Andromeda galaxy more closely. Hubble had resolved the outer regions of the Andromeda galaxy into stars but had not been able to do so for the inner region. Baade decided to make the attempt, feeling that this might help clear up the problems.

At about this time fate stepped in as the United States began to mobilize for World War II. Baade was called in by American officials in order to determine his wartime status. He was a German citizen but had every intention of becoming a U.S. citizen. In fact, he had obtained his initial citizenship papers sometime earlier. When asked to produce these papers, Baade could not; he had lost them. The officials decided they had no choice but to assign him the status of enemy alien and to restrict his movements for the course of the war. This could have been very discouraging for Baade. The bulk of his fellow American scientists were involved in the war effort which, because of his status, he could not join. He must have been quite pleased, though, when the officials declared his region of restriction to be the Mount Wilson area. Not only did he have the 100-inch telescope practically to himself, but the blackout of neighboring Los Angeles during the war made the skies around Mount Wilson darker than they probably ever will be again.

Even with these advantages, Baade knew he would have to use extreme care if he was to succeed in resolving the central regions of the Andromeda galaxy where even Hubble had failed. He estimated that he would need to expose the photographic plates for 9 hr to achieve his goal. But during this extended exposure, a number of factors could potentially ruin the needed resolution. There is always a faint glow in the sky from the aurora. It is usually not visible to the eye but is sufficient to fog a plate during a 9-hr exposure. Baade partially compensated for this by using plates that were less sensitive to the bluish light from the aurora and more sensitive to red light. During the 9-hr exposure, Baade carefully monitored the focusing of the telescope. It was an exacting, tiring job which took many hours to master. We have already discussed the fact that the seeing in the atmosphere is excellent only for short, unexpected instants (see Section 10.5). During such an instant, when the atmosphere was temporarily still, the star Baade used to judge the focusing would appear clearly. He then had to make an instantaneous decision as to whether or not the image was out of focus and, if so, in which direction. Then, at the same instant, he had to turn the focusing knob in the right direction by the precise amount. Baade's work was a display of devotion and patience reminiscent of that which we have observed in many of the great observers of the past.

By 1943, Baade had seen all his precautions pay off. "I had just managed to get in under the wire, with nothing to spare," he said. There on the plate was the evidence of Baade's success, the central regions of the Andromeda galaxy just barely resolved into tens of thousands of individual stars.

Baade Discovers Stellar Populations Baade studied the newly resolved stars and came to realize that they seem different, as a group, from the stars in the outer regions of the Andromeda galaxy. The revelation which emerged from Baade's work is that there are basically two "populations" of stars.

The stars in the central regions of spiral galaxies, along with those found in globular clusters and the halos of spiral galaxies have been classified as **population II** stars. They appear red when viewed as a group. For this reason, it was fortunate that Baade had used red-sensitive photographic plates when trying to resolve the central regions of the Andromeda galaxy.

The stars in the arms of spiral galaxies, including those in open clusters, as well as the sun and the stars in the sun's vicinity, are classified as **population I** stars. Viewed in bulk, regions of galaxies where population I stars reside appear considerably bluer than the regions where population II stars reside. As an aid to keeping the two populations straight, keep in mind that population I stars were the *first* to be carefully studied by astronomers living on a planet near the sun.

Studies of the spectra of the two populations of stars have revealed a major difference between them. Population II stars consist almost entirely of hydrogen and helium; elements heavier than helium amount to less than 1 percent of their composition. Population I stars are strongly contrasted in this respect. Their heavy-element content is roughly 2 to 4 percent.

Further study has revealed that Baade's two populations are two extremes based on location and heavy-element content. A continuous variation in stellar populations has been revealed between the two extremes. Some astronomers now find it convenient to divide all stellar populations into five gradations of population.

Astronomers have interpreted the two extreme populations of stars as representing two stages of star formation in a galaxy. Population II stars are thought to have formed in the earlier stages of a galaxy's life when it was more nearly spherical in shape. Population II stars are old and thus reddish when seen in bulk because of the high percentage of K and M stars, those with the longest lives. The bluish O and B stars died long ago. Furthermore, many of these stars are in the old, very luminous red giant stage.

Population I stars are thought to have formed more recently, after the spinning of the galaxy had flattened its shape. They are young and bluish when seen in bulk because the brightest ones are bluish main sequence stars. They have a higher heavy-element content because they were formed from hydrogen mixed with heavy elements produced in earlier supernovas (see Section 14.11). The discovery of populations of stars constitutes partial verification of the theory of stellar evolution; K and M stars live much longer than O and B stars. Also, heavy elements are formed in stars and were not present in the early stages of galaxy formation, a verification of the theory of the origin of the elements.

Figure 16.8 The two types of Cepheid variable stars. As shown by the dotted lines, a population I Cepheid is brighter than a population II Cepheid having the same period. If you were to observe a population I Cepheid but mistake it for a population II Cepheid, you would incorrectly infer a luminosity for it that is too small. This would lead you to infer that the object is closer than it really is.

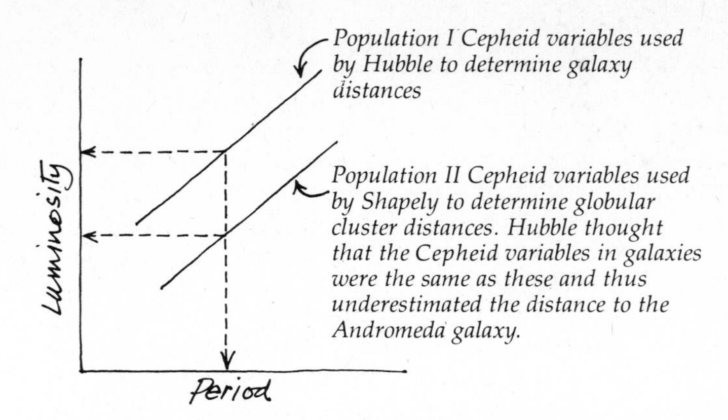

Refining and Correcting the Cepheid Variable Method Baade followed up his discovery and in 1952 came to a very important conclusion concerning stellar populations and Cepheid variables. He found that Cepheid variables, formerly thought to be of only one type, could also be separated into two populations. The Cepheid variables Shapley studied in globular clusters (Section 15.3) are population II stars. However, the Cepheid variables Hubble studied in the Andromeda and other galaxies are population I stars, the only type bright enough for him to see. The crucial point, Baade found, was that when he compared a population I Cepheid with a population II Cepheid with the same period, the two stars did not have the same luminosity. In fact, the population I Cepheid is about four times more luminous. Another way to describe this discovery is to say that, on a graph of luminosity versus period, the Cepheid variables fall on two distinct lines according to their population class (Figure 16.8).

Now when Hubble, in the early 1920s, measured the distance to the Andromeda galaxy, he had been observing the brighter population I Cepheids without realizing it. But he used Shapley's graph of luminosity versus period for Cepheid variables to determine the luminosities of the Andromeda Cepheids. Without realizing it at the time, Shapley had used dimmer population II Cepheids, and as a result Hubble had underestimated the luminosities of the Andromeda Cepheids. An underestimation of luminosity leads to an underestimation of distance, as well. (If a star is dim, it must be close in order to look as bright as it does.)

The upshot of Baade's discovery is this. After correction for interstellar absorption, Shapley's determination of the size of our galaxy, 100,000 ly, is correct. Hubble's method of determining the distances to galaxies, using Cepheid variables, underestimated their distances by a factor of 2. The Andromeda galaxy is more nearly 2 million ly away. This adjustment results in an increase in the calculated size of the Andromeda galaxy. If it is twice as far away as was thought, it must be twice as large as was thought. The diameter of the Andromeda galaxy now works out to be roughly 100,000 ly, which is comparable to the estimated 100,000-ly diameter of our own galaxy. The newly determined distance to the Andromeda galaxy also explained why the globular clusters surrounding it were formerly thought to be four times too dim. Since the new analysis showed that they are twice as far away, the globular clusters must be $2^2 = 4$ times as luminous as formerly estimated.

By 1952, all the distance and size measurements we have been discussing in this and the previous chapter had been brought into harmony. The Milky Way galaxy and the Andromeda galaxy are both about 100,000 ly in diameter and separated by 2 million ly, center to center. In a scale model of the two galaxies, placing one on the East Coast of the United States and the other on the West Coast, the two galaxies would have a diameter of roughly 200 km (125 mi), about the size of a small New England state. For comparison imagine a scale model having the sun on the East Coast and Alpha Centauri on the West Coast. Each star would be only about 20 cm (8 inches) in diameter.

As we stand watching the Andromeda galaxy on a dark night, we are looking at another gigantic galaxy like our own. The billions of stars in the galaxy combine to make only a faint cloud of light visible to our naked eye. It is amazing that the galaxy is visible at all across the immense gulf of 2 million ly. When one finds the Andromeda galaxy with the naked eye, one is viewing the most distant object the naked eye can see. The light entering one's eye has been traveling for the last 2 million years. Hard to imagine, but thus saith the Cepheid variables.

16.8 HUBBLE CLASSIFIES THE GALAXIES

As Russell obtained data on the stars, he found a wide variety of types, cool–dim, cool–bright, hot–bright, etc. Then he found order in the data, an order he displayed in the Hertzsprung-Russell diagram (see Section 12.4). Similarly, as Hubble accumulated data on the galaxies, he originated an ingenious diagram which brought order to what was at first a bewildering variety of sizes and shapes. In another sense, Hubble's idea is somewhat analogous to the scheme devised to classify the stars into various categories (O,B,A,F,G,K,M, Section 12.3). Hubble classified nearly all galaxies into three broad categories, categories showing a gradual progression from one type to the next.

The three categories of galaxies are displayed in Figure 16.9. This figure, nicknamed "the tuning fork," shows normal spiral galaxies, barred spiral galaxies, and elliptical galaxies.

Normal Spiral Galaxies The first type of galaxy is found on the upper right of the tuning fork. Called **normal spiral galaxies** (labeled "S"), they are composed of two main features, the nucleus and the arms. Two principal arms are often attached to the nucleus at diametrically opposite points. Many normal spirals have smaller, shorter arms as well. The normal spirals are further classified into subgroups a, b, and c.

Figure 16.9 Hubble's "tuning fork" classification of galaxy structures.

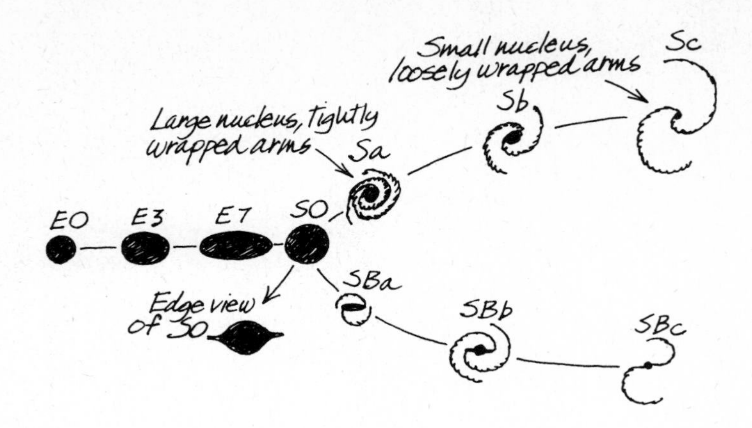

Sa galaxies have very large nuclei, consisting (as do all galactic nuclei) chiefly of old, dim stars. In Sa galaxies, the nucleus constitutes a large fraction of the whole galaxy. The arms of Sa galaxies, consisting (as do all galaxy arms) of the full range of bright and dim stars, dust, and gas, are tightly wound around the nucleus. At the other end of the scale, Sc galaxies have relatively tiny nuclei and loosely wound, open arms (see Figures 16.10 through 16.12). The Andromeda galaxy (Figures 16.5 and 16.7) is an Sb galaxy.

Barred Spiral Galaxies A related type of galaxy is displayed on the lower right of the tuning fork. These are **barred spiral galaxies** (labeled "SB") and are similar to S galaxies except that either their nuclei are elongated or two distinct bars of stars emerge in opposite directions. Each of the two principal arms of the SB galaxy is attached to one end of the bar. Again, the subgrouping a, b, and c is used with the same meaning (see Figures 16.13 and 16.14).

Elliptical Galaxies The third main class of galaxy is found on the left side of Hubble's tuning fork, the **elliptical galaxies** (labeled "E"). These are the plainest looking of the galaxies; they exhibit no details in structure. They are so named because they are elliptical in outline. They seem to consist of nothing but stars scattered throughout the volume of the galaxy. The stars are more densely packed near the center, but are not situated in any detectable arm arrangement (see Figure 16.15).

There seems to be an almost complete absence of gas and dust in E galaxies, another factor contributing to their plain appearance. Spiral galaxies all exhibit quantities of dust and gas in their outer regions, which obscure the stars behind them. This contributes to their interesting, patchy appearance. Spirals that can be observed nearly edge-on usually exhibit a narrow, strong concentration of dark dust and gas along their central plane (see Figure 16.16). On the other hand, the nuclei of spiral galaxies are rather similar to those of elliptical galaxies in that they are flattened, quasispherical regions of stars having little dust and

Figure 16.10 The "Sombrero galaxy," M104, an Sa galaxy. This splendid galaxy features an enormous dust-free nucleus surrounded by a thin disk of dust, gas, and stars. The arms are tightly wound and not easy to see. M104 is about 40 million ly from us and 26,000 ly in diameter.

gas. Some think of an elliptical galaxy as the nucleus of a spiral galaxy lacking arms, dust, and gas.

E galaxies are subgrouped under the numbers 0 to 7, according to the amount of flattening (or eccentricity) observed in their outline. An E0 galaxy is circular in outline, while an E7 is the most flattened type of elliptical galaxy.

At the point of the tuning fork, where all three classes of galaxies meet, Hubble placed the S0 systems (see Figures 16.17 and 16.18). These are like E galaxies in that they lack gas, dust, and arms, but like S (or SB) galaxies in that they resemble the nuclei of S or SB galaxies in outline. Some astronomers believe that an S0 system was once a normal spiral galaxy that collided with another galaxy. The interaction could have swept all the gas and dust out of the galaxy, spelling the end of star formation. Since spiral arms are marked out principally by bright, hot, young stars, none can be formed.

Figure 16.11 M81, an Sb galaxy. This is one of the best views available of a spiral galaxy. Bright, young stars trace out delicate arms which wind around a moderate-sized nucleus. M81 is only about 10 million ly from us, which explains how such a detailed photograph is possible. It is quite large, having a diameter of about 50,000 ly.

Irregular Galaxies The usefulness of Hubble's scheme of galactic classification is demonstrated by the fact that, since he proposed it in 1925, nearly all galaxies discovered fit nicely into one or another of the categories. Of the relatively few galaxies that do not fit Hubble's scheme, most are relatively formless groupings of stars which are not symmetrical and lack a well-formed nucleus, but which usually have large amounts of gas and dust. These are known as **irregular galaxies** (labeled "Irr").

During the round-the-world trip by Magellan's fleet in 1522, the ships traveled far into the southern hemisphere. There they observed two large clouds of light in the unfamiliar skies, clouds appearing to be about as bright as the Milky Way. Now known as the Magellanic clouds, these are two irregular galaxies which are companions of our

Figure 16.12 M33, an Sc galaxy. This galaxy is the third largest in the local group, the cluster of galaxies in which our galaxy resides (see Section 16.12). It has a small nucleus and loosely wound arms. It is roughly 2 million ly from us and about 50,000 ly in diameter.

galaxy. The Large Magellanic cloud (Figure 16.19) is about 150,000 ly away and the Small Magellanic cloud is about 200,000 ly away. The shapes of the clouds seem to show evidence that these irregular galaxies are distorted in shape by the gravitational pull of the huge, nearby Milky Way galaxy. The diameters of the two clouds are about 50,000 ly and 20,000 ly, smaller than that of the Milky Way galaxy but by no means tiny. These two very close galaxies have been extensively studied. They are sufficiently close that many individual objects, such as stars, clusters, and nebulas, can easily be observed in them. As we have seen, one of the greatest astronomical breakthroughs of the century occurred as a result of Leavitt's studies of the Cepheid variables in these galaxies (Section 12.9).

When the *observed* galaxies are counted, one finds that about 70

Figure 16.13 An SBb barred spiral, NGC 1530.

Figure 16.14 An SBc barred spiral, NGC 7479.

percent of them are spiral and barred spiral galaxies, about 20 percent are elliptical, and about 5 percent are irregular. These figures undoubtedly reflect an effect of observational selection, however, because spiral nebulas are generally larger and more luminous than elliptical galaxies. It is felt that there probably are very many dwarf elliptical galaxies in space, which are too dim and too distant to be observed. Of the galaxies near ours, those in the local group of galaxies (see also Section 16.12), about 60 percent are small, elliptical galaxies. However, some of the largest, most massive galaxies known are the rare giant elliptical galaxies. M87 (Figure 14.15) has roughly half the diameter of our own galaxy but perhaps 50 times more mass.

Some of the most interesting looking galaxies are considered peculiar because they do not fit well into any of Hubble's categories. Some of these galaxies are shown in Figures 16.20 and 16.21.

Figure 16.15 A giant elliptical galaxy, M87, type E0. This galaxy is known to have about 10,000 globular clusters surrounding it. Can you see any of them close to the galaxy? M87 is about 40 million ly from us and about 40,000 km in diameter. Recent research has disclosed the possibility that M87 may have a giant black hole (See section 17.5) at its center.

Figure 16.16 A spiral galaxy seen edge-on. This is NGC 4565, an Sb galaxy. It is a splendid illustration of the thin disk of gas and dust running through the central plane of such a galaxy while the nuclear bulge protrudes at the center. Our galaxy seen edge-on might look much like this one.

Figure 16.17 An S0 galaxy, NGC 5866. Possibly this was a spiral galaxy which was stripped of its gas and dust.

Figure 16.18 A peculiar S0 galaxy, NGC 2685. What are we to make of this unusual galaxy which has two axes of symmetry? Could another galaxy have collided or had a near miss with it, leaving it in this unusual condition?

Figure 16.19 The Large Magellanic cloud. Our galaxy's largest near companion in space, this is usually classified as an irregular galaxy, although some feel that it may be a barred galaxy which has been severely distorted by our galaxy's gravitation.

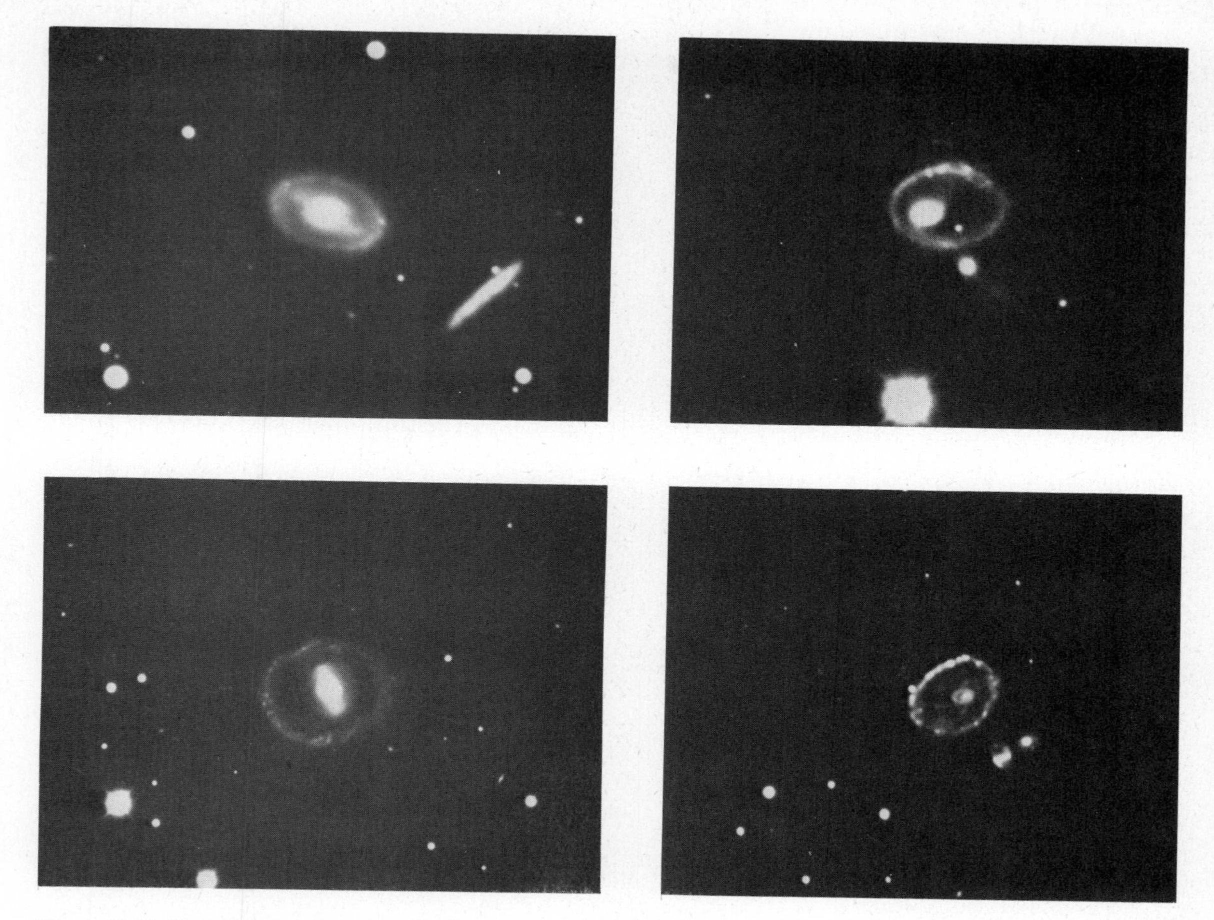

Figure 16.20 Four ring galaxies. Not all galaxies fit neatly into Hubble's scheme. These rare ring galaxies, like many other peculiar galaxies, may be the products of a collision or a near miss between galaxies.

16.9 THE ROTATION OF GALAXIES

As Kant hypothesized (Section 15.1) and Oort demonstrated (Section 15.6), our galaxy rotates. When a spectroscope is used to observe galaxies we see nearly edge-on (as in Figure 16.16), it reveals that one side is moving toward us and the other is receding. It is clear that all galaxies rotate. If they didn't, gravitation would cause them to collapse.

In the outer regions of the disk of our galaxy, a region such as the one in which the sun resides, our galaxy exhibits **differential rotation;** the farther a star lies from the galaxy's center, the more slowly it moves and the longer it takes to complete one orbit.

Nearly spherical galaxies, particularly E0 through E3 galaxies, exhibit a different pattern. In such galaxies there is little or no organized rotation. Each star follows an orbit around the galaxy's center, but the stars' orbits are oriented randomly. The stars and globular clusters in

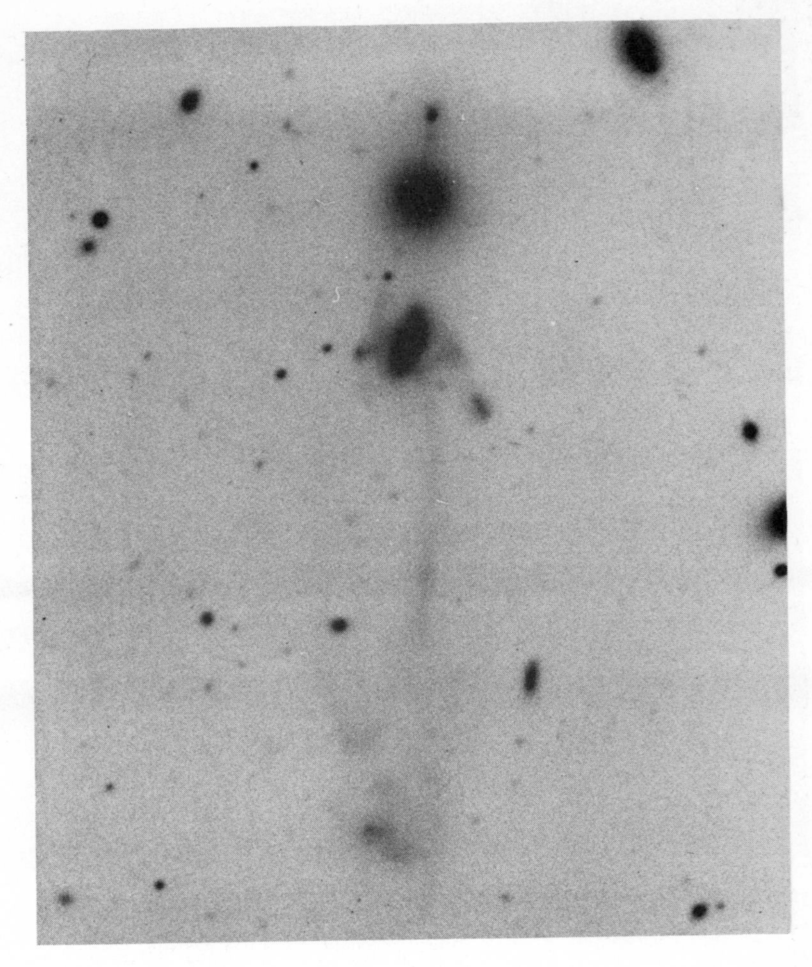

Figure 16.21 A peculiar system of galaxies, NGC 3561. This is a negative print, which shows more detail. On such a print, regions of light appear dark, and vice versa. Near the top is a small galaxy connected to a larger one which is in turn connected to a spiral. The spiral apparently has a long, straight string of material attached to it, which ends in a confused muddle.

the spherical halo of our galaxy orbit in a similar way (see Figure 15.20). The same must be true of the stars within globular clusters themselves.

Rotation and Spiral Arms When Rosse discussed spiral galaxies (Section 16.2), he said that their spiral shape must indicate rotation. This is still generally believed. But beyond that, the spiral arms present an unsolved puzzle. Astronomers are fairly certain that they rotate much more slowly than the stars that make them up. In other words, a typical star in the outer regions of a spiral galaxy catches up to and passes through the arms of the galaxy. The arms must rotate slowly compared to the orbital motions of the stars; otherwise, in the typical lifetime of a galaxy, they would wind around the galaxy many, many times instead of the two or three times observed. (Recall that the sun has made about 20 revolutions of our galaxy so far; see Section 15.6.)

Many hypotheses have been proposed to explain the spiral arms.

None has found universal acceptance, but one of the most discussed is the density wave hypothesis. This hypothesis suggests that a galactic arm represents a region where matter has been compressed to about 10 percent above normal. This compression induces stars to form there, and the newly formed stars outline the arm's position. Some of the other hypotheses invoke galactic magnetic fields or near collisions of galaxies. It is a fascinating unsolved problem.

16.10 THE DISTANCES TO THE GALAXIES

The Method of Cepheid Variables Hubble was the first to measure the distance to the Andromeda galaxy in a convincing manner. His method, we recall, was to observe a Cepheid variable star in the galaxy and from its period obtain its luminosity L. Then he determined the variable's apparent brightness b. These are the two quantities required to determine the distance of the star and thus the distance of the galaxy in which it resides. We use the equation $L = bR^2$, where R is the distance.

The method has two limitations. The accuracy of the result is largely limited by the accuracy of the determination of the star's corrected apparent brightness, that is, the apparent brightness b it would have if there were no interstellar absorption. The star looks dimmer than it should because of interstellar absorption of its light by the dust in our own galaxy. Astronomers try to allow for this effect, but the amount of absorption varies in our galaxy along differing lines of sight. That is to say, the *average* amount of correction needed is known but, because of undetected concentrations of dust in some regions of our galaxy, the needed correction in a given direction may be more than average by an unknown amount. A greater difficulty is that there is no way to correct for absorption in the *other* galaxy. Another limitation of the method is that, if a galaxy lies more than about 7 million ly away, the Cepheids are too distant to be seen as individuals. Because the overwhelming majority of all galaxies lie beyond this distance, astronomers began to search for other methods of determining the distances to galaxies. Once again, they exercised their ingenuity in answering the question asked so many times in the history of astronomy, How far away is it?

The Two Basic Classes of Distance Determinations Modern methods of determining galactic distances beyond 7 million ly fall into two basic classes. In each case, astronomers look for an object in the galaxy that is sufficiently bright to be seen clearly. The first class of methods involves estimating the luminosity of the object by some means and then measuring its apparent brightness. Then $L = bR^2$ may be used to calculate the distance R. The procedure employing Cepheid variables fits into this class of methods. The other class of methods involves somehow estimating the *actual* diameter of some object in the galaxy and then

When determining the distance to a galaxy, as in so much of astronomical research, the assumption is made that similar-looking objects in other galaxies are actually similar to those in ours with respect to size, luminosity, and the like. This assumption is reasonable, but it may be wrong. However, it is necessary to make this assumption if any progress is to be made.

measuring its *angular* diameter. One then computes how distant the object would have to be in order to have such an angular diameter. This class of methods is not affected by interstellar absorption but has problems of its own. (See below.) We next discuss a few of these methods in more specific terms.

Globular Cluster Method Beyond the distance from which Cepheid variables are visible, one can try to observe a galaxy's associated globular clusters (see Section 15.3). Most globular clusters in our galaxy have diameters somewhat less than 100 ly. One assumes that this is also true for the globular clusters of other galaxies (see Figure 16.15). The observer then measures the angular diameters of the galaxy's globular clusters and computes their distance. The major source of error here is the difficulty in accurately measuring the angular diameter of a dim, distant globular cluster. The limiting distance of this method is about 50 million ly.

Supergiant Star Method We next discuss another method with about the same distance limit as the globular cluster method. The first step is to measure the apparent brightness of the brightest stars in the galaxy that are not variable stars. It is assumed that, as in our galaxy, these are the rarest, largest, most luminous giant stars, the supergiants. In our galaxy, most supergiants have luminosities between about 300,000 and 800,000 suns. A problem with this method is that at such great distances very bright stars may be confused with very bright emission nebulas, the H II regions (discussed in Box 15.3).

Supernova Method Beyond 50 million ly, no ordinary individual objects in a galaxy can be observed. One then can hope to observe a supernova in the galaxy. Supernovas become so luminous that, at their brightest, they can be observed up to distances of 300 million ly. However, supernovas occur in any given galaxy at an average rate of about one every 50 years. This method is not very accurate because the maximum luminosities of individual supernovas vary quite widely. This is unfortunate, perhaps, but if a galaxy is that far away, no other method of observing individual objects in the galaxy is known.

Whole Galaxy Methods The 200-inch mirror at Mount Palomar can photograph uncounted galaxies lying beyond a distance of 300 million ly. How distant are these galaxies? This question can be answered only by obtaining a rough estimate based on the appearance of such a distant galaxy as a whole. It may be reasonable to assume that the distant galaxy has a total *luminosity* roughly equivalent to that of the average nearby galaxies whose total luminosities have been measured. Then use $L = bR^2$. An alternative is to assume that the distant galaxy has a *diameter* roughly equivalent to the average diameter of nearby galaxies and then measure the galaxy's angular diameter. Both of these methods give only a rough idea of the distance but still are better than nothing. In the next three sections we describe some of the interesting discoveries the above methods have made possible.

16.11 THE NUMBER OF KNOWN GALAXIES

How many galaxies are there? The answer has become more and more staggering as ever larger telescopes have been used to seek out galaxies. A very large telescope can photograph such a tiny portion of the sky at one time that it is impractical to survey the entire sky and count individual galaxies. Instead, an observer photographs sample regions of the sky, counts the galaxies seen in these regions, and then estimates the total number that would be visible to the telescope in the whole sky. The reader will recognize this method as similar to the star gauges Herschel performed (see Section 15.2).

Edwin Hubble, who demonstrated that spiral nebulas are actually galaxies, performed a galaxy count using the 100-inch (254-cm) telescope on Mount Wilson. He estimated that it would require several thousand years to photograph the entire sky with the 100 inch. His galaxy count was made from about 1300 phtotgraphs of selected regions of the sky. Hubble found about 44,000 galaxies in these regions and estimated that about 100 million galaxies would be counted in a complete survey with the 100 inch. This amounts to 400 galaxies in each area in the sky the angular size of the full moon.

The 200-inch (508-cm) telescope at Mount Palomar has about four times the light-collecting surface area of the 100 inch and, as a result, can see many more galaxies. When the 200 inch is used to photograph a region of the sky away from the obscuring lane of dust in the Milky Way, the result is awesome. A large number of tiny points of light are seen on the photograph; these are foreground stars in our own galaxy. Also seen are faint, fuzzy blobs of light even more numerous than the foreground stars; these are incredibly numerous, incredibly distant galaxies. Many of the original photographs of nearby galaxies reproduced in this book show this same phenomenon. When the original plates are studied with a magnifying lens, numerous faint background galaxies are visible. Unfortunately, in the process of reproducing astronomical photographs in a book, most of the faint background detail is lost. A count of the galaxies visible in the region of the sky defined by the bowl of the Big Dipper yields an estimated one million galaxies. The estimated number of galaxies visible to the 200 inch is over one billion. If a large telescope were placed in space, perhaps on the moon, away from the earth's atmosphere and lights, many billions more galaxies probably would be visible. As larger optical telescopes have been built, there has been no indication that we have seen the most distant galaxies.

16.12 CLUSTERS OF GALAXIES

The Local Group As data on galaxies have been collected, their arrangement in space has become more apparent. The Milky Way galaxy has two companions, the irregular galaxies known as the Magellanic clouds, both roughly 200,000 ly from it. The Andromeda galaxy, about 2 million ly from us, is the nearest galaxy with a size comparable to ours. It has two small elliptical galaxies as close companions. M33 is a spiral galaxy also about 2 million ly from us. Its diameter is roughly one-half that of the Andromeda galaxy, and it is about 500,000 ly from the Andromeda galaxy. This comes to seven galaxies in our galactic neighborhood, counting our own.

Over the years about ten other small galaxies have been discovered in this neighborhood. These additional galaxies added to the seven already mentioned comprise what is called the **local group** of galaxies, seventeen galaxies clustered in a roughly spherical volume of space having a diameter of about 2.5 million ly. Of the ten additional galaxies that have been added to the original seven in the local group, two are small, irregular galaxies comparable in size to the Small Magellanic cloud; that is, they have diameters of about 10,000 ly. The eight remaining galaxies are all small elliptical galaxies ranging in diameter from about 1000 to 5000 ly and ranging in distance out to about 850,000 ly. About a dozen more minor galaxies have been observed whose distances are so poorly known that one cannot be sure whether or not they belong to the local group.

Other Clusters of Galaxies As other galaxies outside the local group were studied, it became more and more apparent that they are probably members of groups as well. In 1967, Lick Observatory published a survey of the sky made with a special 20-inch (51-cm) telescope designed to photograph large areas of the sky. Each plate covers a square of the sky 6° on a side. On these plates, galaxies so dim that they have an apparent brightness 60,000 times less than that of the dimmest star the naked eye can see are just barely observable. A statistical analysis of these galaxies proved what had long been suspected, that it is normal for galaxies to be found in clusters (see Figure 16.22).

The nearest large cluster of galaxies is the Virgo cluster (Figure 16.23) found in the constellation Virgo. Many of the galaxies in this cluster are visible using small telescopes. Messier (see Section 14.1) observed 16 of them and gave them M numbers. Altogether about 3000 members of the Virgo cluster are known, and doubtless many more exist which are too faint to see. The center of the Virgo cluster is about 78 million ly from us. The cluster is so huge that some astronomers suspect the local group to be a small grouping on the fringe of the Virgo cluster.

Other clusters of galaxies are more distant, but some are even more spectacular. The Coma cluster (Figure 16.24) is about 370 million ly distant but contains at least 11,000 *visible* member galaxies. One of the

Figure 16.22 A small cluster of four galaxies. The largest galaxy is NGC 3185.

Figure 16.23 A small portion of the Virgo cluster of galaxies. The Virgo cluster is too large and too near to appear in its entirety in this photograph. This cluster is about 78 million ly from us and contains roughly 3000 galaxies.

most distant clusters of galaxies, cluster 1410, lies at a distance estimated to be 4.5 billion ly. A count of sample regions of the sky away from the obscuring regions of the Milky Way was made at Mount Palomar to estimate the numbers of visible clusters of galaxies. An average of about one cluster was found in each square degree of the sky, which implies a total of over 40,000 clusters within reach of the 200-inch telescope.

Figure 16.24 A portion of the Coma cluster of galaxies. The bright, overexposed object with diffraction spikes (artifacts produced by the telescope) is a star in our own galaxy. The small, fuzzy patches scattered throughout this photograph are galaxies in the Coma cluster.

Twentieth-century science has produced a fascinating picture of nature as a series of clusters of ever-increasing size: the nucleus of the atom, a tiny cluster of protons and neutrons; the atom, a cluster of electrons surrounding a nucleus; molecules, clusters of atoms; planets, clusters of molecules; solar systems, clusters of planets around a central star; galaxies, clusters of stars many of which in turn are arranged in globular and open clusters; and finally, clusters of galaxies. Research is presently going on at both ends of this magnificent picture. At one end, evidence is beginning to emerge which seems to indicate that protons are clusters of smaller particles called quarks. At the other end, some astronomers speculate that the clusters of galaxies are themselves arranged in superclusters. The evidence for this is still controversial.

16.13 HUBBLE'S LAW AND THE EXPANDING UNIVERSE

Slipher Discovers the Galactic Red Shift We discussed some of Slipher's work with the spectroscope in Section 16.4 and Box 16.2. However, it is his work on the spectra of galaxies for which he is best known today. As early as 1899 spectroscopes had become good enough to reveal that spiral nebulas have spectra consisting of a continuous background of light crossed by black lines. The spectra are similar to that of the sun, only much less distinct.

The task of photographing the spectra of galaxies was a difficult one. Galaxies are faint, for one thing, and Slipher was working with a rather small telescope, by today's standards. Furthermore, the spectroscope spreads out the light of the galaxy into a spectrum, so that the light arriving at each place on the photographic plate is all that much fainter. Slipher found that, in order to achieve a sufficiently bright spectrum, he had to expose the photographic plate for more than one night. Some spectra required 60 hr of exposure time. In the course of such a long exposure, the telescope must be made to follow the galaxy across the sky with great precision. Slipher found that the controls of his telescope were too crude for such precision. He compensated for this by leaning against the telescope to make fine adjustments.

One of the first successes Slipher achieved was to measure the radial velocity of the Andromeda galaxy as a whole. In 1913 he published the result that the Andromeda galaxy has a blue shift (see the Doppler effect, Section 7.6) indicating a relative velocity of approach. Today it is recognized that most of this measured speed is caused by the motion of the sun in its orbit around the center of our galaxy. The residual speed of approach of the Andromeda galaxy itself is about 50 km/sec (30 mi/sec). We need not fear a collision of these two galaxies, because if they are aimed directly at one another, it will require about 12 billion years for them to cover the 2 million ly gap separating them. Moreover, the two galaxies may very well have a relative tangential (sideways) motion as well as a radial motion, which would cause them to miss each other altogether.

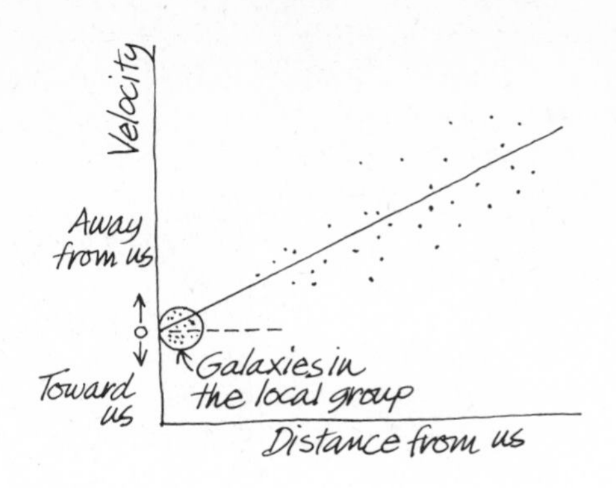

Figure 16.25 Hubble's law: velocity versus distance for galaxies. On this graph, each dot represents the velocity and distance of one galaxy. We see that the more distant a galaxy is from us, the faster it tends to move away from us. The galaxies in our local group do not seem to be participating in the general expansion. Roughly as many are approaching us as are receding. All speeds in the local group are relatively small.

Slipher went on to obtain the radial velocities of other galaxies. He found one that exhibits a blue shift, indicating a relative velocity of approach. But from then on, to his and all astronomers' surprise, every galaxy showed a red shift. After the radial velocities of 45 galaxies had been measured, 39 of them by Slipher, only the original two showed a blue shift. The largest radial velocity Slipher found for any galaxy was a speed of recession of 1800 km/sec (1100 mi/sec). This speed may be compared with the 250 km/sec speed of the sun in its orbit in our galaxy.

Hubble's Law These unusual results intrigued Edwin Hubble at Mount Wilson. Would the 100-inch reflector find that most galaxies are moving away from us if it were used to measure the radial velocities of galaxies too faint for Slipher's 24 inch to record? The answer was pursued by one of Hubble's Mount Wilson associates, Milton Humason, who had undertaken the time-consuming task of extending Slipher's work to more galaxies and fainter galaxies. And Slipher's pattern held. Almost all galaxies are moving away from us. The greatest speed of recession Humason found at the time was an amazing 40,000 km/sec (25,000 mi/sec).

Hubble began to inspect the data on radial velocities and to compare each galaxy's speed with his own determination of the galaxy's distance. Gradually a pattern began to emerge. The galaxies in the local group are drifting relatively slowly in a random way, and all the radial velocities are rather small. Six of the local galaxies show small red shifts, and the other ten show small blue shifts. However, all galaxies *outside* the local group (with one or two possible exceptions) show red shifts only. The overwhelming majority of all galaxies observed are moving away from us, some at enormous speeds.

Next Hubble produced a graph using these data on galaxies, a graph that has since acquired an importance in astronomy equal to that of the Hertzsprung-Russell diagram. He graphed the velocity of recession of galaxies versus their distance from us (Figure 16.25). The graph shows a definite trend for galaxies outside the local group: The more distant a galaxy, the greater its velocity of recession (Figure 16.26). The data

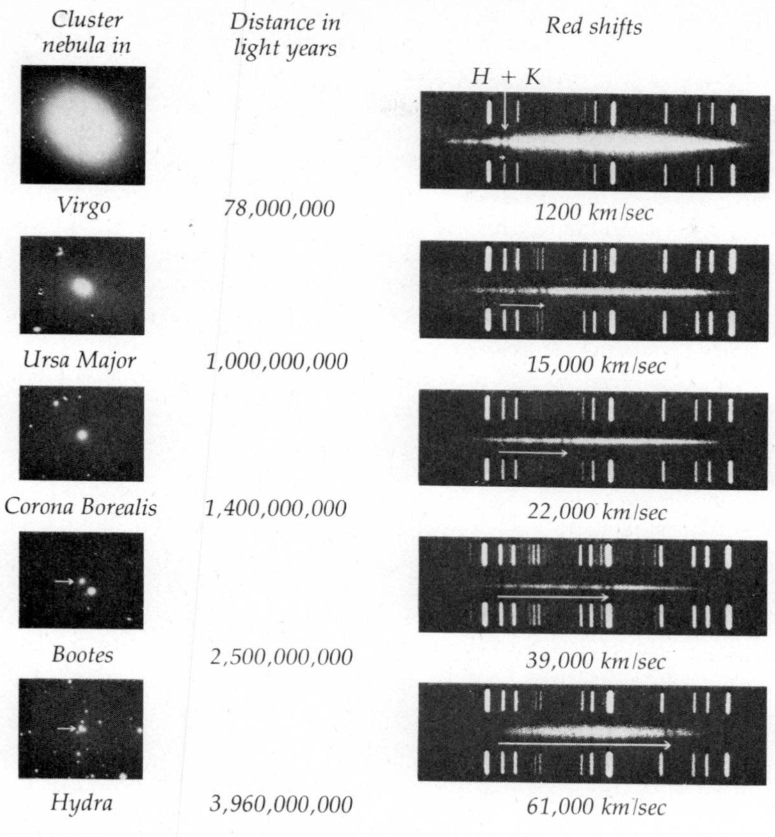

Figure 16.26 Galaxies in selected clusters and the red shift in their spectra: Hubble's law. In each spectrum, red is on the far right. Above and below each galaxy's spectrum is a comparison spectrum.

Hubble used were subject to relatively large amounts of error, particularly in the distance determinations, but it seemed to him that the graph could be approximated by a straight line. This would mean that, if galaxy A were twice as far from us as galaxy B, its velocity of recession would also be twice as great. If galaxy C were three times as distant, its velocity would be three times as great, and so on.

The relationship between distance R and velocity of recession V can be expressed as a graph (the straight line in Figure 16.25) or as an equation:

$$V = H \times R$$

This equation and the equivalent graph are known as the **red shift law** or as **Hubble's law.** In the equation, H is a number known as **Hubble's**

constant. Its numerical value is determined from the graph of velocity versus distance. The original value of H obtained by Hubble is now known to be inaccurate. He was in error largely because he used the wrong period-luminosity relation for Cepheid variables, which made his distances too small (see Section 16.7). Nevertheless, since Hubble's time more sophisticated techniques of measurement have all upheld the form of his law: The more distant a galaxy, the proportionately greater its velocity of recession. Only the numerical value of H has been reevaluated. A modern determination of Hubble's constant is $H = 17$ km/sec per 1 million ly. This indicates that, for galaxies outside the local group, each galaxy is moving away from us at a speed of 17 km/sec multiplied times the number of millions of light years it is from us. For example, if a galaxy is 100 million ly away, its radial velocity of recession is $17 \times 100 = 1700$ km/sec.

One significant aspect of Hubble's discovery is that, if H is accurately known, Hubble's law provides another way to determine the distances to galaxies. One photographs the galaxy's spectrum, measures its Doppler shift, and calculates its velocity of recession V. Then, using Hubble's law, one uses V and H to compute R, the galaxy's distance. This method has allowed astronomers to probe the distances of the most distant objects known, a most valuable tool.

Besides being a valuable tool, the red shift law is a remarkable phenomenon. Think about it; all the distant galaxies are moving away from us. The more distant the galaxy, the faster it is moving. This idea is sometimes called "the expanding universe." It is a puzzle. Why is the universe expanding? Moreover, why does it seem to be expanding away from *us?* We postpone a discussion of these questions until the next chapter.

16.14 THE RADIO TELESCOPE AND GALAXIES

Radio Sources in Our Galaxy A large variety of objects emit radio waves strong enough to be detected with the largest radio telescopes. The sun is a weak source except during the time of a solar flare. The moon is close enough that its very weak thermal radio radiation can be detected. Jupiter emits radio waves. A number of the emission nebulas in our galaxy can be detected. Many of the supernova remnants in our galaxy, including the Crab nebula, are easily observed with a radio telescope. The record for greatest apparent brightness recorded by a radio telescope is held by Cassiopeia A, a supernova remnant. In Section 15.8 we saw how important the 21-cm radiation of hydrogen in our galaxy is.

Conspicuously absent among the thousands of radio sources appearing in radio telescopes are ordinary stars. The sun is detectable only because it is so near. Only recently have very sophisticated techniques made it possible to detect a few of the nearer stars with radio equipment. Some neutron stars, however, are observed as pulsars.

Figure 16.27 The 43-meter-diameter (140 ft) radio telescope of the national radio astronomy observatory. The total moving weight of the telescope is 2600 tons. This telescope specializes in high-precision measurements at shorter radio wavelengths.

On the other hand, most of the objects appearing in the catalogs of radio objects are thought to be located outside the Milky Way galaxy. Of the brighter objects in these catalogs about one-third have not been identified, about one-third are galaxies, and about one-third are quasars, the subject of Section 16.15.

Radio Telescopes Achieve Poor Resolution Once a radio object has been discovered by a radio telescope, its identification is a difficult, time-consuming task. The principal hindrance to such work is the long wavelength of radio waves. To understand why this is so, recall (see Section 6.4) that the objective lens or mirror of a light telescope should have as large a diameter as feasible. The larger the diameter, the more light collected. But, more to the point here, the larger the diameter, the greater the resolution obtainable. (The better the resolution obtained by a telescope, the better the fine details of objects may be observed.) The quality of the resolution is also determined by the wavelength of the incoming radiation. The smaller the wavelength, the better the resolution obtained with a given diameter objective. Light has such a small wavelength that the human eye can obtain a resolution of about 2 minutes of arc. An 8-inch-diameter (20 cm) optical telescope can easily produce a resolution of 1 second of arc.

In most radio telescopes, the function of the objective mirror is performed by a large, metallic, dish-shaped antenna. The antennas of radio telescopes are huge compared to light telescopes (Figure 16.27). The largest antenna to date is the Arecibo dish in Puerto Rico, which has a diameter of about 300 meters (1000 ft) (see Figure 15.22). Such antennas are large enough to collect sufficient power from radio waves

Figure 16.28 The twin 27-meter-diameter (89 foot) radio telescopes of the Owens Valley radio interferometer.

so that the many faint radio objects in the sky can be observed. This is especially so because the radio signals can be amplified and processed by electronic equipment and computers.

The principal limitation of a radio telescope is the resolution obtainable. Since radio waves have wavelengths ranging from millions to billions of times longer than light waves, the resolution obtained with a radio telescope is very poor. Suppose a radio telescope were designed to obtain about the same resolution the human eye does. If it were designed to collect radio waves having a wavelength of $\frac{1}{2}$ meter, a reflecting dish having a diameter of about 1 km (3300 ft) would be required. This would be more than three times larger than the Arecibo telescope.

In the early days of radio telescopes it was often very difficult to identify radio objects because of this poor resolution. Radio telescopes could determine only that an object lay somewhere inside a relatively large region of the sky. Astronomers using optical telescopes were often uncertain which of the many objects visible in that region was the radio source.

Radio Interferometry Achieves Good Resolution This handicap of radio telescopes has been at least partially overcome by a technique called **radio interferometry.** The technique uses at least two radio telescopes separated by a large distance (Figure 16.28) but pointed in the same direction. The collected radio waves are combined electronically, resulting in a compounded signal which is then analyzed using a computer. This signal contains information on the size of the radio object and its location in the sky, information much more precise than that obtainable using only one of the radio telescopes. One can think of the two combined telescopes separated by a distance L as being two pieces of a much larger single radio telescope of diameter L (see Figure 16.29). The two combined smaller radio telescopes collect much less radio energy than the large one would but obtain about the same resolution. In one instance, two radio telescopes, one in West Virginia and one in Sweden, were used to distinguish details of radio objects as small as 0.0005 second of arc (much better than optical telescopes have achieved) and positions of objects accurate to about 1.0 second of arc, a vast improvement over a single radio telescope.

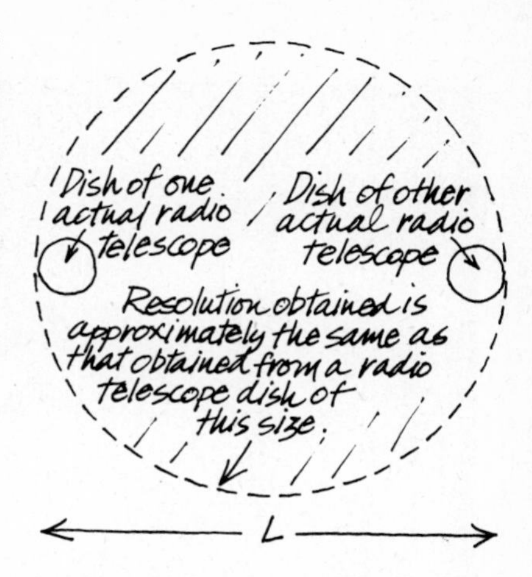

Figure 16.29 A radio interferometer.

Radio Galaxies The results of the search of the heavens with radio telescopes has turned up fascinating results. Most galaxies emit over 98 percent of their radiation as visible starlight. Only a few nearby, normal galaxies are observable using a radio telescope; examples are the Milky Way galaxy and the Andromeda galaxy. **Radio galaxies,** galaxies that appear in catalogs of radio objects and which have been optically identified, are for the most part peculiar objects radiating much more radio energy than ordinary galaxies. There are a number of types of radio galaxies, some of which we mention here.

Some radio galaxies are spherical galaxies of Hubble type E0 (see Figure 16.30). They appear peculiar in that they have a central band of absorbing gas and dust, a feature absent in ordinary E0 galaxies.

Other radio galaxies emit their radio signals from two or more enormous regions lying outside the visible boundary of the galaxy. Cygnus A is a good example. It has two radio-emitting regions, one on either side of the visible galaxy. Each of the two regions is about as large as the visible galaxy itself (see Figure 16.31). It is speculated that these regions originated in some major explosion in the galaxy, which expelled two huge clouds; fast-moving electrons in the clouds emit synchrotron radiation (see Section 14.8). This is the source of the radio waves we receive. The details of the mechanism by which the clouds were produced are far from clear.

Many examples of peculiar radio galaxies are known. Figure 16.32 is a photograph of M87, a galaxy in the Virgo cluster. When photographed using a short exposure time, a blue jet may be seen.

The sources of the energy that produce the powerful radio emissions of radio galaxies are still unexplained and delightfully puzzling. But the objects discussed in the next section are even more bewildering and controversial.

Figure 16.30 Centaurus A, NGC 5128, An E0 peculiar galaxy. This is the nearest known "violent" galaxy. It is a powerful radio and x-ray source. It is about 14 million ly from us and about 50,000 ly in diameter. Among its peculiarities is that it is an elliptical galaxy yet has a dense lane of dust across it.

16.15 QUASARS

The Discovery of Quasars As mentioned above, most ordinary stars do not reveal their presence to a radio telescope. However, in the late 1950s several objects in the sky were called **radio stars.** The term was used because certain radio sources had been identified as ordinary-looking, blue, starlike objects.

In 1960, the 200-inch Mount Palomar telescope was used to photograph the spectrum of one of these radio stars, and the result was totally baffling. Instead of the normal adsorption lines of hydrogen, calcium, etc., found in ordinary stars, this spectrum had emission lines. Furthermore, the positions of the emission lines did not fit the pattern of any known element. Perhaps, it was suspected, radio stars are not actually stars.

Then, in 1962, a group of radio astronomers in Australia used a 64-meter-diameter (210 ft) radio telescope to obtain a very precise posi-

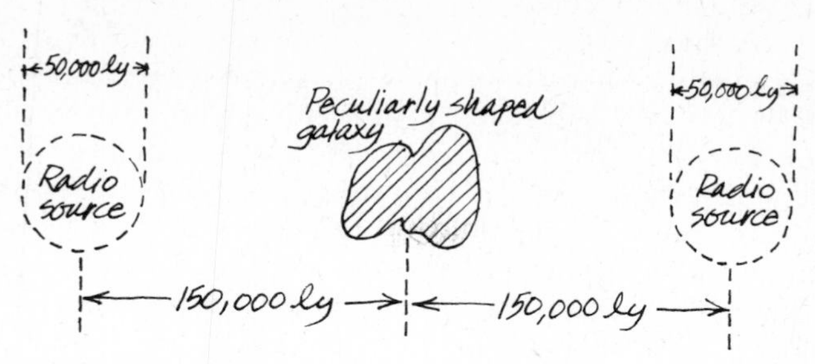

Figure 16.31 The galaxy and associated radio sources known as Cygnus A. The second strongest radio source in the northern sky, Cygnus A, can be detected even by some amateur radio telescopes. Walter Baade was the first to identify this strange-looking galaxy with the radio source, which is estimated to be 500 million ly from us. Its appearance led earlier astronomers to suspect that it was two galaxies colliding, but this hypothesis has now been dropped. Perhaps the galaxy has a belt, much like Centaurus A (Figure 16.30), which gives it the appearance of being double. The origin of the two radio sources is still a matter for speculation.

tion of a radio source known as 3C 273. Ordinarily a single radio telescope of this size would not be able to achieve great accuracy, but the Australians made use of a fortunate occurrence. Occasionally the moon passes directly in front of this object. By measuring precisely when the moon's edge cut off the radio signals from 3C 273, and by knowing the position of the moon at that time, an excellent determination of the location was obtained.

Maarten Schmidt had received his education in the Netherlands and had later moved to the United States. Upon receiving the information concerning the precise location of 3C 273, he decided to study the object visible at that location using the Mount Palomar facilities. Optically, 3C 273 is a tiny blue object which looks like a star. It had been photographed many times in sky surveys and passed over because of its ordinary appearance. Schmidt obtained its spectrum and found that it shows an emission spectrum having bright lines in no recognizable pattern. He pondered the meaning of this spectrum as he made his measurements, but he remained puzzled.

Figure 16.32 A short-exposure photograph of M87 showing its peculiar jet of light. M87 also appears in Figure 16.15 in a long-exposure photograph designed to show some of its many globular clusters. M87 is a powerful radio source.

By early 1963, he was ready to publish a report on his data concerning 3C 273. As he worked on this paper, he suddenly had one of those wonderful moments of insight that we have noted from time to time in this book. Aha! Schmidt suddenly realized that the lines in the spectrum are ordinary lines due to ordinary atoms, but that they are shifted in the spectrum by such a large amount that no one before him had recognized their nature. One must recall that the usual shifts of the lines in the spectra of ordinary stars are very tiny. Astronomer's minds were simply not prepared to look for these particular lines at this place in the spectrum of a starlike object. Schmidt found that the lines are those produced by hydrogen, oxygen, and magnesium, but that they have been shifted by a relatively large amount toward longer wavelengths, an unprecedented red shift. Schmidt then checked the spectra of other radio stars and found a large shift present in each of their spectra. Since then the term "radio star" has been dropped for these objects, and the term "quasar" has generally replaced it. A **quasar** is an object that appears to be a star in a telescope but has a spectrum of bright (emission) lines that are strongly red-shifted.

Today several hundred quasars are known. Many objects have been found that fit the definition of a quasar, except that they are not detectable with a radio telescope. Such objects are also usually included under the term "quasar."

After Schmidt's breakthrough, the next task was to interpret the new discovery. For openers, it was required to explain the large red shifts quasars exhibit. The first thing most people think of is the Doppler effect, since quasars are moving away from us at high speed. The more the lines in the quasar's spectrum are shifted toward longer wavelengths, the faster the object is presumed to be moving. The results of this interpretation, if it is correct, are amazing. 3C 273, the quasar Schmidt first recognized, would be moving at 15 percent of the speed of light (often written $0.15c$). Another quasar was checked by Schmidt, and its speed came out to be $0.30c$. Since then, the speed record has been broken frequently as more quasars have been discovered. In 1970 the record was $0.88c$, and today several quasars are known with red shifts indicating speeds just over $0.90c$, over 90 percent of the speed of light. Then another surprising discovery was made.

Quasars Are Small Not long after the initial discovery of quasars, evidence began to accumulate that they must be surprisingly small. This was indicated by the fact that many quasars are variable, that is, they change their apparent brightness. This was interesting enough, but the real surprise was the rapidity of the variations. One quasar doubled its brightness in 24 hr.

The rapidity of the variation gives one an idea of the maximum possible size of the quasar. It cannot be much larger than the distance a light beam would travel in the time the variation took place. In other words, the more rapid the variation observed, the smaller the object.

The small sizes of some quasars have been verified by studies using radio interferometry.

When this idea is applied to the variations of quasars, their diameters come out to be roughly $\frac{1}{10}$ ly; some are larger, and some are smaller. This is a totally unexpected result. Even relatively small objects like the Ring nebula (see Figure 13.17) are about 1 ly in diameter.

The Cosmological Interpretation of Quasars There are two schools of thought concerning the interpretation of data about quasars. The **cosmological interpretation** accepts that the large red shifts observed in quasars are caused by their rapid motion away from us, which produces a Doppler shift. It is only natural to follow this up by assuming that Hubble's law also applies. In other words, the greater the quasar's speed, the greater its distance. Using Hubble's law then gives the distances to quasars. The quasar known as 3C 273 is about 2 billion ly away. The quasar moving at $0.30c$ is roughly 5 billion ly away. The distance of quasars moving at about $0.90c$ is roughly 14 billion ly. (For comparison, one of the most distant galaxies known has a red shift indicating a speed of $0.36c$ and a distance of about 6 billion ly.)

But at this point the cosmological interpretation runs into a severe problem. If quasars are as distant as Hubble's law suggests, they must have an enormous luminosity; this is because they have an apparent brightness so great that they were earlier mistaken for stars in our own galaxy. With the use of $L = bR^2$, it is estimated that a quasar as distant as Hubble's law suggests would produce between 10 and 100 times as much radiation as the combined radiation from all of the 100 billion stars in the Milky Way. Furthermore, all this radiation must be generated from a region only about $\frac{1}{10}$ ly in diameter. So far, no satisfactory mechanism for producing such overwhelming amounts of energy in such a small region has been conceived.

The Local Hypothesis A minority of astronomers have rejected the cosmological interpretation. They do so principally because of the enormous luminosities the reasoning leads to. These scientists have proposed various forms of the **local hypothesis.** They suggest that quasars are not very distant, implying that their luminosities are *much* smaller than the majority view holds. This would make the task of explaining the source of their energy less difficult.

The Local Hypothesis: Why Do Quasars Show Large Red Shifts? But if quasars are relatively close, perhaps even in our local group of galaxies or neighboring clusters of galaxies, why do they have such large red shifts? There are two schools of thought on this question. The first school points out that, according to Einstein's general theory of relativity, as a light beam travels outward from a massive object, the effect of the gravitational pull on the light is to shift its wavelength to longer wavelengths. This prediction has been experimentally verified. A very tiny red shift has been detected in a beam of electromagnetic radiation traveling upward in a building. Some astronomers were led to speculate that quasars might not have high speeds but rather might be enormously massive, compact objects with an associated, very strong gravitational field. The light from a quasar, according to this idea, is red-shifted by a large **gravitational red shift** as it leaves the quasar.

This suggestion has not received much support. Wouldn't a star having enough mass to produce a red shift soon become a black hole (see Section 17.5) and disappear? A larger, more massive object might be able to exist, but it is hard to see how it could have an emission spectrum, an indication of a low-density gas. Although the gravitational red shift hypothesis has not been totally discounted, few astronomers currently accept it.

Figure 16.33 Stephan's Quintet. The red shifts of the four smaller galaxies imply a recession of about 6000 km/sec. The largest galaxy, however, has a red shift indicating a speed of recession of 800 km/sec. The standard analysis is that the larger galaxy is actually much closer to us and not associated with the other four. Yet some of the other indications of the distances of all five galaxies tend to place them at the same distance. The question is not settled.

Does Hubble's Law Apply to Quasars? The other school of thought that adheres to the local hypothesis accepts that large red shifts do indeed indicate a large speed of recession but rejects the proposal that Hubble's law applies to quasars. In other words, quasars are moving away from us rapidly but are not nearly as distant as Hubble's law implies.

To assert that Hubble's law may not apply to quasars is enough of a challenge to the establishment, but Halton Arp has gone further. He has become one of the world's leading experts on peculiar galaxies. Figure 16.21 shows one of the many examples of peculiar groups of galaxies he has photographed. Among the most interesting photographs are those showing two or more galaxies seemingly side by side.

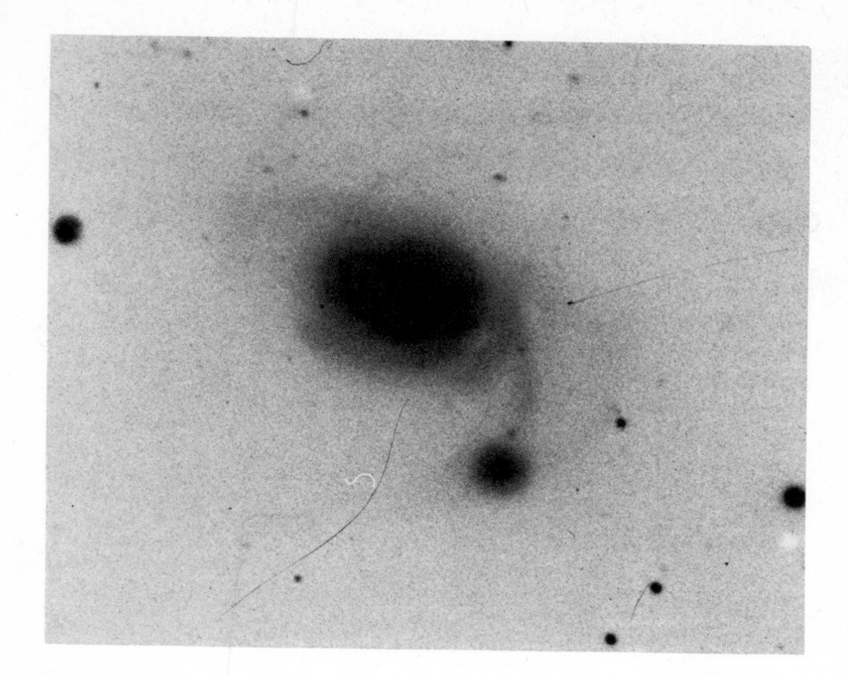

Figure 16.34 Two apparently connected galaxies with different red shifts, NGC 7603. These two galaxies have red shifts that differ radically; one is about twice the other. (The lines coming out of the galaxy are defects in the print.)

Arp and others have found surprisingly large numbers of such cases in which one of the galaxies in the group has a different red shift than the rest (see Figure 16.33). The standard answer has been that the odd galaxy is only in the same line of sight as the other or others and is really well behind or in front of the others, at its proper place according to Hubble's law. Arp has countered this by finding examples of pairs of galaxies with differing red shifts and with evidence of connections between them (see Figure 16.34). If the connection is real, the two galaxies must both be the same distance from us.

This argument for the local hypothesis goes on to say that, if Hubble's law doesn't apply to some galaxies, why should it apply to all quasars? In fact, Arp has found some examples of what appear to be connecting links between a galaxy and a quasar, each object having a different red shift (see Figure 16.35).

The reactions to Arp's work have ranged from emotional upset to delight in finding a new puzzle. At any rate, his results lend some support to the relatively few who favor the local hypothesis.

However, the local hypothesis also has certain problems with energy production. Some interpreters of this theory hold that at least some quasars are objects blasted out of a nearby galaxy, perhaps even our own. We see no blue shifts in quasars because the ones that were originally approaching us have long since passed us and are now retreating.

But where does the energy come from to produce such a rapidly moving object as a quasar? Wouldn't such a violent explosion rip any object apart rather than producing a quasar? Furthermore, the total

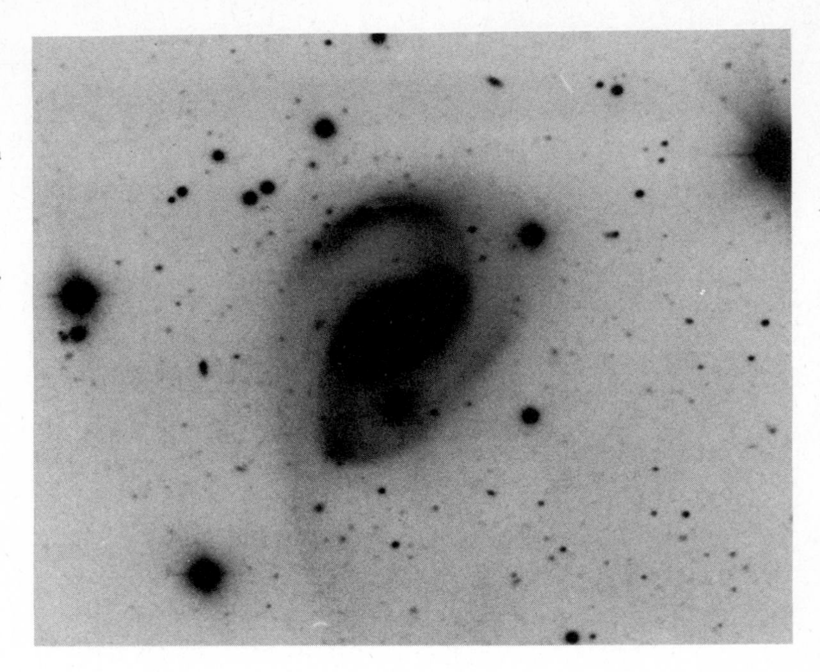

Figure 16.35 A quasar (Markian 205) and a galaxy (NGC 4319). If these two objects (the quasar is the black dot below the galaxy's nucleus) are connected by a bridge of matter, then the two objects are the same distance from us. Yet, they have very different red shifts, indicating that the quasar is receding from us at a speed that is 19,000 km/sec (43 million mi/hr) faster than the speed of recession of the galaxy!

energy required to produce the motions of the quasars may well be as great as the energy production required by the cosmological interpretation. What process could possibly produce so much energy?

Quasars are one of the most fascinating and baffling phenomena of twentieth-century astronomy. The data are still coming in. Some observers say that they can glimpse a faint structure around one or two quasars as if the quasar were the very bright nucleus of a galaxy. It will be intriguing to watch as astronomers try to come to grips with the mysterious quasars.

SUMMING UP

A great deal of puzzlement was resolved when Hubble showed that spiral nebulas are enormous galaxies lying at least 1 million ly away. Baade's findings concerning the two stellar populations of stars refined our knowledge of the distances to galaxies. Further studies revealed that galaxies commonly exist in clusters and that the more distant a galaxy the greater its speed away from us.

Radio telescopes have revealed radio galaxies and quasars. It has not yet been possible to explain how such objects produce such enormous quantities of energy. The study of quasars has led to an unresolved confrontation in which one side questions the universal applicability of Hubble's law to all objects.

In the next chapter, we investigate the properties of the universe taken as a whole.

EXERCISES

1. Define the following terms: local group of galaxies, radio interferometry, quasar, gravitational red shift.

2. Write several paragraphs contrasting the opinions of Kant and Laplace concerning the nature of the Andromeda nebula.

3. Hubble settled the question of the existence of galaxies in 1923. List the pre-1923 evidence for and against their existence.

4. What discovery led to the first estimate of the distance to the Andromeda galaxy? Why did this result lead to uneasiness in some quarters when the Andromeda galaxy and the Milky Way galaxy were compared with respect to size and the luminosity of their globular clusters?

5. Contrast population I stars with population II stars.

6. Explain how the discovery that there are two populations of Cepheid variables led to a revision of the estimated distance to the Andromeda galaxy.

7. Make a sketch of each of these types of galaxies: Sa, Sb, Sc, SBa, SBb, SBc, E0, E3, E7.

8. List six methods of determining the distance to a galaxy. (Don't omit Hubble's law.)

9. State Hubble's law in words. If a galaxy is 300 million ly from us, what is its approximate speed of recession according to Hubble's law?

10. List three types of radio galaxies.

11. What is the evidence that quasars are remarkably small? Roughly how small are they?

12. Answer these questions on the basis of the cosmological interpretation of quasars: (a) Does Hubble's law apply to all quasars? (b) Are quasars near or very distant from us? (c) Why does their distance pose a problem concerning their luminosity?

13. Answer these questions on the basis of the local hypothesis concerning quasars: (a) Does Hubble's law apply to all quasars? (b) What is the possible evidence favoring the answer to question a? (c) Are quasars relatively close or very distant? (d) Why does the answer to question c make the luminosity problem arising in the cosmological interpretation less severe? (e) What is the problem concerning the energy of motion of quasars in this hypothesis?

READINGS

A biography of Hale which is well worth reading is

Explorer of the Universe, by Helen Wright, Dutton, New York, 1966.

Much information on Messier, his work, and his objects is to be found in

Messier's Nebulae and Star Clusters, by Kenneth Glynn Jones, American Elsevier, New York, 1969.

Two good references on galaxies are

Galaxies, by Harlow Shapley, revised by Paul W. Hodge, Harvard University Press, Cambridge, Mass., 1972.

Exploring the Galaxies, by Simon Mitton, Scribner, New York, 1976.

Part of the pleasure in studying galaxies derives from looking at them. One should not miss the photographs and commentary in

The Hubble Atlas of Galaxies, by Alan Sandage, Carnegie Institute of Washington, Washington D.C., 1961.

The manner in which astronomers came to grips with the galaxies is well described on pp. 101–174 of the book by Berendzen et al. cited in Chapter 15. See also the book by Asimov cited in Chapter 13.

Quasars are featured in the excellent book

Black Holes, Quasars, and the Universe, by Harry L. Shipman, Houghton Mifflin, Boston, 1976, chaps. 1–12.

COSMOLOGY, THE EDGE OF THE UNIVERSE, AND LIFE

I have been conducting an informal poll for some years, asking people whether they have ever thought about the edge of the universe. It is surprising to find that the majority have. Let's put the question this way: What would you discover if you sought the edge of the universe by traveling in a straight line for a very long time at high speed? Here are some of the answers that might come to mind (see also Figure 17.1).

1. You might come to an impassable boundary of some sort. Perhaps it would be a force barrier or an impenetrable wall. Perhaps beyond the barrier space itself would end; there would be less than nothing into which you could not move (whatever *that* means).

2. You might come to the end of matter, but space would continue forever. In other words, if you traveled far enough, you would eventually pass no more galaxies. You would from then on see only empty space ahead. Behind you, all the matter in the universe would appear to be within a shrinking ball of light.

3. You might continue to move past new galaxies forever. That is, space is infinite in extent and so is matter.

Of these, the first seems very unlikely, although such a theme has been used in science fiction (we live in a giant game preserve, for example). The second and third answers seem plausible at first, but some people find infinity, even an infinite expanse of empty space, hard to accept. We have already quoted Kepler's protest, "The infinite is unthinkable!" in Section 6.5. These three answers do not exhaust the possibilities, as we shall see in Section 17.6.

In this chapter we study this and other questions of a similarly immense scope. The first six sections deal with topics drawn from the branch of astronomy known as **cosmology,** the study of the universe, its origin, its present state, and its future. It is a subject for scientists who enjoy thinking about space on the grandest scale imaginable—the universe, everything that is. As we shall see, it is also a field that attracts scientists who have developed the art of pushing new, seemingly bizarre, concepts to the outer limits of human thought. The seventh and eighth sections deal with the possibility of finding intelligent life elsewhere in the universe.

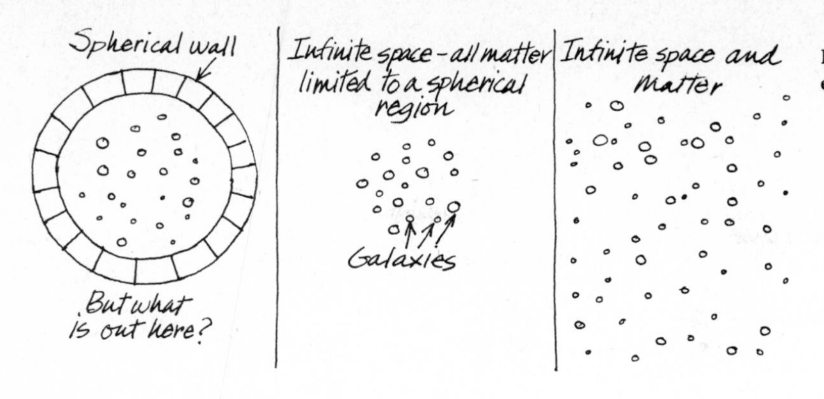

Figure 17.1 Three concepts of the edge of the universe.

17.1 A GEOCENTRIC UNIVERSE AFTER ALL?

One of the central themes in the history of astronomy has been the search for the place of humankind in the universe. We saw in Chapter 2 that for thousands of years astronomers interpreted the motions in the night sky to mean that the earth is at the center of the universe. Copernicus' revolutionary idea that the sun is at the center was difficult to accept at first because people's minds had been prepared, one might say prejudiced, to believe that we are at the center. We have seen how Einstein's theory of relativity restored a kind of democracy between the choices; when working with his theory one may conceive of the center of the universe at any location and still obtain correct answers. Some places may merely be more convenient to use than others.

William Herschel felt he had uncovered evidence that we are at least close to the center of the universe; he found data which seemed to show that the solar system is very close to the center of the Milky Way galaxy (Section 15.2). This was countered by Shapley's measurements of globular clusters. These measurements could most simply be interpreted to mean that we are near the outskirts of the Milky Way galaxy (Section 15.3).

In Section 16.13 we discussed Hubble's law which declares that, except for those in the local cluster, all galaxies appear to be moving away from us and that, the more distant the galaxy, the faster it is traveling away from us. The obvious implication seems to be that, after all, we are at the center, the center of the expanding universe. This implication was doubted from the first. It seems fair to say that astronomers of the twentieth century have a prejudice about such things, a prejudice which is the opposite of the prejudice of the early astronomers: If any evidence seems to imply that we are in the center, there must be some other explanation.

Does Hubble's Law Imply That We Are at the Center? In the case of Hubble's law, careful thought reveals that, if we lived in *any* galaxy, Hubble's law would still hold. To understand this, refer to Figure 17.2.

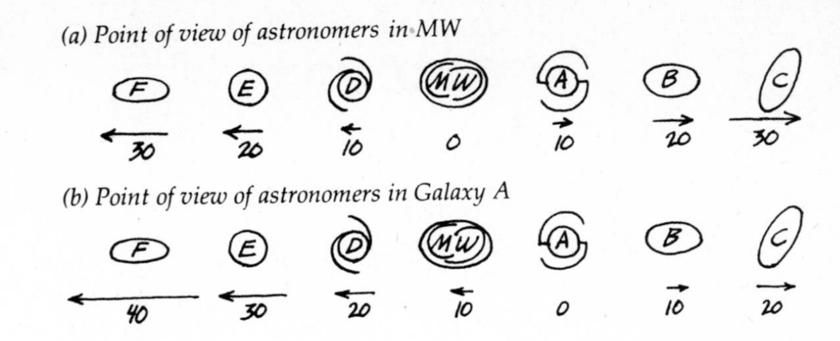

Figure 17.2 Hubble's law illustrated from two points of view.

Figure 17.2a represents an idealized set of galaxies all of which lie on a straight line. "MW" represents our Milky Way galaxy. The arrow below each galaxy (except MW) represents the radial speed of the galaxy as measured by astronomers in our galaxy. (These speeds are determined using the Doppler effect.) The numbers below the arrows indicate the speed of the galaxy measured in arbitrary units. Notice that the speeds conform with Hubble's law: All galaxies are moving away from us, and the more distant galaxies are moving faster.

Now, if we could somehow establish communication with astronomers in galaxy A, we might be surprised at first to hear them declare that Hubble's law also holds for them. Their representation of the same galaxies is shown in Figure 17.2b. This seeming contradiction with our own representation disappears if we realize that the Doppler effect does not measure absolute speeds but only relative speeds. We now see that, in Figure 17.2a, we made the *assumption* that we are at rest. The residents of galaxy A have similarly made the assumption that they are at rest. All that either group, MW or A, can really say with certainty is that MW and A are separating from each other at a speed of 10. It is an unwarranted assumption on our part to decide that *we* are the ones "actually" at rest.

You can see that the two representations of the galaxies' speeds in figure 17.2 are actually equivalent. For example, consider galaxies C and F. According to the astronomers residing in the Milky Way galaxy, galaxy C is moving at a speed 30 to the right and galaxy F is moving at a speed 30 to the left. This means that the *relative speed of separation* of C from F is 30 + 30 = 60. Now calculate the relative speed of separation of C from F using the data gathered by the astronomers of galaxy A. You find the same relative speed of separation, 20 + 40 = 60. The reader should check that a calculation of the relative speed of separation of any pair of galaxies in the figure yields the same result no matter whose data we use.

Thus, we cannot conclusively determine from Hubble's law that we are at rest and at the center of the universe. No matter which galaxy we lived in, we would get the same results as embodied in Hubble's law. In light of this insight, a better wording of Hubble's law is:

All galaxies are moving away from all other galaxies and, the greater the distance between any pair of galaxies, the greater their relative speed of separation.

17.2 THE BIG BANG THEORY

The Cosmic Egg Hubble's law indicates that the galaxies are presently moving farther apart. This leads reasonably enough to the idea that in the past they were closer together. If one imagines a movie of the universe running backward, one would expect to see all the matter in the universe eventually clump together. This is the basis of the concept that at one time in the past all this matter was condensed into a ball. This ball is sometimes known by the colorful term, the **cosmic egg.** We say that our universe began when the egg exploded.

How big was the cosmic egg? The calculations led to a strange conclusion. No known force would be able to withstand the grip of gravity in the cosmic egg. Once formed, theory proposed, the egg would have continued to skrink until it became a mathematical point or, in more technical terms, a *singularity*. The idea of all the matter in the universe occupying a volume infinitely smaller than the point of a pin, is, to say the least, amazing.

The Big Bang The second main feature of this cosmological conception is that at one point in time the cosmic egg erupted in a colossal explosion which caused its contents to expand. According to this idea, we see the universe still expanding today (Hubble's law) as a result of this initial outburst. This theory has come to be known as the **big bang theory.**

How long ago did the big bang occur? There are a number of ways to estimate this. One method depends on Hubble's law, which gives the speeds of separation of the galaxies at the present time. Using this, you calculate how long ago the galaxies would have been clumped together. The result obtained depends critically on the measured value of H, Hubble's constant in Hubble's law: $V = HR$. The smaller H is, the slower the galaxies are moving now, so the longer the time since the big bang. Hubble's own determination of H implied a universe about 2 billion years old—this is the time from the big bang until now. This estimate ran into a serious conflict with the findings of geologists, a conflict reminiscent of the conflict between Helmholtz and the geologists in the previous century (Section 9.7). By the late 1920s geologists had studied the earth and had found reason to believe that it is about 5 billion years old (see Section 10.1). The question, of course, was, How could a 5-billion-year-old earth exist in a 2-billion-year-old universe?

The conflict was resolved when Baade discovered that Hubble had been using an incorrect luminosity graph for Cepheid variables when he determined the distances to galaxies (Section 16.7). A corrected determination of H pushed the estimated time of the big bang back to about 5 billion years ago.

Since then, increasingly sophisticated techniques have derived lower and lower values for H. As a result, the astronomical estimate for the age of the universe is currently about 20 billion years.

Pioneering theoretical studies of the explosion known as the big bang were undertaken by a team led by George Gamow (see also Section 14.7). Working with assumed conditions during the explosion of the cosmic egg, Gamow calculated that, in the first hour of the explosion, all chemical elements heavier than hydrogen now found in the universe were formed. As explained in Section 14.7, this idea was later abandoned. However, it is still believed that substantial quantities of helium were formed at this stage.

As explained in Section 14.11, elements heavier than helium are now thought to originate in stars.

The Early History of the Universe According to the Big Bang Theory Gamow and his associates produced a detailed history of the early universe, that is, for the era just after the explosion of the cosmic egg. We discuss some of their findings in terms that have been updated in light of more recent discoveries in physics.

According to the big bang theory, it is important to consider the relative amounts of energy in radiation and in matter in the early universe. According to Einstein, both electromagnetic radiation and matter are forms of energy (see Section 9.7). In the very early universe, the amount of energy present as radiation was far greater than that of matter—radiation dominated matter. As a result, the radiation was so intense that it would not allow the particles of matter to combine in any way.

As time went on, the early universe expanded. This had the effect of lowering the temperature of the universe which, according to the laws of physics, reduced the amount of energy in the radiation. Thus, as the universe expanded, radiation dominated matter less and less.

The calculations concerning the early stages of the universe have become very detailed in recent years. Physicists feel that they can compute with confidence what the universe was like even during the first few seconds of the explosion. Only the events in the first $\frac{1}{1000}$ sec are thought to be uncertain! After $\frac{1}{100}$ sec the temperature of the universe was about 1×10^{11} or 100 billion K. At 1 sec, the temperature had already dropped to 1×10^{10} or 10 billion K.

A key event is thought to have occurred after about 4 min, when the temperature had dropped to about 1 billion K. At this point, the protons and neutrons could combine because the radiation no longer had enough energy to prevent them from doing so. This is the stage at which the original helium in the universe was formed.

Another important event happened when the temperature of the universe had dropped to 5000 K, which occurred about 500,000 years after the explosion of the cosmic egg. Below this temperature the radiation was no longer able to prevent the electrons in the universe from joining up with the hydrogen and helium nuclei. Whole, un-ionized atoms came into being at last. After this, the matter in the universe was the dominant form, and galaxies began to form.

We recall from Wien's law (Section 9.4) that, at each temperature, an object emitting thermal radiation has a principal wavelength. This same law can be applied to the original radiation from the big bang. As the universe expanded, the principal wavelength of the radiation increased to ever longer wavelengths. At the 500,000-year mark, when the temperature was about 5000 K, the principal wavelength of the radiation was about 5800 Å or 5.8×10^{-7} meter. Gamow and his associates calculated that, as the universe continued to expand, the principal wavelength of the radiation continued to lengthen until today it should be about 1 cm or $\frac{1}{100}$ meter. With the use of Wein's law, one calculates that this corresponds to a temperature of about 25 K. When the calculation is repreated today in light of recent discoveries, a temperature nearer 3 K results. This prediction, that the universe would be bathed in radiation left over from the big bang, was forgotten by almost everyone for about 40 years. (But we shall meet it again.)

We discuss the big bang theory further in succeeding sections. First, however, let us consider an ingenious alternative.

17.3 THE STEADY STATE THEORY: AN ALTERNATIVE PROPOSAL

The basic ideas of the big bang theory explained the observed universe so well that it was quickly accepted by most astronomers. The question seemed settled in the minds of the majority until the late 1940s when a small group of cosmologists proposed another theory which challenged the big bang theory. Among this group was Thomas Gold. It was Gold who later proposed that pulsars are neutron stars (Section 14.15). The new theory, called the **steady state theory,** was taken up by Fred Hoyle (see Box 17.1) who became the foremost spokesman for the theory.

If it was the intention of the authors of the steady state theory to shock cosmologists out of a too easy acceptance of the big bang theory and to make them rethink their ideas and seek new data, they were successful.

Two Cosmological Principles Hoyle's presentation of the steady state theory ran along these lines. Astronomers have based much of their thinking on an idea we shall call the **cosmological principle:** The universe looks the same in every direction as viewed from every location. In other words, if you choose any two locations in the universe from which to view the surroundings out to, say, several billion light years, the views will be essentially the same. You will see approximately the same number of galaxies in each region and will find them spaced by roughly the same average distances. Furthermore, Hubble's law will hold in the same form as we know it as viewed from any location. This is an extension of the discussion surrounding Figure 17.2. Boiled down, this cosmological principle says that there are no unusual places in the universe.

According to Hoyle, the point of departure for the steady state theory was that there is no evidence that the matter in the universe was

BOX 17.1

FRED HOYLE

I do not demand that an idea I am following turn out to be right in the end, just that it interest me.

Fred Hoyle

Fred Hoyle (born in Bingley, England, in 1915) has been at the forefront of some of the most exciting areas of research in astronomy for many years. During World War II he did research on radar and by 1945 was a lecturer in mathematics at Cambridge University. The flood of books and research papers from Hoyle soon reached a crest that has not subsided to this day. In 1946 he contributed to work on the formation of the solar system by suggesting that a companion star to the sun had exploded. He proposed that the material from the star formed the planets. Then, in 1948 with Gold and Bondi, he became one of the principal proponents of the steady state theory. He also contributed the idea that the fusion of other elements besides hydrogen could occur in a star, once the temperature and pressure in the star becomes high enough. This work in conjunction with Geoffrey Burbidge, E. Margaret Burbidge, and William A. Fowler led in 1957 to a theory of nucleosynthesis, in which the origin and abundance of atoms heavier than hydrogen are explained (see Section 14.11).

Hoyle has produced work in many other areas, much of it strikingly original. He is also the author of numerous books on astronomy which can be recommended to readers of this text. It is a rare pleasure to be able to read comprehensible, popular science writing by one of the outstanding figures in the field.

Not everyone agrees that the perfect cosmological principle is indeed "perfect." The name indicates something of the spirit of the steady state theorists.

ever crushed to a high density and high temperature. Surely, Hoyle reasoned, the effects of so drastic a condition would still be observable. Thus, he asserted, we must seek a theory that does not require a cosmic egg or a big bang. The steady state theory solves the problem by proposing what Hoyle called the **perfect cosmological principle,** which he maintained was more aesthetically pleasing and intellectually satisfying than the cosmological principle. It states: The universe looks the same in every direction as viewed from every location and *when viewed at any time.* The italicized phrase is meant to suggest that just as there are no unusual places in the universe, including our own location, there are also no unusual times in the universe, including the time we happen to be alive. This small additional phrase, if correct, has far-reaching implications for cosmology, as we shall explore.

To understand the differences between the two principles, let us choose an average region of space, as shown in Figure 17.3a. In this region there are five major galaxies separated by a certain average distance. What if we were somehow able to come back and view the same region of space many billions of years later? Figure 17.3b represents the prediction of the big bang cosmological principle. The galaxies have continued to move away from each other, and there are only two galaxies remaining in the region. The average distance between galaxies has increased dramatically. The perfect cosmological principle, on the other hand, predicts that the view will be essentially the same at any later time, as represented by Figure 17.3c. Here we see that there are still five galaxies in the region and that the average distance be-

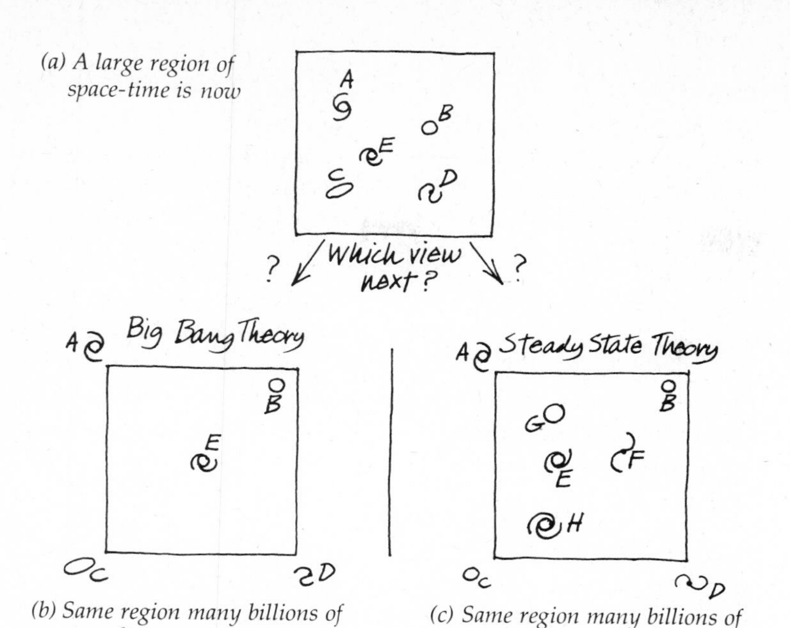

Figure 17.3 Predictions of the big bang and steady state theories.

tween galaxies is still roughly the same as in Figure 17.3a. This follows from the idea that the region will not look markedly different at any time.

We notice in Figure 17.3c that the expansion of the universe has occurred. Galaxies A, C, and D have left the region, but the region still has nearly the same appearance because of the presence of three new galaxies, F, G, and H. How can these new galaxies be explained?

Continuous Creation We explore the answer to this question by listening to an imaginary discussion between two cosmologists. Ben speaks for the big bang theory and Stella for the steady state theory.

Ben: I admit that the perfect cosmological principle has a certain appeal aesthetically, but it leads you to require new galaxies (Figure 17.3c). Where did these mysterious galaxies appear from?

Stella: The answer is simple. We require that the creation of matter, perhaps in the form of neutrons, be taking place all over the universe at a gradual rate. Each neutron breaks up into an electron and a proton. These then combine to form a hydrogen atom. As the old galaxies move apart, the new hydrogen atoms collect gravitationally and gradually form new galaxies in the space between.

Ben: You must be joking! The creation of matter? From what?

Stella: Matter is created from nothing. One minute there is nothing there, and the next minute—bingo—a brand new neutron appears.

It was not formed from other matter or from energy; it was created from nothing.

Ben: I can hardly believe my ears. I suspect that you are being deliberately outrageous. This sounds more like religion than science. How can a scientist accept this idea?

Stella: Honestly, I see no problem. If an idea reminds you of religion, why must you reject it for that reason? That is merely prejudice. Be reasonable. If the perfect cosmological principle is correct, then it follows that the creation of matter is always taking place.

Ben: I would rather take it the other way around. Since I can't accept that creation is always taking place, I conclude that the perfect cosmological principle is wrong.

Stella: But you can't just say, "I don't buy it." You have to find something wrong with the theory.

Ben: Alright. You must admit that, during the various studies of matter that have gone on all over the world, no one has ever observed a neutron appearing out of nowhere.

Stella: True, but that is because neutrons are created only infrequently at any location. The galaxies move apart quite slowly, so matter is created slowly to take up the empty space. My calculations indicate that if you watched the entire volume of the Empire State Building for the creation of a neutron, you would have to wait 100 years before one was created. And even then, would you be likely to detect so insignificant an object? Put it this way. In the 5-billion-year history of the earth, only about 4 grams of matter have been created inside the earth. This is barely enough matter to make a coin.

Ben: Yes, but your idea violates the principle of the conservation of mass and energy (Section 9.7), which says that the total amount of mass plus energy in the universe does not change.

Stella: Yes, we do violate it, but not very much. Besides, the conservation of energy is only a hypothesis on a par with our principle; it is unprovable. Energy is *almost* always conserved, we assert.

Ben: Well now, one more thing. The big bang theory implies an age of the universe of about 20 billion years. How can you figure an age for the universe using your theory?

Stella: We see the universe as infinite in time. It never began and it never will end. The creation of matter and the expansion of the universe have gone on for all time in the past and will go on for all time in the future. This very nicely eliminates the problems your big bang theorists have trying to explain away the singularity involved with your cosmic egg. All you have to do is get used to a new idea, Ben. Creation is still going on. After all, where did your cosmic egg come from? Was it created out of nothing all at once? I find *that* hard to accept.

The steady state theory was a great deal more sophisticated and impressive than can be conveyed here. The basic ideas were embodied in the elegant mathematical framework of Einstein's theory of relativity (Section 17.4), which has great beauty. But the theory also engendered irritation and outrage in some astronomical quarters and caused a great deal of extremely useful controversy and rethinking on both sides of the issue. Astronomers went back to their telescopes and spectroscopes seeking evidence that would confirm one theory or the other. All in all, the steady state theory was a healthful shot in the arm for astronomy.

The 3° Blackbody Radiation The battle for evidence was more or less a draw until 1965 when an accidental discovery was made with a radio telescope constructed at Bell Telephone Laboratories. Two scientists, Arno Penzias and O. C. Wilson, discovered radio waves reaching the earth from all directions. After a thorough study by many researchers, it was concluded that this radiation has a principal wavelength indicative of a temperature of about 3 K. This phenomenon is called **3° blackbody radiation.** It was quickly interpreted as radiation left over from the early stages of the big bang when the universe was very hot and radiation dominated matter. This, of course, confirms the prediction Gamow had made about four decades earlier. Penzias and Wilson received the Nobel Prize in 1978 for their work.

The discovery is in some ways similar to Jansky's discovery of radio waves coming from the center of our galaxy (Section 15.8). In both cases, the work was originated at Bell Laboratories and both involved an, at first, unwanted signal which later turned out to have great astronomical significance.

Here at last was evidence that could be interpreted as a sign that the universe indeed *was* very hot long ago, a contradiction to the basis of the steady state theory that the universe was never in a compressed state. The steady state theory is unable to account for 3° blackbody radiation in a natural way. Once the nature of the newly discovered radiation was confirmed, Hoyle dropped his theory. One cosmologist, Dennis Sciama, lamented the passing of the steady state theory by writing, "For me, the loss of the steady state theory has been a cause of great sadness. The steady state theory [had such] a sweep and beauty. . . ."

17.4 GENERAL RELATIVITY AND GRAVITATION

While in college, Einstein was taught that certain laws of physics take different forms depending on whether the observer is at rest, is moving from one place to another, is rotating, or is moving and rotating. This disturbed Einstein, and he spent a great deal of time thinking about it. He believed that this was an indication that the laws of physics were somehow incomplete. His sense of scientific beauty insisted that the laws of physics have the same form no matter how the observer is moving.

Special Relativity By 1901 he was employed in a Swiss patent office and used his spare time to work on this problem, as well as several others. He started with a special case of the problem. He began by assuming that the laws of physics would be the same for any two nonrotating observers traveling in a straight line at a constant speed relative to each other. This, plus the assumption that the speed of light is a universal constant, led to what is known as the **special theory of relativity,** which Einstein published in 1905.

The two assumptions Einstein started with led him to many seemingly bizarre conclusions which we have mentioned earlier in this book (see Boxes 6.6 and 7.1, Section 9.7, and Box 12.1). He found that matter and energy are two forms of the same thing, mass-energy ($E = mc^2$), that the speed of light is the maximum speed possible in the universe, that time is relative, that one cannot hold the concept of absolute rest, and much, much more. His conclusions have been experimentally verified and incorporated into contemporary physics.

The General Theory of Relativity All this had proceeded from a special case of the problem of the laws of physics. Next Einstein turned his intellectual energies to solving the problem in its entirety. He assumed that the laws of physics would have the same form for any two observers *no matter how* they were moving relative to each other, whether rotating, traveling in a curved path, speeding up, or slowing down.

As Einstein followed the logical consequences of his idea, the theory grew more and more complex, abstract, and mathematically involved. He once again saw a simple initial assumption grow and flower until his theory became one of the most beautiful products of the human mind. He began to think of space, which to us is the mere nothingness between matter, as a phenomenon having properties all its own. He found, to his surprise, that his study of motion was evolving into a theory of gravitation which went far beyond anything dreamed of by Newton. After 10 years of some of the most creative and involved thinking ever performed, he published what is known as the **general theory of relativity** in 1915 (see Box 17.2).

Einstein's theory of gravitation answers some of the earlier objections leveled at Newton's theory by such thinkers as Leibniz (Section 6.10). Recall that, in reply to requests for an explanation of how gravity works, Newton replied, "I do not make hypotheses."

The Difficulty in Grasping General Relativity For several reasons, the theory is difficult to grasp. Among these is that Einstein found it necessary to think in four dimensions, and this forced him to use a branch of mathematics known as tensor calculus. This presents a predicament in nonmathematical books such as this one. The only way to state Einstein's ideas correctly without distortion is to use the language of tensor calculus. We have only two choices. We could ignore general relativity, but that would force us to omit several extremely fascinating topics in later sections. Instead, we will try to describe the basis of the general theory of relativity, a theory of gravitation, using a simple analogy. This analogy is not an exact one, and if pushed too far will lead to inaccuracies. But at least it gives one the flavor of the idea.

We shall use such phrases as "thinking in four dimensions" and "four-dimensional space." The reader should know that these are merely shorthand phrases which we find handy. An expert on relativity would not talk this way and might say that such phrases are not technically correct. I hope such phrases will be helpful in forming a preliminary idea of Einstein's thoughts.

BOX 17.2

EINSTEIN'S GREATEST BLUNDER

As Einstein was pondering his general theory of relativity, he came to a crucial point at which cosmology entered the scene. He had arrived at the basic equation of general relativity. Next he had to decide whether or not a constant that appeared in the equation had the value zero. Like all great creators of theories, Einstein was guided by his intuition and sense of beauty as much as by his knowledge of the actual behavior of the universe. Now he was presented with a dilemma. His aesthetic sense told him that the constant must be zero, for then the theory would have much more beauty, symmetry, and simplicity. On the other hand, if the constant were zero, it would imply that the universe was expanding or contracting. This distressed Einstein, because at that time astronomers assumed that the universe was static—permanent. If so, the constant must not be zero.

Which way should he go? Should he follow his inner sense of beauty or the facts as he knew them? Earlier, in the special theory of relativity, he had predicted effects that seemed preposterous until experiments had confirmed them. After anguishing over the question, he decided, for once in his life, to go against his intuition; he published his theory with a nonzero constant and therefore a static universe.

Readers who have studied Chapter 16 know that, about 14 years after Einstein published his results, Hubble published his red shift law which implies that the universe is indeed expanding. Einstein called the decision not to follow his intuition, even when at odds with the "facts," "the greatest blunder of my life."

Einstein could think in four dimensions. We are used to three. As we read Einstein's theory, we find it hard to conceptualize his idea of gravitation. For the purposes of an analogy let us imagine that a colony of ants can think in only two dimensions: back and forth, right and left. We assume that ants cannot conceive of up and down. This means, for example, that they can picture squares and circles, two-dimensional figures, but are totally unaware of cubes and spheres, figures requiring three dimensions. The ant interprets every surface it crawls on to be a flat two-dimensional surface, like a flat field, a plane. Even if the surface has hills and valleys, the ant, which cannot imagine up and down, sees the surface as flat. The analogy is this: Einstein is to us as we are to these ants.

Let us imagine that a colony of ants lives on a large, flat mattress, as in Figure 17.4a. They find that they can walk around on their mattress with ease. An ant named Newtant finds that a marble rolls in a straight line on the mattress.

Now suppose we place a bowling ball near the center of the mattress as shown in Figure 17.4b. The ants now find that the bowling ball causes some interesting phenomena. The ants must brace their legs or else they will fall and tumble toward the bowling ball. Newtant does some experiments and finds that, the closer one gets to the bowling ball, the more one must brace one's legs. He also finds that more massive ants are more strongly affected. He rolls marbles sideways with respect to the bowling ball and finds that, given the proper speed, the

Figure 17.4 **Two-dimensional ants on a mattress illustrate an idea from the theory of general relativity.**

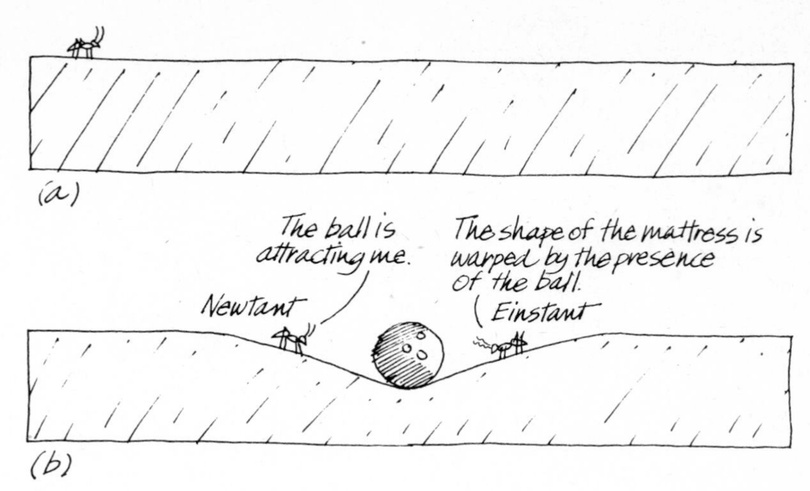

marble rolls around the bowling ball rather than rolling into it—orbital motion. He then announces his law of bowling ball gravitation: "The bowling ball attracts all matter. The closer you get to the ball, the harder it pulls. The greater your mass, the greater the pull. If the bowling ball had more mass, it would pull harder then the present one does." When Newtant's students ask him how the bowling ball is able to attract something without reaching out to it Newtant replies, "I do not make hypotheses."

Some time later, another ant named Einstant comes along. He has thought about the puzzle of the bowling ball and its mysterious gravitational pull. He proposes a new point of view. He tells the other ants that they must learn to think in three dimensions. He says that the presence of mass (the bowling ball) changes the shape of the space (the mattress) in the region of the mass. The reason you feel a pull toward the bowling ball is not that the ball pulls directly on you, but rather that it causes the mattress to slope downward (a three-dimensional concept) toward the ball. The closer you get to the bowling ball, the more the mattress is distorted from its original flat shape, and the more you must brace your legs. The other ants stare at him in wonder, but we humans can see what he means.

General Relativity and Gravitation Now we are in a position to consider Einstein's idea. He imagined space to be a four-dimensional concept. In the absence of matter (mass), space is "flat." However, the presence of matter causes space to be distorted in the region around the matter so that it forms a four-dimensional valley. A small object placed near the matter tends to fall down the four-dimensional slope. We have called this tendency the "force of gravity." The nearer we come to an object, the more the four-dimensional space slopes and the stronger the pull we say that we experience. A more massive object distorts the space around itself more strongly, and so we say we experience a stronger pull.

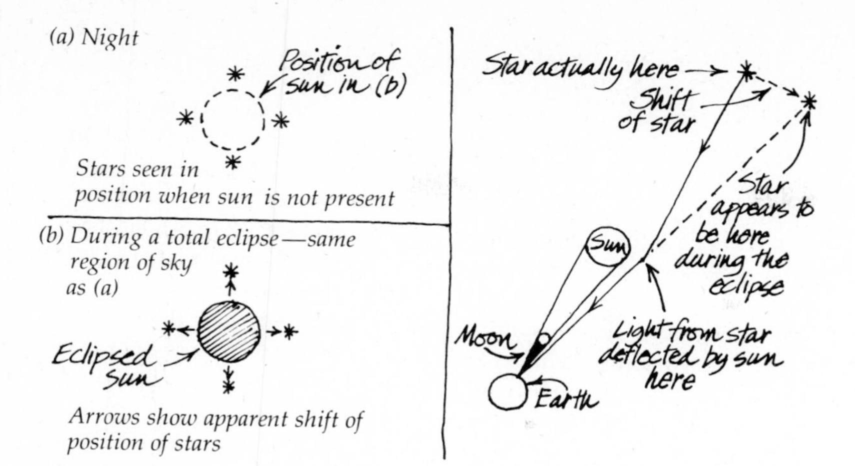

Figure 17.5 A test of general relativity. Astronomers calculated in advance the sun's position during a total eclipse. They then photographed the stars in that region of the sky at night (a). Months later, on the day of the eclipse, they photographed the stars near the position of the eclipsed sun in the sky. They then measured the change in the position of the stars (b), that is, the angle between the position of each star before and during the eclipse (frame two). They found that the angles matched the angles predicted by general relativity and not those predicted using Newton's theory of gravitation. The reason for the apparent change in position of a star is that the light from the star curves in the region near the sun where the distortion of space by the sun is most noticeable. The angles shown here are grossly exaggerated for clarity. The largest angular shift predicted by Einstein was just under 2 *seconds* of arc.

Three Tests of General Relativity When Einstein expressed these ideas in mathematical form, he found that his results generally agreed with the results of Newton's theory of gravitation but only in situations where gravitation is weak. However, in regions where gravitation is strong, for example, near very massive objects, he predicted that his theory would yield results different from those of Newton's theory. Einstein described experiments that could be performed to test his theory against Newton's.

For example, he stated that, as light passes near the sun, the distortion of space near the sun causes it to curve by a certain amount. This was tested in 1919 when an expedition was made to a region where a total solar eclipse was to occur. [(Eddington (see Box 13.2), one of the first to comprehend Einstein's ideas and work for their acceptance, was on this expedition.] There, astronomers measured the positions of stars appearing near the sun during the eclipse and found that the light from these stars had indeed been curved by approximately the amount predicted by Einstein (see Figure 17.5). This experiment was repeated using radio waves produced by the *Viking* lander on Mars during the conjunction of Mars near the end of 1976. The results agreed even more closely with Einstein's predictions than the eclipse results did. We have already discussed another confirmation of Einstein's theory, the gravitational red shift, in Section 16.15. Another attempt at confirmation is in a state of confusion at the present time (see Box 17.3 and Figure 17.6).

BOX 17.3

GRAVITY WAVES

One of the predictions of the general theory of relativity is that, when mass is accelerated (undergoes a change of velocity), it produces gravity waves. This effect is analogous to that produced when electrons are accelerated; they produce electricity waves, better called electromagnetic radiation. The theory predicts that the gravity waves will be very weak—much, much weaker than electromagnetic radiation—and thus extremely difficult to detect.

Until 1969, no one even attempted to search for gravity waves because they were expected to be so weak. That year, Joseph Weber of the University of Maryland began an experiment to detect these waves, the results of which had astronomers buzzing with excitement. Weber's instrument consisted of a large metal cylinder to which sensors were attached to detect any vibration of the cylinder (see Figure 17.6). The sensors were so sensitive that they could detect a vibration of the cylinder back and forth over a distance as small as the diameter of the nucleus of an atom. The idea was that passage through the cylinder of a gravity wave from outer space would cause the cylinder to "ring" like a bell at the natural frequency of vibration of the cylinder.

Almost at once, Weber began detecting about six events per day. His detector had the ability to determine the general direction from which the gravity waves came. He concluded that the ones he was detecting came from the center of our galaxy. This led astronomers to speculate how these gravity waves might be caused. The result was great puzzlement because, since gravity waves are extremely weak, something very violent must be producing them if Weber's instrument could detect them. Furthermore, the events Weber detected were so frequent that some violent ongoing process had to be involved. One theoretician proclaimed that the explanation might require the complete annihilation of the mass of 10,000 stars per year! At that rate, the galaxy would be depleted of stars in a mere 10 million years.

Other researchers then built gravity wave detectors in order to gather more data. One person claimed to detect them, but the great surprise was that none of the other detectors ever registered any gravity waves. Some of the detectors were quite similar to Weber's, and others were radically different, but they did not yield positive results.

Thus, another mystery of modern astronomy hangs in the balance. Do gravity waves exist? The consensus is that they probably do but that Weber was not detecting them but rather some other disturbance. If that is true, what caused Weber's carefully constructed experiment to show positive results? No one can explain this to the satisfaction of all concerned.

We now turn to one of the most exciting predictions to have come out of general relativity, black holes.

17.5 BLACK HOLES

After Einstein's general theory of relativity appeared in 1915, not all physicists were able to immediately comprehend his profound new ideas, nor were all prepared to deal with the mathematics involved. One of the notable exceptions was Karl Schwarzschild ("Shvarts'-shilt"), a German physicist, who studied the theory, understood it, and began to work out some of its consequences. One of his greatest discoveries involved the bending of space by highly concentrated matter.

Figure 17.6 Joseph Weber and his gravity wave detector.

We have seen that the general theory of relativity asserts that the presence of matter distorts the four-dimensional shape of space, creating a four-dimensional valley. The more concentrated mass is in a region, the more distorted the surrounding space. Schwarzschild found that, if the mass concentration in a region exceeds a certain value, space becomes so distorted that it is curved into a four-dimensional sphere.

The analogy using ants on a mattress may be helpful here. Suppose we replace the ball on the mattress with one so massive that it distorts the surface of the mattress into an enclosed cavity, as shown in Figure 17.7. An unfortunate ant has been trapped inside the cavity. Recall our assumption that ants can only conceive of two dimensions. The trapped ant still pictures the surface of the mattress as flat. She thinks she

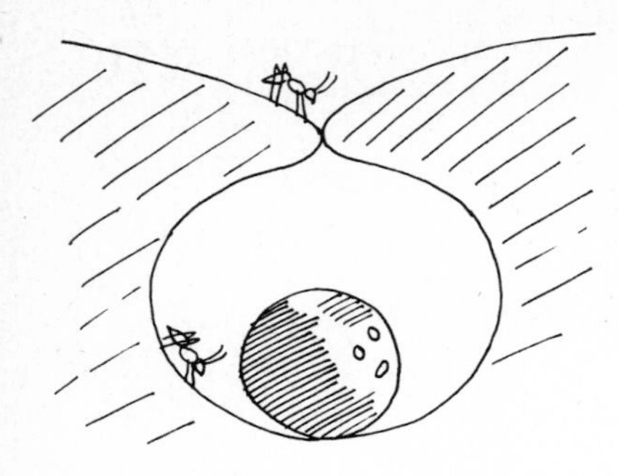

Figure 17.7 A very massive bowling ball creates a cavity.

feels a very strong pull from the ball. Suppose she tries to get away from it by crawling in a straight line ("straight" to her two-dimensional mind) directly away from the ball. To her amazement, she does not crawl very far before she finds the ball dead ahead of her once more. No matter which direction she takes away from the ball, she must return to it if she travels in a "straight" line. The puzzled ant cannot conceive that she is inside a three-dimensional closed cavity. We three-dimensional creatures see that she is traveling inside a three-dimensional sphere.

Now consider the ant outside the cavity. He notices a pull as he wanders near the edge of the cavity. If he wanders too close to the edge, he will fall in and find himself trapped forever, unable to leave. He will have left his large, fairly flat universe and entered a small, highly curved universe dominated by a massive ball.

Schwarzschild's discovery was that, if mass is sufficiently concentrated in a region, space will curve around it into a four-dimensional sphere, a **black hole.** According to this theory, a human in a rocket ship inside a black hole would be unable to escape. If she tried to escape by traveling directly away from the central massive object in a straight line ("straight" to her three-dimensional mind) she would soon find that the massive object was dead ahead. She would not observe a barrier or wall keeping her in the black hole, but she would not be able to get very far from the central massive object. She would not be able to conceive that she is in a four-dimensional closed cavity. We are used to light beams traveling in straight lines as viewed in three dimensions. However, inside a black hole, three-dimensional straight lines are curved in four dimensions so that they travel on a four-dimensional sphere. Thus, light beams are also unable to leave the black hole.

How would a black hole appear from the outside? Since no light can leave the black hole, it appears black. As a space traveler encountering a distant black hole, you would find it an eerie experience. Your spaceship would be attracted by the black hole's powerful gravitational pull, but you would be unable to see any object causing the attraction. If you wandered too close to the black hole, you would enter it, never to return. Even before entering the black hole, you would be killed because your feet, being closer to the hole, would be attracted more strongly than your head. You would be pulled to pieces. There is a boundary around every black hole. Outside that boundary, the gravitational field is indistinguishable from that of an ordinary star. Inside that boundary, you are trapped forever. You should not think that a black hole inevitably pulls all objects outside the boundary into itself. A planet could orbit a black hole outside the boundary just as the earth orbits the sun.

How Could a Black Hole Be Formed? Thus, Einstein's theory of general relativity predicts that black holes *could* exist. But what process could form a black hole? Schwarzschild derived a formula giving the radius r of the sphere into which a mass must be squeezed in order to become a black hole. This is the famous Schwarzschild radius, $r = 1.5 \times 10^{-27}$ times the mass of the object. In this formula, mass is in kilograms and r is in meters. It says that the earth would become a black hole if it were squeezed into a ball of radius 1 cm (0.39 inch). The sun would have to be compressed into a radius of 3 km (1.8 mi), and the Milky Way galaxy would have to be compressed into a radius of about 0.1 ly to form a black hole. Clearly, a great crushing force is required if a black hole is to be formed.

It has been proposed that some stars may become black holes as they die. Recall that in Chapter 13 we related that a dying star having a mass less than the Chandrasekhar limit, 1.4, suns, becomes a white dwarf, a star that opposes the crush of gravity by means of electron pressure. Stars with masses between 1.4 and 5 suns *may,* it has been proposed, shed enough mass to get under the limit and also become white dwarfs (see Section 13.9). Other, even more massive stars may become supernovas, and some of them may form a neutron star at their center. A neutron star is able to withstand the grip of gravity by means of neutron pressure.

In 1939, the same year he worked out a theory based on general relativity showing that neutron stars might exist, Oppenheimer also published a paper concluding that, if the core of a star collapses and if its mass is too large, even neutron pressure will be insufficient to stop the collapse. Gravitation will crush the star to ever smaller sizes without an end. According to more recent calculations, the most massive neutron star possible would have a mass of about 3 suns. It has been speculated that the core of a very massive star might bypass the white dwarf and neutron star stages and eventually be crushed by gravitation so powerfully that it shrinks to less than the Schwarzschild radius, thereby becoming a black hole. Thus, black holes might exist.

Searching for Black Holes But *do* black holes exist? It is not hard to imagine the dismay of astronomers setting out to search for objects that, by their very nature, emit no light!

In 1970, an x-ray telescope named *Uhuru* was launched into orbit around the earth. X-rays do not pass through the earth's atmosphere. This was the first major opportunity astronomers had to view the sky for a prolonged time in the x-ray region of the electromagnetic spectrum. Among the most fascinating discoveries was a source of x-rays named Cygnus X-1. This x-ray source is very near a star that is a spectroscopic binary. The exciting fact is that the data on the unseen companion indicate that it has a mass no smaller than 4 suns, well above the 3-sun limit.

The most common interpretation of this observation is that Cygnus X-1 consists of an ordinary star and a black hole. The two orbit each other at a relatively small distance. Gas from the star is falling into the black hole. As this gas plunges into the black hole, and before it reaches the boundary of no return, it is violently compressed and reaches such a high temperature that it emits x-rays in accordance with Wien's law (see Section 9.4). The gas then passes into the black hole and is gone forever.

We have seen that, in the past, an attractive theory has often been challenged by alternative theories. One does not want to accept an exciting theory until every other, less extraordinary possibility has been eliminated. In the case of Cygnus X-1, the devil's advocates have been at work devising other possible theories to explain the situation. Not all astronomers agree that we have found even one black hole. Kip Thorne, an expert on the question, puts it this way, "The evidence makes me . . . about 90 percent certain that in the center of Cygnus X-1 there is indeed a black hole." Others are much less certain. Nevertheless, most astronomers agree that, since black holes are such an attractive result of general relativity, they *ought* to exist.

17.6 THE EDGE OF THE UNIVERSE RECONSIDERED

Earlier we posed the question, What would you discover if you sought the edge of the universe by traveling in a straight line for a very long time at high speed? We listed three possibilities: you might find an impassable barrier, you might find the end of matter but not of space, or you might discover that matter and space continue on without limit. Einstein's research into the nature of time and space, known as the theory of general relativity, led him to the opinion that a fourth possibility was the most likely. We discuss Einstein's opinion in this section but, before reading on, the reader should ponder the content of the two preceding sections and try to anticipate Einstein's idea.

The Concept of a Closed Universe According to the general theory of relativity, the presence of matter curves the surrounding space. The more mass in a region, the more curved the space. When Einstein applied this concept to the universe as a whole, he concluded that the total effect of all the matter in the universe is to curve the space in the universe as a whole. The more concentrated the mass in the universe, the more curved the universe. As he pondered this idea, he came to the conclusion that the most simple, beautiful, satisfying circumstance would be that the mass in the universe is sufficiently concentrated so that the universe as a whole is curled up into a four-dimensional sphere. Currently, cosmologists refer to this concept as the **closed universe.** Somewhat loosely put, Einstein proposed that we are living inside a gigantic black hole.

The Consequences of a Closed Universe If Einstein is correct, the consequences are staggering. In a closed universe, if you could travel in a seemingly perfectly straight line long enough, you would return to your starting point after having traversed a four-dimensional sphere. Light beams, if they were able to travel long enough, would return to their source. (This is the idea behind the joke about the astronomer who looked into a telescope to check the haircut on the back of his neck.) There are technical reasons why neither the space traveler nor the light beam could actually complete the trip, but the concept illustrates Einstein's striking idea. In a closed universe, there is no boundary at which you can say, "The universe ends here," but, on the other hand, you cannot say that the universe goes on forever. The universe is limited in extent.

Imagine a rocket ship traveling at top speed. Suppose it leaves our galaxy and passes by galaxy after galaxy. Each galaxy it reaches is moving away from our galaxy faster than the one before. Eventually, all the galaxies in front of the rocket will be moving away from our galaxy faster than the rocket is. Thus, the expansion of the universe will prevent our rocket from actually going completely around the universe.

If Einstein is right and the universe is closed, then it seems that a redefinition of the word "universe" may be in order. Just as a black hole might exist in our universe and be observed by us who are outside it, isn't it possible that our universe exists in a larger universe which might also contain other closed universes? And might not that larger universe in turn be closed and imbedded in a still larger universe? The mind reels. Perhaps we should redefine the word "universe" to mean "all that exists in our own four-dimensional sphere," rather than "all that exists."

Is the Universe Actually Closed? Startling though Einstein's concept of the edge of the universe might be, many cosmologists find a closed universe to be by far the most satisfying answer. To begin with, they find the alternative, an open universe, infinite in extent, even harder to accept. Furthermore, the mathematical development of the general theory of relativity is most beautiful and elegant when applied to a universe that is closed. Einstein believed the universe is closed because his sense of beauty led him to that opinion. Einstein, to whom striving for simplicity was a lifelong mission, said, "From the standpoint of the theory of relativity, the condition for a closed surface is very much simpler than the corresponding . . . [open] universe."

But is the universe actually closed? It should be emphasized that the general theory of relativity does not demand either a closed or an open universe. Einstein's preference for a closed universe was based on aesthetic feelings. Granted, conclusions reached on the basis of Einstein's aesthetic feelings are not to be taken lightly; nevertheless, cosmologists have persisted in a search for evidence to reinforce or contradict this deeply held belief.

The Critical Density The whole question hinges on the average concentration, that is, the average density, of the matter in the universe. If the average density is less than a critical value, the universe is open. If the average density is higher than this critical value, the universe is closed. One test to determine which situation actually exists, consists of surveying the number of grams of matter in the observable universe. One then divides this number by the number of cubic centimeters in the observable universe, thereby computing the average density (number of grams per cubic centimeter) of the universe.

This discussion of the critical density is correct only if Einstein was right and the constant mentioned in Box 17.2 is zero. In some versions of relativity, other physicists take the constant to be nonzero.

One must also determine the numerical value of the critical density to which the actual density is to be compared. The critical density depends on the measured value of Hubble's constant (see Section 16.13). An error in the determination of this constant would lead to an error in the determination of the critical density and might change the entire outcome of the determination. Based on a modern determination of Hubble's constant, the critical density is calculated to be about 6×10^{-30} gram per cubic centimeter. This density would be achieved in a universe in which each cubic volume of space 82 cm (2.7 ft) on a side contained one hydrogen molecule. In the actual universe, matter is not spread out uniformly; some is clumped into stars, some is in planets, and some lies between the stars as dust and gas. It is the average density that matters.

Measuring the Mass of the Observable Universe Measuring the amount of mass in the observable universe is not an easy task. The easiest mass to account for is that in the stars within galaxies, because stars advertise their presence by emitting light. Even this measurement is not simple. One must judge the masses of galaxies by various indirect means. The results of such measurements indicate that, if matter existed only in galaxies, the density of the universe would be roughly 0.2×10^{-30} gram per centimeter, about one-thirtieth of the critical density.

Thus, cosmologists who believe that the universe is closed are required to search for evidence of matter lying within the clusters of galaxies and between the individual galaxies themselves. One approach is too assume that the clusters of galaxies are **gravitationally bound,** that is, that enough gravitational force is produced by the matter in each cluster to prevent the individual galaxies within the cluster from escaping. The determination is a difficult one to make. So far, the results seem to make it unlikely that enough matter can be inferred between the galaxies within a cluster to close the universe. Worse yet, there may be evidence that at least some clusters of galaxies are not gravitationally bound at all.

The search thus turns to the vast regions of space between the clusters of galaxies. Could there be individual stars, dim dwarf galaxies, or incredibly thin gas in this region? How can we know? Only one thing is certain; as one cosmologist puts it, there is "plenty of space [between the clusters] for all the missing mass we need." But is it there?

In their desire to find the "missing mass," some cosmologists have speculated that there might be a vast amount of mass in black holes formed from collapsed stars or even from entire collapsed galaxies. Perhaps there is enough mass-energy in the gravity waves (if they exist) to supply the need. Perhaps the halos of galaxies contain huge amounts of unobservable mass. Perhaps, perhaps, perhaps.

In 1974, James E. Gunn and several colleagues surveyed the available data and concluded that the density of the universe is so low that the universe is open. But Gunn reported, "The evidence for the open universe is not conclusive because each of the [lines of logic indicating an open universe] by itself has loopholes. But we feel that openness is the most reasonable conclusion from data now at hand." It will be exciting in coming years to watch the continuing drama. Will the accumulating data ever verify Einstein's opinion? Only the brave would risk a prediction as to the eventual outcome.

The Fate of the Universe At present the galaxies are receding from each other. Yet their mutual gravitational attraction must be gradually slowing the speed of recession. Will this gravitational brake be strong enough to halt the expansion of the universe some day and then cause the galaxies to begin to move back toward each other?

The general theory of relativity proposes a fascinating set of alternatives. If the universe is open, it predicts that the universe is too gently curved (in Newtonian terms: the gravitational pull is too weak) to halt the expansion. The galaxies will recede, ever more slowly, but never stop. This is analogous to the case of a rocket launched from earth at a speed so rapid that it is able to escape from the earth. On the other hand, if the universe is closed, the theory predicts that the universe is sufficiently curved (the gravitational pull is sufficiently large) to stop the expansion of the universe eventually and then to reverse it.

Cosmologists who accept the big bang theory and who also believe the universe is closed have painted an awesome picture for us. A tiny cosmic egg explodes, sending matter streaming out in all directions. The matter clumps up into galaxies and further clumps into stars. Humans come along and observe the process. Hubble discovers the expansion of the universe. Einstein predicts the eventual end of the expansion. The expansion *does* stop, and the galaxies begin to approach each other. (How many billions of years in the future this might occur is at present difficult to compute.) The contraction of the universe accelerates, and the galaxies begin to touch. Matter and radiation heat up tremendously in the contraction. Gravitational forces grow immense. All matter is compressed until the universe is only 1 billion ly across—then 1 million, 100, 10 ly across. Finally, in the twinkling of an eye, the universe is crushed down into a tiny ball, perhaps returning to the singularity from which it began. In one sentence, the history of the universe may consist of: big bang–expansion–contraction–big crunch.

And what then? Will the universe remain a tiny ball forever? There are those who speculate that the tiny ball may be another cosmic egg which may again explode. This concept known as the **oscillating universe,** envisions an endless series of expansions and contractions of the universe, one of the most intriguing visions yet proposed by thinkers in the twentieth century.

17.7 LIFE IN THE UNIVERSE

We close this book with one of the oldest and yet most current and exciting questions in astronomy, Is there life in the universe? Phrased this way, the question is one of the oldest trick questions in the world.

The answer is, Yes—on earth. The answer, at once both obvious and profound, leads to the next question, Is there life *elsewhere* in the universe? We may be a bit closer to an answer than were the ancients who speculated that each heavenly body is populated, but we just don't know. A great deal of thought and energy is being expended to answer this question, and an answer could turn up at any time. Until then, there is intellectual pleasure to be had from pondering this awesome question. Ultimately, we will not know the answer until we detect a definitive example of extraterrestrial life or until we inspect every nook and cranny of the universe and find it barren.

Some Recent Views Concerning Extraterrestrial Life In recent years a number of books have appeared presenting the case that ancient astronauts visited the earth. The authors of these books allege that ancient monuments, including Stonehenge, the stone faces on Easter Island, and the Pyramids in Central America and Egypt were built by visitors from outer space. As exciting a prospect as this might be, scientists find the evidence unconvincing. It is usually easy to think of alternate explanations for these accomplishments in ordinary human terms. Sadly, some of these books contain errors in reporting the facts, misstatements that might make their case seem more persuasive to a nonexpert. The scientific consensus is that there is no compelling evidence that ancient astronauts walked the earth. It could have happened, but we have no clear signs that it did.

A related topic is unidentified flying objects (UFOs). A large literature of sensational books exists on this subject as well. Many of the books on UFOs are merely sensational, superficial reports of sightings. Again, it is fair to say that most scientists find no good evidence among UFO reports that life is visiting us from outer space. Buried among all the mistaken sightings of ordinary astronomical objects and the many hoaxes there *may* lie an actual sighting of an alien craft. If so, the evidence for it has not yet surfaced, most scientists believe.

The Scientific Search for Extraterrestrial Life The *scientific* evidence that life might exist elsewhere in the universe is not nearly so spectacular or monetarily profitable to write about. Yet scientists find it much more *interesting* evidence because it is solid, if not conclusive, evidence. Life does not seem to exist on any of the heavenly bodies we have explored so far. But this is negative evidence. Two discoveries point, if ever so tentatively, in the other direction.

Amino Acids in Meteorites The first of these discoveries occurred in 1970. Scientists working for NASA identified amino acids in a meteorite that had fallen near Murchison, Australia. This discovery was exciting because amino acids are the building blocks of proteins. Proteins, in turn, are the complex molecules that are among the main structural components of living beings on earth. For example, our cell membranes and our muscles are composed of protein. The discovery of amino acids in the Murchison meteorite was followed by a similar discovery in two more meteorites in 1971. Then, in 1974 NASA workers found fatty acids in the Murchison meteorite and one other meteorite. Animal fat is composed of two subunits, fatty acid and glycerol, a type of alcohol. While these discoveries *do not* by any means *prove* that life exists elsewhere than on earth, they do show that some of the principal *components* of life as we know it have been formed elsewhere in the solar system.

Molecules in Space The second discovery in this regard is actually a long series of discoveries. As radio telescope techniques have been refined, radio astronomers have detected among the incoming radio waves an ever-increasing number of spectral lines, radio waves of a single wavelength. The first of these to be discovered was the 21-cm line of hydrogen (see Section 15.8). The origin of these radio waves has been identified. They are emitted by molecules in space. (A molecule is composed of two or more atoms.) The number of molecules identified by their spectral lines has now climbed to nearly 50, including such common molecules as water, carbon monoxide, formaldehyde, ammonia, and methyl alcohol.

Before these discoveries, it had been supposed that such molecules could not form between the stars. It is now hypothesized that these molecules form on dust particles in space. This idea is supported by the fact that many of the molecules are seen in two particularly dusty regions of the sky, the Orion nebula and in the direction of the center of our galaxy.

Among the most exciting molecules found are those known as **amino acid precursors.** For example, two of these, methylamine and formic acid are found in space. On earth these chemicals react to form glycine, the simplest amino acid. Growing numbers of astronomers speculate that planets might be "seeded" by these molecules which are among the materials necessary to begin life. If this is so, life may indeed be abundant in the universe.

Laboratory Experiments Concerning the Beginning of Life Many experiments have been performed in the laboratory to determine if life could have begun on earth without introduction of the necessary materials from outer space. In 1953, Stanley Miller performed the pioneering experiment in this field. He placed an atmosphere of methane, ammonia, hydrogen, and water in a closed container. These materials were theoretically among the chemically active principal components of the atmosphere of the early earth. The atmosphere of Jupiter still has this composition, but the earth's atmosphere has changed since then. Miller passed an electric discharge through the chamber to supply the energy for chemical reactions. He assumed that on the young earth this energy would have been supplied by ultraviolet radiation from the sun. He found that amino acids formed in the container. Similar experiments since then have demonstrated that a wide variety of the complex molecules necessary for life can form spontaneously under conditions similar to those on the early earth. No life has spontaneously formed in the laboratory, but some suppose that it might require a billion years for this to occur.

Radio Telescopes Search for Extraterrestrial Life As provocative as these discoveries are, they do not by any means confirm that life exists elsewhere. Another approach to the problem has been to use radio telescopes to search for radio signals from outer space, signals broadcast by intelligent beings. At first the signals from pulsars were thought to fill the bill, but now the neutron star theory seems to be a much more natural explanation (see Section 14.15). A systematic search for intelligence using a radio telescope is beset with many problems. Which stars do you concentrate on and which wavelengths do you examine? How do you procure time on the busy radio telescopes for such research?

Not much actual work has been done. In the early 1960s, Frank Drake spent several weeks observing the two nearest stars most like the sun. He used a 26-meter-diameter (85-ft) radio telescope and observed at a wavelength of 21 cm. This is the wavelength of the hydrogen spectral line. It was chosen because it might be a "famous" wavelength for other civilizations, just as it is for us. The results were negative. Beginning in 1973, a radio telescope at Ohio State University was used to inspect large regions of the sky for intelligible signals. Again, workers observed at 21-cm. No messages have yet been received.

In contrast to this passive approach, several attempts have been made to send messages to potential extraterrestrial civilizations. Figure 11.4 shows the plaque that was attached to the *Pioneer 10* and *Pioneer 11* space probes which will eventually leave the solar system. The chances of these tiny probes ever being found in the vastness of space are very small.

The first radio signal ever sent specifically to contact extraterrestrial life was emitted on November 16, 1974, to dedicate the refurbished Arecibo radio telescope, the world's largest (see Figure 15.22). Using the telescope in reverse, a 450,000-watt radio transmitter emitted a signal. The resulting beam was aimed at the globular cluster M13, also known as the Hercules cluster. The message, in an easily broken code (it is hoped), starts with the numbers 1 to 10. Then follow, among other things, representations of atoms, of the DNA molecule, of a human body, of the solar system, and of the Arecibo telescope. Since M13 is about 25,000 ly from us, we can expect an answer no sooner than roughly 1300 generations from now.

17.8 THE PLEASURES OF SPECULATING ABOUT EXTRATERRESTRIAL LIFE

After long hours in the observatory or at their desks, astronomers often enjoy leaving the realm of strict facts and evidence behind them in order to speculate, letting their minds roam freely without concern for rigid accuracy. At such times, conversations like the following can occur.

Will Anyone Out There Talk to Us?

Oscar (the optimist): You know, when I look through the telescope and see all the stars out there, I have to believe that there are many other civilizations of intelligent beings.

Patricia (the pessimist): I know what you mean, but the more I think about it the less sure I am. I really doubt that we will ever communicate with any other civilization.

Oscar: How can you say that? Since a round-trip message to the Andromeda galaxy would take 4 million years, let's confine our attention to our own galaxy. Yet, in our galaxy alone there are, let's say, about 100 billion stars.

Pat: Yes, but surely not all stars have societies able and willing to communicate. Let's speculate that binary stars do not have planets. That leaves out at least two-thirds of all stars, since at least about that many stars have a stellar companion. So only about three-tenths of all stars could have planets.

Oscar: OK, I'll go along with that, if you'll agree that all single stars *do* have planets. That leaves 0.3 times 100 billion or 30 billion stars with planets, an impressive number.

Pat: Yes, but we can't applaud yet. Think of this. We do not know for certain of any other star, except the sun, that has a planet roughly the size of the earth. What if the planets of the solar system were formed by some freak accident and planets are actually very rare. But I'll go along with you; let's say 30 billion stars have planets. But then, how do you know that all the stars can support life?

Oscar: All right, you were flexible; I'll be flexible. Of the 100 stars nearest us, I don't know, maybe one in ten is a G star similar to our sun. So say that one-tenth of all stars are G stars. Let's assume that all G stars can support life and that each has one or more planets. That leaves 3 billion G stars with planets.

Pat: I feel like I'm conceding a lot there. *Surely*, one in ten is too high. Besides, I can well imagine that many G stars do not have planets. But, let that go. However, I will *not* concede that every G star has a planet *suitable for life*. Many G stars might have only one planet, and it might be too close or too far from its star to support life. Maybe many G stars have no planets with enough water. Maybe many G stars have nothing but planets covered by oceans and have no continents. I wonder if only one in 100 G stars might have an inhabitable planet.

Oscar: Oh, come on! Only one in 100? You're too stingy. We have nine planets in our system, and at least one is definitely able to support life. Let's at least say that one G star in ten has a planet able to support life. And I feel strongly that that is very pessimistic.

Pat: It goes against my grain but, for now, I'll accept one G star in ten with a habitable planet. That leaves one-tenth of 3 billion or 300 million G stars in our galaxy that have habitable planets.

Oscar: Good. Now, I gave up a lot on that last round, so now I am going to ask you to admit that *every* habitable planet actually will develop life.

Pat: Well, I find that pretty hard to swallow. Yet, we know that there are amino acids and fatty acids in meteors. Life *might* follow. How do we know? Well, suppose I agree that there have been 300 million stars having life-bearing planets. But if I give you that, I will ask you to agree that not all these planets achieve the status of having *intelligent* life. Maybe most life never gets beyond the plant stage. How do we know?

Oscar: We don't. How about making the guess that one in ten planets that bear life eventually bears intelligent life?

Pat: One in ten? Well . . . OK.

Oscar: Fine, then that still leaves 30 million planets that have intelligent life—30 million!

Pat: Wait! I'm not done. If another society is going to communicate with us, it must develop technology. Many societies will not.

Oscar: I can't agree! Surely *any* intelligent species will become technological. Intelligence implies curiosity and the desire to control the environment.

Pat: Now, wait. Dolphins are quite intelligent, but their lack of hands has kept them from developing technology. The high Egyptian culture developed some technology but died out before being able to communicate with extraterrestrials. Maybe very intelligent beings would not desire to control the environment but would rather seek oneness and peace with nature. I'm not sure, but don't some of our religions have an attitude something like that? I'd say only one intelligent species in 100 would develop high technology.

Oscar: I still say that intelligence will always develop technology. Let's settle on one technological society in ten.

Pat: How about one in 50?

Oscar: Well, this is where I'm giving away the most yet, but I'll settle for one in 30. One-thirtieth of the 30 million planets with intelligent life leaves 1 million societies in our galaxy with high technology.

Pat: Reluctantly, I'll take that figure. Now, how many of these million societies will *want* to communicate with others?

Oscar: Now, come on, Pat. You're just saying anything to win your point. The urge to communicate must be *universal* among intelligent beings.

Pat: Not really. Many people argue that we should not try to advertise our own presence to the galaxy. They point out that, if we do contact another intelligent species, they will almost certainly be more advanced than we are. After all, we have been able to communicate with other planets for only about 20 years. Chances are that anyone we contact would be more advanced than that. Then these people go on to point out that, here on earth, whenever a more technologically advanced civilization has come into contact with a lesser one, it was always disastrous for the lesser. Look at Cortez and the Aztecs, for example. The conclusion is that we should not communicate.

Now here is another point of view. Look at our society. We *could* be attempting contact right now. However, very little effort is being made. We could spend, say, $10 billion and build a radio telescope that could easily communicate with any similar radio telescope in our galaxy. But Congress does not seem interested. Very few radio telescopes have ever been used, even briefly, to look for incoming intelligent signals. I submit that most societies will *not* attempt communication.

Oscar: Oh, very well. Let's be ridiculous and say that only one technological society in ten will try to communicate. That still leaves one-tenth of 1 million, or 100,000 societies in our galaxy trying to communicate.

Pat: Not so fast!

Oscar: What now?

Pat: Now we have to consider how long a technological society will last. Look at us. We developed the ability for interstellar communication only about 20 years ago. Now we have an exploding population which threatens the very existence of our civilizations. We are using up our energy and other natural resources at an alarming rate. Our two major powers have nuclear missiles aimed at each other. The touch of a button could pound us back into the dark ages. I will be surprised if our technological society lasts 20 *more* years!

Oscar: Pat, you are a confirmed pessimist, for sure. But you must recall that we are also intelligent. Intelligence means the ability to solve problems. We are rapidly becoming aware of our dwindling resources. We are seeking ways to use solar and nuclear energy. There are already signs of a reduction in the world's birthrate. We have already seen arms reduction agreements signed by the major powers. I am sure that our technological society will get through this crisis and last thousands of years.

Pat: I wish I had your faith. I see mostly decadence and corruption in our society. Nevertheless, for the sake of argument I will grant you that the average technological society that wants to communicate lasts for as long as 100,000 years, a figure I very much disbelieve.

Oscar: You grant that much? Fine. Then we have 100,000 societies each lasting 100,000 years and all of them trying to communicate. Then how can you say that you doubt that we will ever communicate with even one of them?

Pat: The reason is that our galaxy is roughly 10 billion years old. Now, if your 100,000 societies spring up at random during that time as stars are born and die, then in 10 billion years there would be, on the average, *one* such society each 100,000 years, because 10 billion divided by 100,000 is 100,000. Don't you see? That means that, on the average, there is only one society willing to communicate in our galaxy at any one time. On the average, while one society is dying out, another will be building up. Well, now you know why I'm so pessimistic!

Oscar: I was tricked! I shouldn't have agreed that one society in 30 will develop technology, and what about. . . .

Pat (interrupting): Well, if you take that back, I must insist on shorter lengths of time for technological societies to exist. Maybe 200 years is more like it. . . .

The conversation lasted well into the night. No firm conclusions were reached, except that more data were needed. Then they went outside to look at the stars in this, our astounding universe.

SUMMING UP

Properly understood, Hubble's law does not require that we live at the center of the universe. However, most astronomers agree that the law does imply that the universe began with a big bang. The 3° blackbody radiation is seen as strong confirmation of this theory.

Einstein's general theory of relativity explains gravitation as the effect of the curvature of space due to the presence of mass. It predicts that black holes might exist, and some black hole candidates are under active observation. General relativity also predicts that, if the universe is dense enough, it will eventually collapse, reversing the present expansion caused by the big bang.

Although most astronomers suppose that life exists elsewhere in our vast, amazing universe, no convincing evidence of extraterrestrial life has yet come to light. Humans have gone to the moon, but will we eventually expand into the universe? Will we meet other beings there? All we who are alive now know is that the universe is there for us to view, contemplate, and enjoy.

EXERCISES

1. Define the following terms: cosmology, cosmic egg, black hole, open universe, closed universe, critical density for closure, oscillating universe, extraterrestrial intelligence.

2. Does Hubble's law imply that our galaxy is at the center of the universe? Restate Hubble's law in light of this insight.

3. Summarize the major events in the first 300,000 years of the universe according to the big bang theory. How old is the universe according to this theory?

4. (a) State the cosmological principle. (b) What is the perfect cosmological principle? (c) What is the age of the universe according to the steady state theory? What does the steady state theory have to say about creation? (d) What discovery led to the demise of the steady state theory? Explain why the steady state theory could not survive this discovery.

5. How, according to the general theory of relativity, are gravitational effects produced? List two tests of this theory that tend to confirm it.

6. What is the current status of the search for gravity waves?

7. How might a black hole be formed as a star dies? What is the most persuasive evidence known that a black hole might exist? Explain the proposed mechanism by which this possible black hole makes its presence known.

8. Which of the three possible answers to the question at the beginning of the chapter about the edge of the universe appeals most to you? What further proposition did Einstein advance to answer this question? Does it appeal to you more than any of the first three possibilities?

9. What is the ultimate fate of the universe if it is closed? If it is open?

10. Study the argument in Section 17.8 and supply numbers that seem reasonable (or, perhaps, "pleasing") to *you*. Then calculate how many civilizations are attempting communication in our galaxy at any one time.

READINGS

Books that cover the topics in this chapter are

Cosmology Now, edited by Laurie John, Taplinger, New York, 1973, by many experts in the field, including R. Penrose on black holes.

The Red Limit, The Search for the Edge of the Universe, by Timothy Ferris, Morrow, New York, 1977, an exciting coverage of cosmology.

Man Discovers the Galaxies, by Richard Berendzen, Richard Hart, and Daniel Seeley, Science History Publications, New York, 1976, pp. 175–215, an historical account.

The Universe, by Isaac Asimov, Walker, New York, 1971.

Black Holes, Quasars, and the Universe, by Harry L. Shipman, Houghton Mifflin, Boston, 1976, chap. 13–16.

A detailed description of the current big bang theory is very well done in

The First Three Minutes, A Modern View of the Universe, by Steven Weinberg, Basic Books, New York, 1977.

A good introduction to relativity is to be found in

The Relativity Explosion, by Martin Gardner, Random House (Vintage Books), New York, 1976.

Responsible replies to the ancient astronaut books are

The Past is Human, by Peter White, Taplinger, New York, 1976, by an archeologist.

The Space-Gods Revealed, A Close Look at the Theories of Erich von Daniken, by R. Story, Harper & Row, New York, 1976, which refutes such claims, item by item.

Two interesting books on UFOs are

UFO's—Explained, by Philip J. Klass, Random House, New York, 1974, which is useful if perhaps a bit too passionate.

The UFO Experience, by J. Allen Hynek, Regnery, Chicago, 1972, which is perhaps a bit too credulous.

Of the many books on extraterrestrial life, among the most interesting are

Intelligent Life in the Universe, by I. S. Shklovsky and Carl Sagan, Dell, New York, 1966, in which some of the information is now dated but which is still very thought-provoking.

The Galactic Club, Intelligent Life in Outer Space, by Ronald N. Bracewell, Freeman, San Francisco, 1975, a brief, well done, nontechnical survey.

APPENDIX I

THE ENJOYMENT OF THE CONSTELLATIONS

Many people find an abiding pleasure in knowing at least the easier, brighter constellations. The groupings of the stars grow to be familiar and never cease to entertain as they move in their seasonal rhythms.

I became interested in learning the constellations as a youngster after reading *The Stars* by H. A. Rey (Houghton Mifflin, Boston, 1970), an entertaining book which facilitates finding and remembering the constellations. Rey shows how easy it is to locate the star Arcturus (see my Figure 2.1). I tried it, it worked, and I was hooked.

In this appendix, there are five simple star maps. On any evening, one of the maps will be appropriate. Each map features a distinctive grouping of stars which should be rather easy to identify. Not all stars visible to the naked eye are shown; only the brightest and those most important to the featured region are shown.

My hope is that you will try out one of the maps, be encouraged to obtain a copy of *The Stars* (or other star charts), and become hooked too.

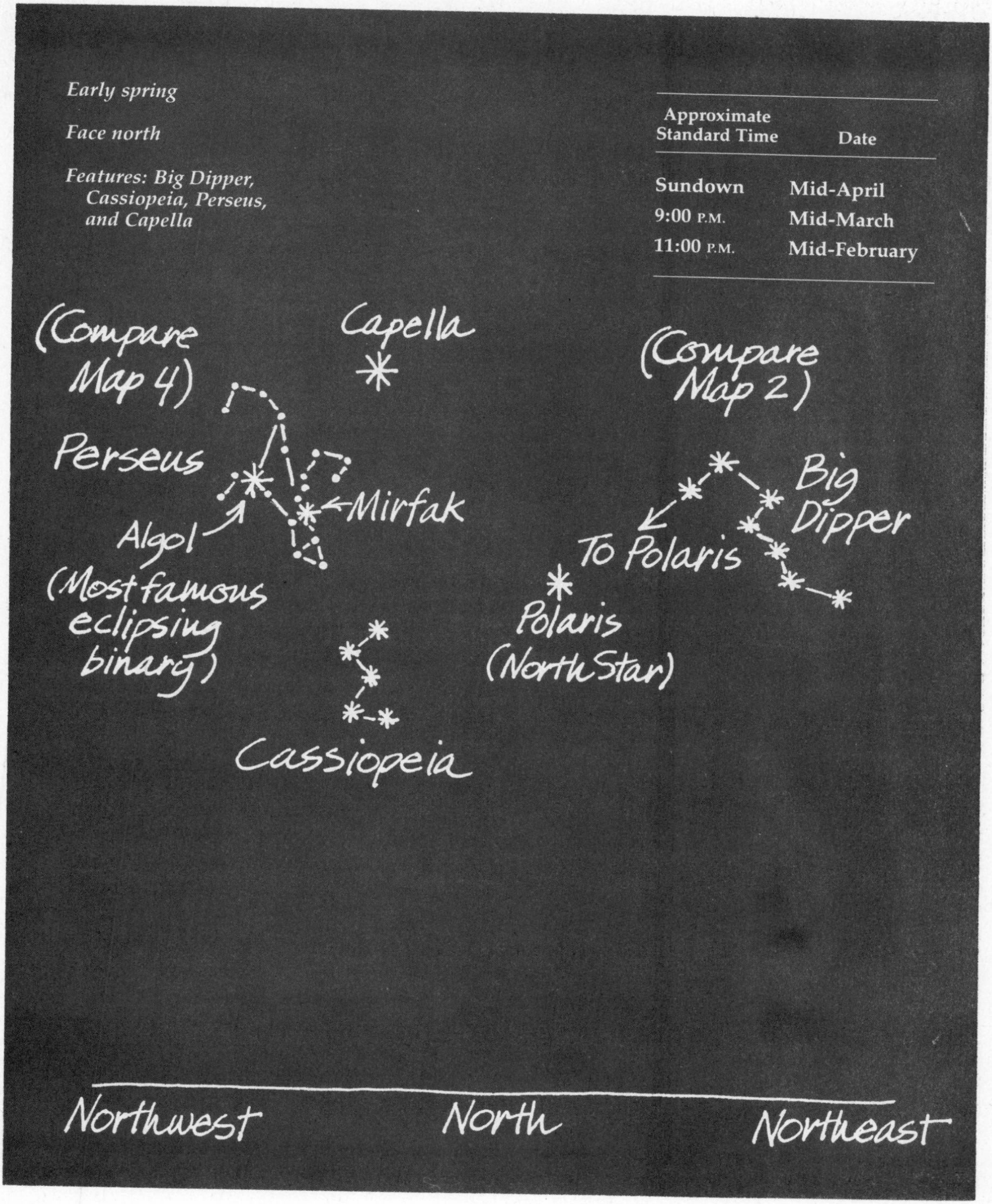

Early spring

Face north

Features: Big Dipper, Cassiopeia, Perseus, and Capella

Approximate Standard Time	Date
Sundown	Mid-April
9:00 P.M.	Mid-March
11:00 P.M.	Mid-February

(Compare Map 4)

Capella

(Compare Map 2)

Perseus

Mirfak

Algol

(Most famous eclipsing binary)

Big Dipper

To Polaris

Polaris (North Star)

Cassiopeia

Northwest

North

Northeast

Map 1

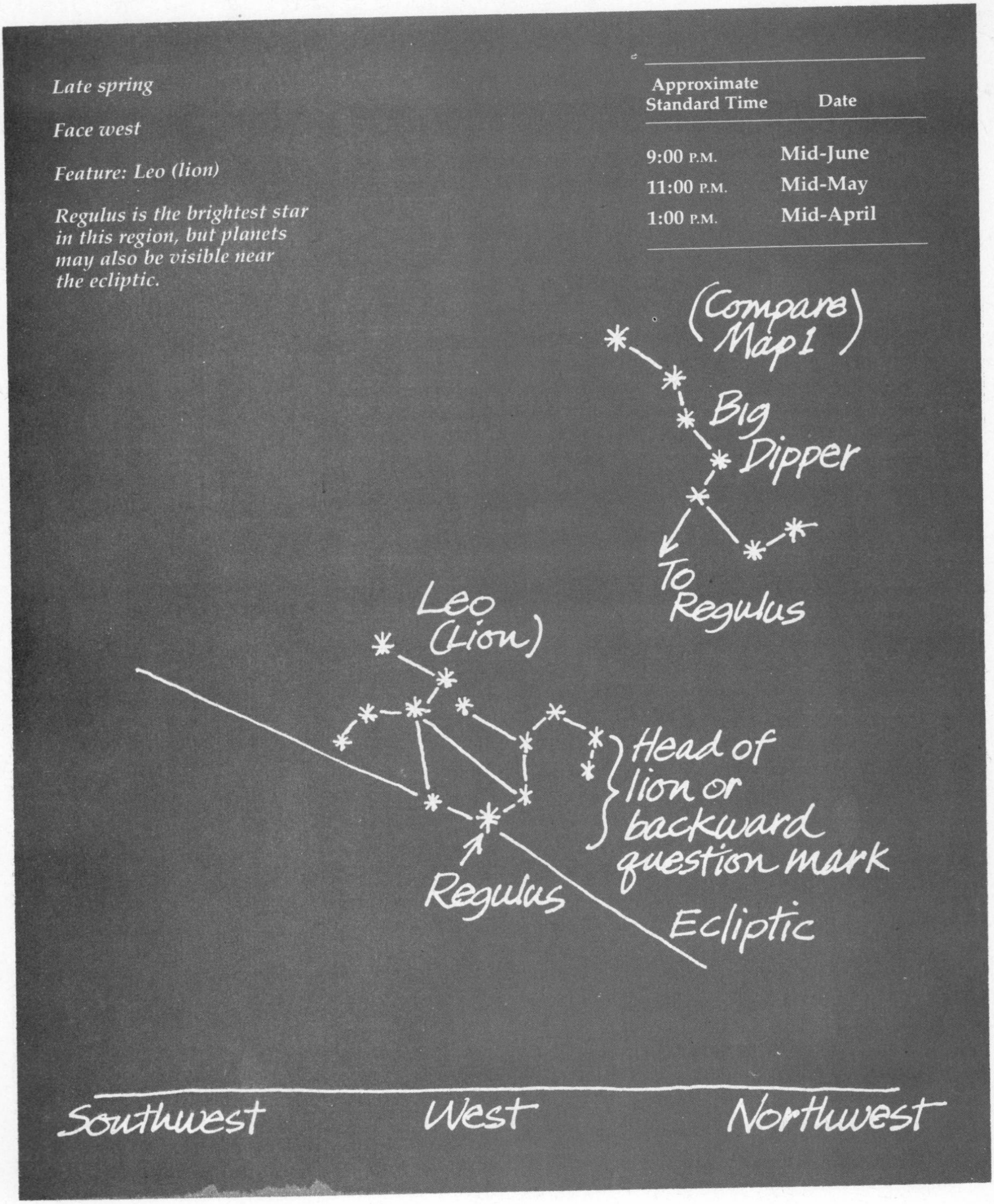

Late spring

Face west

Feature: Leo (lion)

Regulus is the brightest star in this region, but planets may also be visible near the ecliptic.

Approximate Standard Time	Date
9:00 P.M.	Mid-June
11:00 P.M.	Mid-May
1:00 P.M.	Mid-April

(Compare Map 1)
Big Dipper
To Regulus
Leo (Lion)
Head of lion or backward question mark
Regulus
Ecliptic
Southwest
West
Northwest

Map 2

Summer

Lie on your back, feet toward south.

Feature: Summer Triangle

Approximate Standard Time	Date
7:00 P.M.	Mid-September
9:00 P.M.	Mid-August
11:00 P.M.	Mid-July

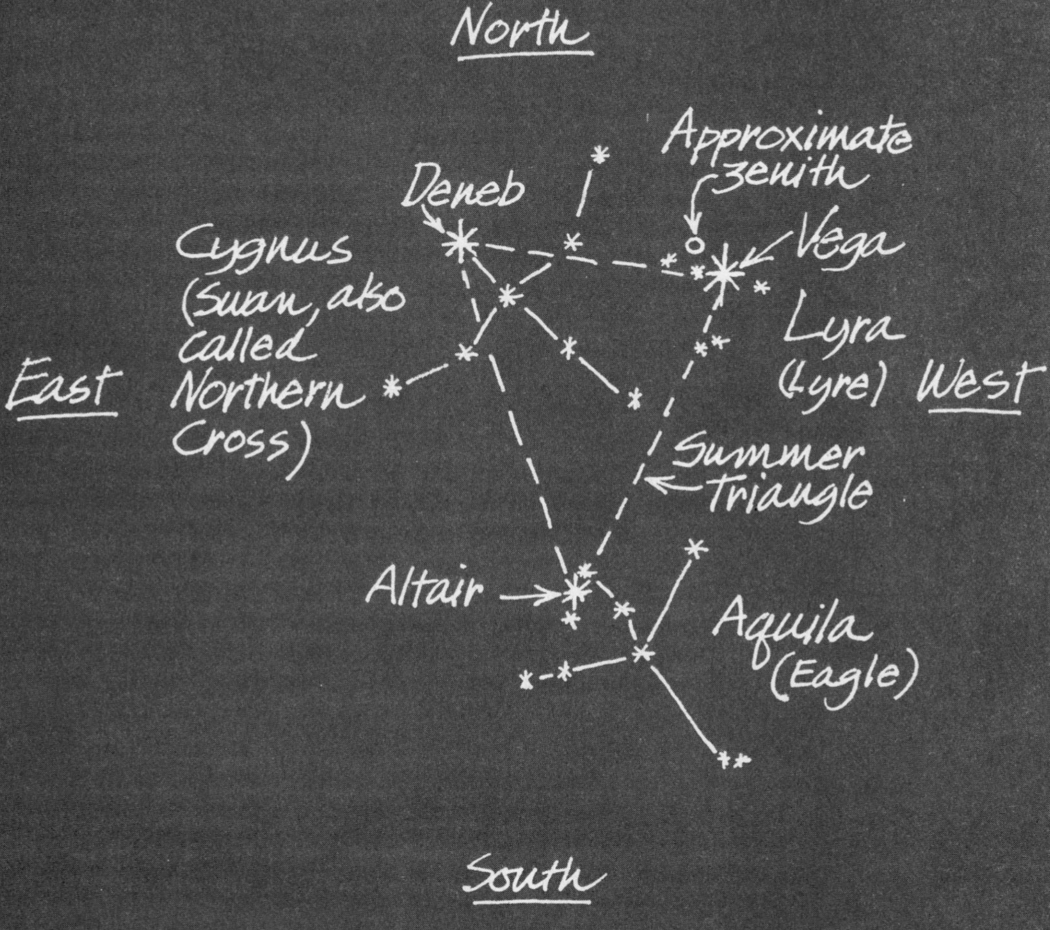

North

Approximate zenith

Deneb

Vega

Cygnus (Swan, also called Northern Cross)

Lyra (Lyre)

East

West

Summer Triangle

Altair

Aquila (Eagle)

South

Map 3

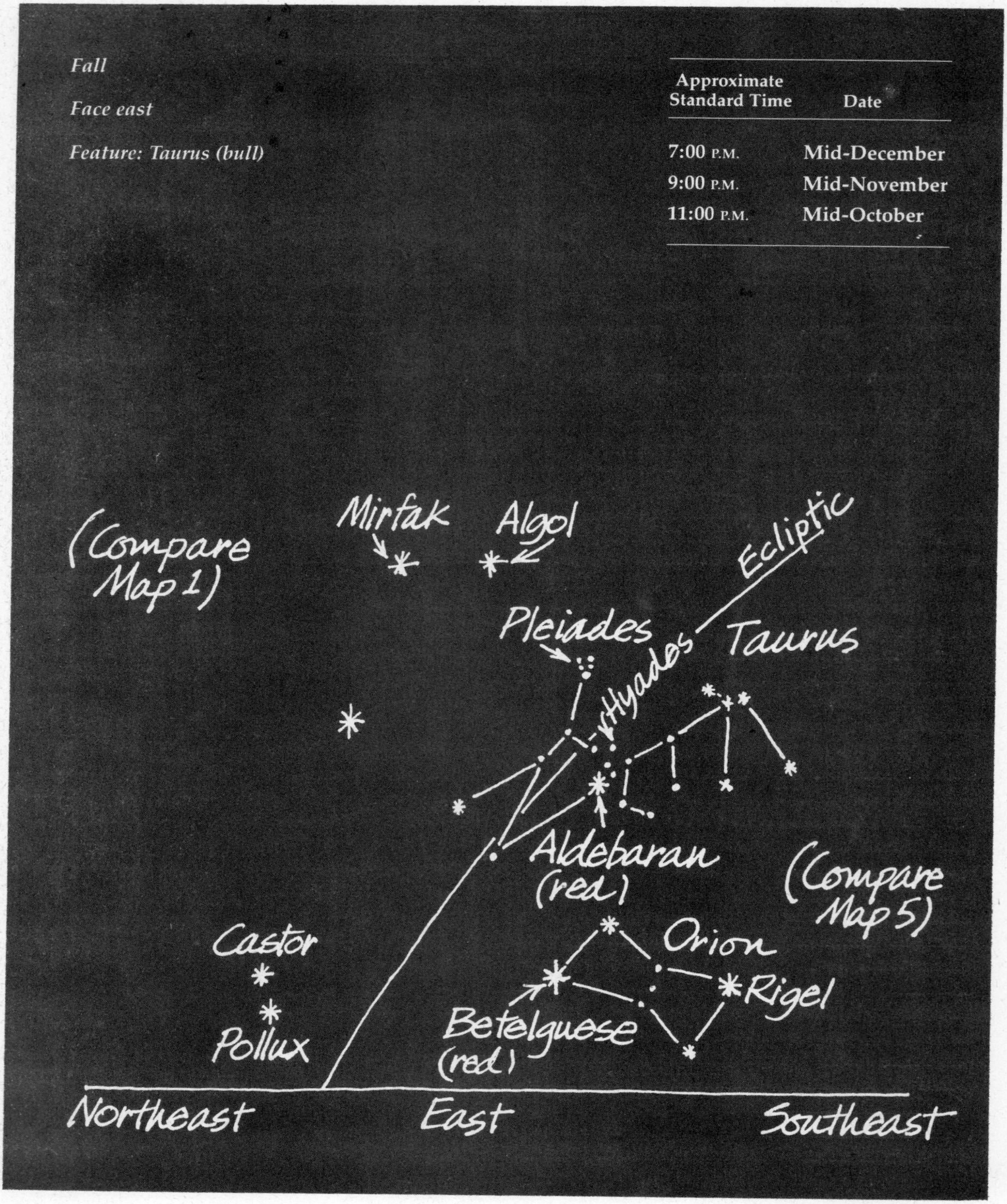

Fall

Face east

Feature: Taurus (bull)

Approximate Standard Time	Date
7:00 P.M.	Mid-December
9:00 P.M.	Mid-November
11:00 P.M.	Mid-October

(Compare Map 1)
Mirfak
Algol
Ecliptic
Pleiades
Hyades
Taurus
Aldebaran (red)
(Compare Map 5)
Castor
Pollux
Orion
Rigel
Betelguese (red)
Northeast
East
Southeast

Map 4

Winter

Face south

Features: Winter Triangle and Orion

Approximate Standard Time	Date
Sundown	Mid-March
9:00 P.M.	Mid-February
11:00 P.M.	Mid-January

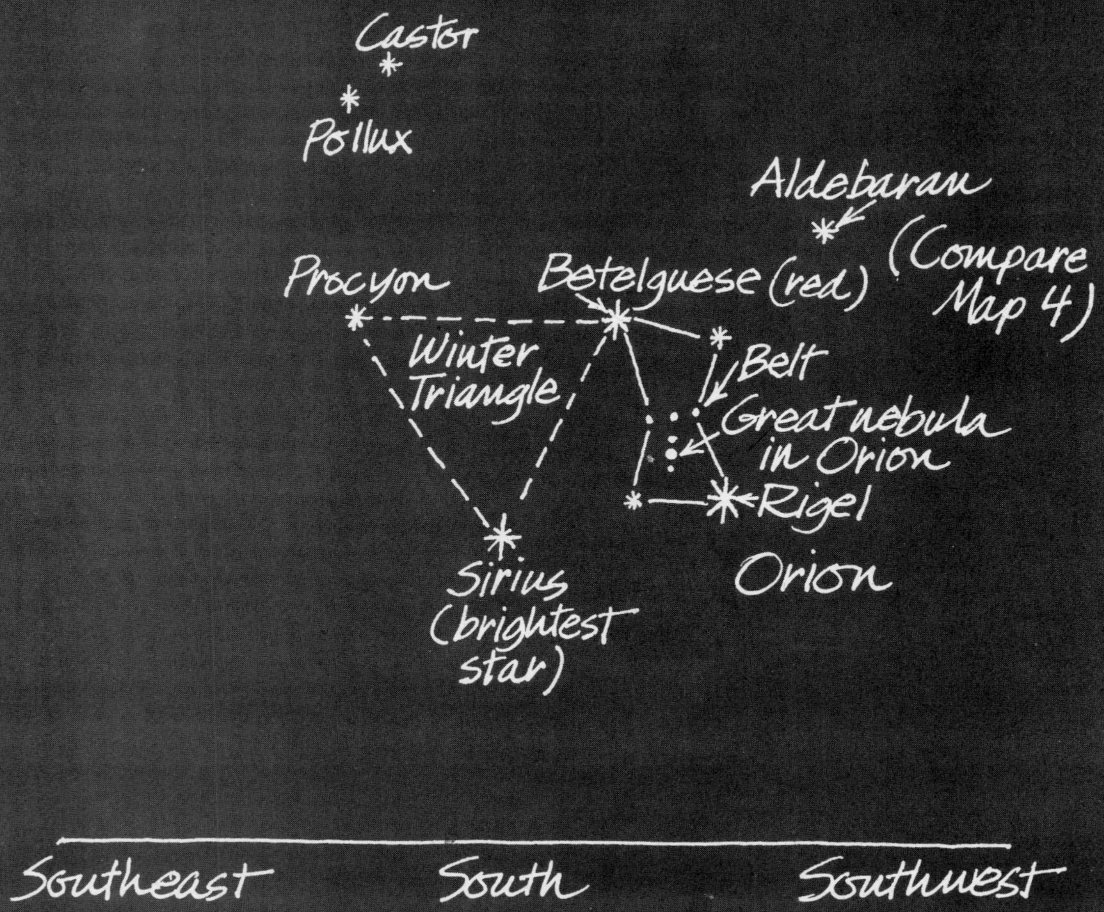

Castor
Pollux
Aldebaran
(Compare Map 4)
Procyon
Betelguese (red)
Winter Triangle
Belt
Great nebula in Orion
Rigel
Orion
Sirius (brightest star)

Southeast South Southwest

Map 5

APPENDIX 2

SOLUTIONS TO PUZZLES

Puzzle 2.1 The puzzle asks the reader to picture the view on the left in Figure P2.1, looking out a huge semicircular window at the sky. This window faces north, which places west on the left and east on the right. The bottom straight line of the window represents the horizon. The reader knows that the star moves and that Polaris does not. What path will the star follow? The reader also knows that the angular separation of the star and Polaris does not change. Thus the star cannot move to the left or right. The only way for the star to move without retreating from or approaching Polaris is in a circle. The center of the circle is at Polaris. Since all stars move generally upward in the east and downward in the west, the direction of the star's motion on its circular path is as shown on the right. As viewed by a person living in the northern hemisphere and facing north, the stars appear to trace out circles in a counterclockwise direction. You can further deduce that each star follows its path over and over. Finally, note that the moving star in the figure is circumpolar. It never sets because it never reaches the horizon. The reader is encouraged to observe the motions of stars near Polaris.

Figure P2.1

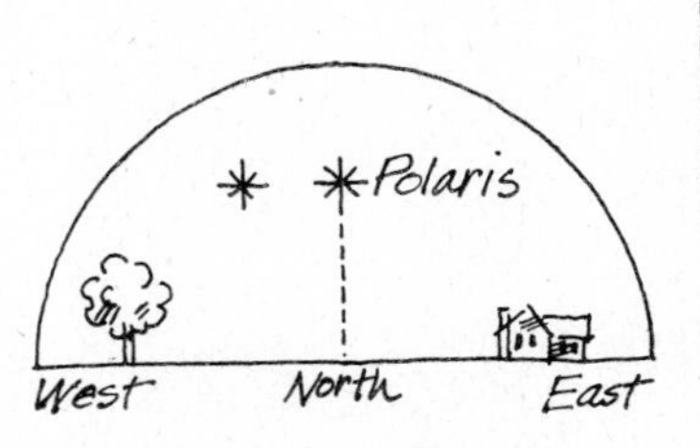

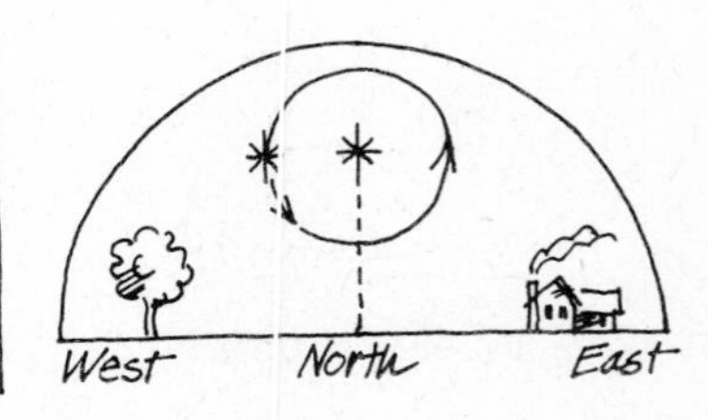

Puzzle 2.2 Which star, A or B, will win a race back to the starting line (see Figure P2.2.)? In a horse race one might bet that the horse with the shorter track would win, as shown in (b). It does not work this way for stars. Such a finish would violate the idea that the angular separation of all stars remains constant. If either star were to win the race, it would leave the other behind, thereby increasing the angular separation. The race must always end in a tie. Each star traces its own path completely in the same length of time as all other stars. Think about it this way. Suppose stars A and B represent the pointers of the Big Dipper. If (b) were correct, A and B would no longer point at Polaris.

Figure P2.2

(a) The race begins.

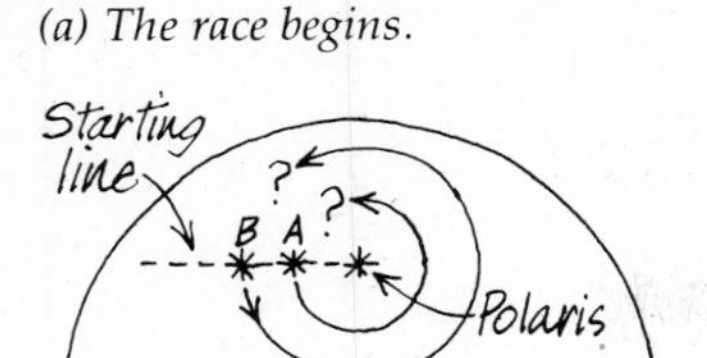

(b) The end of the race will not *be like this.*

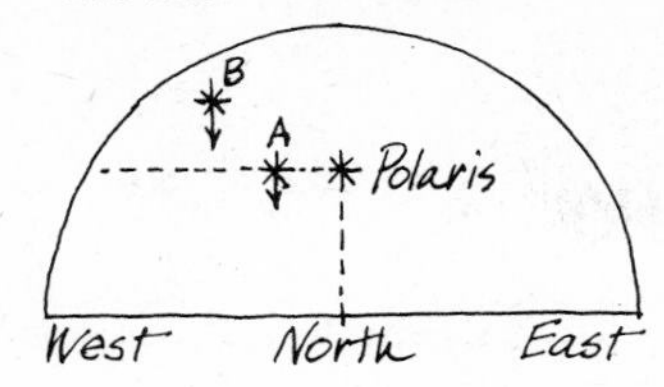

(c) The end of the race—a tie.

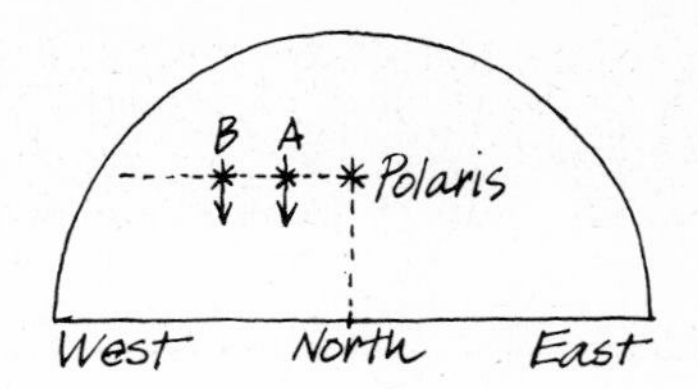

As shown in (c) the pointers still point correctly at the end. (The race takes almost, but not quite, 24 hr to complete, as explained in the text.)

Puzzle 2.3

Figure P2.3

Tonight at 11:00 P.M.

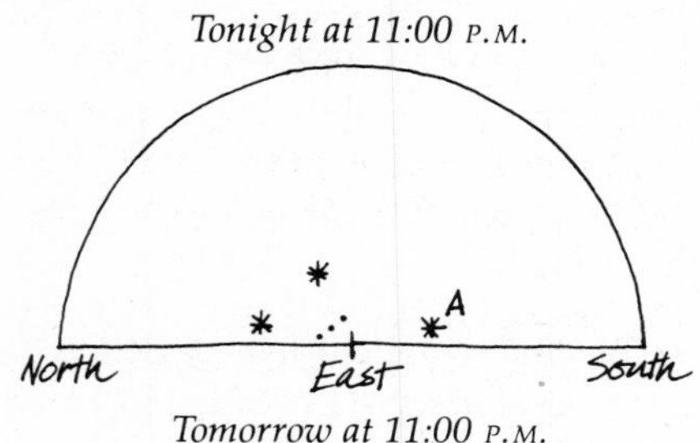

Star A is just above the eastern horizon at 11:00 P.M. Where will it be tomorrow at 11:00 P.M.?

Tomorrow at 11:00 P.M.

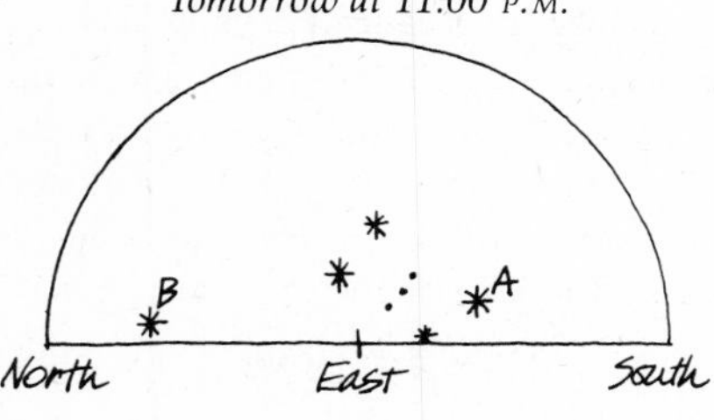

Each star takes about 23 hr, 56 min to travel its path once and return to the same place relative to the horizon. Thus, star A will rise tomorrow 4 min before 11:00 P.M., at 10:56 P.M. This means that by 11:00 P.M. tomorrow, star A will be farther above the horizon than it was today at the same time. We see that new stars, such as star B, are visible at 11:00 P.M.

Position of star A at 11:00 P.M. on four successive days.

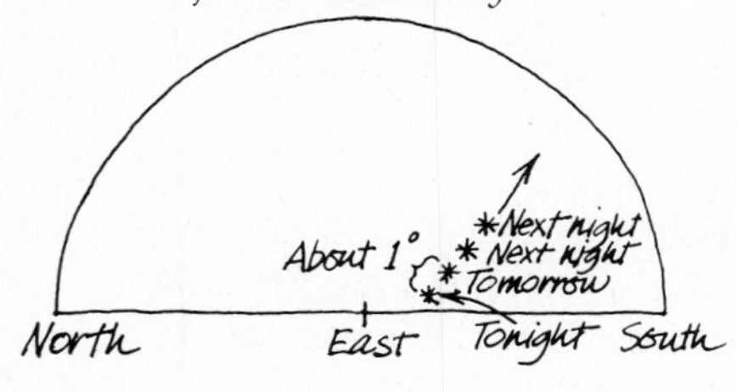

Each day at the same time each star has traveled 360° (one full circle) plus approximately one more degree, the angle it moves in 4 min. The result is that the star appears higher in the east each day at the same time. For clarity, the illustration exaggerates the star's daily advance.

Puzzle 2.4 The star returns to the meridian 4 min before the sun does. One result is that the sun falls behind the star by an angle slightly less than 1°. Figure P2.4 shows that, by losing the race, the sun has shifted to a new position further to the east (left) of the star. (We have agreed to imagine that the stars are visible by day.)

Figure P2.4

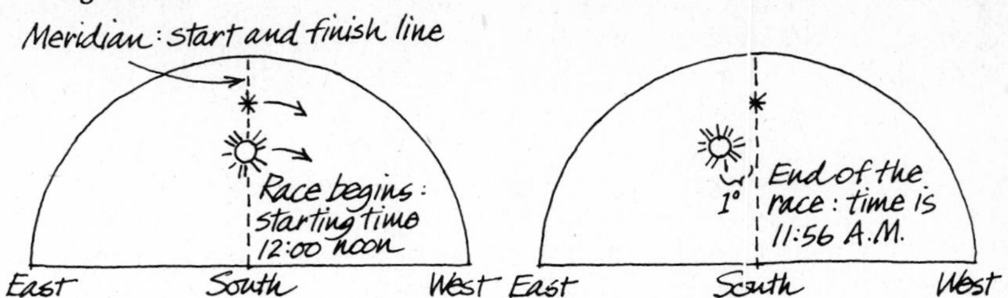

The sun and a star run a one-lap race from the meridian. Which one returns to the meridian first?

End of the race. The star takes 23 hr, 56 min to go around once. The sun takes 4 min longer, or 24 hr. Thus the star wins the race at 11:56 A.M.

Figure P2.5

The end of the race

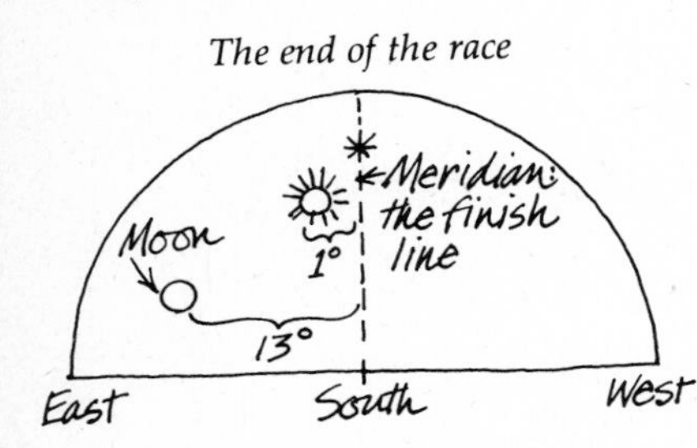

Puzzle 2.5 The beginning of the race (not shown) saw the star, sun, and moon on the meridian (see Figure P2.5). All three next set in the west and then rose in the east. The result of the race is that the star, which took 23 hr, 56 min to run one lap, won. The sun, which required 24 hr to run one lap, lost by 4 min. The moon, which required about 24 hr, 50 min to run one lap, lost to the star by about 54 min. The moon falls behind the fixed stars (to the east) by about 13° per day.

Puzzle 3.1 Figure P3.1a shows an improvement over Figure 3.3. Here the observer can see more of the celestial sphere because the earth is drawn smaller than in the previous figure. Figure P3.1b shows further improvement. The earth is shown very small, so small that it is not possible to show a dot for the observer. Now the observer can see almost exactly half of the celestial sphere. Conclusion: The earth is very small in comparison to the celestial sphere. In other words, the distance to the stars is very great compared to the size of the earth.

Figure P3.1

(a) This illustration has a double purpose. The first is discussed in the caption. The second is to show that the axis of the earth, extended into space, becomes the axis of the celestial sphere. Similarly, the extended equator of the earth becomes the celestial equator.

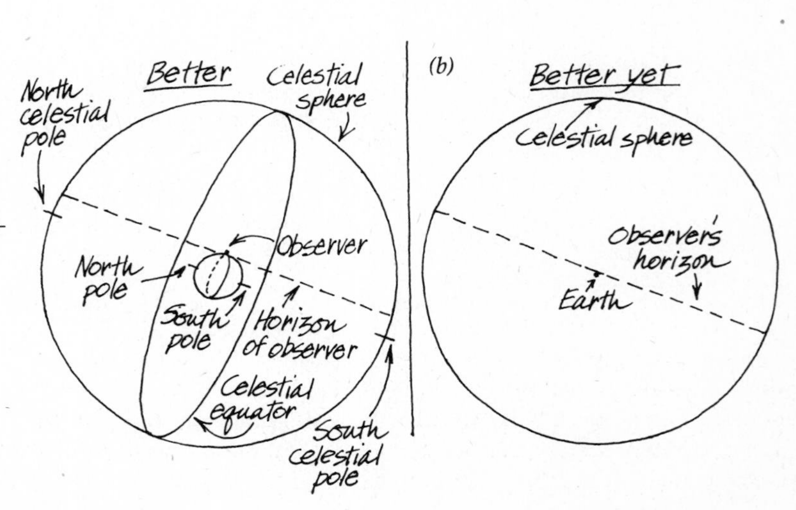

Puzzle 3.2 Figure P3.2 is another example of the view of the motion of a celestial object as viewed from high above the earth's north pole. As in all such pictures, there are two ways to think about the meaning of the picture.

Figure P3.2

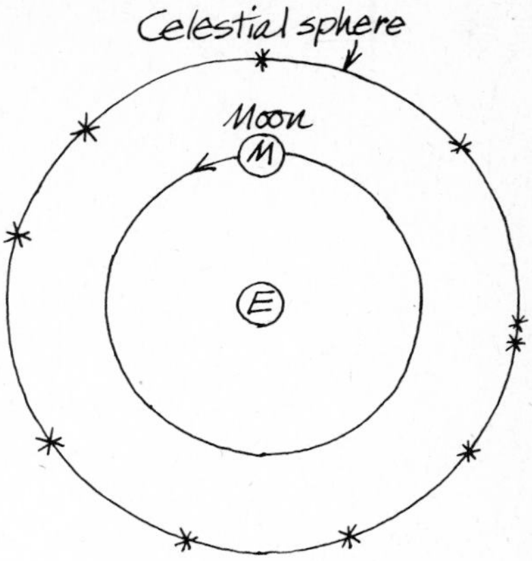

1. Mentally keep the stars stationary. It is known that the moon appears to travel among the fixed stars from west to east, 13° per day. The moon appears to return to a place among the fixed stars in one sidereal period, twenty-seven days, 8 hr. Thus, in the figure, you should imagine the moon moving as shown by the arrow, 13° per day, and taking twenty-seven days, 8 hr to travel its path once. This explains the motion of the moon with respect to the fixed stars.

2. You can also imagine the whole figure, except the stationary earth, turning clockwise around the earth once every 23 hr, 56 min, carrying the moon along. The moon simultaneously crawls along the paper 13° per day counterclockwise. This explains the motion of the moon with respect to an observer on earth.

Puzzle 4.1 The race begins when the observer is at position A in Figure P4.1. The sun and the star are lined up on the meridian. At B, 23 hr, 56 min later, the earth has rotated once on its axis and brought the observer (and his meridian) back to the fixed star. The star has won the race. The sun now appears to be to the east (left) of the star by 1°. The time is now 11:56 A.M. Four more minutes must elapse until, at C, the sun again crosses the meridian. If we were to assume that the earth does not revolve around the sun but merely that the earth rotates in place, the outcome of the race would always be a tie.

Figure P4.1

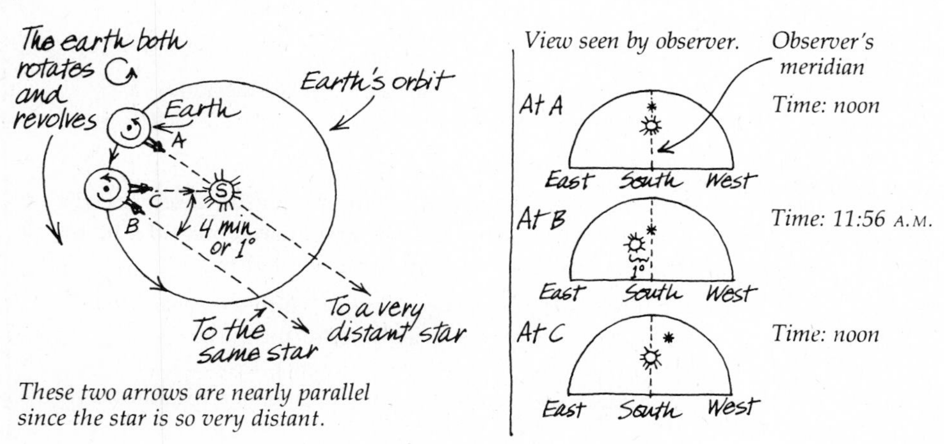

These two arrows are nearly parallel since the star is so very distant.

Puzzle 5.1 Both stars have their altitudes increased by atmospheric refraction, but the lower star will be affected more than the upper one. Thus, these stars appear closer together than they actually are. The angular separation is actually *more* than 5°. Tycho corrected all his observations for atmospheric refraction if they were made far from the zenith. This is just one instance of the extraordinary care he exercised in his zeal to measure the skies accurately.

Puzzle 5.2 The planet's sidereal period is $P = 8$ years. Then $P^2 = P \times P = 8 \times 8 = 64$. Now if $P^2 = 64$, what is R^3? Easy, if you know Kepler's third law: $P^2 = R^3$. Thus $R^3 = 64$, also. We wish to find R, the average distance of the planet from the sun. We now know that $R^3 = R \times R \times R = 64$. What number cubed gives 64? After some trial and error we discover that $4^3 = 4 \times 4 \times 4 = 64$. Thus $R = 4$ AU. The result is that the new planet will be found at an average distance of 4 AU from the sun. Notice that no difficult parallax measurements were required to determine this distance.

Puzzle 6.1.

Figure P6.1

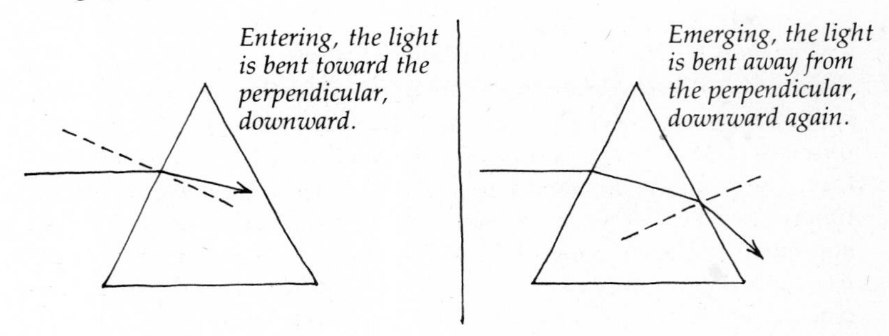

Puzzle 10.1

1. The same side of the moon always faces the earth, and so does the observer (see Figure P10.1).

2. See Figure P10.1.

3. The time from sunrise to sunrise for the observer is the same as from the first quarter moon to the first quarter moon for us. A complete day on the moon lasts about one earth month.

4. Our observer always sees the earth overhead. According to Figure P10.1, as viewed from the moon, the earth does not appear to move around in the sky. Actually, the moon rocks back and forth a little, so the earth seems to move back and forth a bit around its average position.

5. Yes. The earth rotates once every 24 hr, in which time the moon moves only 13° in its orbit.

6. The two sets of phases are exactly opposite. At full moon the earth looks new, etc.

7. Midnight on the moon's near side has the full earth in the sky making the night quite bright. Midnight on the moon's far side is darkest.

Figure P10.1

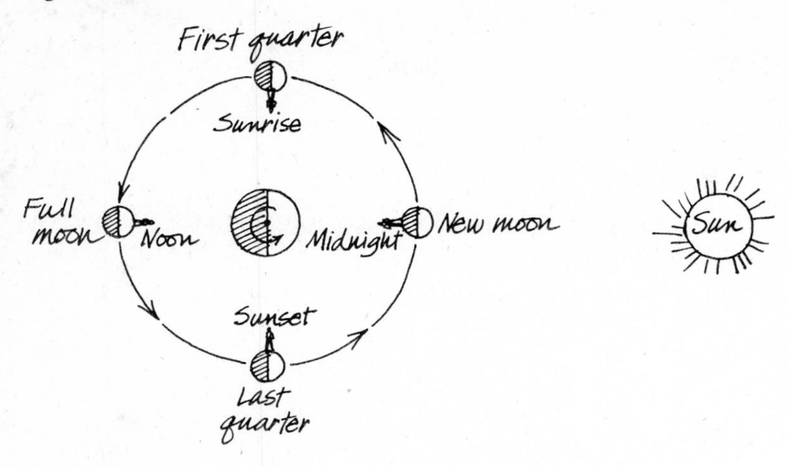

Puzzle 11.1 A blue shift, a shift toward shorter wavelengths, indicates motion toward the observer. Furthermore, the faster the motion, the greater the blue shift. For the side of Saturn's rings approaching us, a greater blue shift of the inner edge indicates that it is moving faster than the outer edge. The conclusion is that the rings consist of individual orbiting particles.

Puzzle 12.1

1. $L = bR^2 = \frac{1}{3} \times 3^2 = \frac{1}{3} \times 3 \times 3 = \frac{1}{3} \times 9 = 3$ suns.

2. $L = bR^2 = \frac{1}{3} \times 9^2 = \frac{1}{3} \times 81 = 27$ suns. We may observe that the star in question 1 looked just as bright as this one but, since it is three times farther away, it is actually $3^2 = 9$ times brighter.

3. $L = bR^2 = \frac{1}{9} \times 9^2 = \frac{1}{9} \times 81 = 9$ suns. Compare question 2. The two stars are the *same distance* from us. Thus, if one appears to be one-third as bright, it actually is one-third as bright.

4. $L = bR^2$ becomes $8 = \frac{1}{2} \times ?$ The number the question mark represents must be 16, since one-half of 16 is 8. Thus $R^2 = R \times R = 16$ or $R = 4$ AU. We will find in section 12.9 that one can determine certain stars' luminosities before knowing their distance. One can then follow the procedure used here to find the distance of such a star.

Puzzle 12.2 The letters refer to the questions given in the puzzle.

1. (A) Using the H-R diagram you have drawn, you can see that W and the sun have the same surface temperature. (B) According to Stefan's law, the hotter star would emit more radiation from an equal area of its

surface. Here the temperatures are equal. Thus, the two stars have equal brightness, area for area. (C) Because it lies above the sun on the H-R diagram, star W is brighter taken as a whole. (D) Although these two stars are equal in brightness area for area, W is brighter as a whole. Thus, it must have more area than the sun. Consequently, W is the larger star.

2. (A) X is hotter. (B) X. (C) They have equal overall brightness. (D) One square kilometer of X is brighter than 1 km^2 of the sun. The sun achieves an equal overall brightness by having more square kilometers of surface area. The sun is larger than X.

3. (A) The surface temperatures are equal. (B) They are equal. (C) The sun. (D) The sun is larger.

4. (A) The sun. (B) The sun. (C) The luminosities are equal. (D) Z is larger than the sun.

Puzzle 13.1

1. The mass of a B star is roughly 10 suns and that of a K star roughly 0.7 sun. The B star has 10/0.7 = 14.3 times more fuel.

2. The B star has a luminosity of about 800 suns; the K star's luminosity is about 0.2 sun. The B star uses its fuel 800/0.2 = 4000 times faster.

3. The B star has somewhat more fuel but uses it very much more rapidly than the K star. The K star should last much longer than the B star on the main sequence. The B star will leave the main sequence first.

Puzzle 13.2

1. Whereas the *contraction* of a gas tends to cause it to heat up, the *expansion* of the star's envelope causes the surface temperature to decrease somewhat. Answer: Half.

2. By Wien's law, the star's principal wavelength increases as the surface temperature drops. The star gets redder.

3. A cooler object radiates less per unit area. Answer: Decrease.

4. A larger star has more surface from which to emit radiation. Answer: More luminous.

5. If the surface area increases 10,000 times while the surface temperature merely drops by half, one would guess that the effect of the surface area dominates. Detailed calculations indicate that this is correct. The luminosity increases as a star leaves the main sequence. The sun, for example, will become several hundred times more luminous during this transition.

6. The decreasing surface temperature sends the star to the right, and the increased luminosity sends the star upward on the H-R diagram.

7. The star's surface is becoming cooler, redder, and brighter. It is becoming a red giant.

APPENDIX 3

THE PLANETS

Planet	Average distance from the sun: Millions of km	AU	Sidereal period (earth years)	Synodic period (earth days)	Average orbital velocity (km/sec)
Mercury	57.9	0.387	0.241 (88.0 days)	115.8	47.9
Venus	108.2	0.723	0.615 (225 days)	583.9	35.1
Earth	149.6	1.000	1.000 (365.3 days)	—	29.8
Mars	227.9	1.524	1.881	779.9	24.1
Jupiter	778.3	5.203	11.86	398.9	13.1
Saturn	1427	9.539	29.46	378.1	9.65
Uranus	2870	19.18	84.03	369.7	6.80
Neptune	4497	30.06	164.8	367.5	5.43
Pluto	5913	39.53	248.0	366.7	4.74

[a] S, Solid; C, Clouds.

Eccentricity of orbit	*Inclination of orbit to earth's orbit (degrees)*	*Mass (earth = 1)*	*Average density (gm/cm³)*	*Surface gravity (earth = 1)*	*Equatorial sidereal period of rotation*	*Inclination of orbit to equator (degrees)*	*Equatorial radius (km)*	*Average temperature at surface, (°C)[a]*
0.206	7.00	0.055	5.44	0.37	59 days	Roughly 2	2,439	350 (S) day, −170 (S) night
0.007	3.39	0.815	5.24	0.89	243 days	3	6,055	−30 (C), 480 (S)
0.017	0	1	5.50	1	23 hr, 56 min	23.45	6,378	22 (S)
0.093	1.85	0.108	3.97	0.38	24 hr, 37 min	23.98	3,395	−23 (S)
0.048	1.31	317.8	1.33	2.60	9 hr, 55 min	3.08	71,400	−150 (C)
0.056	2.49	95.2	0.68	1.1	10 hr, 14 min	26.73	60,400	−180 (C)
0.047	0.77	14.7	1.1	1.1	12 hr?	97.9	25,900	−200 (C)
0.009	1.78	17.2	1.7	1.3	16 hr?	28.8	24,500	−230 (C)
0.249	17.17	0.002?	?	?	6 day, 10 hr	?	1,500?	?

APPENDIX 4

THE SATELLITES OF THE SOLAR SYSTEM

Parent planet	*Satellite*	*Average distance from center of planet (thousands of km)*	*Orbital sidereal period (earth days)*
Earth	Moon	384	27.32
Mars	Phobos	9.4	0.319
	Deimos	23.5	1.26
Jupiter	5 Amalthea	182	0.498
	1 Io	422	1.77
	2 Europa	671	3.55
	3 Ganymede	1,070	7.16
	4 Callisto	1,880	16.7
	13 Leda	11,100	239
	6 Himalia	11,500	251
	7 Elara	11,800	260
	10 Lysithea	11,900	264
	12 Ananke	21,200	631
	11 Carme	22,500	692
	8 Pasiphe	23,500	739
	9 Sinope	23,700	758
	14 (Unconfirmed)	—	—
Saturn	Janus	160	0.749
	Mimas	186	0.942
	Enceladus	238	1.37
	Tethys	295	1.89
	Dione	378	2.74
	Rhea	527	4.52
	Titan	1,220	15.9
	Hyperion	1,480	21.3
	Iapetus	3,560	79.3
	Phoebe	12,900	530
Uranus	Miranda	130	1.41
	Ariel	191	2.52
	Umbriel	265	4.14
	Titania	436	8.71
	Oberon	589	13.5
Neptune	Triton	356	5.88
	Nereid	5,570	359
Pluto	Charon	19	6.4

[a] R, Retrograde revolution.
[b] Date of discovery is shown in parentheses.

Inclination of orbit to planet's equator (degrees)[a]	Radius of satellite (km)	Discovered by[b]
18–29	1,740	—
1.0	14	Hall (1877)
1.3	8	Hall (1877)
0.4	100	Barnard (1892)
0	1,800	Galileo (1610)
0	1,500	Galileo (1610)
0	2,600	Galileo (1610)
0	2,500	Galileo (1610)
26.7	10?	Kowal (1974)
27.6	80	Perrine (1904)
24.8	20?	Perrine (1905)
29.0	55	Nicholson (1938)
33 R	10?	Nicholson (1951)
16 R	10?	Nicholson (1938)
35 R	10?	Melotte (1908)
27 R	10?	Nicholson (1914)
—	—	Kowal (1975)
0.0	100	Dolfus (1966)
1.5	200	W. Herschel (1789)
0.0	250	W. Herschel (1789)
1.1	500	Cassini (1684)
0.0	400	Cassini (1684)
0.4	800	Cassini (1672)
0.3	2,900	Huygens (1655)
0.4	100	Bond (1848)
14.7	700	Cassini (1671)
30 R	100	W. Pickering (1898)
3.4	200	Kuiper (1948)
0	700	Lassell (1851)
0	500	Lassell (1851)
0	900	W. Herschel (1787)
0	800	W. Herschel (1787)
20 R	1,900	Lassell (1846)
27.4	300	Kuiper (1949)
0	600?	Christy (1978)

APPENDIX 5

TYPICAL MAIN SEQUENCE STAR DATA

Spectral class	*Luminosity (sun = 1)*	*Principal wavelength (Å)*	*Surface temperature (K)*	*Radius (sun = 1)*	*Mass (sun = 1)*	*Average density (gm/cm³)*
O	500,000	700	40,000	20	30	0.01
B	800	2,000	16,000	5	10	0.1
A	20	3,000	10,000	2	3	0.5
F	2.5	3,600	8,000	1.3	1.5	1
G	1	5,100	5,700	1	1	1.4
K	0.2	7,300	4,000	0.8	0.7	2
M	0.008	11,000	3,000	0.3	0.2	10

APPENDIX 

TWENTY SELECTED STARS

Name	*Visual apparent magnitude*	*Distance (ly)*	*Luminosity (sun = 1)*	*Spectral class*	*Approximate radius (sun = 1)*	*Mass (sun = 1)*
Sun	−26.8	0.000016	1	G2	1	1
α Cen A (Alpha Centauri, also Rigil Kentaurus)	0.0	4.3	1.5	G2	1.2	1.1
Barnard's star	9.5	5.9	0.0004	M5	0.3	—
α CMa A (Sirius)	−1.5	8.6	23	A1	1.8	2.3
α CMa B (Sirius B)	8.7	8.6	0.002	White dwarf	0.02	1.0
61 Cyg A	5.2	11	0.08	K5	0.7	0.63
α CMi A (Procyon)	0.37	11	7.2	F5	1.7	1.8
Kruger 60A	9.8	13	0.001	M3	0.5	0.3
40 Eri B	9.6	16	0.003	White dwarf	0.02	0.43
α Lyr (Vega)	0.04	27	53	A0	3	—
α Aql (Altair)	0.77	17	11	A7	2	—
β Gem (Pollux)	1.15	35	33	K0	30	—
α Boo (Arcturus)	−0.06	36	100	K2	25	—
α Tau (Aldebaran)	0.86	68	160	K5	45	—
α Car (Canopus)	−0.72	195	1,500	F0	60	—
α Sco A (Antares)	0.94	500	9,000	M1	600	—
α Ori (Betelguese)	0.41	520	15,000	M2	700	—
β Ori (Rigel)	0.14	900	53,000	B8	40	—
α Cyg (Deneb)	1.25	1600	63,000	A2	50	—
ζ Pup (Naos)	2.2	2000	50,000	O5	20	—

GLOSSARY

In this glossary, most of the important terms used in this book are defined. Each entry is accompanied by the section number in which the item is first discussed. For many entries, further references may be found in the index.

absolute zero (9.2) The lowest possible temperature. Zero on the Kelvin scale of temperature.

absorption, interstellar (15.7) The weakening of starlight as it passes through interstellar dust.

absorption spectrum (9.5) A spectrum made up of dark lines on a colored, bright background.

acceleration (6.7) The rate at which an object's speed and/or direction of motion changes.

altitude (of a heavenly object) (2.1) The angular separation between any heavenly object and the point on the horizon directly below it.

amino acid (17.7) One of the building blocks of protein.

amplitude of a (water) wave (7.2) The distance from its maximum height to its maximum depth.

angle (2.1) A measure of the space between any two straight lines that intersect.

angular diameter (2.1) The angular separation between any two oppositely situated points on a round object.

angular separation (of two objects) (2.1) The angle between the two lines of sight from the observer to the objects.

annual motion of the sun (2.2) The motion of the sun about 1° per day eastward with respect to the fixed stars.

aphelion (5.6) The point on an orbit around the sun at which the orbiting body is farthest from the sun.

apparent brightness (9.1) How bright an object appears; used as opposed to an object's luminosity or actual brightness.

apparent motion (intro. to Chap. 2) The observed motion of an object, described without regard to whether the motion is actually occurring as described.

astrometric binary star (12.6) A binary star whose nature is revealed by the wavy, periodic path it traces out in the course of its proper motion.

astronomical unit (4.8) A unit of distance equal to the average distance of the earth from the sun.

atmospheric refraction (5.2) The bending in the earth's atmosphere of light from a celestial object, which causes the object to have a greater altitude than it would in the absence of an atmosphere.

aurora (9.6) Light emitted in the earth's upper atmosphere by atoms of the atmosphere stimulated by charged particles showering down into the atmosphere.

barred spiral galaxy (16.8) Similar to a normal spiral galaxy except that either the nucleus is elongated in shape or two distinct bars of stars emerge from the nucleus in opposite directions. The arms are attached to the bars.

binary star (12.6) An object composed of two stars each of which orbits permanently in the gravitational field of the other. See also "double star."

black dwarf (13.6 and 13.9) Used in two ways. (1) A star that does not have enough mass for fusion to be initiated. (2) A star that has left the white dwarf stage and cooled so much that it no longer emits a significant amount of light.

black hole (17.5) An object which, on the basis of the general theory of relativity, is possible if a mass becomes sufficiently compacted that the space surrounding it becomes so curved that the mass is enclosed in a four-dimensional sphere from which no matter or radiation can escape.

blue shift (also violet shift) (7.6) A change in wavelength toward shorter wavelengths of electromagnetic radiation, produced by the relative motion of the source toward the observer. See also "Doppler effect."

bolometer (9.1) An instrument that measures the total amount of energy striking it in the form of electromagnetic radiation.

bright points on the sun (9.8) Intense regions of activity on the sun, which produce strong x-rays and which last an average of 8 hr.

celestial equator (2.1) An imaginary circle on the celestial sphere, which divides the celestial sphere into a north half and a south half.

celestial pole (2.1) Either of the two locations on the celestial sphere about which the objects in that region of the sky appear to travel in circles.

celestial sphere (2.1) The sphere that appears to surround the earth and to which the stars may be thought to be attached when one considers their apparent motions.

Cepheid variable star (12.9) A type of pulsating variable, named for the first member of its class to be discovered, Delta Cephei. Very useful as such a star's distance is easily determined without use of the method of parallax.

Chandrasekhar's limit (13.9) The largest mass a white dwarf can have, about 1.4 suns.

chromatic aberration (6.6) The disturbance of color produced in a refracting telescope.

chromosphere (8.2) A layer of pink gas on the sun. It lies just above the photosphere.

circumpolar star (2.1) Any star that is always above one's horizon.

closed universe (17.6) If the average density of matter in the universe is sufficiently great, space will be curved back upon itself forming a four-dimensional sphere known as a closed universe. Such a universe will eventually collapse.

coma of a comet (11.8) A halo of dust and gas that has escaped from the nucleus of the comet as the nucleus is warmed by the sun and which surrounds the nucleus in a spherical region.

conjunction (2.4) When the angular separation of two objects has decreased to a minimum, they are said to be in conjunction. (This is only a rough definition. The exact technical definition does not concern us in this book.)

continuous creation (17.3) A feature of the steady state theory: that the production of matter out of nothing is an ongoing process in the universe.

continuous spectrum (9.5) A complete spectrum of all colors of light gradually shading from one wavelength to the next. No wavelengths are omitted.

convection (9.6) A type of vertical circulation in which warmer, less dense material rises and cooler, more dense material falls.

core of a star (13.6) The central region of a star where the pressure and temperature are greatest and where fusion occurs during various stages of a star's life.

corona (8.2) The tenuous, hot region of gas extending far above the sun's chromosphere.

coronal holes (of the sun) (9.8) Regions on the sun in which x-ray emission is weak, possibly the sites of origin of the solar wind.

cosmic egg (17.2) The extremely dense ball of matter that, according to the big bang theory, contained all matter and exploded to initiate our universe.

cosmic ray (14.6) A nucleus of an atom moving through space at a speed which, in many cases, closely approaches the speed of light.

cosmological principle (17.3) The principal assumption of the big bang theory: that the universe looks the same in every direction as viewed from any location.

cosmology (intro. to Chap. 17) The study of the universe, its origin, its present state, and its future.

crater (10.2) A hole or depression in a body caused by the impact of another body or by volcanic action.

critical density (of the universe for closure) (17.6) A particular average density. If the average density of the universe exceeds this value, it is a closed universe. Otherwise it is an open universe.

cube (of a number) (5.8) The result produced when a number is multiplied by itself and then by itself once more. For example, $5^3 = 5 \times 5 \times 5 = 125$.

daily motion (of the sun) (2.2) The sun's apparent motion which brings it back to the meridian once every 24 hr on the average.

data reduction (5.2) The process of applying corrections to raw data so that they will be useful for others.

declination, the sun's motion in (2.2) The apparent motion of the sun which carries it north and then south of the celestial equator during each year.

deduction (6.7) The act of inferring a specific fact from general principles or axioms.

deductive method (of obtaining a scientific law) (Box 6.5) A method involving the act of logically inferring a law using as a starting point previously known laws, principles, or axioms.

deferent (3.5) A circle having the earth at or near its center.

degenerate matter (13.9 and 14.12) (1) Electron degenerate matter: matter supported against the force of gravity by means of its electron pressure, as in a white dwarf. (2) Neutron degenerate matter: matter supported against the force of gravity by means of its neutron pressure, as in a neutron star.

degree (angular) (2.1) An angle equal to one-ninetieth of a right angle.

density (10.1) The mass of an object or sample of a substance divided by its volume.

differential rotation (of a galaxy) (16.9) A type of rotation, principally of the outer regions of a spiral galaxy, characterized by the fact that, the more distant a star is from the center of the galaxy, the longer its orbital period.

differential rotation (of a gaseous sphere) (8.1) The type of rotation that proceeds at different periods at different distances from the object's equator.

differentiation (of a planet) (10.1) The separation of the substances of a planet during the liquid phase of its existence. The denser the substance, the nearer to the planet's center it settles.

diffuse nebula (15.4) A cloud of interstellar matter.

direct motion of a planet (2.4) Apparent motion from west to east with respect to the fixed stars.

dispersion of light (6.6) The act of breaking up a light beam into its component colors by refracting each particular wavelength through a different angle.

Doppler effect (7.6) A source of waves emitting a particular wavelength and traveling radially relative to an observer is seen to emit a longer wavelength if the source is receding and a shorter wavelength if the source is approaching.

double star (12.6) This term is used in a number of ways. (1) A binary star. (2) Any two stars that appear in nearly the same line of sight in a telescope. (3) Two stars having a small angular separation, that is, nearly the same line of sight, but which are too far apart to physically influence each other significantly.

dust (in interstellar space) (15.4) Tiny, solid particles which lie chiefly near the central plane of spiral galaxies and which strongly absorb the light passing through them.

dwarf star (13.2) A term often used in other books for any star on the lower main sequence. This book reserves the term for use in the phrase "white dwarf star," however.

dynamics (6.1) The study of nature seeking the reasons for and underlying principles of observed motions.

earthshine (10.2) The phenomenon in which light reflected from the earth illuminates the side of the moon currently turned away from the sun.

eccentricity of an ellipse (5.6) A measure of the degree of flattening of an ellipse. It is a number between zero and one. The smaller the number the more nearly circular the ellipse.

eclipse (8.2) The partial or total disappearance of one celestial object as it passes behind another.

eclipsing binary star (12.6) A binary star whose nature is revealed by variations in its apparent brightness. Also known as an eclipsing binary or a photometric binary.

eclipsing variable star (12.6) See "eclipsing binary star."

ecliptic (2.2) The apparent path the sun annually traces out among the fixed stars.

electromagnetic radiation (2.3) Any radiation of any of the following subclasses: gamma radiation, x-rays, ultraviolet radiation, visible light, infrared radiation, microwaves, radio waves.

electron (9.3) A negatively charged subatomic particle. Electrons form the outer electron cloud of an atom.

element, chemical (9.5) A substance composed entirely of one kind of atom.

ellipse (5.5) A geometric figure that can be drawn by placing a loop of string over two thumbtacks and holding the string taut against a moving pencil.

elliptical galaxy (16.8) A galaxy that is elliptical in outline and lacks the arms and dust lanes of a spiral galaxy.

elongation (2.4) The angular separation of an object from the object it orbits. Often applied to the angular separation of a planet from the sun.

emission nebula (15.4) See "H II region."

emission spectrum (9.5) A spectrum made up of bright lines on a dark background.

empirical method of obtaining a scientific law (Box 6.5) A method that involves searching the data produced by observations to find any patterns in them.

envelope of a star (13.6) The region of a star between its visible surface (photosphere) and its core.

epicycle (3.5) A circle upon which a planet travels and which has an imaginary point at its center.

equinox (2.2) One of the two times of the year at which the center of the sun lies on the celestial equator.

eruptive variable star (12.8) A star that suddenly and unexpectedly increases in apparent brightness.

expanding universe (16.13) An interpretation of Hubble's law, which infers that each galaxy is rushing away from all other galaxies.

extraterrestrial intelligence (17.7) Any rational being that does not live on or come from the earth.

eyepiece (6.4) The lens in a telescope near to which the eye is placed. This lens aids in magnifying the image produced by the objective lens or mirror.

facula (8.1) A bright patch on the sun which precedes a sunspot by a few days and lasts longer than the sunspot.

fatty acid (17.7) One of the two subunits of the fat of animals.

filament on the sun (8.2) A prominence seen against the brighter surface of the sun.

fission (13.8) The production of atomic nuclei by splitting apart a more massive nucleus.

fixed star (2.1) A star that does not change its angular separation with respect to other stars. Actually, all stars change their relative positions, but only very slowly. See "proper motion."

flare, solar (9.6) A violent outburst near the surface of the sun. A flare looks white in a telescope and produces intense electromagnetic radiation and spews plasma into space.

flash spectrum (of the sun) (9.5) Bright lines observed in a spectroscope just after the sun's photosphere has been completely eclipsed.

fluorescence (7.3) The ability of some substances to absorb ultraviolet radiation and emit visible light in response.

focal length (6.4) The distance from the center of a lens or mirror to its focal point.

focal plane (6.4) A plane passing through the focal point (focus) of a lens or mirror and parallel to the lens or mirror.

focal point or focus (of a lens or mirror) (6.4) The point at which light from a very distant point source on the lens or mirror axis is brought together (focused) by a lens or mirror.

focus of an ellipse (Box 5.2) Either of the two points marked by a thumbtack in the definition of an ellipse. See "ellipse."

follower (or following) spot (8.1) The second largest spot in a sunspot group. It often is located to the rear of the group with respect to the direction of the sun's rotation.

fusion (9.7) The production of atomic nuclei by means of the joining together of nuclei of lesser mass.

gamma radiation (also gamma rays) (7.3) Electromagnetic radiation having a wavelength shorter than that of x-rays.

gauge, star (15.2) A count of the stars in a particular region of the sky.

geocentric (intro. to Chap. 3) Earth-centered.

globular cluster (15.3) A typical globular cluster consists of 100,000 stars existing in a spherical region having a diameter of 100 ly.

granules (on the sun) (9.6) Brighter spots resembling grains of rice in rice soup, which are due to the convection of the upper regions in the photosphere of the sun.

gravitationally bound cluster of galaxies (17.6) A cluster of galaxies within which the gravitational force on each galaxy due to the combined attractions of the other masses in the cluster is sufficient to prevent most of the galaxies from escaping from the cluster.

gravitational red shift (16.15) The lengthening of wavelength that occurs in a beam of electromagnetic radiation as it leaves a massive object.

gravity wave (17.4) A variation in gravitation due to the acceleration of a mass.

greenhouse effect (10.2) A phenomenon that causes the trapping of heat in an atmosphere. Sunlight passes freely through the atmosphere, is absorbed by the planet's solid surface, and reradiated as longer-wavelength radiation to which the atmosphere is opaque.

guest star (intro. to Chap. 14) A term used by ancient Chinese astronomers for what is now called a nova.

H II region (15.4) Also called an emission nebula. A cloud of dust and gas. The gas emits radiation by fluorescence due to the ultraviolet radiation received from neighboring stars. The gas is predominantly ionized hydrogen.

halo, galactic (15.8) A large spherical region surrounding the disk of the Milky Way galaxy and having the same center as the center of the disk. It is the region in which globular clusters and an unknown number of individual stars reside.

heat (9.2) The energy stored in the random motion of the atoms in an object.

heliocentric (4.3) Sun-centered.

Hertzsprung gap (13.2) A region on the Hertzsprung-Russell diagram between the red giants and O stars in which few stars are found.

Hertzsprung-Russell diagram (12.4) A graph displaying the luminosities or other equivalent characteristics of stars versus their color, spectral class, or surface temperature. Also known as an H-R diagram.

highlands (on the moon) (6.5) The higher, lighter-colored, rough regions of the moon. See also "mare."

horizon (2.1) The circle at which the sky appears to meet the ground.

inferior conjunction of an inferior planet with the sun (2.4) A conjunction occurring as the planet appears to cross from the east side to the west side of the sun.

inferior planet (2.4) A planet that never appears to go into opposition with the sun. There are two: Mercury and Venus.

interferometer (12.5) A device in which two beams of electromagnetic radiation are combined.

interferometry, radio (16.14) A method of obtaining a high degree of resolution by means of the combined radio waves gathered from two widely separated radio telescopes.

inertia (6.3) The sluggishness of any object; the property that makes it difficult to start moving and difficult to stop or deflect from a straight path once it is moving. The more massive an object, the more inertia it is said to have.

infrared radiation (7.3) Electromagnetic radiation having a wavelength longer than that of visible light but shorter than that of microwaves.

ion, ionized, un-ionized (13.9) An ion is an atom that has had one or more of its electrons removed. In such a condition, the atom is said to be ionized. A whole atom is said to be un-ionized.

irregular galaxy (4.8) A relatively formless galaxy, not symmetrical and usually containing large amounts of gas and dust.

isotope (10.2) Any two atoms having the same number of protons but differing numbers of neutrons in their nuclei are said to be isotopes.

joule (9.7) A unit of energy equal to that gained by a 1-kg mass when it is raised a distance of 10.2 cm.

jovian planet (11.2) Any of the planets of the solar system beyond Mars (excluding Pluto) characterized by their large size, large mass, low density, and gaseous or combination gaseous-rock-ice compositions.

Kelvin scale of temperature (9.2) A means of expressing temperature in which absolute zero equals zero kelvins (0 K). If T_k is the temperature on the Kelvin scale, T_c is the same temperature on the Celsius scale, and T_f is the same temperature on the Fahrenheit scale, then $T_k - 273 = T_c = \frac{5}{9}(T_f - 32)$.

kilogram (6.8) A unit of mass equal to 1000 grams. An object weighing 1 newton (0.225 lb) near the surface of the earth has a mass of 0.102 kg. See also "newton."

kilometer (8.3) One thousand meters or 0.6214 mi.

latitude (3.1) A measure of the overland distance from the earth's equator to one's location.

leader (also, preceding) spot (8.1) The largest spot in a sunspot group. It often is located to the front of the group with respect to the direction of the sun's rotation.

light curve (12.8) A graph of apparent brightness versus time for any luminous object.

light year (12.1) The distance light travels in 1 year; about 10 million million km.

line (in a spectrum) (7.5) A vertical stripe in a spectrum on which radiation of one particular wavelength is focused or at which radiation is weak compared to its immediate surroundings.

local group (of galaxies) (16.12) A name given to the cluster of roughly 17 galaxies of which the Milky Way galaxy is a member.

luminosity (9.1) How bright an object actually is as opposed to its apparent brightness (how bright it looks). It may be expressed as energy emitted per second (watts) (see Section 9.1) or in suns (see Section 12.2).

magnitude (12.2) A measure of the apparent brightness of an object. The smaller the numerical value of the object's magnitude, the greater its apparent brightness. A star just barely visible to the naked eye has a magnitude of about 6.

main sequence (12.4) The diagonal band of stars on the Hertzsprung-Russell diagram reaching from upper left to lower right.

main sequence star (12.4) A star situated on the main sequence on the Hertzsprung-Russell diagram. Also known as "a star on the main sequence."

mare (6.5) A low, dark-colored, flat, often round region on the moon. The plural of "mare" is "maria."

mass (6.7) The quantity of matter in an object.

mass-energy (9.7) A term deriving from Einstein's recognition, embodied in the equation $E = mc^2$, that mass and energy are equivalent.

mass-luminosity relation (12.7) The finding that the more luminous a main sequence star the greater its mass.

meridian (2.1) An imaginary line on the celestial sphere passing through both poles and the zenith.

Messier object, M number, or M object (16.1) The number or the object having that number in a list of nebulas published by Messier.

meteor (11.9) A rock that was previously orbiting the sun and which is now plunging through the earth's atmosphere. See also "meteoroid" and "meteorite."

meteorite (11.9) A piece of rock that formerly orbited the sun and which has survived, at least partially, a trip through the atmosphere of the earth and now lies on or near the surface of the earth. See also "meteor" and "meteoroid."

meteoroid (11.9) A rock in orbit around the sun. See also "meteor" and "meteorite."

micrometeoroid (11.9) A dust-sized fragment of rock in orbit around the sun.

microwaves (7.3) Electromagnetic radiation having a wavelength shorter than that of radio waves but longer than that of infrared radiation.

Milky Way (6.5) An irregular band of faint white light in the sky, which reaches in a circle completely around the celestial sphere.

Milky Way galaxy (15.2) The huge disk of stars in which the sun resides. Sometimes referred to as "our galaxy."

minute of arc (2.1) An angle equal to $\frac{1}{60}$°. Sixty minutes of arc equal 1°.

moon effect (or moon illusion) (10.2) The optical illusion that the moon and other celestial objects appear to be abnormally large when seen near the horizon.

multiple star (12.6) Three or more stars each of which orbits permanently in the gravitational fields of the others.

neutrino (9.7) A tiny particle of subatomic matter which can only exist when traveling at the speed of light and which is able to penetrate great thicknesses of matter with ease.

neutron (9.7) A subatomic particle that has no charge. Protons and neutrons constitute the nucleus of an atom.

neutron star (14.12) A star composed chiefly of neutrons, which is roughly 20 km in diameter yet has a mass of about 1 sun.

nova (12.8) Literally "new star." A type of eruptive variable that has experienced an explosion which makes the star appear suddenly much brighter and which throws off a shell of matter.

nucleosynthesis (14.11) The theory that elements heavier than helium originate inside stars.

nucleus of an atom (9.7) The tiny,

dense cluster of protons and neutrons at the center of an atom.

newton (unit of force) (6.8) A unit of force equivalent to 0.225 lb.

observational selection (12.4) A phenomenon inherent in some types of data collection in which certain kinds of data are overemphasized or underemphasized.

objective lens or mirror (6.4) The component of a telescope that gathers light, thereby producing a brighter image.

occultation (11.4) The passage of one astronomical object in front of another from the point of view of an observer. Applied especially to the passage of the moon in front of planets or stars and to the passage of planets in front of stars.

open cluster (13.7) A loose, irregular concentration of stars in space roughly 10 ly across and containing as many as several hundred stars or fewer than 10.

open universe (17.6) If the average density of the matter in the universe is sufficiently small, the universe is said to be open. See also "closed universe."

opposition of a planet with the sun (2.4) When a planet and the sun are in opposite regions of the sky, at or near an angular separation of 180°, they are said to be in opposition.

optical pulsar (14.17) A pulsar that has been detected to emit rapid pulses of visible light.

optical telescope (6.4) A telescope designed to collect visible light.

orbit (4.4) The path of a planet around the sun, of a satellite around a planet, or, in general, of any object moving under the influence of a gravitational force.

oscillating universe (17.6) A closed universe which expands and contracts in a possibly endless cycle.

parallax angle (5.3) The angle formed by two lines of sight directed at the same object, one from each of two observation points.

parallax, concept of (5.3) The apparent motion of a nearby object with respect to a distant background as the observer moves sideways.

parallax, method of (5.3) A method of determining the distance to an object one cannot reach.

parsec (12.1) A distance equivalent to 3.26 ly. A star at this distance exhibits a standard parallax angle of 1 second of arc.

peculiar galaxy (16.14) A galaxy having very unusual or even unique features.

penumbra of a sunspot (8.1) The less darkly colored outer region surrounding the umbra of a sunspot.

perfect cosmological principle (17.3) The principal assumption of the steady state theory: that the universe looks the same in any direction as viewed from every location and when viewed at any time.

perihelion (5.6) The point on an orbit around the sun at which the orbiting body is nearest the sun.

period of a variable star (12.8) The length of time required for the pattern of variation of a variable star to occur once.

photometer (12.2) A device for measuring apparent brightness.

photometric binary (12.6) See "eclipsing binary star."

photon (7.1) A particle of light or other electromagnetic radiation in the modern theory of light.

photosphere (8.2) The visible surface of the sun.

plage (8.1) See "facula."

planetary nebula (13.9) A shell of gas having a typical diameter of 0.6 ly surrounding a central star which is dim but very hot.

plasma (9.6) A form of matter consisting of electrons and partial atoms from which electrons have been torn.

plate tectonics or continental drift (10.1) A theory concerning the earth's surface, involving the motion of regions or plates on the surface and the interactions of the plates where they meet.

populations of stars (16.7) The discovery that the stars in a galaxy can be categorized into two groups or populations on the basis of their color, heavy-element content, and location in the galaxy.

pore (on the sun) (8.1) A small, black dot on the sun roughly 1000 km in diameter, from which a sunspot develops.

principal wavelength (9.4) The wavelength of the most intense or strongest electromagnetic radiation emitted by an object.

prominence (8.2) A huge plume of pink gas projecting upward from the sun's surface.

proper motion (12.1) The apparent sideways motion across the celestial sphere of a star with respect to other stars.

proton (9.7) A positively charged subatomic particle. Protons and neutrons constitute the nucleus of an atom.

protostar (13.5) A newly formed star that emits radiation using energy derived from contraction.

pulsar (14.14) A source of very rapid pulsations of electromagnetic radiation. The pulsations have an extremely accurately repeated period. Pulsars are thought to be rotating neutron stars.

pulsating variable star (12.9) A variable star that expands and contracts with the same period as its light variations.

qualitative (3.8) An adjective referring to the aspects of a theory that can be expressed in words without using mathematics.

quantitative (3.8) An adjective referring to the aspects of a theory requiring numbers and mathematics for a clear description.

quasar (16.15) A celestial object that looks like a star in a telescope and has a spectrum of bright (emission) lines which are strongly red-shifted.

radial velocity (7.6) The part of the relative motion of an object that is directed toward or away from an observer.

radiation pressure (13.6) The force per unit area exerted by electromagnetic radiation when it strikes matter.

radio galaxy (16.14) Any galaxy that has been observed using a radio telescope; the term is sometimes reserved for galaxies that emit an unusually large quantity of radio waves.

radio telescope (14.14) A telescope designed to collect radio waves.

radio waves (7.3) Electromagnetic radiation having a wavelength longer than that of microwaves.

ray (connected with a crater) (10.2) One of a number of bright streaks on a body's surface originating at or near a crater.

red shift (7.6) A change in wavelength toward longer wavelengths of electromagnetic radiation produced by the relative motion of the source away from the observer. See also "Doppler effect."

reflecting telescope (6.6) A telescope that employs an objective mirror.

reflection (6.4) The rebounding or throwing back of a light beam upon striking a polished or other smooth surface.

reflection nebula (15.4) A cloud of dust that reflects the light of neighboring stars. It has a spectrum similar to that of a star.

refracting (telescope) (6.4) A telescope that uses an objective lens.

refraction (6.4) The bending of a light beam as it passes from one medium into another.

residence time (on the main sequence) (13.6) The length of time a star remains a main sequence star.

resolution (of a telescope) (6.4) The ability to distinguish the details of distant objects. The finer details a telescope reveals, the better the resolution.

resolve an object, to (16.1) To distinguish details in an object. Often used to indicate that individual stars have been seen in an object, as in the phrase "to resolve the object into stars."

retrograde motion (of a planet) (2.4) Apparent motion from east to west with respect to the fixed stars.

revolution (4.5) The motion of one celestial body around another.

rotation (4.5) The spinning of an object upon its axis.

satellite or moon (6.5) Any object that orbits a planet.

scarp (10.3) A cliff on the surface of a planet or satellite.

second of arc (2.1) An angle equal to $\frac{1}{60}$ minute of arc. Sixty seconds of arc equal 1 minute of arc.

shadow bands (8.2) Fleeting patterns of bright and dark lines in agitated motion often seen on surfaces just before and just after totality during a total eclipse of the sun.

sidereal day (2.1) The time required for one rotation of the celestial sphere.

sidereal period, lunar (2.3) The length of time required for the moon to appear to traverse the celestial sphere once, that is, the time required to start apparently near one star and return to that star again.

sidereal period of a planet (4.8) In the heliocentric system, the length of time required for a planet to complete one orbit.

solar constant (9.1) The amount of energy per second striking a square meter of area lying just above the earth's atmosphere.

solar cycle (8.1) The gradual increase and decrease in activity on the sun's surface, which follows a roughly 11-year cycle. Near the peak of a solar cycle the sun is marked by many more sunspots than average.

solid body rotation (16.9) A type of rotation characterized by the fact that each particle of the rotating object completes one circuit in the same amount of time.

solstice (2.2) One of the two times during a year when the sun appears to lie either as far north or as far south of the celestial equator as possible.

spectral analysis (9.5) The discovery of the properties of an object by means of a study of the spectrum produced by the light emitted by the object.

spectral classification (12.3) A method of assigning stars to categories based on the appearance of their spectra.

spectroscope (7.5) An instrument that employs the phenomenon of dispersion to separate the wavelengths of incoming electromagnetic radiation by refracting each wavelength through a different angle.

spectroscopic binary star (12.6) A binary star whose nature is revealed by a periodic shifting of the lines in its spectrum.

spectrum (7.3) An array of the various wavelengths in an incoming beam of electromagnetic radiation in which the wavelengths of the radiation are sorted according to increasing wavelength.

speed (6.7) The rate at which an object's position changes.

spicule (9.6) A jet of gas extending upward at the top of the sun's chromosphere some 3000 km.

spiral galaxy (normal) (16.8) A galaxy exhibiting a nucleus and arms.

spiral nebula (16.2) A nebula ("cloud of light") having an appearance resembling a whirlpool. This term gave way to the term "spiral galaxy" in the mid-1920s.

square of a number (5.8) The result produced when a number is multiplied by itself. For example, $5^2 = 5 \times 5 = 25$.

stellar evolution (13.2) The study of the "life" of a star, the processes by which it is formed, changes, and reaches its final state.

stellar structure, theory of (13.3) A theory that deduces the conditions in the interiors of stars based on data obtained about them from their surface conditions and their masses.

sunspot (8.1) A region of temporary magnetic activity on the sun, which is cooler than its surroundings and therefore appears darker.

sunspot group (8.1) A collection of sunspots in one region of the sun's surface.

sunspot number, mean annual (8.1) A measure of the sunspot activity that takes into account both individual sunspots and sunspot groups.

superior conjunction (of an inferior planet with the sun) (2.4) A conjunction occurring as the planet appears to cross from the west side to the east side of the sun.

supernova (14.3) A star that dies by exploding and spewing much of its contents into space and producing an enormous outpouring of light.

supernova remnant (14.4) An expanding cloud of gas left over from a supernova explosion.

synchronous rotation (10.2) A type of rotation in which there is a simple relationship between the object's orbital period and period of rotation. In ordinary cases of synchronous rotation the two periods are equal, so that the orbiting object always keeps one face toward the parent object.

synchrotron radiation (14.8) Electromagnetic radiation produced by very rapidly moving electrons spiraling in a magnetic field.

synodic period (2.2 and 2.3) (1) For the moon and superior planets, the time between conjunctions with the sun. (2) For inferior planets, the time from one inferior conjunction with the sun to the next.

systematic error (5.2) Any error in measurement that introduces a consistent deviation from the actual amount.

tangential velocity (7.6) The part of the relative motion of an object that is directed at right angles to a line connecting the object and the observer.

terminator (6.5) The line on the moon or other object which, at any instant, divides the day side from the night side.

terrestrial planet (11.2) Any of the four inner planets of the solar system characterized by their small size, low mass, high density, and rocky composition. See also "jovian planet."

thermal radiation (9.3) Electromagnetic radiation emitted by a substance by virtue of the thermal agitation (heat) of its atoms.

three-degree blackbody radiation (17.3) Radiation received from all regions of the sky, which has a principal wavelength indicative of a temperature of about 3 K.

transit (8.3) The passage of a smaller body in front of a larger one.

ultraviolet radiation (7.3) Electromagnetic radiation having a wavelength longer than that of x-rays but shorter than that of visible light.

umbra (in a shadow) (8.2) The region in a shadow produced by an extended source of light, which no light from the source can reach by traveling in a straight line.

umbra (of a sunspot) (8.1) The dark central region of a sunspot.

uniformitarianism (8.1) The assumption that, lacking evidence to the contrary, processes have proceeded in the past in the same fashion as they now do.

variable star (12.8) A star that changes its apparent brightness.

visual binary star (12.6) A binary star that can be resolved (seen as two stars) through a telescope.

volcanism (also vulcanism) (10.1) Phenomena such as volcanoes on the surface of a planet due to melted rock rising near the planet's surface.

volume (6.7) The quantity of space occupied by an object.

watt (12.2) A unit of power (energy per second).

wavelength (of a wave) (7.2) The distance from one crest to the next; alternatively, the distance from one trough to the next.

weight (6.8) The force on an object due to gravitational attraction.

weightlessness (6.2) A condition in which one cannot sense the gravitational attraction of the earth on one's body. This condition occurs when one is falling freely.

wind, solar (9.6) A plasma emitted by the sun at an average speed of about 400 km/sec.

year (2.2) The time required for the sun apparently to return to a particular place on the celestial sphere, a place known as the vernal equinox.

ylem (14.7) The matter in the early universe just after the explosion of the cosmic egg in Gamow's theory, a theory that attempted to explain the origin of the elements.

zenith (2.1) The imaginary point in the sky directly overhead.

zodiac (2.2) Generally, the constellations through which the ecliptic passes.

zone of avoidance (15.3) A band along and some distance on either side of the Milky Way in the sky in which few globular clusters and practically no galaxies are found.

INDEX

Please refer to the Glossary for definitions of terms.

80 81 82 83 84 6 5 4 3 2